PURE AND APPLIED MATHEMATICS

A Program of Monographs, Textbooks, and Lecture Notes

Executive Editors — *Monographs, Textbooks, and Lecture Notes*

Earl J. Taft
Rutgers University
New Brunswick, New Jersey

Edwin Hewitt
University of Washington
Seattle, Washington

Chairman of the Editorial Board

S. Kobayashi
University of California, Berkeley
Berkeley, California

Editorial Board

Masanao Aoki
University of California, Los Angeles

Glen E. Bredon
Rutgers University

Sigurdur Helgason
Massachusetts Institute of Technology

G. Leitman
University of California, Berkeley

W. S. Massey
Yale University

Irving Reiner
University of Illinois at Urbana-Champaign

Paul J. Sally, Jr.
University of Chicago

Jane Cronin Scanlon
Rutgers University

Martin Schechter
Yeshiva University

Julius L. Shaneson
Rutgers University

LECTURE NOTES
IN PURE AND APPLIED MATHEMATICS

1. *N. Jacobson*, Exceptional Lie Algebras
2. *L.-Å. Lindahl* and *F. Poulsen*, Thin Sets in Harmonic Analysis
3. *I. Satake*, Classification Theory of Semi-Simple Algebraic Groups
4. *F. Hirzebruch, W. D. Neumann,* and *S. S. Koh*, Differentiable Manifolds and Quadratic Forms
5. *I. Chavel*, Riemannian Symmetric Spaces of Rank One
6. *R. B. Burckel*, Characterization of C(X) among Its Subalgebras
7. *B. R. McDonald, A. R. Magid,* and *K. C. Smith*, Ring Theory: Proceedings of the Oklahoma Conference
8. *Yum-Tong Siu*, Techniques of Extension of Analytic Objects
9. *S. R. Caradus, W. E. Pfaffenberger,* and *Bertram Yood*, Calkin Algebras and Algebras of Operators on Banach Spaces
10. *Emilio O. Roxin, Pan-Tai Liu,* and *Robert L. Sternberg*, Differential Games and Control Theory
11. *Morris Orzech* and *Charles Small*, The Brauer Group of Commutative Rings
12. *S. Thomeier*, Topology and its Applications
13. *Jorge M. López* and *Kenneth A. Ross*, Sidon Sets
14. *W. W. Comfort* and *S. Negrepontis*, Continuous Pseudometrics
15. *Kelly McKennon* and *Jack M. Robertson*, Locally Convex Spaces
16. *M. Carmeli* and *S. Malin*, Representations of the Rotation and Lorentz Groups: An Introduction
17. *George B. Seligman*, Rational Methods in Lie Algebras

Other volumes in preparation

RATIONAL METHODS IN LIE ALGEBRAS

George B. Seligman
Department of Mathematics
Yale University
New Haven, Connecticut

MARCEL DEKKER, INC., New York and Basel

COPYRIGHT © 1976 by MARCEL DEKKER, INC. ALL RIGHTS RESERVED.

Neither this book nor any part may be reproduced or transmitted in any form or by any means, electronic or mechanical, including photocopying, microfilming, and recording, or by any information storage and retrieval system, without permission in writing from the publisher.

MARCEL DEKKER, INC.

270 Madison Avenue, New York, New York 10016

LIBRARY OF CONGRESS CATALOG CARD NUMBER: 76-13414

ISBN: 0-8247-6480-3

Current printing (last digit):
10 9 8 7 6 5 4 3 2 1

PRINTED IN THE UNITED STATES OF AMERICA

Foreword

These notes grew out of seminars and lectures at Yale University over a period of years. The first objective has been to make accessible by direct methods and explicit display a number of central results of recent years in the theory of semisimple (adjoint) algebraic groups over arbitrary ground fields (of characteristic zero), by concentrating on the adjoint representation and the structure of the Lie algebra. Mainly, this undertaking involves replacing considerations of algebraic geometry and galois descent by techniques from linear algebra and the theory of various classes of linear algebras. None of the principal results presented here is really new; Chevalley, Mostow, Borel, Tits, Satake and others have seen to that. What I have hoped for is that those, like my hearers and myself, who seek a substitute for backgrounds adequate to the methods of some of these authors, will find here a more elementary route to understanding of some important results.

Except for Chapter I, which principally exposes older work of Chevalley and Mostow, there is probably no part of the monograph that does not derive largely from ideas of Jacques Tits. Specific references should make this debt clear, but I wish to point out particularly the case of Chapter III, where the principal achievement may be viewed as a version of an introduction of coordinates into "Tits geometries", as performed in [TB]. The methods used as a substitute for Tits' geometrical considerations are developed directly from earlier work of Tits aimed particularly at the construction of certain exceptional Lie algebras. Thus his influence is doubly decisive in this case.

I regret that the intrusion of other responsibilities has prevented me from giving what should have been the finishing touches to the explicit description of simple Lie algebras of relative rank one. There are several workers making rapid progress along these lines, notably Allison and Faulkner, and I have tried to give a summary of current knowledge, with some ventures beyond that, in Chapter V.

I have attempted to give the reader who is mainly interested in finding central results, and in understanding what these say without reading the whole monograph, a fighting chance to do so. For this spiritual fellow of mine, "Reference Guide" precedes the Table of Contents. May it spare me some of the curses I have bestowed on other writers for their shortcomings on this score!

Students and colleagues, who have participated in seminars and suffered through lectures, have often added insights and corrections to improve the presentation. Over the years, Tsuneo Tamagawa, Jim Humphreys, John Faulkner, Rob Gordon, Robert Wilson, Bruce Allison, Brian Parshall and Georgia Benkart have earned mention on this score. Chapters IV and V, in particular, contain abundant evidence of Allison's influence in more recent times.

To complete the list of what is not original, there is the matter of my title. It was used forty years ago for a paper in the Annals of Mathematics, a paper that is an ancestor to almost all algebraic efforts in this area. Apart from its being appropriate to describe the approach in these notes, it seemed particularly suitable as a means of acknowledgeing from the outset my debt for guidance and encouragement, over more than half of those forty years, to its author, Nathan Jacobson.

The fact that the monograph is a continuing record of evolution, rather than a polished account of the status of a field, may be blamed for some minor discrepancies in notation and for an occasional failure to make a relatively recent publication (more recent than the writing of the notes) the primary reference, even though it would be the best one. Thus one finds [B], [Bo], [BO] used in various places to refer to the Bourbaki volume on root-systems, and finds this excellent work cited secondarily to Steinberg's Yale notes in Chapter I (begun in 1968 and completed in early 1969).

Steady summer support from the National Science Foundation, most recently through Grant no. NSF-GP-33591X, has been of assistance in the development of these notes. In addition, the Eugene Higgins Trust has supported the work during the academic years, both at Yale and at the Institute for Advanced Study. Donna Belli has managed to keep a good humor while typing from masses of messy manuscript and enduring corrections from a rather intolerant author. She deserves all our thanks.

George B. Seligman

Reference guide

All derivations of a semisimple Jordan algebra are inner	333, 334
Bruhat decomposition for the group of automorphisms	32
Cartan decomposition $(K + P)$	291
Commutator formulas for exponentials of root vectors	35–39
Compact real forms: Existence	289–290
Uniqueness	292
Conditions for the even Clifford algebra in 5 dimensions to be a division algebra (Tamagawa's theorem)	340, 341
Conjugacy: of maximal nil subalgebras	27
of maximal solvable ("Borel") subalgebras	28
of maximal split tori	17, 22, 27
of splitting Cartan subalgebras	28
Construction of canonical split subalgebra	54, 55
Coördinatization: of relative types D, E	64
of relative type A_1	65
of relative type $A_r (r \geq 2)$	69
of relative type $B_r (r \geq 3)$	74
of relative type F_4	84
of relative type G_2	86–87
of relative type $C_r (r \geq 3)$	101–102
of relative type $C_2(B_2)$	111–112
of relative type $BC_r (r \geq 3)$	120–121
of relative type BC_2	140
of relative type BC_1	143
Decomposition: of almost algebraic Lie algebra	24
of almost algebraic solvable Lie algebra	22
of simple Lie algebra relative to canonical split subalgebra	56–58
of the center $Z(L_0)$ of the centralizer of a maximal split torus relative to a Weyl reflection	148–149
Derivation-simple algebras are simple	B3
Determination: of $\text{Hom}_S(M_1 \otimes M_2, M_3)$ for split simple S and certain irreducible M_i	59–61
of some biadditive mappings to the symmetric elements of an involutorial division algebra	335, 336
Equivalence of anisotropy and compactness for real semisimple Lie algebras	292
Extension: of isomorphisms as modules for subalgebras to isomorphisms of (simple) Lie algebras	156
of isomorphisms from centralizers of maximal split tori to simple Lie algebras	152, 153, 161, 164, 166
of subalgebra-isomorphisms of modules to isomorphisms	159
of the symmetric part of an antihermitian form to an antihermitian form	204, 336–339
Freudenthal's formula for weight multiplicities	306
Identification of the centroid for simple Lie algebras	172, 176

Jacobson-Morozov theorem	4
Maximal nil subalgebras	18
Maximal tori are Cartan subalgebras (in completely reducible Lie algebras)	2
Multiplicities of certain modules in certain tensor products (for split simple algebras)	58
Nondegeneracy of trace form of completely reducible algebraic Lie algebra	19
Realizations: of compact simple Lie algebras	302
of simple algebras of type $A_r (r \geq 3)$	177
over reals	292
of simple algebras of type A_2	177, 180-182
over reals	292
of simple algebras of type A_1	177, 184, 189-190, 198, 203-204
over reals	298
of simple algebras of type $B_r (r \geq 3)$	185
over reals	293-294
of simple algebras of type B_2	188-189
over reals	296
of simple algebras of type C	198, 211-212
over reals	295, 296
of simple algebras of type D	188
over reals	292-293
of simple algebras of type E_6, E_7, E_8	196
over reals	293
of simple algebras of type F_4	190-191
over reals	294
of simple algebras of type G_2	193-194
over reals	294
of simple algebras of type BC	198, 204, 232, 257, 285-288
over reals	296, 298, 302
Realizations of simple groups associated with simple Lie algebras	
type A	179-180, 183, 184-185, 190
type B	187-188, 189
type C	211, 212-213
type D	188
type E	196
type F	192-193
type G	195
type BC	211, 233, 244, 249, 251, 252, 253, 256, 266, 283
Semisimple subalgebras (L^P) associated with a set P of simple roots	145-147
Simplicity of groups generated by exponentials of nilpotents (the groups G')	53
Stability under derivations of the radical of a Jordan algebra	333
Structure of the group associated with a closed set of positive roots	40, 42
Summary of properties of root-systems	11, 12, 15
Tables of multiplicities of weights	308, 309, 311-316, 318-322
Tits systems, elementary properties of	43-47
Wedderburn-Artin theorem for semisimple Jordan algebras	332, 333

CONTENTS

I. STRUCTURE OF REDUCTIVE LIE ALGEBRAS — 1
  1. Introduction; Maximal split tori — 1
  2. Abstract root-systems; behavior of the root spaces — 10
  3. Questions of conjugacy — 16

II. THE AUTOMORPHISM GROUPS — 30
  1. Generalities — 30
  2. The structure of U^+ — 34
  3. (B,N)-pairs — 43
  4. Applications to automorphism groups — 48

III. CONSTRUCTIONS FOR SIMPLE LIE ALGEBRAS. — 54
  1. Decompositions relative to split subalgebras. — 54
  2. Algebras of types A_1, D, E. — 61
  3. Algebras of relative type $A_r (r \geq 2)$. — 65
  4. Algebras of relative type $B_r (r \geq 2)$. — 69
  5. Algebras of types G_2 and F_4. — 74
  6. Algebras of type $C_r (r \geq 2)$. — 87
  7. Algebras of type $BC_r (r \geq 3)$. — 113
  8. Algebras of type BC_2. — 124
  9. Algebras of type BC_1. — 140

IV. DATA FOR ISOMORPHISM. — 144
  1. Introduction: necessary conditions. — 144
  2. On certain canonical subalgebras. — 145
  3. The criteria for isomorphism. — 149
  4. Lemmas on the extension of isomorphisms. — 156
  5. Proof of the isomorphism theorem: Types A_1, BC_1, D, E. — 161
  6. Proof of the isomorphism theorem: the remaining reduced cases. — 163
  7. Conclusion of the proof: the remaining non-reduced cases. — 165

V. REALIZATIONS — 167
  1. Centroids — 167
  2. Algebras of type A — 176
     a) Derived algebras of simple associative algebras — 176
     b) The exceptional case of rank two; Tits' second construction — 180
     c) The case of rank one; the Tits-Koecher construction — 184
  3. Algebras of types B and D; Skew transformations relative to quadratic forms — 185
     a) Algebras of type $B_r (r \geq 3)$ — 185
     b) Algebras of type $D_r (r \geq 4)$ — 188
     c) Algebras of type B_2 — 188
     d) Some algebras of type A_1 ("type B_1") — 189
  4. Algebras of type F_4 — 190
  5. Algebras of type G_2 — 193
  6. Algebras of type E — 196

7.	Algebras of types C,BC	196
	a) General case; skew elements of simple associative algebras with involution	196
	b) The exceptional case of type C_3	211
	c) The exceptional case of type BC_2	213
	d) Algebras of type BC_1; the remaining cases	234
	i) $A = E^+$, E an associative division algebra of degree ≥ 2	234
	ii) The case where A is of degree two.	239
	iii) The case where $A = Z(A)$ is of degree one.	258
8.	Simple Lie algebras over the reals.	289
Appendix A.	Computation of multiplicities	305
A-1.	Multiplicities of dominant weights	307
A-2.	Decomposition of certain tensor products	323
Appendix B.	Some facts about Jordan algebras	332
1)	The radical is derivation-stable	332
2)	When all derivations are inner	333
3)	Jordan division algebras arising in connection with Lie algebras of type G_2	334
4)	A theorem on some mappings of involutorial algebras	336
Appendix C.	A Theorem of Tamagawa	340
Bibliography		343
Index		345

CHAPTER I

STRUCTURE OF REDUCTIVE LIE ALGEBRAS

§1.   Introduction; Maximal split tori.

Let F be a field of characteristic zero, V a finite dimensional vector space over F. We consider Lie algebras L of endomorphisms of V, acting completely reducibly in V. By [J], Theorem II.11 and Corollary 2 of §III.5, $L = Z \oplus [LL]$, where Z is the center of L and where the derived algebra $[LL]$ is semi-simple. Moreover, every element of Z is a semisimple endomorphism of V.

We shall on occasion wish to make the assumption that L is algebraic. Rather than entering here into a discussion of the Lie algebras of algebraic linear groups, we define this notion for L as above, in self-contained terms which are shown by Chevalley ([C II], [C III]) to be equivalent to being the Lie algebra of an algebraic group of endomorphisms of V:

Definition. L is algebraic if for every $T \in Z$ and field extension K of F containing all characteristic roots of T, and for every basis $u_1,\ldots,u_n$ of V_K relative to which T acts diagonally, say $u_i T = \lambda_i u_i$, $1 \le i \le n$, Z_K also contains every endomorphism S of V_K satisfying the following conditions:

1) S acts diagonally relative to $u_1,\ldots,u_n$, say $u_i S = \mu_i u_i$, all i.

2) If $m_1,\ldots,m_n$ is any set of rational integers with

$$\sum_{i=1}^n m_i \lambda_i = 0, \text{ then } \sum m_i \mu_i = 0 .$$

The significance of this assumption for some of our arguments will appear at the time those arguments are presented. For now, we shall make no use of it.

Let T be a subalgebra of L which is maximal with respect to the two properties:

(a) T is commutative;

(b) Every $T \in T$ is semisimple and has all its characteristic roots in F.

Any T which satisfies (a) and (b) is called a <u>split toral algebra</u> (sometimes a "split torus"). Thus our T is a <u>maximal split toral subalgebra</u> ("maximal split torus").

A <u>toral subalgebra</u> ("torus") H is one which is commutative and consists of semisimple elements; thus Z is toral in our setting. Let H be a maximal toral subalgebra of L containing the maximal split torus T above.

Since the associative algebra of endomorphisms of V generated by commuting semisimple endomorphisms consists of semisimple endomorphisms (cf. [J], §III.7, proof of Theorem 11, or [C II], Proposition 4, page 69), it is clear that if $T \in L$ is semisimple, $[T, H] = 0$, then $T \in H$. In particular, $Z \subseteq H$, $H = Z \oplus (H \cap [LL])$, and $H \cap [LL]$ is a maximal toral subalgebra of $[LL]$.

Likewise, if $T \in L$ is semisimple and has all its characteristic roots in F, and if $[T, T] = 0$, then $T \in T$.

<u>Lemma 1.</u> If $T \in T$, then ad T is semisimple and has all its characteristic roots in F.

For if $u_1, \ldots, u_n$ is a basis for V with $u_i T = \lambda_i u_i$, and if E_{ij} are the matrix units relative to this basis: $u_k E_{ij} = \delta_{ik} u_j$, then $u_k [E_{ij}, T] = \delta_{ik} \lambda_j u_j - \lambda_i \delta_{ik} u_j = (\lambda_j - \lambda_i) u_k E_{ij}$, or $[E_{ij}, T] = (\lambda_j - \lambda_i) E_{ij}$. Thus ad T acts diagonally on End (V), hence on every (ad T) - stable subspace, especially on L.

<u>Proposition 1.</u> H is a Cartan subalgebra of L.

<u>Proof.</u> We must show H is its own normalizer, or that $H \cap [LL]$ is its own normalizer in the semisimple algebra $[LL]$. By the argument of Lemma 1, for each $T \in H$, ad T is a diagonal endomorphism of L_K for some field extension K of F, so ad T is semisimple. Thus if $W \in L$, $[[WT]T] = 0$ implies $[WT] = 0$. In particular, if W normalizes H, we have $[WT] \in H$, hence $[[WT]T] = 0$, and therefore $[WT] = 0$, for all $T \in H$. Hence we must show H is its own

centralizer, and the remarks above show we may assume L is <u>semisimple</u>.
As a completely reducible Lie algebra, L is <u>splittable</u> ("almost
algebraic" - cf. [J], p. 100) in the sense that if $B \in L$, then there
are $S, N \in L$, with S semisimple and N nilpotent, such that
$[S,N] = 0$, $B = S + N$. Moreover, S and N may be written as polynomials in B without constant term. In particular, if B centralizes
H so do S and N, and hence $S \in H$. Let C be the centralizer
of H in L. If $B = S + N$ is in C, and if $A \in C$, then
$A(\text{ad } B) = A(\text{ad } N)$, $A(\text{ad } B)^k = A(\text{ad } N)^k$, and this is zero for
$k \geq 2r - 1$, where $N^r = 0$. Thus ad X is a nilpotent transformation
of C for every $X \in C$, so that Engel's theorem implies that C is
a <u>nilpotent</u> Lie algebra. If $Y \in L$ normalizes C, then for $T \in H$,
$[[YT]T]] \in [C T] = 0$, so that $[YT] = 0$ and $Y \in C$. Thus C is its
own normalizer, hence is a <u>Cartan subalgebra</u> of L. {So far we have
only used that L is splittable.} Since L is assumed semisimple,
C is commutative and is a non-singular subspace with respect to the
Killing form of L, by [J], §IV.1 - using extension of the base
field. Now if $N \in C$ is nilpotent, so is ad N, acting in L;
hence ad T ad N for all $T \in C$ are nilpotent by the commutativity
of C, from which N is in the radical of C with respect to the
Killing form of L. But this means $N = 0$, so that if $B = S + N$
as above we have $B = S \in H$, $C = H$, and the proof of Proposition 1
is complete.

    Now ad T is a diagonalizable transformation of L for every
$T \in T$, and T is commutative. Therefore we have a family of linear
functions α on T (the weights of the adjoint representation of
T on L) such that $L = \Sigma \oplus L_\alpha$, where $L_\alpha = \{X \in L | [XT] = \alpha(T)X$
for all $T \in T\}$. Those $L_\alpha \neq 0$ will be called <u>root-spaces</u> relative
to T. In particular, $L_0 \supseteq H$, a Cartan subalgebra of L, and
$[L_\alpha L_\beta] \subseteq L_{\alpha+\beta}$. Thus each L_α is stable under ad H, indeed under
ad L_0. Clearly both Z and $[LL]$ are stable under ad L and
$Z \subseteq L_0$, so that $[LL] = (L_0 \cap [LL]) \oplus \Sigma_{\alpha \neq 0} L_\alpha$.

    In $[LL]$, let $X_\alpha \in L_\alpha$, $Y_\beta \in L_\beta$; then ad X_α ad Y_β
maps L_γ into $L_{\gamma+\alpha+\beta} \neq L_\gamma$ unless $\alpha + \beta = 0$. It follows that
L_α and L_β are orthogonal with respect to the Killing form of
$[LL]$ unless $\alpha + \beta = 0$. Thus we have:

Lemma 2. If $L_\alpha \neq 0$, then $L_{-\alpha} \neq 0$, and if $\alpha \neq 0$, L_α and $L_{-\alpha}$ are totally isotropic and dual to one another with respect to the Killing form, hence have the same dimension. The restriction of the Killing form to $L_0 \cap [LL]$ is non-singular.

If $\alpha \neq 0$ and $0 \neq X_\alpha \in L_\alpha$, there is $T \in \mathcal{T}$ with $\alpha(T) \neq 0$, and $[X_\alpha T] = \alpha(T) X_\alpha$, from which X_α is <u>nilpotent</u> by [J], p.44 (Lemma 4). By [J], p. 100 (Theorem 17), there are elements $Y, Z \in L$ (indeed, in $[LL]$) such that $[X_\alpha, Z] = 2X_\alpha$, $[YZ] = -2Y$, $[X_\alpha, Y] = Z$. Write $Z = \Sigma_\beta Z'_\beta$, $Z_\beta \in L_\beta$, $Y = \Sigma_\gamma Y_\gamma$, $Y_\gamma \in L_\gamma$; from $[X_\alpha, Z] = 2X_\alpha$ we have $[X_\alpha, Z_0] = 2X_\alpha$, $[X_\alpha, Z_\beta] = 0$ if $\beta \neq 0$. From $[X_\alpha Y] = Z$ we have $Z_0 = [X_\alpha, Y_{-\alpha}]$. By [J], pp. 98-99 (Lemma 7) there is $W \in L$ with $[WZ_0] = -2W$, $[X_\alpha, W] = Z_0$; writing $W = \Sigma W_\beta$, we see from $[X_\alpha, W] = Z_0$ that $[X_\alpha, W_{-\alpha}] = Z_0$ (so $W_{-\alpha} \neq 0$), and from $[WZ_0] = \Sigma_\beta [W_\beta Z_0] = -2 \Sigma W_\beta$ that $[W_\beta Z_0] = -2W_\beta$ for all β, in particular for $\beta = -\alpha$. Thus we may replace Z by Z_0, Y by $W_{-\alpha}$ to assume $Z \in L_0$, $Y \in L_{-\alpha}$. Now the representation theory ([J], §IV.2) of the 3-dimensional algebra $\{X_\alpha, Y, Z\}$ shows that <u>all characteristic roots of</u> Z <u>are integers</u>. Since $Z \in L_0$, it follows that $Z \in \mathcal{T}$, $\alpha(Z) = 2$. Thus:

Lemma 3. Let $0 \neq X \in L_\alpha$, $\alpha \neq 0$. Then there is $Z \in \mathcal{T}$, $Y \in L_{-\alpha}$, with $[XZ] = 2X$, $[YZ] = -2Y$, $[XY] = Z$.

We insert here a brief summary of the representation theory of the split 3-dimensional algebra S with basis E, F, H, $[EH] = 2E$, $[FH] = -2F$, $[EF] = H$. One sees easily that the algebra is simple, so that every S-module (S written on the right, as in [J],) is completely reducible ([J], p. 79, Theorem 8). By [J], §III.8, every irreducible S-module has the following form:

There is a basis $v_0, v_1, \ldots, v_m$, with:

$v_i F = v_{i+1}$, $i < m$; $v_m F = 0$;

$v_0 E = 0$, $v_{i+1} E = -(i+1)(m-i) v_i$, $0 \leq i \leq m$;

$v_i H = (m-2i) v_i$, $0 \leq i \leq m$.

Thus the dimension of an irreducible S-module determines the module to within isomorphism, and there is an irreducible S-module of each positive dimension. In irreducible modules of odd dimension, $2n+1$, H acts diagonally with <u>even</u> integral eigenvalues $0, \pm 2, \ldots, \pm 2n$, while in irreducible modules of even dimension $2n+2$, H acts diagonally with odd integral eigenvalues $\pm 1, \pm 3, \ldots, \pm 2n+1$. Thus the assertion about characteristic roots of Z used in the proof that $Z \in T$ in Lemma 3 is clear.

Now let j be a characteristic root of H in the S-module M, and let $0 \neq y \in M$, $yH = jy$. Let M be written as a direct sum of irreducible summands, so that y has in each summand a component which belongs to the eigenvalue j of H. It follows from the discussion above that y can have non-zero components only in summands of dimensions m satisfying $m \equiv j + 1 \pmod{2}$ and $m \geq |j|$. Let q be maximal with $yE^q \neq 0$. Then it follows from the fact that E annihilates only v_0 in the displayed irreducible module, and from the fact that E maps the i-eigenspace for H into the $(i+2)$-eigenspace, that yE^q lies in the sum of the irreducible summands of highest dimension containing a non-zero component of y. Call this dimension $d + 1$, so that $d \equiv j \pmod{2}$, and $yE^{q+1} = 0$ means that yE^q belongs to the eigenvalue d of H: $yE^qH = dyE^q$. But remarks above show that $yE^qH = (j+2q)yE^q$, from which $d = j + 2q$.

Applying similar considerations with r maximal for $yF^r \neq 0$ gives $-d = j - 2r$, from which we obtain by addition:

$$j = r - q$$

That is, <u>knowing that</u> $0 \neq y$ <u>is an eigenvector for</u> H, <u>the corresponding eigenvalue may be found as</u> $r - q$, <u>where</u> r <u>is maximal with</u> $yF^r \neq 0$, q <u>maximal with</u> $yE^q \neq 0$.

We also note that if $j > 0$ is an eigenvalue of H, then E maps the eigenspace belonging to $j - 2$ onto that belonging to j, and that this mapping is one-one if and only if no irreducible summands of dimension $j - 1$ are present. In fact, the kernel of the restriction of E to the $(j-2)$-eigenspace is just the subspace generated by the vectors "v_0" in such summands.

Now return to our setting; let $0 \neq \alpha$, $0 \neq X \in L_\alpha$, $Y \in L_{-\alpha}$, $Z \in T$, $Z = [X\, Y]$, $\alpha(Z) = 2$. Let β be any root (possibly zero) relative to T, and let $M = \sum_{i \in \mathbb{Z}} L_{\beta + i\alpha}$. Then M is stable under $E = \mathrm{ad}\, X$, $F = \mathrm{ad}\, Y$, $H = \mathrm{ad}\, Z$, and S is as above.

<u>Lemma 4.</u> Let $q \in \mathbb{Z}$ be maximal with $L_{\beta+q\alpha} \neq 0$, and let $r \in \mathbb{Z}$ be maximal with $L_{\beta-r\alpha} \neq 0$. Then q and r are non-negative and

$\beta(Z) = r - q$. Moreover, every $\beta + i\alpha$, $-r \leq i \leq q$, is a root.

Proof. Let $0 \neq x \in L_{\beta+q\alpha}$, so $xH = (\beta(Z) + 2q)x$. Let $xF^k \neq 0$, $xF^{k+1} = 0$. Since $x E \in L_{\beta+(q+1)\alpha} = 0$, we have $k = \beta(Z) + 2q$.

On the other hand, we see that the characteristic roots of H are symmetric about zero, so that the smallest one, $\beta(Z) - 2r$, is the negative of the largest one $\beta(Z) + 2q$. Thus $\beta(Z) = r - q$, $k = r + q$, so that $xF^j \neq 0$, $0 \leq j \leq r + q$. But $xF^j \in L_{\beta+(q-j)\alpha}$, and $q - j$ runs from $-r$ to q. This completes the proof.

Corollary. There is a unique element $H_\alpha \in T$ with $\alpha(H_\alpha) = 2$, $H_\alpha = [UV]$, $U \in L_\alpha$, $V \in L_{-\alpha}$. This H_α is the element Z of Lemma 3.

For let $Z' = [UV]$ be as in the statement of the Corollary. For each root β, $\beta(Z') = r - q = \beta(Z)$ by Lemma 4. Hence $Z - Z' \in T \cap [LL]$, $\beta(Z-Z') = 0$ for all roots β, so that $Z - Z'$ is central in $[LL]$, which is semisimple, and $Z' = Z$.

Lemma 5. The elements H_α generate $T \cap [LL]$.

Proof. By remarks above, we may choose roots $\alpha_1, \ldots, \alpha_r$ whose restrictions form a basis for the dual space of $T \cap [LL]$. Then if $T \in T$, we have, for each root $\alpha \neq 0$, $T = \lambda H_\alpha + T'$, where $\alpha(T') = 0$. We consider the value of the Killing form at (H_α, T'). If $T_1, T_2 \in T$, we have $(T_1, T_2) = \Sigma_\beta (\dim L_\beta) \beta(T_1) \beta(T_2)$; now if β is a root, $\Sigma_{i \in \mathbb{Z}} L_{\beta+i\alpha}$ is stable under $\text{ad } T'$ and under the adjoint representation of $\{X, Y, H_\alpha\}$. The trace on this space of $(\text{ad } H_\alpha)(\text{ad } T')$ is $\Sigma_i (\dim L_{\beta+i}) \beta(T')(\beta + i\alpha)(H_\alpha)$, since $\alpha(T') = 0$. But this is $\beta(T')(\text{Trace of } (\text{ad } H_\alpha)$ in this space), and $\text{ad } H_\alpha = [\text{ad } X, \text{ad } Y]$, has trace zero. Thus $(H_\alpha, T') = 0$, as a sum of such traces. On the other hand,

$$(H_\alpha, H_\alpha) = \Sigma_\beta (\dim L_\beta) \beta(H_\alpha)^2 = \Sigma_\beta (\dim L_\beta) r_{\beta\alpha}^2,$$

where $\beta(H_\alpha) = r_{\beta\alpha} = r - q$ as before. Since $0 \neq H_\alpha \in T \cap [LL]$, $\beta(H_\alpha) \neq 0$ for some β, so (H_α, H_α) is a <u>positive integer</u>.

Now we have $(T, H_\alpha) = (\lambda H_\alpha + T', H_\alpha) = \lambda(H_\alpha, H_\alpha) = \frac{1}{2} \alpha(T)(H_\alpha, H_\alpha)$, or the usual formula $\alpha(T) = \frac{2(T, H_\alpha)}{(H_\alpha, H_\alpha)}$. Since $\dim(T \cap [LL]) = r$, it suffices for our Lemma to show that if $\alpha_1, \ldots, \alpha_r$ are as above, then $H_{\alpha_1}, \ldots, H_{\alpha_r}$ are linearly independent. Thus assume

$\Sigma_{i=1}^{r} \lambda_i H_{\alpha_i} = 0$, from which for all $T \in \mathcal{T}$,

$$0 = (T, \Sigma \lambda_i H_{\alpha_i}) = \Sigma \lambda_i (T, H_{\alpha_i}) = \frac{1}{2} \Sigma \lambda_i (H_{\alpha_i}, H_{\alpha_i}) \alpha_i(T).$$

That is, $\frac{1}{2} \Sigma (H_{\alpha_i}, H_{\alpha_i}) \lambda_i \alpha_i = 0$, from which all $\lambda_i = 0$, and Lemma 5 is proved.

Now we show that $\mathcal{T} = (\mathcal{T} \cap \mathcal{Z}) + (\mathcal{T} \cap [LL])$, from which it follows that <u>restriction to</u> $\mathcal{T} \cap [LL]$ <u>of the subspace of</u> $\mathcal{T}*$ <u>generated by the roots is an isomorphism of this subspace onto</u> $(\mathcal{T} \cap [LL])*$. Thus let $T \in \mathcal{T}$, and write $T = Z + X$, $Z \in \mathcal{Z}$, $X \in [LL]$. By Lemma 1, ad $X =$ ad T has all its characteristic roots in F. Now $[LL]$ is semisimple, and may also be viewed as a Lie algebra of linear transformations of itself in the adjoint representation. Since ad X is semisimple some maximal toral subalgebra of ad $[LL]$ contains ad X, and Proposition 1 yields a Cartan subalgebra K of $[LL]$ containing X.

Now let M be any $[LL]$-module (of finite dimension). We claim <u>all eigenvalues of</u> X <u>in</u> M <u>lie in</u> F. Since ad $X =$ ad T annihilates $\mathcal{T}$, this will show $X \in \mathcal{T}$, hence $Z \in \mathcal{T}$ as well, and will complete the argument. If Ω is the algebraic closure of F, K_Ω has a basis of elements $H_{\alpha_1}, \ldots, H_{\alpha_r}$ as above (here $K_\Omega = K_\Omega \cap [L_\Omega L_\Omega]$), where $\alpha_1, \ldots, \alpha_r$ are independent roots. Then $\alpha_i(X)$ is an eigenvalue of ad X, so is in F for every i. Writing $X = \Sigma_i \lambda_i H_{\alpha_i}$, we have

$$(\ddagger) \qquad \alpha_j(X) = \Sigma_i \lambda_i \alpha_j(H_{\alpha_i})$$

for all j, and the $\lambda_i \in \Omega$ are unique by the linear independence of the H_{α_i}. Since the α_j are also linearly independent, the integral matrix $(\alpha_j(H_{\alpha_i}))$ is non-singular. Thus the system $(\ddagger)$ of linear equations in the λ_i has a unique solution in F, and $\lambda_i \in F$. Since the H_{α_i} commute and have integral eigenvalues, all eigenvalues of X are in F.

Having chosen a basis $H_{\alpha_1}, \ldots, H_{\alpha_r}$ for $\mathcal{T} \cap [LL]$ as above, we may introduce an ordering of the roots as follows: $\gamma > \beta$ if the first index i for which $\gamma(H_{\alpha_i}) \neq \beta(H_{\alpha_i})$ has $\gamma(H_{\alpha_i}) > \beta(H_{\alpha_i})$ (these values being integers). From remarks above, this is a linear ordering of the roots, indeed of the rational space

they generate, or even of the real linear space of real functions on
the set $\{H_{\alpha_1},\ldots,H_{\alpha_r}\}$ generated by the (restrictions of the) roots,
Sums of positive elements are positive, as are positive scalar multiples
of positive elements.

Having fixed this ordering, we call a root γ **simple** if γ
is positive, but is not the sum of two positive roots. Let $\beta_1,\ldots,\beta_s$
be the distinct simple roots; then no $\beta_i - \beta_j$ is a root for $i \neq j$,
so $\beta_i(H_{\beta_j}) = r - q = -q \leq 0$. Moreover, every positive root is
easily seen to be a sum of simple roots (consider a minimal counter-
example!), and one of the non-zero roots γ, $-\gamma$ is positive. Since
the roots generate $(T \cap [LL])^*$, so do the simple roots (more properly, their restrictions do). Thus one sees by the proof of Lemma 5
that the H_{β_i}, β_i simple, generate $T \cap [LL]$.

Lemma 6. The simple roots β_i (restricted to $T \cap [LL]$) form a
basis for $(T \cap [LL])^*$.

Proof. By the remarks above, it suffices to show the β_i are
linearly independent. Now for $\lambda_1,\ldots,\lambda_s \in F$, $\Sigma \lambda_i \beta_i = 0$ is equivalent
with $\Sigma_i \lambda_i \beta_i(H_{\beta_j}) = 0$ for all j, H_{β_j} as above. Thus the β_i
are linearly dependent over F if and only if the rank of the
integral matrix $(\beta_i(H_{\beta_j}))$ is less than s, in which case there are
rational λ_i, not all zero with $\Sigma_i \lambda_i \beta_i(H_{\beta_j}) = 0$ for all j,
hence with $\Sigma \lambda_i \beta_i = 0$. Thus it suffices to show the β_i are linearly
independent <u>over the rationals</u>.

Now a shortest relation of rational dependence among the β_i
may be taken in the form $\beta_1 = \Sigma_{i=2}^{a} \lambda_i \beta_i + \Sigma_{i=a+1}^{b} \lambda_i \beta_i$, where
$\lambda_2,\ldots,\lambda_a$ are <u>positive</u>, $\lambda_{a+1},\ldots,\lambda_b$ are <u>negative</u>. We cannot have
all $\lambda_i \leq 0$, since then $\beta_1 = \Sigma_{i=a+1}^{b} \lambda_i \beta_i \leq 0$, although $\beta_1 > 0$.
Thus $a \geq 2$, and minimality of our relation gives $\Sigma_{i=2}^{a} \lambda_i \beta_i \neq 0$;
since this is the same linear function on $T \cap [LL]$ as is given by
forming the Killing scalar product with $H = \Sigma_{i=2}^{a} 2\lambda_i (H_{\beta_i}, H_{\beta_i})^{-1} H_{\beta_i}$,
we have $H \neq 0$. Now $\beta_1(H) \leq 0$ since $\beta_1(H_{\beta_i}) \leq 0$ for $2 \leq i \leq a$,
while for similar reasons

$\Sigma_{i=a+1}^{b} \lambda_i \beta_i(H) \geq 0$. Finally, $\Sigma_{i=2}^{a} \lambda_i \beta_i(H)$

$= \Sigma_{i,j=2}^{a} \lambda_i \lambda_j 2(H_{\beta_j}, H_{\beta_j})^{-1} \beta_i(H_{\beta_j})$

$= \Sigma_{i,j=2}^{a} \lambda_i \lambda_j 2(H_{\beta_j}, H_{\beta_j})^{-1} 2(H_{\beta_i}, H_{\beta_i})^{-1} (H_{\beta_i}, H_{\beta_j})$

$= (\Sigma_i \lambda_i 2(H_{\beta_i}, H_{\beta_i})^{-1} H_{\beta_i}, \Sigma_j \lambda_j 2(H_{\beta_j}, H_{\beta_j})^{-1} H_{\beta_j})$

$= (H,H) = \Sigma_\gamma (\dim L_\gamma) \gamma(H)^2$, and $\gamma(H)$ is rational.

Since $0 \neq H \in T \cap [LL]$, some $\gamma(H) \neq 0$, so $(H,H) > 0$. Thus $0 \geq \beta_1(H) = \Sigma_{i=2}^{b} \lambda_i \beta_i(H) > 0$, a contradiction.

<u>Corollary</u>. The H_{β_i}, for β_i simple as above, form a basis for $T \cap [LL]$.

§2. _Abstract root-systems; behavior of the root spaces._

Let Σ be the set of (non-zero) roots, X_Σ the rational subspace of $(T \cap [LL])^*$ generated by Σ. Then $s = r$ above ($= \dim(T \cap [LL])$) and $\beta_1, \ldots, \beta_r$ are a basis for X_Σ. If $\beta \in \Sigma$, then β is uniquely of the form $\beta = \pm \Sigma\, n_i \beta_i$, the n_i being non-negative integers (assume β is a minimal positive counter-example to deduce this). If β and γ are roots, $\gamma \neq 0$, $\beta(H_\gamma) = r - q$ as before, then $-q \leq \beta(H_\gamma) \leq r$, so that $\beta - \beta(H_\gamma)\gamma \in \Sigma$.

Moreover, we have seen above that the restriction of the Killing form of L to the rational space X_Σ^* generated by the H_β is positive definite, and every H_γ is in this space, satisfying $\beta(H_\gamma) = 2(H_\beta, H_\beta)^{-1}(H_\beta, H_\gamma)$ for every root $\beta \neq 0$. Setting $H_\beta^c = 2(H_\beta, H_\beta)^{-1} H_\beta$ for such β, we thus see that H_β^c is the unique element Z of X_Σ^* such that $(Z, T) = \beta(T)$ for all $T \in X_\Sigma^*$.

Now we introduce in X_Σ a positive definite scalar product by duality: $(\phi, \psi) \underset{\text{def}}{=} (Z_\phi, Z_\psi)$, where $\phi, \psi \in X_\Sigma$ and where $Z_\phi, Z_\psi \in X_\Sigma^*$ are defined by $(Z_\phi, T) = \phi(T)$ for all $T \in X_\Sigma^*$. Making the assumption that we have discarded 0 from Σ, further we have that, if $\alpha, \beta \in \Sigma$,

(1) $\quad 2(\alpha, \beta)(\beta, \beta)^{-1} = 2(H_\alpha^c, H_\beta^c)(H_\beta^c, H_\beta^c)^{-1} = 2(H_\alpha, H_\beta)(H_\alpha, H_\alpha)^{-1}$

$\quad\quad\quad = \alpha(H_\beta)$ _is an integer_ $r_{\alpha\beta}$,

(2) $\quad \alpha - r_{\alpha\beta} \beta \in \Sigma$, and every $\alpha - j\beta$, j an integer between 0 and $r_{\alpha\beta}$, is either in Σ or is zero.

We may equally well embed X_Σ in $X_\Sigma \otimes_\mathbb{Q} \mathbb{R}$, where (ϕ, ψ) extends to a positive definite scalar product in unique and obvious fashion. The set Σ (without 0) still generates this real space. Now Σ is a "_system of roots_" in X_Σ, or in $X_\Sigma \otimes \mathbb{R}$, in at least one current sense. We choose to formulate this for $X_\Sigma \otimes \mathbb{R} = V$: V has a positive definite scalar product, Σ generates V, and we have (1) and (2). By (2), reflection in the hyperplane orthogonal to $\beta \in \Sigma$ maps Σ into Σ (this reflection sends $\phi \in V$ into $\phi - 2(\phi, \beta)(\beta, \beta)^{-1} \beta$, as one easily sees). The (finite) group of orthogonal transformations of V generated by these reflections w_β (sometimes also written σ_β or S_β) is the _Weyl_ group W of the system of roots.

We now refer to Steinberg's "Appendix on Finite Reflection Groups" from the notes of his Yale 1967-68 lectures on Chevalley groups, abbreviated [SA]. The set $\Pi = \{\beta_1,\ldots,\beta_r\} \subseteq \Sigma$, now called the set of "roots", is a <u>simple system</u> in the sense of [SA], p.2. By (12), p. 4 of [SA], replacing P/α by $P/$ (all positive multiples of α) here and in (11) it follows that if P is the set of positive roots in our ordering, and if P' is the set of positive roots with respect to another ordering on $X_\Sigma \otimes R = V$, with Π' the unique ((9), p. 2 of [SA]) simple system contained in P', then there is $w \in W$ with $P'w=P$, $\Pi' w = \Pi$. It further follows that every root is an <u>integral</u> combination of Π' (since this holds for Π). As we shall see below, the only positive multiples of $\alpha \in \Sigma$ which can be in Σ are $\alpha, \frac{1}{2}\alpha, 2\alpha$, and hence at most one of $\frac{1}{2}\alpha, 2\alpha$. Since $w_\alpha = w_{2\alpha}$, W is generated by the w_α, where $\alpha \neq 2\beta, \beta \in \Sigma$. Such α are called primitive; all elements of Π are primitive, and for discussion of W, we may replace Σ by the set of primitive roots. By (16), p.5 of [SA], W is generated by the w_α for $\alpha \in \Pi$, and by (24) of [SA], W is simply transitive on the positive systems, also on the simple systems. One will also note the results (28) and (36) of [SA] pertaining to subgroups of W, which may be appealed to later. Furthermore (38) of [SA] shows that W is defined by the generators w_α and the relations $(w_\alpha w_\beta)^{n(\alpha,\beta)} = 1$ for $\alpha, \beta \in \Pi$. See also Bourbaki [Bo], Chap. VI,§1), where all the results here quoted from [SA] may also be found.

Now if $\alpha,\beta \in \Sigma$ and if $(\alpha,\beta) = 0$, then w_α and w_β commute, or $n(\alpha,\beta) = 2$ (of course $n(\alpha,\beta) = 1$ if one of α,β is a multiple of the other). If $\alpha \neq \beta$ are in Π, we have seen that $(\alpha,\beta) \leq 0$, and for further consideration of $n(\alpha,\beta)$ we may assume $(\alpha,\beta) < 0$. Then $(\alpha,\beta)^2 \leq (\alpha,\alpha)(\beta,\beta)$ gives $1 \leq \frac{2(\alpha,\beta)}{(\beta,\beta)} \frac{2(\alpha,\beta)}{(\alpha,\alpha)} < 4$, with equality on the right ruled out since α and β are linearly independent. By (1), we have $1 \leq r_{\alpha,\beta} r_{\beta,\alpha} < 4$, the middle number being the product of two negative integers, which are therefore $\{-1, -1\}, \{-1, -2\}$ or $\{-1, -3\}$, and one shows directly that in these cases $n(\alpha,\beta)$ is equal to 3, 4 or 6, respectively, with the usual representation by diagrams: $\cdot_\alpha \quad \cdot_\beta$ if $n(\alpha,\beta) = 2$;

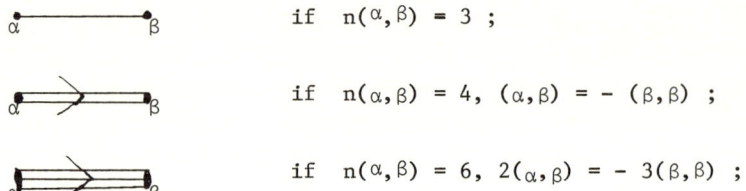

$\quad$ if $n(\alpha,\beta) = 3$;

$\quad$ if $n(\alpha,\beta) = 4$, $(\alpha,\alpha) = -(\beta,\beta)$;

$\quad$ if $n(\alpha,\beta) = 6$, $2(\alpha,\alpha) = -3(\beta,\beta)$;

we associate a <u>Dynkin diagram</u> with Π, and obtain a classification of these as in [J], §IV.5. There is, however, one refinement which must be taken into account: Namely, let $0 \neq \alpha$ be a root, and let $\gamma = j\alpha \neq 0$ be a root linearly dependent on α ; we may assume $j > 0$. Then $r_{\alpha\gamma} = \frac{2}{j}$ and $r_{\alpha\gamma} = 2j$ are integers; from the second of these, $j = \frac{k}{2}$, k an odd integer, or $j = k$, an integer. From the first, $\frac{4}{k}$ is an integer in the first case, and $k = 1$, $j = \frac{1}{2}$, while in the second case, $\frac{2}{k}$ is an integer, $k = 1$ or 2, and $j = 1$ or 2. Thus $j = \frac{1}{2}, 1$ or 2. Replacing α by $\frac{1}{2}\alpha$ if this is in Σ, we see that not both of $\frac{1}{2}\alpha$, 2α can be roots. A root α is called <u>primitive</u> if $\frac{1}{2}\alpha$ is not a root. Clearly simple roots are primitive. Now check the list of (indecomposable) Dynkin diagrams of [J], §IV.5 to see when a simple root α may be such that 2α is a root: If a diagram contains a segment α———β, we have $r_{\beta,2\alpha} = -\frac{1}{2}$, which is impossible; likewise for $\alpha\!\Rightarrow\!\beta$, $\alpha\!\Rightarrow\!\beta$. For $\alpha\!\Leftarrow\!\beta$, we have $r_{\beta,2\alpha} = -\frac{3}{2}$, also impossible. Thus the only indecomposable simple systems in which 2α may be a root are as follows: A_1: $\cdot_\alpha$

B_n: •——•——• ... •——•⇒•$_\alpha$

We distinguish these cases by coloring (in our imagination) α red. With care as to this distinction, the argument numbered XVI in [J], p. 122 goes over to show that the (possibly colored) diagram of Π determines the set Σ. (See [Bo], Chap. VI, §4.)

$\quad$ If $\beta \neq 0$ is a root (now relative to T) and if α is a root such that $\alpha(H_\beta) > 0$, we take $M = \Sigma_{i \in \mathbb{Z}} L_{\alpha+i\beta}$, S a 3-dimensional algebra with basis $X_\beta \in L_\beta$, $Y_{-\beta} \in L_{-\beta}$, $H_\beta = [X_\beta Y_{-\beta}] \in T$ as before, and consider the adjoint representation of S on M. Here L_α is the space belonging to the characteristic root $\alpha(H_\beta) > 0$ of ad H_β, and our observations on the representations of S have shown that ad X maps the subspace of M belonging to the characteristic root $\alpha(H_\beta) - 2$ of ad H_β onto L_α; i.e.

$[L_{\alpha-\beta}X_\beta] = L_\alpha$. On the basis of these assertions, and noting from Lemma 3 that any $X_\beta \neq 0$ in L_β can play the role above, we have

Lemma 7. Let $\beta \neq 0$ be a root, α a root, $\alpha(H_\beta) > 0$. Then for any $X_\beta \neq 0$ in L_β, $[L_{\alpha-\beta}X_\beta] = L_\alpha$. In particular, dim. $L_{\alpha-\beta} \geq$ dim. L_α, dim. $L_\beta >$ dim. $L_{2\beta}$. If $T \cap [LL] = H \cap [LL]$, in which case $[LL]$ is split relative to $H \cap [LL]$ ([J], Chap. IV), we have dim. $L_\beta = 1$, $L_{2\beta} = 0$, so that the only roots $k\beta$ are $0, \pm \beta$.

Proof. We need to support the assertions beginning with "dim $L_\beta >$ dim $L_{2\beta}$". This is clear if $L_\beta = 0$, and otherwise dim $L_\beta \geq$ dim $L_{2\beta}$, $Y \to [YX]$ being a linear mapping of L_β onto $L_{2\beta}$: but the kernel contains $X_\beta \neq 0$, from which the assertion follows. Now suppose $T \cap [LL] = H \cap [LL]$, which we have seen is a Cartan subalgebra of $[LL]$; then $L_0 = Z + (H \cap [LL]) = Z + (T \cap [LL])$, and $L_\beta = [L_0 X_\beta] = [TX_\beta] = F X_\beta$ is one-dimensional, from which $L_{2\beta} = 0$ by the inequality on dimensions. The last assertion now follows from our considerations on root-systems.

To see a non-split example of the above phenomena, let V be a 3-dimensional complex vector space with basis e_1, e_2, e_3 and a hermitian scalar product (x,y) whose matrix relative to this basis is

$$\begin{pmatrix} 0 & 0 & 1 \\ 0 & 1 & 0 \\ 1 & 0 & 0 \end{pmatrix}$$

. Let L be the set of endomorphisms X of V which are skew with respect to (x,y), i.e., $(uX,v) = -(u,vX)$ for all $u,v \in V$. Then L is a Lie algebra over $\mathbb{R}$ (not over $\mathbb{C}$) of endomorphisms of V, regarded as a 6-dimensional space over $\mathbb{R}$. Relative to the basis e_1, e_2, e_3 for V over $\mathbb{C}$, the matrices of elements of L have the form:

$$\begin{pmatrix} \xi & \eta & \rho i \\ \zeta & \tau i & -\overline{\eta} \\ \sigma i & -\overline{\zeta} & -\overline{\xi} \end{pmatrix}, \quad \begin{array}{l} \rho, \sigma, \tau \in \mathbb{R}, \\ \xi, \eta, \zeta \in \mathbb{C}, \ i^2 = -1. \end{array}$$

Let T be the subalgebra (always over $\mathbb{R}$) with basis diag $\{-1, 0, 1\}$. The centralizer H of T in L consists of all diagonal matrices diag $\{\xi, \tau i, -\overline{\xi}\}$ in L, and one sees that T is a maximal split toral subalgebra of L. Relative to T, the roots are $0, \pm \alpha, \pm 2\alpha$, with root-spaces as follows: $L_0 = H$,

$$L_\alpha = \left\{ \begin{pmatrix} 0 & \eta & 0 \\ 0 & 0 & -\overline{\eta} \\ 0 & 0 & 0 \end{pmatrix} \middle| \; \eta \in \mathbb{C} \right\}, \; L_{-\alpha} = \left\{ \begin{pmatrix} 0 & 0 & 0 \\ \zeta & 0 & 0 \\ 0 & -\overline{\zeta} & 0 \end{pmatrix} \middle| \; \zeta \in \mathbb{C} \right\},$$

$$L_{2\alpha} = \left\{ \begin{pmatrix} 0 & 0 & \rho i \\ 0 & 0 & 0 \\ 0 & 0 & 0 \end{pmatrix} \middle| \; \rho \in \mathbb{R} \right\} \; L_{-2\alpha} = \left\{ \begin{pmatrix} 0 & 0 & 0 \\ 0 & 0 & 0 \\ \sigma i & 0 & 0 \end{pmatrix} \middle| \; \sigma \in \mathbb{R} \right\}$$

where $\alpha(\text{diag}\{-1,0,1\}) = 1$. One readily verifies that L acts irreducibly on the real space V, and that the center Z of L consists of the diagonal matrices $\text{diag}\{\tau i, \tau i, \tau i\}$, $\tau \in \mathbb{R}$. The derived algebra $[LL]$ is thus semi-simple of dimension 8. (In fact $[LL]$ is normal simple, and becomes isomorphic to the complex Lie algebra of 3 by 3 matrices of trace zero upon extension of the base field to the complex numbers - see [J], pp. 300-301.) We shall later consider the "semi-split forms" of split simple Lie algebras; in that terminology, $[LL]$ is the real semi-split form of A_2.

All the information about roots and root-spaces may fail to characterize L, even when L is simple. For instance, given a symmetric nondegenerate bilinear form of Witt index zero on $\mathbb{Q}^3$ there is no skew diagonalizable endomorphism of $\mathbb{Q}^3$ other than zero. Thus if L is the Lie algebra of skew endomorphisms of $\mathbb{Q}^3$, a normal simple Lie algebra of type B_1 ($= A_1$), one has $T = 0$, $L_0 = L$. Yet two such L's are isomorphic if and only if the associated forms are cogredient - see [J], p. 13 - and not all forms of Witt index zero on $\mathbb{Q}^3$ are cogredient (i.e., if $f(x,y)$ and $g(x,y)$ are the forms, there is an automorphism N of $\mathbb{Q}^3$ and a rational $\rho \neq 0$ such that $q(x,y) = \rho f(xN, yN)$ for all x,y), as one easily sees by comparing the forms with matrices $\text{diag}\{1,1,1\}$ and $\text{diag}\{1,-2,3\}$ - that the latter has Witt index 0 is an easy exercise in elementary number theory. In the case where the ground field is the reals and L is normal simple, the condition "$T = 0$" will, together with the "type" of L, determine L to within isomorphism as the unique "compact real form" of its type - for the existence of such a form see [J], §IV.7, and for uniqueness Pontrjagin, <u>Topological Groups</u>, 2nd ed., §§65 and 66.

14

For later reference, we include here the following:

Lemma 8. Let $\Sigma \subset V$ be a system of roots, in the sense of (1) and (2). Let P be the set of positive roots in some ordering, $\Pi \subset P$ the simple roots. Let Q be a subset of Σ satisfying:

(a) If $\alpha, \beta \in Q$, $\alpha+\beta \in \Sigma$, then $\alpha+\beta \in Q$; and

(b) If $\alpha \in \Sigma$, then either $\alpha \in Q$ or $-\alpha \in Q$.

Then there is $w \in W$ with $P \subseteq Q^w$ (image of Q under w).

Proof. If $w \in W$, evidently Q^w satisfies (a) and (b). Choose $w \in W$ to maximize $|Q^w \cap P|$. If $P \not\subseteq Q^w$, and if $\alpha \in P$ is chosen minimal not in Q^w, it follows from (a) that $\alpha \in \Pi$. By (b), $-\alpha \in Q^w$, and if $2\alpha \in \Sigma$, it follows by (a) applied to $-\alpha$ and 2α that $2\alpha \notin Q^w$. Since α, 2α are the only possible multiples of α in P, we see that $Q^w \cap P = Q^w \cap P'$, $P' = P -$ (multiples of α). By (11) of SA we have $P'^{w_\alpha} = P'$, so $(Q^w \cap P)^{w_\alpha} = (Q^w \cap P')^{w_\alpha} = Q^{ww_\alpha} \cap P'$ has $|Q^w \cap P|$ elements. Moreover $\alpha = (-\alpha)^{w_\alpha} \in Q^{ww_\alpha} \cap P$, from which $|Q^{ww_\alpha} \cap P| > |Q^w \cap P|$, a contradiction, and Lemma 8 is proved. (cf. [Bo], Prop. 20, p. 161.)

Exercise. Analyze the real Lie algebra L of all endomorphisms of $\mathbb{C}^n (n \geq 2)$ which are skew, as in the case $n = 3$ already studied, where the hermitian form on $\mathbb{C}^n$ has matrix

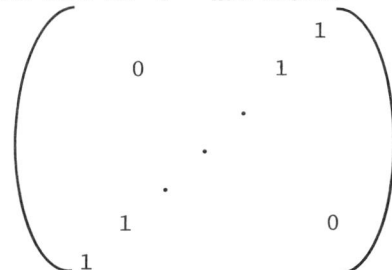

That is, show our hypotheses apply to L, find a maximal split torus T and a root-decomposition relative to T and find the Dynkin diagram for a simple system of restricted roots (the cases n even, n odd will probably have to be distinguished.)

§3. **Questions of Conjugacy**.

Let T be a maximal split toral subalgebra of L as above, and likewise assume an ordering as above on the restricted roots relative to $T \cap [LL]$. We use $\alpha, \beta, \ldots$ for roots relative to T. Set $N = \Sigma_{\alpha > 0} L_\alpha, P^+(T) = L_0 + N$.

The following considerations seem to have originated with Mostow (cf. [M]; see also [C III]).

As centralizer of the completely reducible Lie algebra T, L_0 is completely reducible on V ([J], pp. 100 ff), and thus is almost algebraic. Moreover, if $0 \neq N \in L_0$ is nilpotent, we obtain as in §1 an $H \in L_0$, semisimple, all characteristic roots of H being integers, with $[NH] = 2N$. But then $[H,T] = 0$ and H is diagonalizable over F, so $H \in T$, and $[NH] = 0$ since $N \in L_\theta$. This contradiction combines with the almost-algebraicity of L_0 to show that L_0 consists of semisimple endomorphisms of V.

Now for $\alpha \neq 0$, $X \in L_\alpha$ is nilpotent, and if α, β are positive, then $[L_\alpha L_\beta] \subseteq L_{\alpha+\beta}$ where $\alpha+\beta > 0$ shows that $\bigcup_{\alpha > 0} L_\alpha$ is a "weakly closed set" of nilpotent endomorphisms of V, in the sense of [J], Chapter II, §1. By [J], Theorem 1, p. 33, N consists of nilpotent endomorphisms of V.

Lemma 9. $P^+(T)$ *is almost algebraic; if* $B \in P^+(T)$ *is semisimple, then there are* $N_1, \ldots, N_k \in N$ *such that conjugation by* $\exp(N_1) \ldots \exp(N_k)$ *carries* B *into* L_0.

Proof. Clearly N is an ideal in $P^+(T)$. Now let $X \in P^+(T)$. Then we can write uniquely $X = U + N$ where $U \in L_0, N \in N$. By the remarks above, U is semisimple, hence ad U is also semisimple. Relative to ad U, we have the Fitting decomposition $N = N_0 \oplus N_1$, where $[N_0 U] = 0, [N_1 U] = N_1$. Thus we can write $N = N_0^{(1)} + [N_1 U]$, $N_0^{(1)} \in N_0, N_1 \in N_1$, so $X = U + N_0^{(1)} + [N_1 U]$. But now $(U+N_0^{(1)}) \exp(-(\text{ad } N_1)) \equiv U + N_0^{(1)} + [N_1 U] \pmod{[NN]}$; thus since $\exp(\text{ad } N_1)$ stabilizes $[NN]$ we have $X \exp(\text{ad } N_1) \equiv U + N_0^{(1)} \pmod{[NN]}$.

Now we proceed inductively: Suppose we have found $N_1, N_2, \ldots, N_{j-1}$, and $N_0^{(j-1)} \in N_0$, with $N_i \in N^i (= [NN^{i-1}])$ such that $X \exp(\text{ad } N_1) \ldots \exp(\text{ad } N_{j-1}) \equiv U + N_0^{(j-1)} \pmod{N^j}$.

16

Decompose the difference of the two sides of this congruence according to the Fitting decomposition of N^j relative to ad U to get:

$$X \exp(\text{ad } N_1)\ldots\exp(\text{ad }N_{j-1}) = U + N_0^{(j-1)} + M_0 + [M_1 U],$$

where $M_0 \in (N^j)_0$, $M_1 \in (N^j)_1$. Then let $N_j = M_1 \in (N^j)_1$ to get as before

$$X \exp(\text{ad } N_1) \cdots \exp(\text{ad } N_j) \equiv U + N_0^{(j-1)} + M_0 \pmod{N^{j+1}}.$$

Setting $N_0^{(j)} = N_0^{(j-1)} + M_0$, we have extended our procedure.

By Engel's theorem N is nilpotent, so that $N^{j+1} = 0$ for some j. In this case the procedure above gives

$$X \exp(\text{ad } N_1) \ldots \exp(\text{ad } N_j) = U + N_0, \text{ where } N_0 \in N, U \in L_0,$$

$[UN_0] = 0$. Thus U is the semisimple part and N_0 the nilpotent part of $U + N_0$. Now it is easy to verify that for N nilpotent, $X \exp(\text{ad } N) = \exp(-N) X (\exp N) = (\exp N)^{-1} X (\exp N)$. Setting $Y = (\exp N_1) \ldots (\exp N_j)$ above, we have $Y^{-1} XY = U + N_0$, $Y^{-1} P^+(T) Y = P^+(T)$, $Y^{-1} N Y = N$, and $Y^{-1} T Y$ is semisimple resp. nilpotent with T. Thus $X = Y U Y^{-1} + Y N_0 Y^{-1}$ is the decomposition of X into semisimple and nilpotent parts, and these lie in $P^+(T)$: $P^+(T)$ is almost algebraic. If X is semisimple, $X = Y U Y^{-1}$, or $Y^{-1} X Y = U \in L_0$, completing the proof of the Lemma.

Remark. If $X \in P^+(T)$ is nilpotent, then $Y N_0 Y^{-1} = X$, which is therefore in N. Thus N is the set of nilpotent elements of $P^+(T)$.

Corollary. If T' is a second maximal split toral subalgebra of $P^+(T)$, then T' is conjugate to T by an element $(\exp N_1) \ldots \exp(N_k)$ as in Lemma 9.

Proof. Considering the adjoint action of T' on $P^+(T)$, the roots of T' in $P^+(T)$ are linear functions, so we can choose $X \in T'$ such that no non-zero root vanishes at X. By Lemma 9, X has a conjugate in L_0, which must be in T since it is semisimple with characteristic roots in F. Let this conjugate be $Y^{-1} X Y$. Comparing centralizers, we have $C(Y^{-1} T'Y) = C(Y^{-1} X Y) \supseteq T$. Thus $Y^{-1} T'Y + T$ is commutative, generated by endomorphisms diagonalizable over F, so is itself diagonalizable. By the maximality of T, we have $Y^{-1} T' Y \subseteq T$. But then $T' \subseteq Y T Y^{-1}$, also a split toral subalgebra, from which $Y^{-1} T' Y = T$ by the maximality of T'.

17

Lemma 10. *Let* $N = N^+(T)$ *and* $P = P^+(T)$ *be as above. Then* P *is the normalizer of* N *in* L.

Proof. Let K be the normalizer of N in L. Clearly $P \subseteq K$.
Now $T \subseteq K$, so K decomposes relative to $\mathrm{ad}\,T$ into root-spaces, which are necessarily of the form $K \cap L_\beta$. If $K \cap L_\beta \neq 0$ for some β with $\beta < 0$, we have by Lemma 3 an $X_\beta \in K \cap L_\beta$, $Y_{-\beta} \in L_{-\beta} \subseteq N$, and $[X_\beta Y_{-\beta}] \neq 0$ in T, so that X_β does not normalize N. Thus $K \subseteq P$, and Lemma 10 is proved.

Lemma 11. $N = N^+(T)$ *is a maximal nil (i.e., consisting of nilpotent elements) subalgebra of* L.

Proof. Let M be a nil subalgebra of L properly containing N. Apply Engel's theorem to the representation of N on the quotient space M/N induced from the adjoint representation of N on M to obtain $M \in M$, $M \notin N$, $[M\,N] \subseteq N$. Thus $M \in K = P$, and M is nilpotent. But we have seen that the set of nilpotent elements of P is N (remark after the proof of Lemma 9). This contradiction establishes our result.

Lemma 12. *If the trace form* $(X,Y) = \mathrm{Tr}(XY)$ *is nondegenerate on* L, *then* N *is the radical of the restriction to* P *of this form.*

Proof. We must show $N = P \cap P^\perp$. Let $L = L_0 + \sum_{\alpha \neq 0} L_\alpha$ be the decomposition of L into root-spaces relative to T, $V = \Sigma_\lambda V_\lambda$ the decomposition of V into weight-spaces relative to T. From $V_\lambda L_\alpha \subseteq V_{\lambda+\alpha}$ it follows that $(L_\alpha, L_\beta) = 0$ unless $\alpha+\beta = 0$, so that the restriction to L_0 of the trace form is non-degenerate. It actually follows at once that $N = P \cap P^\perp$, but we give the following argument for later reference: N is an ideal in P consisting of nilpotent endomorphisms, from which it follows by [J], Theorem 2, p.35, that N is contained in the radical of the associative algebra of endomorphisms of V generated by P. In particular, NY is nilpotent for every $N \in N$, $Y \in P$, so that $(N,Y) = \mathrm{Tr}(NY) = 0$ and $N \subseteq P^\perp$. But $P = L_0 + N$, $(P^\perp \cap L_0) \subseteq L_0^\perp \cap L_0 = 0$, so $P \cap P^\perp = N + (L_0 \cap P^\perp) = N$.

The following proposition, which follows from more general considerations in [CII], [CIII], is given here as needed in our context:

Proposition 2. *Let* L *be algebraic. Then the trace form* $(X,Y) = Tr(XY)$ *is nondegenerate on* L.

Proof. Let S be the radical of the trace form. By Cartan's criterion ([J], p. 68) S is solvable, so by Lie's theorem, somewhat extended ([J], Theorem 7, p. 74), $[LS]$ consists of nilpotent endomorphisms of V. But L can contain no nil ideal $N \neq 0$, since VN would be an L-stable subspace of V, so $V = VN \oplus U$, with U L-stable, $UN \subseteq U \cap VN = 0$, and $VN = VN^2$. But then $VN = VN^2 = \ldots = VN^k = 0$ for k sufficiently large, by Engel's theorem, and $N = 0$. Thus $[LS] = 0$, and $S \subseteq Z$, the center of L, and it suffices to show that (X,Y) is nondegenerate on Z. Thus let $T \in Z$, $(T,Z) = 0$. By linearity, $(T,Z_K) = 0$, where the trace form is that of endomorphisms of V_K, K any extension field of F. Take K to be an extension containing all characteristic roots of T; let $v_1, \ldots, v_n$ be a basis for V_K with $v_i T = \lambda_i v_i (\lambda_i \in K)$, and so arranged that $\lambda_1, \ldots, \lambda_s$ form a basis for the $\mathbb{Q}$-subspace of K generated by $\lambda_1, \ldots, \lambda_n$. Thus $\lambda_1, \ldots, \lambda_s$ are $\mathbb{Q}$-linearly independent, and for $j > s$, we have $\lambda_j = \sum_{i=1}^{s} r_{ji} \lambda_i, r_{ji} \in \mathbb{Q}$. For $1 \leq i \leq s$, let Z_i be the endomorphism of V_K with $v_\ell Z_i = \delta_{i\ell} v_\ell$, $1 \leq \ell \leq s$; $v_j Z_i = r_{ji} v_j$, $s < j \leq n$. Now if $m_1, \ldots, m_n$ are integers with $\sum_{i=1}^{n} m_i \lambda_i = 0$, substituting for all λ_j, $j > s$, gives $\sum_{i=1}^{s} k_i \lambda_i = 0$, with $k_i = m_i + \sum_{j=s}^{n} m_j r_{ji}$, so that $k_i = 0$ for all i. By our definition of algebraicity in §1, $Z_i \in Z_K$, $1 \leq i \leq s$, so that $(T,Z_i) = 0$.

Now $(T,Z_i) = \lambda_i + \sum_{j=s+1}^{n} r_{ji} \sum_{k=1}^{s} r_{jk} \lambda_k$
$= \sum_{k=1}^{s} (\delta_{ik} + \sum_{j=s+1}^{n} r_{ji} r_{jk}) \lambda_k = 0$,

so that $1 + \sum_{j=s+1}^{n} r_{ji}^2 = 0$ for all i, which is absurd. The only case where no absurdity is involved is that where $s = 0$, i.e., where $T = 0$, which must be the case.

Remark 1. If $T = \text{diag}\{\lambda_1, \ldots, \lambda_n\}$ is a diagonal matrix, those diagonal matrices $S = \text{diag}\{\mu_1, \ldots, \mu_n\}$ such that $\sum_{i=1}^{n} m_i \lambda_i = 0$, $m_i \in \mathbb{Z}$, implies $\sum m_i \mu_i = 0$, are called the __replicas__ of T. We have shown that __if__ $\text{Tr}(TS) = 0$ __for all replicas__ S of T, __then__ $T=0$.

Remark 2. Let D be a space of diagonal matrices over a field F of characteristic zero. Let $T_1, \ldots, T_d$ be a basis for D, and suppose that each T_i has only rational eigenvalues. Then if $T \in D$, D contains all replicas of T. For we may permute positions on the diagonal, relabel, adjust by rational scalar factors, and add rational multiples of one basic T_j to another to assume
$T_j = \text{diag } \{0,0,\ldots,01,0,\ldots,0,r_{d+1,j},\ldots,r_{n,j}\}$, where the r_{kj} are rational, and where this holds for all j, $1 \leq j \leq d$. Then

$T = \Sigma_{j=1}^d \rho_j T_j = \text{diag }\{\lambda_1,\ldots,\lambda_n\}$, where $\lambda_j = \rho_j$, $1 \leq j \leq d$, and where $\lambda_k = \Sigma_{j=1}^d r_{kj} \rho_j$, $k > d$.

Now let $S = \text{diag}\{\sigma_1,\ldots,\sigma_n\}$ be a replica of T. We define $\mu_1,\ldots,\mu_d$ by $\mu_j = \sigma_j$, $1 \leq j \leq d$. Then we claim $S = \Sigma_{j=1}^d \mu_j T_j \in D$. For the two sides evidently agree in the first d positions; for k > d, the entry in the k-th position of $\Sigma_j \mu_j T_j$ is $\Sigma_j r_{kj}\mu_j = \Sigma_j r_{kj} \sigma_j = \sigma_k$ since $\lambda_k = \Sigma_j r_{kj} \rho_j = \Sigma_j r_{kj} \lambda_j$. Thus the assertion of Remark 2 is proved.

Lemma 13. Let L be algebraic, so that the trace form (X,Y)=Tr(XY) is nondegenerate on L. Let N be an arbitrary maximal nil subalgebra of L, P its normalizer. Then $N = P \cap P^\perp$.

Proof. As in the proof of Lemma 12, $N \subset P \cap P^\perp$. Now if $X \in P$, $X = S + N$ the decomposition of X into semisimple and nilpotent parts, we know that $S, N \in L$, and that ad S, ad N are polynomials in ad X, hence normalize N. Thus $S, N \in P$. Now $F N \cup N$ is a weakly closed set of nilpotent endomorphisms of V so that the subalgebra $FN + N$ is nil by [J], Theorem 1, p. 33, and $N \in N$ by maximality. Thus $X \in P \cap P^\perp$ implies $S \in P \cap P^\perp$, and we must show $S \in P \cap P^\perp$, S semisimple, implies $S = 0$.

Now $P \supseteq Z$, and $Z \cap Z^\perp = 0$ by the proof of Proposition 2. Writing $S = Z + T$, $Z \in Z$, $T \in [LL]$, we have Z semisimple, hence T semisimple by [SZ] = 0, $T \in P$, (S,Z) = 0 = (Z,Z), so that Z = 0, $S = T \in [LL]$ (that (Z, [LL]) = 0 is clear from (Z, [LL]) = ([ZL],L) = (0,L)) . Thus we have $S \in [LL]$, $S \in P \cap P^\perp$, S semisimple. Let K be the algebraic closure of F. Then $M = [LL]$ is semisimple, $S \in M \subseteq M_K$; we claim that if $R \in \text{Hom}_K(V_K, V_K)$ has a diagonal matrix which is a replica of that of S, then $R \in M_K$. That is, we know that $S \in M_K$ is semisimple, hence is contained in a maximal toral subalgebra H of M_K; since K is algebraically closed, H is necessarily split, and $M_K = [M_K M_K]$, so that if $\alpha_1, \ldots, \alpha_r$ are roots relative to H, forming a basis for H^*, $H_{\alpha_1}, \ldots, H_{\alpha_r}$ form a basis for H. Now we have $S = \Sigma_{i=1}^r \rho_i H_{\alpha_i}$, $\rho_i \in K$, and there is a basis $v_1, \ldots, v_n$ for V_K relative to which H has diagonal matrices. Since all eigenvalues of H_{α_i} are integral, we have the situation of Remark 2, by which H contains all replicas of S, and therefore M_K does.

Now it is a matter of linear equations to note that P_K is the normalizer of N_K in L_K, so that $S \in P \cap P^\perp$ is orthogonal to P_K with respect to the trace form on L_K (acting in V_K). Now let R be a replica of S, so $R \in L_K$ by the above. We claim $R \in P_K$ from which we shall have $\text{Tr}(RS) = 0$ and $S = 0$ by Remark 1 above. Letting $v_1, \ldots, v_n$ be a basis for V_K with $v_i S = \lambda_i v_i$, $v_i R = \mu_i v_i$, and letting E_{ij} be the corresponding matrix units: $v_k E_{ij} = \delta_{ik} v_j$, we have $[E_{ij}, S] = (\lambda_j - \lambda_i) E_{ij}$, $[E_{ij}, R] = (\mu_j - \mu_i) E_{ij}$, so that if $m_{ij} \in \mathbb{Z}$, $\Sigma_{i,j} m_{ij}(\lambda_j - \lambda_i) = 0$ we have $\Sigma_{i,j} m_{ij}(\mu_j - \mu_i) = 0$. Thus the matrix for ad R, acting in $\text{Hom}_K(V_K, V_K)$, is a replica of that for ad S.

What we wish to show is that $R \in P_K$, or that R normalizes N_K. This will complete the proof, and follows from

<u>Remark 3.</u> <u>Let</u> $A = \text{diag}\{\alpha_1, \ldots, \alpha_n\}$, <u>and let</u> $B = \text{diag}\{\beta_1, \ldots, \beta_n\}$ <u>be a replica of</u> A. <u>Then</u> B <u>may be written as a polynomial in</u> A.

This is really quite obvious; let $\alpha_1, \ldots, \alpha_r$ be the distinct characteristic roots of A and choose $p(X) \in K[X]$ with $p(\alpha_i) = \beta_i$, $1 \leq i \leq r$. Since $\alpha_i = \alpha_j$ implies $\beta_i = \beta_j$, we have $p(A) = B$.

Applying the remark to $A = \text{ad } S$, $B = \text{ad } R$, shows that ad R stabilizes N_K, and completes the proof of Lemma 13.

<u>Lemma 14.</u> <u>Let</u> L <u>be algebraic, so that the trace form</u> $(X,Y) = \text{Tr}(XY)$ <u>is nondegenerate on</u> L. <u>Let</u> T <u>be a maximal split toral subalgebra of</u> L, <u>and let</u> $N^+(T)$, $P^+(T)$ <u>be as above. Let</u> P <u>be an arbitrary maximal nil subalgebra of</u> L, <u>and let</u> P <u>be its normalizer.</u>

Then:

$$P^\perp = N, \quad N^\perp = P,$$
$$P^+(T) = (P \cap P^+(T)) + N^+(T),$$
$$P = (P \cap P^+(T)) + N.$$

Proof: By Lemma 13, $N = P \cap P^\perp$. From the fact that $([XY],Z) = (X,[YZ])$, we see by taking $X, Y \in P$, $Z \in P^\perp$, that $[PP^\perp] \subseteq P^\perp$, hence that $P^\perp$ is stable under ad N. By Engel's theorem there is a minimal k with $P^\perp (\text{ad } N)^k \subseteq N = P \cap P^\perp$, so that if $Y \in P^\perp$, $Z_1, \ldots, Z_{k-1}$, $X \in N$, then $[[[YZ_1] \ldots]Z_{k-1}]X] \in N$. Thus $P^\perp (\text{ad } N)^{k-1}$ normalizes N, so is contained in P. Since this space is also contained in $P^\perp$, we have violated the minimality of k unless $k = 0$. But $k = 0$ means $P^\perp \subseteq N$, from which the first two assertions follow.

For the remaining two assertions, it is sufficient by Lemmas 10 and 11 to prove the following: Let N_1, N_2 be two maximal nil subalgebras, with normalizers P_1, P_2. Then $P_1 = (P_1 \cap P_2) + N_1$. Consider $((P_1 \cap P_2) + N_1)^\perp = (P_1 \cap P_2)^\perp \cap N_1^\perp = (P_1^\perp + P_2^\perp) \cap P_1$
$= (N_1 + N_2) \cap P_1 = N_1 + (P_1 \cap N_2)$. Since N_1 is an ideal in P_1, the set $N_1 \cup (P_1 \cap N_2)$ is a weakly closed set of nilpotent endomorphisms of V, in the sense of [J], §II.1, so that $N_1 + (P_1 \cap N_2)$ is nil by [J], Theorem 1, p. 33. By the maximality of N_1 we have $N_1 + (P_1 \cap N_2) = N_1$, or $((P_1 \cap P_2) + N_1)^\perp = P_1^\perp$, hence $P_1 = (P_1 \cap P_2) + N_1$, and Lemma 14 is proved.

Proposition 3. Let R be a solvable Lie algebra of endomorphisms of V as above. Assume that R is almost algebraic. Then N, the set of nilpotent elements of R, is an ideal, and if T is any maximal toral subalgebra of R, we have $R = T + N$, a vector-space direct sum. Any two maximal toral subalgebras of R are conjugate via a product of elements $\exp(N)$, $N \in N$.

Proof. Since R is solvable, $[RR]$ is nil by Lie's theorem. Let N' be a maximal nil ideal containing $[RR]$, and let $N \in N$. Then $\{N\} \cup N'$ is nil and weakly closed, from which $FN + N'$ is a nil ideal by [J], p. 33, and $N \in N'$. Thus $N = N'$ is an ideal. By the proof of Proposition 1, the normalizer (= centralizer) H

22

of T in R is a Cartan subalgebra, and T is the set of semisimple elements of H. As above, the nilpotent elements of H are an ideal $N"$ in H, and from the fact that H is almost algebraic and T central in H we have $H = T \oplus N"$ (ideal direct sum). In the decomposition of R relative to H, the "Fitting one component" ([J], p. 39) R_1 is contained in $[RR] \subseteq N$. Thus $R = R_0 \oplus R_1 = H + N = (T \oplus N") + N = T + N$. Since $T \cap N$ is evidently zero, our first assertion is proved.

For the question of conjugacy, the proof previously given for Lemma 9 applies to show that if $U \in R$ is semisimple, then U has a conjugate of the required form in T. Now we modify the proof of the Corollary to Lemma 9 as follows: suppose $U \in T'$ is such that T' is the set of semi-simple elements of the centralizer of U. Then for a product Y of elements $\exp(N)$, $N \in N$, $Y^{-1} U Y \in T$, from which $T \subseteq Y^{-1} T' Y$, so that $T = Y^{-1} T' Y$. That such a "U" exists can be seen as follows: The centralizer H' of T' is a Cartan subalgebra of R, is almost algebraic, and the roots of H'_K in R_K vanish on the nilpotent elements (K being the algebraic closure), so are determined by their values on T'_K, therefore by their values on T', by linearity. Since there is an element of H'_K where no non-zero root vanishes, there is such an element U in T' (the base field F being infinite). Then ad U is non-singular on the "Fitting one component" of R relative to H', so that H' is the centralizer of U and T' is the set of semisimple elements of the same.

Now let N_1, N_2 be two maximal nil subalgebras of the completely reducible Lie algebra L of endomorphisms of V, and let P_1, P_2 be their respective normalizers in L. Let $M = P_1 \cap P_2$. By the fact that each P_i is almost algebraic, so is M. By Levi's theorem ([J], p. 91), there is a semisimple subalgebra K of M so that $M = K + R$, R the radical (maximal solvable ideal of M), and K is almost algebraic since it is necessarily completely reducible ([J], pp. 79 and 100). Moreover, R is <u>almost algebraic</u>. For if $X \in R$, then ad X maps M into R, hence so do ad S, ad N, where S, N are the semisimple resp. nilpotent parts of X. Write $S = S' + R$, $S' \in K$, $R \in R$; then if $S' \neq 0$, S' cannot be central in K, so there is $L \in K$,

$[LS'] \neq 0$, so that $[LS']+[LR] = [LS] \in R$, and $[LS'] \in K \cap R = 0$, a contradiction. Hence $S' = 0$, $S = R \in R$, and so $N = X - S \in R$. By Proposition 3, the set N of nilpotent elements of R is an ideal, and $R = T + N$, where T is a maximal torus. By the complete reducibility of the adjoint representation of T on M, we have $M = M_0 \oplus R$, both subspaces being stable under the adjoint action of T. This means that $[M_0 T] \subseteq M_0$; but since $T \subseteq R$, $[M_0 T] \subseteq R$, from which $[M_0 T] = 0$. Letting U be the centralizer of T in M, we have $U = M_0 \oplus (U \cap R)$, $U + R = M$, and $U/U \cap R \cong U + R/R = M/R$ is semisimple. Thus $U \cap R$ is the radical of U, and if $U = K_0 + (U \cap R)$ is a Levi decomposition ([J], p. 91), the subalgebra K_0 being semisimple, we have $M = U + R = K_0 + R = K_0 + T + N$, with $[K_0 T] = 0$. Thus $K_0 + T$ is a subalgebra of M, with K_0 semisimple and T toral and central, hence completely reducible ([J], p. 81). This establishes the following proposition, a slight modification of Proposition 5 on page 144 of [CIII].

<u>Proposition 4</u>. <u>Let</u> M <u>be an almost-algebraic Lie algebra of endomorphisms of</u> V. <u>Let</u> N <u>be the largest nil ideal in</u> M. <u>Then there is a completely reducible subalgebra</u> M_0 <u>of</u> M <u>with</u> $M = M_0 + N$. <u>The radical of</u> M <u>is</u> $T + N$, <u>where</u> T <u>is the center of</u> M_0.

(The only remark which perhaps should be made is that N is necessarily solvable, so is contained in R, whenever N is a nil ideal of M. The uniqueness of N and the rest of the proposition have then been proved.)

Now let T be a maximal split torus in L, and let $N_1 = N^+(T)$, $P_1 = P^+(T)$ be as before, relative to a chosen ordering of the roots relative to T. Let N_2 be a second maximal nil subalgebra of L, P_2 its normalizer.

<u>Lemma 15</u>. <u>There is an element</u> $Y = \Pi_i \exp(N_i)$, $N_i \in N_1$, <u>such that</u> $T \subseteq Y^{-1} P_2 Y$, <u>in fact such that</u> $L_0 = L_0(T) \subseteq Y^{-1} P_2 Y$.

<u>Proof</u>. By Lemma 14, $P_1 = (P_1 \cap P_2) + N_1$, and N_1 is the set of nilpotent elements of P_1. Since each P_i is almost algebraic (any normalizer in L is), we may apply Proposition 4 to $M = P_1 \cap P_2$ to get $M = M_0 \oplus N = [M_0 M_0] + Z_0 + N$, where M_0 is completely reducible, Z_0 its center, and N the maximal nil ideal of M, hence $N \subseteq N_1$. Thus $P_1 = M_0 + N_1$, a direct

sum since M_0 can contain no nil ideals. Now M_0 is almost algebraic and N_1 is the set of nilpotent elements of P_1, from which we see that every element of M_0 is semisimple.

As in the proof of Proposition 3, we choose $Z \in Z_0$ so that its centralizer in P_1 coincides with the centralizer of Z_0 there. By Lemma 9, there is Y, of the desired type with $Y^{-1} Z Y \in L_0$ (= $L_0(T)$). Thus T centralizes $Y^{-1} Z Y$, $Y T Y^{-1}$ centralizes Z, hence Z_0, or T centralizes $Y^{-1} Z_0 Y$, so that $Y^{-1} Z_0 Y \subseteq L_0$. Replacing N_2 by $Y^{-1} N_2 Y$, with normalizer $Y^{-1} P_2 Y$, we see by the fact that conjugation by Y stabilizes N_1 and P_1 that we may assume $Z_0 \subseteq L_0$.

With $Z \in Z_0$ as above, let C be its centralizer in P_1. Since $Z \in L_0$, we have $T \subseteq C$, so that the decomposition of C into root-spaces relative to T yields the decomposition $C = (C \cap L_0) + (C \cap N_1)$, with $M_0 \subseteq C$, $[M_0 \ M_0]$ being semisimple. As a nil ideal, $C \cap N_1$ is clearly in the radical of C, and $C \cap L_0$ is completely reducible since both L_0 and $F.Z \subseteq L_0$ are ([J], p. 102, Theorem 18). It follows that a Levi decomposition of C is given by

$$C = \underbrace{[C \cap L_0, C \cap L_0]}_{\text{semisimple}} + \underbrace{\{ Z(C \cap L_0) + (C \cap N_1) \cdot \}}_{\text{radical}}$$

By the theorem of Malcev and Harish-Chandra ([J], p. 92) there is an element $Y = \Pi \exp(N_i)$, $N_i \in C \cap N_1$, with $Y^{-1}[M_0 \ M_0] Y \subseteq [C \cap L_0$, $C \cap L_0] \subseteq L_0$. Now the N_i centralize Z hence Z_0, so that $Y^{-1} Z_0 Y = Z_0 \subseteq L_0$, from which $Y^{-1} M_0 Y \subseteq L_0$. Now $P_1 = Y^{-1} P_1 Y = Y^{-1}(M_0 + M_1) Y = Y^{-1} M_0 Y + N_1$. But $P_1 = L_0 + N_1$, a vector space direct sum, from which $Y^{-1} M_0 Y = L_0 \supseteq T$. Since $M_0 \subseteq M = P_1 \cap P_2$, we have $L_0 \subseteq Y^{-1} P_2 Y$, and the lemma is proved. (Actually the description of Y obtained from the statement in [J] is not as explicit as ours; however the proof involves, on p. 93, exactly such a choice as our N_i.)

<u>Lemma 16</u>. Let T be as above, N <u>a maximal nil subalgebra of</u> L, P <u>its normalizer, and suppose that</u> $L_0(T) \subseteq P$. <u>Let</u> P <u>be the set of roots</u> $\alpha \neq 0$ <u>relative to</u> T <u>such that</u> $P \cap L_\alpha \neq 0$. <u>Then</u> $P = L_0 + \Sigma_{\alpha \in P} L_\alpha$, <u>and there is</u> $w \in W$, <u>the Weyl group relative to</u> T, <u>such that</u> P^w <u>contains the set of positive roots in our</u>

given ordering.

Proof. Since $T \subseteq L_0 \subseteq P$, we have $P = L_0 + \sum_{\alpha \in \Sigma} (P \cap L_\alpha)$, and $[L_0, P \cap L_\alpha] \subseteq P \cap L_\alpha$. By Lemma 7, L_0 acts irreducibly in L_α for $\alpha \neq 0$, so that $P \cap L_\alpha = L_\alpha$ if $P \cap L_\alpha \neq 0$. Thus $P = L_0 + \sum_{\alpha \in \Sigma_P} L_\alpha$. Now if $\alpha, \beta \in P$, we have $[L_\alpha L_\beta] \subseteq P$, and it follows from the representation theory of a 3-dimensional algebra formed as in Lemma 3 starting with $X_\beta \in L_\beta$ that $[L_\alpha L_\beta] \neq 0$ if $\alpha + \beta \in \Sigma$. Thus $\alpha + \beta \in \Sigma$ implies $\alpha + \beta \in P$, and P satisfies the first condition of Lemma 8. Next let $\alpha \in \Sigma$, $\alpha \notin P$. Since $(L_\alpha, L_\beta) = 0$ if $\alpha + \beta \neq 0$, we have either $-\alpha \in P$ or $(L_\alpha, P) = 0$. In the latter case $L_\alpha \subseteq P^\perp = N \subseteq P$, a contradiction. Thus the second condition of Lemma 8 is satisfied, and our last assertion follows by that Lemma.

Corollary. P^w *is exactly the set of positive roots.*

For if both α and $-\alpha$ are in P, then since both $L_\alpha, L_{-\alpha}$ consist of nilpotent elements we have $L_\alpha \subseteq N$, $L_{-\alpha} \subseteq N$, and in the trace form $(L_\alpha, L_{-\alpha}) \neq 0$, so that $L_\alpha \subseteq N = P^\perp$, a contradiction. Thus exactly one of α, $-\alpha$ is in P for each $\alpha \in \Sigma$, and the same holds for P^w, which therefore contains no negative root.

Lemma 17. *Let $\alpha \in \Sigma$, α assumed primitive. Let $0 \neq X_\alpha \in L_\alpha$, $X_{-\alpha} \in L_{-\alpha}$, $[X_\alpha X_{-\alpha}] = H_\alpha$ as in §1. Let $0 \neq \xi \in F$, and let $A = A(\xi) = \exp(\xi X_{-\alpha}) \exp(\xi^{-1} X_\alpha) \exp(\xi X_{-\alpha})$, an automorphism of V. Define $X^\sigma = A^{-1} X A$ for $X \in L$. Then σ is an automorphism of L, $T^\sigma = T$, and for $\beta \in \Sigma$, $(L_\beta)^\sigma = L_{\beta w_\alpha}$.*

Proof. That σ is an automorphism of L follows since X_α, $X_{-\alpha}$ are nilpotent and in L (in fact in $N^+(T) \cup N^-(T)$ relative to our given ordering on Σ). If $T \in T$, $\alpha(T) = 0$, then one sees at once that $T^\sigma = T$ (i.e., A commutes with T since $X_\alpha, X_{-\alpha}$ do). If $T = H_\alpha$, one verifies directly that $H_\alpha^\sigma = H_\alpha \exp(\text{ad } \xi X_{-\alpha}) \cdot \exp(\text{ad } \xi^{-1} X_\alpha) \exp(\text{ad } \xi X_{-\alpha}) = -H_\alpha$ (cf. [J], p. 275, proof of Proposition 2). Since every element of T has the form $\lambda H_\alpha + T$, with $\alpha(T) = 0$, we have $T^\sigma = T$.

Now let $\beta \in \Sigma$, $E \in L_\beta$. For $T \in T$, $\alpha(T) = 0$, we have $[E, T]^\sigma = [E^\sigma, T^\sigma] = [E^\sigma, T]$, while $[E, T] = \beta(T) E$ gives $[E, T]^\sigma = \beta(T) E^\sigma$. Thus $[E^\sigma, T] = \beta(T) E^\sigma = \beta^{w_\alpha}(T) E^\sigma$ if $\alpha(T) = 0$. Likewise,

$$[E^\sigma, H_\alpha] = -[E^\sigma, H_\alpha^\sigma] = -[E, H_\alpha]^\sigma = -\beta(H_\alpha)E^\sigma$$
$$= (\beta - 2\beta(H_\alpha)\alpha)(H_\alpha)E^\sigma = \beta^{w_\alpha}(H_\alpha)E^\sigma .$$

(cf. (1) of §2 for $r_{\beta\alpha} = \beta(H_\alpha) = 2(\beta,\alpha)(\alpha,\alpha)^{-1}$).

Thus $E^\sigma \in L_{\beta w_\alpha}$; since σ is an automorphism and L is a direct sum of root spaces (including L_0), we must have $(L_\beta)^\sigma = L_{\beta w_\alpha}$.

We can now prove the main conjugacy theorem.

Theorem 1. Let L be a completely reducible algebraic Lie algebra of endomorphisms of V, a finite-dimensional vector space of characteristic zero. Let T be a maximal split toral subalgebra of L, L_0 the centralizer of T, Σ^+ the set of positive roots in some ordering of the set Σ of roots relative to T, $N^+(T)$ the sum of all L_α for $\alpha \in \Sigma^+$. Then $N^+(T)$ is a maximal nil subalgebra of L with normalizer $P^+(T) = L_0 + N^+(T)$, and $N^+(T)$ is the set of nilpotent elements of $P^+(T)$. If N is any maximal nil subalgebra of L, there are elements $N_1, \ldots, N_k \in N^+(T) \cup N^-(T)$ such that if $Y = \Pi_{i=1} \exp(N_i)$, $Y^{-1} NY = N^+(T)$. Hence any two maximal nil subalgebras of L are conjugate by such a Y.

Proof. Let N be a maximal nil subalgebra, P its normalizer. By Lemma 15, Lemma 16 and its corollary we have $Y = \Pi \exp N_i$, $N_i \in N^+(T)$, with $Y^{-1} PY = L_0 + \Sigma_{\alpha \in P} L_\alpha$, and $w \in W$ with $P^w = \Sigma^+$. Let $w = w_{\alpha_1} \cdots w_{\alpha_j}$, $\alpha_i \in \Sigma$ (not necessarily distinct); let $\xi_1, \ldots, \xi_j \in F^*$; then $A_{\alpha_1}(\xi_1) \cdots A_{\alpha_j}(\xi_j)$ is an element B satisfying the requirements imposed on Y in the statement of the theorem, and $B^{-1} TB = T$, $B^{-1} L_\alpha B = L_{\alpha w}$ for all $\alpha \in \Sigma$ by Lemma 17. It follows that $(YB)^{-1} P(YB) = P^+(T)$, and that $(YB)^{-1} N (YB)$, the set of nilpotent elements of $P^+(T)$, is $N^+(T)$ as we have seen in Lemma 11.   Q.E.D.

What has been done above enables us to complete at once the proof of the following:

Theorem 2. Let L be as in Theorem 1. Let S and T be maximal split toral subalgebras of L. Then there exists $Y = \Pi \exp(N_i)$, $N_i \in N^+(T) \cup N^-(T)$, such that $Y^{-1} S Y = T$.

Proof. Let $N^+(S)$, $P^+(S)$ be defined relative to some ordering of the roots relative to S. By the proof of Theorem 1, there is such Y with $Y^{-1} P^+(S) Y = P^+(T)$. Thus $Y^{-1} S Y$ is a maximal split toral subalgebra of $P^+(T)$, and the theorem follows by the Corollary to Lemma 9.

Corollary. Let L be a split semi-simple Lie algebra over F, in the sense of [J], Chapter IV. Let H_1, H_2 be two splitting Cartan subalgebras of L. Let N^+, N^- be respectively the sums of the positive and the negative root-spaces in some ordering of the roots relative to H_1. Then there are $x_1, \ldots, x_k \in N^+ \cup N^-$ such that $H_2 \Pi_i \exp(\mathrm{ad}\, x_i) = H_1$. (one needs only apply Theorem 2 to the adjoint representation of L).

It will be noted that this result is stronger than Theorem 3 of [J], Chapter 9 (p. 273)-cf. also ex. 18, p. 288 of [J]-in that it does not require that F be algebraically closed. It is a little weaker in that the form of our automorphism is somewhat less precise than is his. Using the Bruhat decomposition of the Chevalley group. Steinberg has shown that an exact analogue of the result in [J] holds for splitting Cartan subalgebras over general fields, even in prime characteristic - cf. Steinberg, Pacific Jour. Math. 1961, 1119-1129, or Seligman, Modular Lie Algebras, §III. 4. Actually, it is not hard to show that every $\exp(\mathrm{ad}\, x_i)$ as above may be written as a product of $\exp(\mathrm{ad}\, y_\gamma)$, where y_γ are root-vectors (corresponding to roots of the same "sign" as x_i), and thereby to obtain the full analogue (for characteristic zero) of the result in [J]. This is left as an exercise.

Problem. Apply the considerations above to prove the following:
Let L be a semi-simple Lie algebra over an algebraically closed field of characteristic zero, and let B be a maximal solvable subalgebra of L, T a Cartan subalgebra of L. Then there are $x_i \in N^+(T) \cup N^-(T)$ such that, $B \Pi_i \exp(\mathrm{ad}\, x_i) = T + N^+(T)$. (Consider the adjoint representation of B on L, and apply Proposition 3 to get $B = T_1 + N_1$ as there. Show that B is its own normalizer, and that if U is the normalizer of N_1, $U \neq B$ implies there is $u \in U$, $u \notin B$, with $[ut] = \lambda(t)u$ for all $t \in T_1$. Then $B' = T_1 + (Fu + N_1)$ contradicts the maximality of B. Hence B is the normalizer of

N_1, and N_1 is the set of nilpotent elements of B. Now show N_1 is a maximal nil subalgebra, by using the fact that if M is a nil algebra containing N_1 properly, N_1 is not its own normalizer in M.)

For related questions, see also [W], Chapter 3, and [HU], Chapter 4.

CHAPTER II

THE AUTOMORPHISM GROUPS

Until we have some new lemmas, etc., we shall refer to "Lemma n", meaning Lemma n of Chapter I. Later, this will be referred to as Lemma I.n, and "Lemma m" will be Lemma m of the current chapter. The same scheme will be followed in subsequent chapters.

1. Generalities.

Let L, T, H, $N = N^+(T)$, $P = P^+(T)$, etc., be as in Chapter I; in particular, the roots relative to T are ordered as in §I.1. We further assume L algebraic. If L is semisimple, and if σ is an automorphism of L; or, more generally, if L_1 and L_2 are semisimple and σ is an isomorphism of L_1 onto L_2, then σ maps toral (resp. split toral, resp. nil subalgebras onto subalgebras of the same kind. If S is a split toral subalgebra of L_{1K} for a field extension K/F, so is $S^\sigma \subseteq L_{2K}$, and elements of S and of S^σ with all eigenvalues rational (in any faithful representation, hence in the adjoint representation, hence in all representations) correspond.

To confirm the underlined assertions, we note that nil subalgebras (for any faithful representation) may be characterized as those for which all ad X are nilpotent, a property preserved by isomorphisms. Again the adjoint representation shows that any commutative subalgebra with all ad X semisimple is contained in a Cartan subalgebra; by passing to a splitting field, one sees that the latter is represented by semisimple transformations in any representation of L. Thus such a subalgebra is toral, and this property characterizes toral subalgebras, showing that σ maps toral subalgebras onto toral subalgebras. Split toral subalgebras (relative to any faithful representation) are represented by the same kind in the adjoint representation, and a toral subalgebra T whose image in the adjoint representation is split may be embedded in a maximal split toral subalgebra ad S of ad L. As in Chapter I, ad S is generated by elements ad H_α, where α is a root of ad L relative to ad S, and where (ad X) $\in$ (ad L)$_\alpha$, (ad Y) $\in$ (ad L)$_{-\alpha}$ have [ad X, ad H_α] = 2 ad X, [ad Y, ad H_α] = $-$ 2 ad Y, [ad X, ad Y] = ad H_α. Since the adjoint representation is faithful, we may drop "ad' throughout to see that $\{X, Y, H_\alpha\}$ is the split 3-dimensional algebra, so that all characteristic roots of H_α are integers and S is split (in any representation). Thus S^σ is also split toral, so T^σ is. Finally, if a toral subalgebra S of L is given, elements of S with rational eigenvalues (in some faithful representation) remain so under extension of the base field to a splitting field for some Cartan subalgebra H containing S. But elements of H with rational eigenvalues (in the extended representation, still faithful) have rational eigenvalues in the adjoint representation, so are rational linear combinations of the H_α's as above. By the representation theory of the 3-dimensional algebra, the H_α^σ have rational eigenvalues in all representations, and therefore all rational combinations of them do, too. Combined with the preservation of semisimplicity under extension of the ground field, this completes our argument.

An isomorphism $\sigma: L_1 \to L_2$ of reductive algebraic (linear) Lie algebras, given as acting in V_1 resp. V_2, will be called <u>rational</u> if both σ and σ^{-1} have the underlined properties above, where now 'nil', 'split', 'semisimple' refer to the given actions in V_1 resp. V_2. In particular, σ is rational if σ is of the form $X \mapsto A^{-1} X A$, where $A: V_1 \to V_2$ is an isomorphism of vector spaces.

Let σ be a rational automorphism of L. Then T^σ is again a maximal split toral subalgebra of L, N^σ a maximal nil subalgebra and P^σ its normalizer. By Lemma 15, there is an automorphism τ of L, consisting of conjugation by an element $Y = \Pi \exp(N_i)$, $N_i \in N$, such that L_0 is contained in $P^{\sigma\tau}$ (L_0 = centralizer of T), By Lemmas 16, 17 and Theorem 1, there is an automorphism η, stabilizing T, consisting of conjugation by an element $\Pi \exp N_i$, $N_i \in N \cup N^-(T)$, such that $P^{\sigma\tau} = P^\eta$. Indeed, the chosen element η induces in the roots relative to T an element w of the Weyl group W of L relative to T. Lemma 17 provides the basis for making such a choice for each $w \in W$; we assume one such choice $\omega(w)$

made, with $\omega(w_\alpha)$ as in Lemma 17 for simple roots α, and fixed for each $w \in W$, with $\omega(1) = 1$, and thus have $P^{\sigma\tau} = P\,\omega(w)$ for some $w \in W$. Thus we have $P\,\sigma\tau\,\omega(w)^{-1} = P$, $T\,\sigma\tau\omega(w)^{-1} \subseteq P$. By the Corollary to Lemma 9, there is an automorphism ρ, consisting of conjugation by a product of $\exp(N_i)$, $N_i \in N$, such that $T\,\sigma\tau\,\omega(w)^{-1}\rho = T$, while we still have $P\,\sigma\tau\,\omega(w)^{-1}\rho = P$.

We adopt the following notations: G = rational automorphism group of L; G' = subgroup of G generated by all $\exp(\mathrm{ad}\,X)$, $X \in N \cup N^-(T)$ (= subgroup generated by conjugation by all $\exp N$, $N \in N \cup N^-(T)$); U^+ (or U^-) = subgroup of G' generated by all $\exp(\mathrm{ad}\,X)$, $X \in N$ (or $X \in N^-(T)$); B = stabilizer of P in G; $B' = B \cap G'$; N = stabilizer of T in G; $N' = N \cap G'$; $H = N \cap B$; $H' = N' \cap B' = H \cap G'$; for $\alpha \neq 0$ a primitive root relative to T, $U_{(\alpha)}$ = subgroup of G' generated by all $\exp(\mathrm{ad}\,X)$, $X \in L_\alpha \cup L_{2\alpha}$.

Then our above considerations show that every $\sigma \in G$ can be written $\sigma = \nu\rho^{-1}\omega(w)\tau^{-1}$, $\nu \in H$. Since H stabilizes the set N of nilpotent elements of P, H normalizes U^+, so $\nu\rho^{-1}$ has the form $\mu\nu$, $\mu \in U^+$. Thus $\sigma = \mu\nu\,\omega(w)\tau^{-1} \in U^+ H\,\omega(w)U^+ \subseteq B\,\omega(w)B$, and we have proved the following

<u>Proposition 1.</u> $G = \bigcup_{w \in W} B\,\omega(w)B$, <u>where</u> $\omega(w) \in G'$ <u>is as above.</u> <u>Also</u> $B = U^+ H = H U^+$, <u>and we have in fact</u> $G = \bigcup_w B\,\omega(w)U^+$.

If α is a primitive root relative to T, let $L^{(\alpha)}$ be the subalgebra of L generated by H (a maximal toral subalgebra containing T), $L_{\pm\alpha}$, $L_{\pm 2\alpha}$. Since $T \subseteq H \subseteq L^{(\alpha)}$, we see that $L^{(\alpha)} = \Sigma_\beta(L_\beta \cap L^{(\alpha)})$. Now $(L_0 \cap L^{(\alpha)}) + L_\alpha + L_{2\alpha} + L_{-\alpha} + L_{-2\alpha}$ is a subalgebra of $L^{(\alpha)}$ containing H, $L_{\pm\alpha}$, $L_{\pm 2\alpha}$, so must be $L^{(\alpha)}$. The derived algebra of $L^{(\alpha)}$ contains $L_{\pm\alpha}$, $L_{\pm 2\alpha}$, and the center of $L^{(\alpha)}$ is contained in H, since it centralizes H. Since H is algebraic (Proof of Lemma I.13), $L^{(\alpha)}$ has algebraic center. The radical R of $L^{(\alpha)}$ is stable under the adjoint action of T and $R \cap L_\beta = 0$, $\beta = \pm\alpha, \pm 2\alpha$, by the embeddability of nonzero elements of these L_β in simple 3-dimensional subalgebras of $L^{(\alpha)}$ (Lemma I.3). Thus $R = R \cap L_0$ consists of semi-simple elements (cf. page 23.) Now if $X \in R$, we know that $[L^{(\alpha)} X] \subseteq R$

consists of nilpotent elements, by Lie's theorem, so is zero. That is, R is central in $L^{(\alpha)}$; so $R =$ center of $L^{(\alpha)}$, and Levi's theorem now gives $L^{(\alpha)} = Z(L^{(\alpha)}) \oplus [L^{(\alpha)}, L^{(\alpha)}]$, with $[L^{(\alpha)}, L^{(\alpha)}]$ semi-simple. If $T \in T$, $\alpha(T) = 0$, then $T \in Z(L^{(\alpha)})$, while $H_\alpha, L_{\pm\alpha}, L_{\pm 2\alpha}$ are all contained in $[L^{(\alpha)}, L^{(\alpha)}]$.

We apply Proposition 1 to $L^{(\alpha)}$, with $N = L_\alpha + L_{2\alpha}$, $N^- = L_{-\alpha} + L_{-2\alpha}$, relative to the maximal split torus T. The Weyl group $W^{(\alpha)}$ must transform α to $\pm\alpha$, and an element of $W^{(\alpha)}$ is determined by its effect on α. By the existence of the reflection w_α, we have $|W^{(\alpha)}| = 2$, and the automorphism group $G^{(\alpha)}$ of $L^{(\alpha)}$ has $G^{(\alpha)} = U^+ H \, \omega(w_\alpha) U^+ \cup U^+ H$ (the latter since H normalizes U^+ and $\omega(1) = 1$). Here an element Θ of H stabilizes T and maps $L_\alpha + L_{2\alpha}$ to itself; since Θ permutes root-spaces and $L_{2\alpha}$ is the center of $L_\alpha + L_{2\alpha}$ (Lemma 7), Θ must map $L_{\pm 2\alpha}$ to $L_{\pm 2\alpha}$, $L_{\pm\alpha}$ to $L_{\pm\alpha}$. For later reference, we summarize these comments as

<u>Lemma 1.</u> Let α <u>be a primitive root</u>, $L^{(\alpha)}$ <u>as above. Then</u> $L^{(\alpha)}$ <u>is completely reducible algebraic; if</u> G, H, $\omega(w)$, U^+ <u>are as above for</u> $L^{(\alpha)}$, <u>we have</u> $G = U^+ H \cup U^+ H \, \omega(w_\alpha) U^+$, <u>and the elements of</u> H <u>stabilize each root-space contained in</u> $L^{(\alpha)}$.

One should also keep in mind that $\omega(w_\alpha)$, and, indeed, all elements of the group "G'" associated with $L^{(\alpha)}$, are the restrictions to $L^{(\alpha)}$ of automorphisms, in the subsets designated by the same letters, of the group G' associated with L. As such, we see that every element ρ of G' in the subgroup generated by $U_{(\alpha)}$, $U_{(-\alpha)}$ has a restriction to $L^{(\alpha)}$ of one of the forms $\sigma\Theta$, $\sigma\Theta \omega(w_\alpha)\tau$, where $\sigma, \tau \in U_{(\alpha)}$, and where Θ stabilizes T, $L_{\pm\alpha}$, $L_{\pm 2\alpha}$, $\Theta \in G'$. If $T \in T$, $\alpha(T) = 0$, T is fixed by $U_{(\alpha)}$, $U_{(-\alpha)}$, hence by Θ. Also, we know that if $X_\alpha, Y_{-\alpha} \in L_\alpha, L_{-\alpha}$, with $[X_\alpha Y_{-\alpha}] \in T$, $\alpha([X_\alpha Y_{-\alpha}]) = 2$, then $[X_\alpha Y_{-\alpha}] = H_\alpha$ (Corollary to Lemma 1.4); hence $[X_\alpha^\Theta Y_{-\alpha}^\Theta] \in T^\Theta = T$, $X_\alpha^\Theta \in L_\alpha$, $Y_{-\alpha}^\Theta \in L_{-\alpha}$, $\alpha([X_\alpha^\Theta Y_{-\alpha}^\Theta]) = 2$, so that $H_\alpha^\Theta = [X_\alpha^\Theta Y_{-\alpha}^\Theta] = H_\alpha$, and Θ <u>fixes</u> T. There follows the

<u>Corollary.</u> Let $\rho \in \langle U_{(\alpha)}, U_{(-\alpha)} \rangle \subseteq G'$. <u>Then either</u> $\rho = \sigma\Theta'$ <u>or</u> $\rho = \sigma\Theta' \omega(w_\alpha)\tau$, <u>where</u> $\sigma,\tau \in U_{(\alpha)}$, $\omega(w_\alpha)$ <u>is as before, and</u> $\Theta' \in \langle U_{(\alpha)}, U_{(-\alpha)} \rangle$ <u>fixes</u> T.

For let σ, τ, Θ be as above. Then $\sigma^{-1} \rho \tau^{-1} \omega(w_\alpha)^{-1}$ is in $<U_{(\alpha)}, U_{(-\alpha)}>$ (or $\sigma^{-1}\rho$, in the former case) and, depending on the case, one of these agrees on T with Θ as above, hence fixes T. We need only set $\Theta' = \sigma^{-1}\rho$ resp. $\sigma^{-1}\rho\tau^{-1}\omega(w_\alpha)^{-1}$ to have the corollary.

2.   The structure of U^+.

<u>Lemma 2</u>.   Let α, β <u>be positive roots relative to</u> T. <u>Then one of the following holds</u>:

a) $\alpha + \beta$ <u>is not a root</u>;
b) $\alpha + \beta$ <u>is a root, but neither</u> $\alpha + 2\beta$ <u>nor</u> $\beta + 2\alpha$ <u>is</u>;
c) $\alpha + \beta$ <u>and</u> $\alpha + 2\beta$ <u>are roots, but none of</u> $2\alpha + \beta$, $2\alpha + 2\beta$, $\alpha + 3\beta$ <u>is</u>;
d) c) <u>holds with</u> α, β <u>interchanged</u>;
e) $\alpha + \beta$, $\alpha + 2\beta$, and $2\alpha + 2\beta$ are roots, but none of $2\alpha + \beta$, $\alpha + 3\beta$, $3\alpha + 2\beta$, $2\alpha + 3\beta$ is;
f) e) <u>holds with</u> α, β <u>interchanged</u>;
g) $\alpha + \beta$, $\alpha + 2\beta$, $2\alpha + \beta$ <u>are roots, but none of</u> $2\alpha + 2\beta$, $\alpha + 3\beta$, $3\alpha + \beta$ <u>is</u>;
h) $\alpha + \beta$, $\alpha + 2\beta$, $\alpha + 3\beta$, $2\alpha + 3\beta$ <u>are roots, but none of</u> $2\alpha + \beta$, $2\alpha + 2\beta$, $\alpha + 4\beta$, $3\alpha + 3\beta$, $2\alpha + 4\beta$ <u>is</u>;
i) h) <u>holds with</u> α, β <u>interchanged</u>.

<u>Proof</u>.   We may assume that $\alpha + \beta$ and one of $\alpha + 2\beta$, $\beta + 2\alpha$ is a root, say $\alpha + 2\beta$ - the conclusions of the lemma are symmetric in α, β. Since only $\gamma, 2\gamma$ can be roots among the positive multiples of γ, it follows that α, β are linearly independent, so that $A_{\alpha,\beta} A_{\beta,\alpha} = 4(\alpha,\beta)^2 (\alpha,\alpha)^{-1}(\beta,\beta)^{-1}$ is an integer between 0 and 3. In fact. $A_{\alpha,\beta} < 0$ (hence $A_{\beta,\alpha} < 0$), since $A_{\alpha,\beta} = 2(\alpha,\beta)(\beta,\beta)^{-1} = r_{\alpha\beta}$ in the notation of Chapter I, so that $A_{\alpha,\beta} \geq 0$ would imply that $\alpha - \beta$, $\alpha - 2\beta$ are roots, and one of $\gamma = \alpha - 2\beta$, $\alpha - \beta, \alpha, \alpha + \beta, \alpha + 2\beta$ would have $A_{\gamma,\beta}$ outside the admissible interval $[-3,3]$ (since $A_{\gamma+\beta,\beta} = A_{\gamma,\beta} + 2$). If $A_{\alpha,\beta} = -2$, then not all of $\alpha - \beta, \alpha, \alpha + \beta$, $\alpha + 2\beta$, $\alpha + 3\beta$ can be roots, for these same reasons, so that neither $\alpha - \beta$ nor $\alpha + 3\beta$ is a root; moreover $A_{\beta,\alpha} = -1$. By the fact that $A_{\beta,\alpha} = -1$ and knowing $\beta - \alpha$ is not a root, we see that $\beta + 2\alpha$

is not a root. If $2\alpha+2\beta$ is a root, we have $A_{2\alpha+2\beta,\alpha} = 2A_{\alpha+\beta,\alpha}$ (by the scalar product formula (1) of §I.2), and since neither $2\alpha + \beta$ nor $\beta - 2\alpha$ is a root, $A_{\alpha+\beta,\alpha} = 1$, $A_{2\alpha+2\beta,\alpha} = 2$, from which 2β is a root. If in addition $3\alpha + 2\beta$ were a root, all of $3\alpha + 2\beta$, $2\alpha+2\beta$, $\alpha + 2\beta$, 2β, $-\alpha +2\beta$ would be, and this gives an inadmissible $A_{\gamma,\alpha}$ as above. Also $A_{2\alpha+2\beta,\beta} = 2A_{\alpha+\beta,\beta} = 0$ since $(\alpha+\beta) \pm \beta$ are roots, while neither of $(\alpha+\beta) \pm 2\beta$ is. Since $(2\alpha+2\beta) - \beta$ is not a root, neither is $2\alpha + 3\beta$. Thus $A_{\alpha,\beta} = -2$ implies that we have either case c) or case e).

If $A_{\alpha,\beta} = -1$, then $\alpha - \beta$ is a root since $\alpha + 2\beta$ is, and $\alpha - \beta$, α, $\alpha+\beta$, $\alpha + 2\beta$ must be the complete "β-string through α". Thus $A_{\alpha-\beta,\beta} = -3$, $A_{\beta,\alpha-\beta} = -1$, $A_{\beta,\alpha} = -1, -2$ or -3. If $A_{\beta,\alpha} \leq -2$, we have $\beta - \alpha$, β, $\beta+\alpha$, $\beta + 2\alpha$, $\beta + 3\alpha$ roots, impossible as above. Hence $A_{\beta,\alpha} = -1$, and $\beta + \alpha, \beta + 2\alpha$ are roots, but not $\beta + 3\alpha$. As we have seen above, $\alpha + 3\beta$ is not a root. Finally, we see that if $2\alpha + 2\beta$ were a root, then from $A_{2\alpha+2\beta,\alpha} = 2 = A_{2\alpha+2\beta,\beta}$, so would be 2β, 2α, from which $A_{\alpha,2\beta} = -\frac{1}{2}$ by the scalar product formula, and this is not an integer. Thus we have the case g).

Finally, if $A_{\alpha,\beta} = -3$, then the β-string through α is α, $\alpha+\beta$, $\alpha+2\beta$, $\alpha+3\beta$, and $A_{\beta,\alpha} = -1$. Since $\alpha - \beta$ is not a root, neither is $2\alpha + \beta$. If $2\alpha+2\beta$ were a root, we have $A_{2\alpha+2\beta,\alpha} = 2$, 2β is a root, and $A_{\alpha,2\beta} = -\frac{3}{2}$, a contradiction. Hence $2\alpha + 2\beta$ is not a root. Now $A_{\alpha+3\beta,\alpha} = -1$, so that $2\alpha + 3\beta$ is a root. Since $A_{2\alpha+3\beta,\beta} = 0$ and $2\alpha+2\beta$ is not a root, $2\alpha +4\beta$ is not a root, and $3\alpha + 3\beta = 3(\alpha+\beta)$ cannot be a root. Thus we have case h), and Lemma 2 is proved.

Lemma 3. <u>Let</u> α, β <u>be as in Lemma 2. Let</u> $X \in L_\alpha$, $Y \in L_\beta$. <u>With the same labeling of cases as in Lemma 2, and</u> $(u,v) = u^{-1}v^{-1}uv$, <u>we have</u> $(\exp(\text{ad } X), \exp(\text{ad } Y)) =$

a) 1;

b) $\exp(\text{ad}[XY])$;

c) $\exp(\text{ad}[XY]) \exp(\text{ad}\frac{1}{2}[[XY]Y])$;

d) $\exp(\text{ad}[XY]) \exp(\text{ad}\frac{1}{2}[[XY]X])$;

e) $\exp(\text{ad}[XY]) \exp(\text{ad}\frac{1}{2}[[XY]Y])$;

f) $\exp(\text{ad}[XY]) \exp(\text{ad}\frac{1}{2}[[XY]X])$;

g) $\exp(\text{ad}[XY]) \exp(\text{ad}\frac{1}{2}[[XY]Y]) \exp(\text{ad}\frac{1}{2}[[XY]X]);$

h) $\exp(\text{ad}[XY]) \exp(\text{ad}[[XY]Y]) \exp(\text{ad}\frac{1}{6}[[[XY]Y]Y]) \cdot$
$\exp(\text{ad}-\frac{1}{6}[[[XY]Y]Y]X]);$

i) $\exp(\text{ad}[XY]) \exp(\text{ad}\frac{1}{2}[[XY]X]) \exp(\text{ad}\frac{1}{6}[[[XY]X]X]) \cdot$
$\exp(\text{ad}\frac{1}{3}[[[XY]X]X]Y]).$

<u>Proof</u>. Since N consists of nilpotent transformations, so does the associative algebra A (without unit) generated by all ad N, $N \in N$ (e.g., by Engel's theorem), and both $(\exp(\text{ad } X), \exp(\text{ad } Y))$ and the expression a)- i) above are of the form $1 + A, A \in A$, as is the inverse of each expression. Thus the transformation $(\exp(\text{ad } X), \exp(\text{ad } Y))$, followed by the inverse of a) -i), is of this form. We show below that <u>this quantity fixes</u> L_0; it then follows that it stabilizes each L_α $(\alpha \neq 0)$ and commutes there with the adjoint action of L_0, which is irreducible by Lemma I.7. By Schur's lemma, the restriction of $1 + A$ to L_α belongs to a division algebra, which necessarily then contains the restriction of A as well; since A is nilpotent, $A|L_\alpha = 0$ for all $\alpha \neq 0$. But we have seen that $A | L_0 = 0$, so $A = 0$. Thus the proof rests on showing that $(\exp(\text{ad } X), \exp(\text{ad } Y))$ agrees with the appropriate one of a) - i) on L_0. We let $H \in L_0$. (It is in fact enough to show the expressions agree on T; for then their quotient stabilizes L_0 and maps $H \in L_0$ into $H + N, N \in N$, hence fixes L_0.)

a) Here $[HX] \in L_\alpha$, $[HY] \in L_\beta$, $[[HX]Y] = 0 = [[HY]X]$
$= [XY],$ and $H(\exp(\text{ad } X), \exp(\text{ad } Y))$

$= (H - [HX] + \frac{1}{2}[[HX]X]) \exp(\text{ad } - Y) \exp(\text{ad } X) \exp(\text{ad } Y)$

$= (H - [HX] + \frac{1}{2}[[HX]X] - [HY] + \frac{1}{2}[[HY]Y]) \exp(\text{ad } X) \exp(\text{ad } Y)$

$= (H - [HX] + \frac{1}{2}[[HX]X] - [HY] + \frac{1}{2}[[HY]Y] + [HX] - [[HX]X]$

$+ \frac{1}{2}[[HX]X]) \exp(\text{ad } Y)$

$= H - [HY] + \frac{1}{2}[[HY]Y] + [HY] - [[HY]Y] + \frac{1}{2}[[HY]Y]$

$= H.$ (of course, here X and Y commute, from which the assertion is immediate; however, I have chosen to show the form taken by the argument in the other cases in this simple case as well).

b) Here we have [[HX]Y] commuting with both X and Y, as do [[HY]X], [[HY]Y], [[HX]X]. Expansion as in a) gives

H (exp(ad X), exp(ad Y)) = H + [[HX]Y] − [[HY]X]
= H + [H[XY]] = H exp(ad[XY]) by the remarks above.

c) We have seen in the proof of Lemma 2 that in this case $A_{\beta,\alpha} = -1$, so that 2α cannot be a root; if 2β were a root, then $A_{2\beta,\alpha} = -2$ would imply that $2\alpha + 2\beta$ is a root, a contradiction. Also $A_{2\alpha+3\beta,\beta} = 2$ shows that $2\alpha + \beta$ would be a root if $2\alpha + 3\beta$ were one, and $A_{2\alpha+4\beta,\beta} = 4$, a disallowed value, shows that neither $2\alpha + 3\beta$ nor $2\alpha + 4\beta$ is a root. Thus

H(exp(ad X), exp(ad Y)) = (H − [HX] − [HY] + [[HX]Y] − $\frac{1}{2}$[[[HX]Y]Y]) ·

exp(ad X) exp(ad Y) = (H − [HY] + [[HX]Y] − $\frac{1}{2}$[[[HX]Y]Y]

− [[HY]X])exp(ad Y) = H + [[HX]Y] − [[HY]X]

− $\frac{1}{2}$[[[HX]Y]Y] + [[[HX]Y]Y] − [[[HY]X]Y]

= H + [H[XY]] + $\frac{1}{2}$[H[[XY]Y]] = H exp(ad[XY]) exp(ad $\frac{1}{2}$[[XY]Y]).

d) Applying c) to Y and X, we have (exp(ad X), exp(ad Y))
= (exp(ad Y), exp(ad X))$^{-1}$ = (exp(ad[YX]) exp(ad $\frac{1}{2}$[[YX]X]))$^{-1}$
= exp(ad $\frac{1}{2}$[[XY]X]) exp(ad[XY]), and these last two factors commute by a) since $3\alpha + 2\beta$ is not a root.

e) Here we have seen that 2β is a root, so that 2α is not. Again, $2\alpha + 3\beta$, $2\alpha + 4\beta$ cannot be roots, as in c). If $3\alpha + 4\beta$ were a root, then $A_{3\alpha+4\beta,\beta} = 2$ gives $3\alpha + 2\beta$ a root, and $A_{3\alpha+2\beta,\alpha} = 4$, which is impossible. Now we have

H(exp(ad X), exp(ad Y)) = (H − [HX] − [HY] + $\frac{1}{2}$[[HY]Y]

+ [[HX]Y] − $\frac{1}{2}$[[[HX]Y]Y]) exp(ad X) exp(ad Y)

= (H − [HY] + $\frac{1}{2}$[[HY]Y] + [[HX]Y] − $\frac{1}{2}$[[[HX]Y]Y] − [[HY]X]

+ $\frac{1}{2}$[[[HY]Y]X] − $\frac{1}{2}$[[[[HX]Y]Y]X]) + $\frac{1}{4}$[[[[HY]Y]X]X]) exp(ad Y)

= H + [[HX]Y] − [[HY]X] + $\frac{1}{2}$[[[HX]Y]Y]

− [[[HY]X]Y] + $\frac{1}{2}$[[[HY]Y]X] − $\frac{1}{2}$[[[HX]Y]Y]X]

+ $\frac{1}{4}$[[[[HY]Y]X]X]. On the other hand,

$$H \exp(\text{ad}[XY]) \exp(\text{ad}\tfrac{1}{2}[[XY]Y])$$

$$= H + [H[XY]] = \tfrac{1}{2}[[H[XY]][XY]] + \tfrac{1}{2}[H[[XY]Y]],$$

which is seen to agree with our expression for $H(\exp(\text{ad } X), \exp(\text{ad } Y))$ by $[[H[XY]][XY]] = [[[[HY]X]Y]X] - [[[[HX]Y]Y]X]$ and $[[[[HY]Y]X]X] = [[[[HY]X]YX] + [[[HY][YX]]X] = [[[[HY]X]Y]X] + [[[HY]X][YX]]$
$= 2[[[[HY]X]Y]X]$, using the Jacobi identity and that $2\alpha + \beta$ is not a root.

f)   This follows from e) in the same way that d) follows from c), using the fact that the factors of e) commute.

g)   Here we have $A_{\alpha,\beta} = -1 = A_{\beta,\alpha}$, and $A_{2\alpha+4\beta,\beta} = 6$, $A_{2\alpha+3\beta,\beta} = 4$ shows that none of $2\alpha+3\beta$, $2\alpha+4\beta$, $3\alpha+2\beta$, $4\alpha+2\beta$ can be a root. Nor can 2α or 2β be a root since, e.g., $A_{\alpha,2\beta} = -\tfrac{1}{2}$. Thus $H(\exp(\text{ad } X), \exp(\text{ad } Y)) = (H - [HX] - [HY] + [[HX]Y] - \tfrac{1}{2}[[[HX]Y]Y]) \cdot$
$\exp(\text{ad } X)\exp(\text{ad } Y) = (H - [HY] + [[HX]Y] - \tfrac{1}{2}[[[HX]Y]Y] - [[HY]X]$
$- \tfrac{1}{2}[[[HY]X]X] + [[[HX]Y]X])\exp(\text{ad } Y) = H + [[HX]Y] - [[HY]X]$
$+ \tfrac{1}{2}[[[HX]Y]Y] - [[[HY]X]Y] + [[[HX]Y]X] - \tfrac{1}{2}[[[HY]X]X]$, while
$H \exp(\text{ad}[XY]) \exp(\text{ad}\tfrac{1}{2}[[XY]Y]) \exp(\text{ad}\tfrac{1}{2}[[XY]X]) = H + [H[XY]]$
$+ \tfrac{1}{2}[H[[XY]Y]] + \tfrac{1}{2}[H[[XY]X]]$, which is seen upon expansion as in e) to agree with $H(\exp(\text{ad } X), \exp(\text{ad } Y))$.

h)   Here none of 2α, 2β, nor twice any of the other roots in the list is a root, nor are $3\alpha+4\beta$, $2\alpha+5\beta$, $3\alpha+5\beta$, $3\alpha+6\beta$. Because it is slightly easier to do i) first, we assume i) is proved. Then, as in d) and f), we have in case h),

$$(\exp(\text{ad } X), \exp(\text{ad } Y)) = \exp(\text{ad}\tfrac{1}{3}[[[XY]Y]Y]X]) \cdot$$
$$\exp(\text{ad}\tfrac{1}{6}[[[XY]Y]Y]) \exp(\text{ad}\tfrac{1}{2}[[XY]Y]) \exp(\text{ad}[XY])$$
$$= \exp(\text{ad}\tfrac{1}{2}[[XY]Y]) \exp(\text{ad}[XY]) \exp(\text{ad}\tfrac{1}{6}[[[XY]Y]Y]) \cdot$$
$$\exp(\text{ad}\tfrac{1}{3}[[[XY]Y]Y]X])$$
$$= \exp(\text{ad}[XY]) \exp(\text{ad}\tfrac{1}{2}[[XY],Y]) (\exp(\text{ad}\tfrac{1}{2}[[XY]Y]),$$
$$\exp(\text{ad}[XY])) \exp(\text{ad}\tfrac{1}{6}[[[XY]Y]Y]) \exp(\text{ad}\tfrac{1}{3}[[[XY]Y]Y]X]).$$

Applying b) to the commutator in the middle of this expression, we have

$\exp(\frac{1}{2}\mathrm{ad}[[XY]Y][XY]]) = \exp(\mathrm{ad} - \frac{1}{2}[[[XY]Y]Y]X])$ since $[[[XY]Y]X]=0$.
Substitution gives h).

i) Similar restrictions hold on the list of roots as in h).

$H(\exp(\mathrm{ad}\ X), \exp(\mathrm{ad}\ Y)) = (H - [HX] - [HY] + [[HX]Y])$.

$\exp(\mathrm{ad}\ X)\exp(\mathrm{ad}\ Y) = (H - [HY] + [[HX]Y] - [[HY]X]$
$- \frac{1}{2}[[[HY]X]X] - \frac{1}{6}[[[[HY]X]X]X] + [[[HX]Y]X]$
$+ \frac{1}{2}[[[[HX]Y]X]X])\exp(\mathrm{ad}\ Y)$
$= H + [[HX]X] - \frac{1}{2}[[[HY]X]X] + [[[HX]Y]X]$
$- \frac{1}{6}[[[[HY]X]X]X] + \frac{1}{2}[[[[HX]Y]X]X] - \frac{1}{6}[[[[HY]X]X]X]Y]$
$+ \frac{1}{2}[[[[HX]Y]X]X]Y]$. On the other hand

$H\exp(\mathrm{ad}[XY])\exp(\frac{1}{2}\mathrm{ad}\ [[XY]X])\exp(\frac{1}{6}\mathrm{ad}[[[XY]X]X])$.

$\exp(\frac{1}{3}\mathrm{ad}[[[XY]X]X]Y]) = (H + [H[XY]] + \frac{1}{2}[H[[XY]X]]$
$+ \frac{1}{2}[[H[XY]][[XY]X]])\exp(\frac{1}{6}\mathrm{ad}[[[XY]X]X])$.

$\exp(\frac{1}{3}\mathrm{ad}\ [[[[XY]X]X]Y]) = H + [H[XY]] + \frac{1}{2}[H[[XY]X]]$
$+ \frac{1}{2}[[H[XY]][[XY]X]] + \frac{1}{6}[H[[[XY]X]X]] + \frac{1}{2}[H[[[XY]X]X]Y]]$
$= H + [[HX]Y] - [[HY]X] - \frac{1}{2}[[[HY]X]X] + [[[HX]Y]X]$
$+ \frac{1}{2}[[[[HY]X]X]X]Y]$
$- \frac{1}{2}[[[[HX]Y]X]X]Y] + \frac{1}{2}[[[[HX]Y]X]X] - \frac{1}{6}[[[[HY]X]X]X]$
$+ [[[[HX]Y]X]X]Y] - \frac{2}{3}[[[[HY]X]X]X]Y]$, which agrees with our
expression for $(\exp(\mathrm{ad}\ X), \exp(\mathrm{ad}\ Y))$.

Lemma 4.   Let $X = X_\alpha + X_{2\alpha}$, $X_\beta \in L_\beta$. Then $\exp(\mathrm{ad}\ X) = \exp(\mathrm{ad}\ X_\alpha)$
$\exp(\mathrm{ad}\ X_{2\alpha})$, the factors commuting. If X_α, $Y_\alpha \in L_\alpha$, $\exp(\mathrm{ad}\ X_\alpha)$
$\exp(\mathrm{ad}\ Y_\alpha) = \exp(\mathrm{ad}(X_\alpha+Y_\alpha))\exp(\mathrm{ad}\frac{1}{2}[X_\alpha Y_\alpha])$.

Proof.   The first assertion results from $[X_\alpha, X_{2\alpha}] = 0$. As in the
proof of Lemma 3, it suffices to prove that both sides of the second
relation agree on $H \in L_0$. Here we have

$H\exp(\mathrm{ad}\ X_\alpha)\exp(\mathrm{ad}\ Y_\alpha) = H + [HX_\alpha] + \frac{1}{2}[[HX_\alpha]X_\alpha]$
$+ [HY_\alpha] + \frac{1}{2}[[HY_\alpha]Y_\alpha] + [[HX_\alpha]Y_\alpha]$, while

$H\exp(\mathrm{ad}(X_\alpha+Y_\alpha))\exp(\mathrm{ad}\frac{1}{2}[X_\alpha Y_\alpha]) = H + [H,X_\alpha+Y_\alpha]$

$+ \frac{1}{2}[[H,X_\alpha+Y_\alpha]X_\alpha+Y_\alpha] + \frac{1}{2}[H[X_\alpha,Y_\alpha]]$, the last term being $\frac{1}{2}[[HX_\alpha]Y_\alpha] = \frac{1}{2}[[HY_\alpha]X_\alpha]$. Comparison now gives the result (which might also be obtained directly from the Baker-Cambell-Hausdorff formula ([J], §V.5)). (The same applies to the following lemma).

Lemma 5. The group $U = U^+$ is generated by all $\exp(\operatorname{ad} X_\alpha)$, $X_\alpha \in L_\alpha$, $\alpha > 0$ (indeed, for α primitive).

Proof. Our ordering of the roots (p. 9) was based on the values of the roots at a basis $H_{\alpha_1}, \ldots, H_{\alpha_r}$ for $T \cap [LL]$, in that order. Now let $M = \operatorname{Max}|\beta(H_{\alpha_i})|$, the maximum being taken over all roots β and $1 \le i \le r$. Let m be an integer, $m > 2M$, $H = m^{r-1}H_{\alpha_1} + m^{r-2}H_{\alpha_2} + \ldots + mH_{\alpha_{r-1}} + H_{\alpha_r} \in T$. Then it is easy to see that for β, γ roots, $\beta > \gamma$ if and only if $\beta(H) > \gamma(H)$. (For most of our purposes, it will suffice that all roots do not vanish at H, or certainly that their values at H are non-zero and distinct).

Now clearly all $\exp(\operatorname{ad} X_\alpha)$ are in U. Thus let $\sigma \in U$, so that $H^\sigma = H + \sum_{\beta > 0} Y_\beta, Y_\beta \in L_\beta$. If some $Y_\beta \ne 0$, let β be minimal in our ordering with $Y_\beta \ne 0$; then $\beta(H) \ne 0$, and $H^\sigma \exp(-\beta(H)^{-1}\operatorname{ad} Y_\beta) = H + \sum_{\gamma > \beta} Z_\gamma$. Proceeding this way, we may assume $H^\sigma = H$, hence $L_\alpha^\sigma = L_\alpha$ for all root-spaces L_α. If $Z \in L_0$, we have $Z^\sigma = Z + \sum_{\beta > 0} Z_\beta$, by the form of σ as a product of $\exp(\operatorname{ad} X)$, $X \in \sum_{\beta > 0} L_\beta$. But from $H^\sigma = H$ we have $L_0^\sigma = L_0$, so that $Z^\sigma = Z$, and σ fixes L_0. By the argument of the proof of Lemma 3, $\sigma = 1$, and our Lemma is proved. (The parenthetical assertion follows from $[L_\alpha L_\alpha] = L_{2\alpha}$ (Lemma 1.7) and Lemma 4.)

We call a subset S of the positive roots <u>closed</u> if $\alpha, \beta \in S$, $\alpha+\beta$ a root, implies $\alpha+\beta \in S$; we call S <u>primitive</u> if $2\alpha \in S$ implies $\alpha \in S$.

Lemma 6. Let S be a closed primitive subset of roots. Let $(\alpha_1, \ldots, \alpha_s)$ be an arrangement in sequence of the primitive roots in S, and let U_S be the subgroup of U generated by the $U_{(\alpha)}$, $\alpha \in S$ primitive. Then every element σ of U_S has a unique representation $\sigma = \sigma_1 \ldots \sigma_s$, $\sigma_i \in U_{(\alpha_i)}$.

Proof. We follow Steinberg's notes, Lemmas 17 and 18. First suppose $\alpha_1 < \alpha_2 < \ldots < \alpha_s$. Then $S - \{\alpha_1, 2\alpha_1\}$ is closed since $\beta, \gamma \in S$, $\beta > \alpha_1, \gamma > \alpha_1$, $\beta + \gamma \in S$, implies $\beta + \gamma > 2\alpha_1$. Moreover, $S - \{\alpha_1, 2\alpha_1\}$ is primitive. By induction on $|S|$ (the result for $S = \{\alpha_s\}$ or $\{\alpha_s, 2\alpha_s\}$ being valid by Lemma 4) we may assume that every element of $U_{S - \{\alpha_1, 2\alpha_1\}}$ can be written in the form $\sigma_2 \ldots \sigma_s$, $\sigma_i \in U_{(\alpha_i)}$. It follows that every element of the set $U_{(\alpha_1)} U_{S - \{\alpha_1, 2\alpha_1\}}$ has the desired form; we show this set is U_S by showing it is closed under left multiplication by the $\exp(\mathrm{ad}\, X_{\alpha_i})$, $1 \leq i \leq s$, these elements generating the $U_{(\alpha_i)}$ by Lemma 4 and the parenthetical remark in Lemma 5. The set is clearly closed under left multiplication by $\exp(\mathrm{ad}\, X_{\alpha_1})$, $X_{\alpha_1} \in L_{\alpha_1}$. If $i > 1$, $Y \in L_{\alpha_i}$, we have
$\exp(\mathrm{ad}\, Y) \exp(\mathrm{ad}\, X_{\alpha_1}) \exp(\mathrm{ad}\, X_{2\alpha_1}) \cdot U_{S - \{\alpha_1, 2\alpha_1\}} =$
$\exp(\mathrm{ad}\, X_{\alpha_1}) \underline{\exp(\mathrm{ad}\, Y) (\exp(\mathrm{ad}\, Y), \exp(\mathrm{ad}\, X_{\alpha_1}))} \cdot \exp(\mathrm{ad}\, X_{2\alpha_1}) U_{S - \{\alpha_1, 2\alpha_1\}}$,
and by Lemma 3, the underlined factor is a product of elements $\exp(\mathrm{ad}\, Z)$, $Z \in L_\gamma, \gamma \in S - \{\alpha_1, 2\alpha_1\}$. Repeating the argument, we have $\exp(\mathrm{ad}\, Z) \exp(\mathrm{ad}\, X_{2\alpha_1}) U_{S - \{\alpha_1, 2\alpha_1\}} = \exp(\mathrm{ad}\, X_{2\alpha_1}) U_{S - \{\alpha_1, 2\alpha_1\}}$, and the existence of our representation follows. For the uniqueness in this case, we may assume uniqueness in $U_{S - \{\alpha_1, 2\alpha_1\}}$, hence $\sigma_1 \neq 1$ if $\sigma_1 \ldots \sigma_s = \tau_1 \ldots \tau_s = \sigma$, $\sigma_1 = \exp(\mathrm{ad}\, X_{\alpha_1}) \exp(\mathrm{ad}\, X_{2\alpha_1})$ by Lemma 4, with one of $X_{\alpha_1}, X_{2\alpha_1} \neq 0$. Likewise $\tau_1 = \exp(\mathrm{ad}\, Y_{\alpha_1}) \exp(\mathrm{ad}\, Y_{2\alpha_1})$, and if $H \in T$, $\alpha_1(H) \neq 0$, $H^\sigma \equiv H - \alpha_1(H) X_{\alpha_1} - 2\alpha_1(H) X_{2\alpha_1}$ (mod $\sum_{\alpha \in S - \{\alpha_1, 2\alpha_1\}} L_\alpha$) as well as $H^\sigma \equiv H - \alpha_1(H) Y_{\alpha_1} - 2\alpha_1(H) Y_{2\alpha_1}$. It follows that $X_{\alpha_1} = Y_{\alpha_1}$, $X_{2\alpha_1} = Y_{2\alpha_1}$, $\sigma_1 = \tau_1$, and uniqueness results by induction.

In the general case, let π be a permutation of $\{1, \ldots, s\}$ so that $\alpha_{\pi(1)} < \alpha_{\pi(2)} < \ldots < \alpha_{\pi(s)}$; let $\pi(s) = j$. Then every element of U_S has a unique representation as a product $\tau_1 \ldots \tau_s$, $\tau_i \in U_{(\alpha_{\pi(i)})}$ and $U_{(\alpha_j)} = U_{(\alpha_{\pi(s)})}$ is normal in U_S by Lemmas 3, 4, as is $U_{(\alpha_{\pi(i)})} U_{(\alpha_{\pi(i+1)})} \ldots U_{(\alpha_{\pi(s)})}$ for each i. Thus we have subgroups $U_1,\ldots,U_s$ of U_S and a permutation π such that for each i, $U_{\pi(i)}\cdots U_{\pi(s)}$ is normal in U_S, and $U_S = U_{\pi(1)}\cdots U_{\pi(s)}$ with uniqueness of representation of elements. We show by induction on s the general group-theoretic principle that then $U_S = U_1 \cdots U_s$, with uniqueness. For $U_j = U_{\pi(s)}$ is normal in U_S, so $U_S/U_j = \overline{U}_{\pi(1)} \cdots \overline{U}_{\pi(s-1)}$, the bar denoting the homomorphic image. Once again $\overline{U}_{\pi(i)} \cdots \overline{U}_{\pi(s-1)} = \overline{U_{\pi(i)} \cdots U_{\pi(s-1)} U_{\pi(s)}}$ is normal in U_S/U_j for each i, and if $\overline{\sigma}_1, \overline{\sigma}_2 \cdots \overline{\sigma}_{s-1} = \overline{\tau}_1 \overline{\tau}_2 \cdots \overline{\tau}_{s-1}$, $(\sigma_i, \tau_i \in U\pi(i))$, we have $\sigma_s, \tau_s \in U_{\pi(s)}$ with $\sigma_1 \cdots \sigma_{s-1} \sigma_s = \tau_1 \cdots \tau_{s-1} \tau_s$, or $\sigma_i = \tau_i$, all i. By inductive hypothesis, $U_S/U_j = \overline{U}_1 \cdots \overline{U}_{j-1} \overline{U}_{j+1} \cdots \overline{U}_s$, with unique representation. Thus every element of U_S has the form $\sigma_1 \cdots \sigma_{j-1}\sigma_{j+1} \cdots \sigma_s \sigma_j$, $\sigma_i \in U_i$. Since U_j is normal, every element has the form

$$\sigma_1 \cdots \sigma_{j-1} \sigma'_j \sigma_{j+1} \cdots \sigma_s, \quad \sigma'_j = (\sigma_{j+1} \cdots \sigma_s)\sigma_j(\sigma_{j+1} \cdots \sigma_s)^{-1}$$

in the above. Now if $\sigma_1 \cdots \sigma_s = \tau_1 \cdots \tau_s$, we have $\overline{\sigma}_1 \cdots \overline{\sigma}_{j-1}\overline{\sigma}_{j+1} \cdots \overline{\sigma}_s = \overline{\tau}_1 \cdots \overline{\tau}_{j-1}\overline{\tau}_{j+1} \cdots \overline{\tau}_s$, from which $\overline{\sigma}_i = \overline{\tau}_i$, all i, or there are $\mu_j, \nu_j \in U_j$, $\sigma_i \mu_j = \tau_i \nu_j$, from which $\sigma_i = \tau_i$ by the case first treated (since $i = \pi(k)$, $k < s$). Then it follows that $\sigma_j = \tau_j$ as well, and the proof is complete.

<u>Corollary 1.</u> <u>If</u> S <u>is the set of all positive roots</u>, $U_S = U$ (Lemma 5) <u>and if</u> $(\alpha_1,\ldots,\alpha_s)$ <u>is a sequence containing each primitive positive root exactly once, every element of</u> U <u>has a unique representation</u> $\sigma = \sigma_1 \cdots \sigma_s$, $\sigma_i \in U_{(\alpha_i)}$.

<u>Corollary 2.</u> <u>If</u> $w \in W$, <u>the Weyl group, let</u> S <u>and</u> T <u>be the sets of positive roots mapped respectively to negative and positive roots by</u> w^{-1}. <u>Then</u> S <u>and</u> T <u>are closed and primitive, and</u> $U = U_S U_T = U_T U_S$, <u>with uniqueness of representation in each case.</u>

In particular, if $\alpha \in \Pi$, the simple system, we have T = all positive roots except $\alpha, 2\alpha$, $S = \{\alpha, 2\alpha\}$, $U_S = U_{(\alpha)}$ in the above. Moreover, we have for $w \in W$, $S = S(w)$, $T = T(w)$ as above, $UH\omega(w)U = UH\omega(w)U_T U_S$, and $\omega(w)U_T\omega(w)^{-1} = U_{T^w} \subseteq U \subseteq UH$, so that $UH\omega(w)U = UH\omega(w)U_{S(w)}$, and $G = \bigcup_w UH\omega(w)U_{S(w)}$.

§3. (B,N) - pairs.

It is an effective and unifying procedure to cast the study of some aspects of the automorphism groups in the language of (B,N)-pairs (or Tits systems). The results here presented are to be found in papers of Tits (Comptes Rendus Paris 254 (1962), 2910-2912; Annals of Math. 80 (1964), 313-329; see also [Bo], Chapter IV, §2).

Let G be a group, B and N subgroups such that:

1) B and N generate G;
2) H = B ∩ N is normal in N;
3) The group W = N/H is generated by a set R of elements of order 2, such that: if $\{\omega(w) | w \in W\} \subseteq N$ is a set of coset representatives, then

a) $\omega(r) B \omega(r) \neq B$ for $r \in R$, and
b) $\omega(r) B \omega(w) B \subseteq B \omega(w) B \cup B \omega(rw) B$ for all $w \in W$, $r \in R$.

(Note: In the applications of this formalism in §4, the groups "G,B,N" will not be exactly those we have denoted by these letters in §1, but rather large subgroups of those.)

If $\omega'(w)$ is a second coset representative of $w \in W$, then $\omega'(w)\omega(w)^{-1} \in H \subseteq B$, so $B \omega'(w) = B \omega(w)$, $\omega'(w) B = \omega(w) B$, and $B \omega'(w) B = B \omega(w) B$ may be denoted simply BwB; likewise, wB and Bw make sense. In particular, $\bigcup_{w \in W} BwB = BWB$ is inverse-closed, contains B and N and is stable under left multiplication by B and by all $\omega(r)$, $r \in R$ (cf. 3b)), hence by N. That is, G = BWB. By the same argument, if $S \subseteq R$, W_S the subgroup of W generated by S, then $BW_S B = G_S$ is a subgroup of G. In particular, if $r \in R$, B ∪ BrB is a subgroup.

The two double B-cosets B, BrB either coincide or are disjoint. If BrB = B, then $\omega(r) \in B$, and $\omega(r) B \omega(r) = B$, contrary to 3a). Thus B ∪ BrB is a disjoint union. In fact, every B ∩ B $\omega(w)$B = ∅ if $w \neq 1$, since otherwise $\omega(w) \in B \cap N = H$, which implies w = 1. More generally, BwB = Bw'B implies w = w'; for if $\ell(w)$ is the length of a shortest representation for w as a product of elements of R, we are done if $\ell(w) = 0$, so may assume the assertion for pairs (w,w') with w of shorter length, and that w = rw", $r \in R$, $\ell(w") + 1 = \ell(w)$. Then $\omega(r)\omega(w)$, hence $\omega(rw) = \omega(w")$, is in BrBBwB = BrBBw'B ⊆ Bw'B ∪ Brw'B by 3b). Thus either Bw"B = Bw'B or Bw"B = Brw'B. We cannot have Bw"B = Bw'B = BwB,

43

since then $w'' = w$ by induction, and this is absurd. Hence $Bw''B = Brw'B$, $w'' = rw'$ by induction. But $w'' = rw$, so $w = w'$, and our proof is complete. That is, G is the disjoint union $\bigcup_{w \in W} BwB$.

__Lemma 7.__ __Let__ $r \in R$, $w \in W$, $\ell(rw) \geq \ell(w)$. __Then__ $rBwB \subseteq BrwB$.

__Proof.__ Use induction on $\ell(w)$, the result being trivial if $\ell(w) = 0$; thus we may assume $\ell(w) \geq 1$, $w = w's$, $s \in R$, $\ell(w') + 1 = \ell(w)$. If the result fails, then by 3b), $rBwB \cap BwB \neq \emptyset$, $rBw \cap BwB \neq \emptyset$ and $rBw' \cap BwBs \neq \emptyset$. Now $\ell(rw') < \ell(w')$ would give $\ell(rw) = \ell(rw's) < \ell(w') + 1 = \ell(w)$, a contradiction so $\ell(rw') \geq \ell(w')$ and by induction $rBw' \subseteq Brw'B$, from which $Brw'B \cap BwBs \neq \emptyset$. By taking inverses in 3b), we have $BwBs \subseteq BwB \cup BwsB = BwB \cup Bw'B$. Thus $Brw'B$ is one of these two double cosets, and cannot be $Bw'B$. Therefore $Brw'B = BwB$, $rw' = w$, and $rw = w'$ has shorter length than has w, a contradiction.

__Lemma 8.__ __If__ $w' = rw$, __then either__ $B \subseteq rB\, wBw^{-1}$ __or__ $B \subseteq rB\, w'Bw'^{-1}$.

__Proof.__ If $\ell(w') \leq \ell(w) = \ell(rw')$, we have by Lemma 7, $rBw'B \subseteq Bw'B, B \subseteq rBw'Bw'^{-1}$. The case $\ell(w') \geq \ell(w)$ yields the other conclusion in identical fashion.

__Lemma 9.__ __If__ $\ell(rw) \leq \ell(w)$, __then__ $rBw \cap BwB \neq \emptyset$.

__Proof.__ By 3), $rBr \subseteq BrB \cup B$, with $rBr \neq B$, so $rBr \cap BrB \neq \emptyset$. Multiplying on the right by rw, $rBw \cap BrBrw \neq \emptyset$. By Lemma 7, $rBrw \subseteq Br(rw)B = BwB$, so $rBw \cap BwB \neq \emptyset$, as required.

__Lemma 10.__ __If__ $r \in R$, $w \in W$, __then__ $\ell(rw) \neq \ell(w)$; if $\ell(rw) < \ell(w)$, __then__ $rB \subseteq BwBw^{-1}B$.

__Proof.__ If $\ell(rw) = \ell(w)$, combining Lemmas 7 and 9 gives $BrwB \cap BwB \neq \emptyset$, so $rw = w$, which is absurd. If $\ell(rw) < \ell(w)$, we have by Lemma 9 that $rBw \cap BwB \neq \emptyset$, or $BrB \cap BwBw^{-1} \neq \emptyset$. Hence $BrB \subseteq BwBw^{-1}B$, from which the second assertion is clear.

__Lemma 11.__ __If__ $w \in W$, $\ell(w) = \ell$, __and if__ $w = r_1 \ldots r_\ell$, $r_i \in R$, __then__ $wBw^{-1}B$ __generates__ $G_{\{r_1,\ldots,r_\ell\}}$.

Proof. By the minimality of our expression for w, $\ell(r_1 w) = \ell - 1 < \ell(w)$. By Lemma 10, $r_1 B \subseteq BwBw^{-1}B \subseteq \langle wBw^{-1}B \rangle$; then $r_2 B \subseteq \langle r_1 wB(r_1 w)^{-1}B \rangle \subseteq \langle wBw^{-1}B \rangle$, and by induction all $\omega(r_i)$, $1 \leq i \leq \ell$, are in $\langle wBw^{-1}B \rangle$, so that $G_{\{r_1,\ldots,r_\ell\}} \subseteq \langle wBw^{-1}B \rangle$. The reverse inclusion is clear.

Proposition 2. *The subgroups of G containing B are exactly the G_S, $S \subseteq R$.*

Proof. Let $P \supseteq B$ be such a subgroup. Then P is a union of double cosets BwB. But $BwB \subseteq P$ implies $\omega(w) \in P$, hence $\omega(w)^{-1} \in P$, hence $wBw^{-1}B \subseteq P$, so that by Lemma 11, P contains all BrB, where $r \in R$ enters into a minimal representation for w. Thus $P \subseteq G_S$, where $S = \{r \in R \mid BrB \subseteq P\}$. But that $G_S \subseteq P$ is clear.

Lemma 12. *For $S \subseteq R$, G_S is its own normalizer in G.*

Proof. If P is the normalizer, P is a union of double cosets BwB. But by normality of G_S in P, $wBw^{-1}B \subseteq G_S$, $\langle wBw^{-1}B \rangle \subseteq G_S$, $\omega(w) \in G_S$, so $BwB \subseteq G_S$, and $P \subseteq G_S$.

Lemma 13. *R is a minimal set of generators for W. If $S, S' \subseteq R$, $G_S = G_{S'}$ only if $S = S'$, $W_S = W_{S'}$ only if $S = S'$; we have $G_S \cap G_{S'} = G_{S \cap S'}$, $W_S \cap W_{S'} = W_{S \cap S'}$.*

Proof. If $r \in R$, $r = r_1 r_2 \ldots r_\ell$ a minimal expression for r in terms of $R' = R - \{r\}$, we apply Lemma 11 to $G_{R'}$ to conclude that $\langle rBrB \rangle = G_{\{r_1,\ldots,r_\ell\}}$. But 3) shows that $B \cup BrB$ is a group, so that $\langle rBrB \rangle = B \cup BrB = G_{\{r_1,\ldots,r_\ell\}}$, $Br_i B \subseteq B \cup BrB$, which implies that $r_i = r$, a contradiction. Thus R is minimal. It follows that if $W_S = W_{S'}$ then $S = S'$. But if $G_S = G_{S'}$, then $W_S = W_{S'}$ from the double-coset decomposition, so $S = S'$. We thus have that if $B < P$, $P = G_S$, where S is uniquely determined as $\{r \in R \mid BrB \subseteq P\}$. By the uniqueness of the double-coset decomposition, $G_S \cap G_{S'} = BW_S B \cap BW_{S'}B = B(W_S \cap W_{S'})B = BW_T B = G_T$, where $T = \{r \mid BrB \subseteq G_S \cap G_{S'}\} = S \cap S'$, and $W_T = W_S \cap W_{S'}$.

45

Proposition 3. Let G' be a subgroup of G, $B' = G' \cap B$, $N' = G' \cap N$. Assume that $B = HB'$. Then (B',N') is a (B,N)-pair for G', with Weyl group isomorphic to W_S, where $S \subseteq R$. If G' is normal in G, $S = R$.

Proof. We have $G = BNB = B'HNHB'$ (by taking inverses, $B = B'H$) = $B'NB'$. Thus $G' = B'N'B'$, as one sees by representing $g \in G'$ in the form anb, $a,b \in B'$, $n \in N$. Now consider the set $HG' = HB'N'B' = BN'B' = B'HN'B' = B'N'HB' = B'N'B = G'H$. From $G'H = HG'$ it follows that HG' is a group P, $B = HB' \subseteq P$, so that $P = G_S$ for some S, $S \subseteq R$ (Proposition 2). Clearly $H' = H \cap N'$ is normal in N', and $N'/_{H'} \cong HN/_H$. But from $BHN'B = P$, we see that HN' is the union of the H-cosets in N corresponding to elements of W_S, or that $HN'/_H \cong W_S$. If $r \in S$, $w \in W_S$, $rB'wB'$ (defined using representatives in N' and the isomorphism above)
$\subseteq G' \cap (rBwB) \subseteq G' \cap (BwB \cup BrwB) = (G' \cap BwB) \cup (G' \cap BrwB)$. But $g \in G'$, $g = anb$, $a,b \in B'$, $n \in N'$ as before, is in BwB only if $nH = w$, in other words, only if $nH' \longleftrightarrow w$ in the isomorphism $N'/_{H'} \cong W_S$ above. Thus $G' \cap BwB = B'wB'$ in this identification, and 3b) is satisfied by (B',N') in G' relative to W_S. For $r \in S$, we have $rBr = rHB'r = HrB'r$, where we may define $rB'r$ using a coset representative in N' for r. Now $rB'r = B'$ implies $rBr = HB' = B$, which is impossible. Hence 3a) is satisfied, and we have a (B,N)-pair. Now if G' is normal, N normalizes G_S, $\emptyset \neq rBr \cap BrB \subseteq$ $\subseteq G_S \cap BrB = BrB$ for all r. Thus $G_S = G$, and $S = R$.

Proposition 4. Let G' be a normal subgroup of G. Let $S = \{r \in R | G' \cap BrB \neq \emptyset \}$. Then $G'B = G_S$, and $W = W_S \times W_{R-S}$ (direct product).

Proof. $G'B$ is a subgroup, since G' is normal; moreover, $B \subseteq G'B$, so that $G'B = G_S$, where $S = \{r \in R | BrB \subseteq G'B\}$. Since $G'B = BG'B$, it is clear that $S = \{r \in R | BrB \cap G' \neq \emptyset\}$. Now if $r \in S$, $r' \in R-S$, we have $r'BrBr' \cap G' \neq \emptyset$ (G' is normal in G). However, $Br'B \cap (BrB \cup B) = \emptyset$, so $Br'r \cap (BrB \cup B) = \emptyset$ ($BrB \cup B$ is a subgroup), $r'Br \cap (BrB \cup B) = \emptyset$. However, $r'Br \subseteq BrB \cup Br'rB$ by 3)b), so that $r'Br \subseteq Br'rB$. Using inverses to get the "right-handed" version of 3)b), we have $r'BrBr' \subseteq Br'rBr' \subseteq Br'rB \cup Br'rr'B$. Hence either $r'r$ or $r'rr'$ is in W_S. Since $r'r \in W_S$ implies $r' \in S$, we have

46

$r'rr' \in W_S \cap W_{\{r,r'\}} = W_{S \cap \{r,r'\}} = W_{\{r\}}$. Thus $r'rr' = r$ or 1, and the latter is impossible, so $r'r = rr'$ and S and $R - S$ commute elementwise. Then $W = W_S W_{R-S}$ and $W_S \cap W_{R-S} = W_{S \cap (R-S)} = 1$.

Proposition 5. <u>Let</u> U <u>be a solvable normal subgroup</u> <u>of</u> B, <u>such that</u> $B = HU$; <u>let</u> G' <u>be the (normal) subgroup of</u> G <u>generated by all conjugates of</u> U. <u>If</u> R <u>does not decompose into commuting proper subsets, and if</u> G' <u>is its commutator group, then every subgroup of</u> G <u>normalized by</u> G' <u>either contains</u> G' <u>or is contained in</u> B.

Proof. Let K be such a subgroup, which we may assume is not contained in B. By Proposition 3, since $B = HU = H(G' \cap B) = H(KG' \cap B)$, both the subgroups G' and KG' of G have (B,N)-pair structures relative to their intersections with B and N, and Weyl groups $W_{S'}, W_S$ respectively, where $S' \subseteq S \subseteq R$. Since G' is evidently normal in G, we have $S' = R$, hence $S = R$, and KG' satisfies all our hypotheses on G, plus having K as a normal subgroup. (That G' is generated by all conjugates of U in KG' follows since $G = HG'$ by $S' = R$ and the proof of Proposition 3, and since $H \subseteq B$ normalizes U.) Thus we may assume $G = KG'$. Applying Proposition 4 to K instead of G', we see since $K \not\subseteq B$ that $KB = G$ (using the indecomposability of R). But the normality of K and the normality of U in B give that KU is normal in $KB = G$, hence contains all conjugates of U, hence contains G'. Thus $KU = KG'$. But now $U/K \cap U \cong KU/K = KG'/K \cong G'/K \cap G'$, the member on the left being solvable and that on the right being its own commutator group. Hence they are trivial, $G' = K \cap G'$, and Proposition 5 is proved.

§4.   Applications to Automorphism Groups.

We return to the setting and notations of §§1,2. We first note that both larger and smaller sets of generators for G' can be given, namely:

i) $\{\exp(\mathrm{ad}N) \mid N \in L, N \text{ nilpotent}\}$, or

ii) $\{\exp(\mathrm{ad}X) \mid X \in L_\alpha, \alpha \in \Pi \cup -\Pi\}$.

For i), we may let M be a maximal nil subalgebra containing N, and know by Theorem I.1 that there is $\sigma \in G'$ with $N^\sigma = M$, hence $N = X^\sigma$ for $X \in N$, $\mathrm{ad}X^\sigma = \mathrm{ad}N = \sigma^{-1}(\mathrm{ad}X)\sigma$, and $\exp(\mathrm{ad}N) = \sigma^{-1}\exp(\mathrm{ad}X)\sigma \in G'$. For ii), let β be a primitive root relative to T; by results cited in §I.2, there is $w = w_{\alpha_1} \ldots w_{\alpha_t}$, $\alpha_i \in \Pi$ such that $\beta = \alpha^w$, where $\alpha \in \Pi$. Then $\omega(w_{\alpha_1}) \ldots \omega(w_{\alpha_t}) = \nu$ is in the group generated by ii), $L_\alpha^\nu = L_\beta$, and if $X_\beta = Y_\alpha^\nu$ is in L_β ($Y_\alpha \in L_\alpha$), $\exp(\mathrm{ad}X_\beta) = \nu^{-1}\exp(\mathrm{ad}Y_\alpha)\nu$ is in the group generated by ii). Combined with Lemma 5 (applied to both U and U^-), this yields the assertion. Since i) generates G', we see that G' is normal in G. Use of the smaller set of generators ii) will make it easier to show that certain subgroups exhaust G', or other results depending on a convenient generating set.

Let H_0 be the subgroup of G fixing T. Then H_0 stabilizes L_0 and every root-space, hence P, and $H_0 \subseteq H$. Clearly H_0 is normal in H, and if $\theta_1, \theta_2 \in H$ induce the same permutation of Π, then $\theta_1\theta_2^{-1}$ stabilizes each root-space, so maps a pair $X_\alpha, Y_{-\alpha}$ with $[X_\alpha Y_{-\alpha}] = H_\alpha$ onto a second such pair (using uniqueness of H_α as in the Corollary to Lemma I.4). Thus H_α is fixed for every α, so that $T \cap [LL]$ is fixed; in particular T is fixed if L is semisimple, $\theta_1\theta_2^{-1} \in H_0$, and $(H : H_0)$ is finite if L is semisimple. If $\theta_1,\ldots,\theta_s$ are a set of coset representatives in H for H/H_0, we have $G = \bigcup_{w \in W} UH\,\omega(w)U = \bigcup_{w \in W} HU\,\omega(w)U = \bigcup_{i=1}^{s}\theta_i \bigcup_{w \in W} H_0 U\,\omega(w)U = \bigcup_{i=1}^{s}\theta_i \bigcup_{w \in W} UH_0\,\omega(w)U = \bigcup_{i=1}^{s}\theta_i G_0$, where G_0 is the subgroup of G generated by U, U^- and H_0. Thus $(G : G_0) < \infty$ if L is semisimple, this index being bounded by the number of automorphisms of the Dynkin diagram of Π. Since G is generated by H, U, U^-, and since $U, U^- \subseteq G_0$ and H normalizes H_0, U, U^-, we see that G_0 is normal in G. We now study the structure of G_0. Namely, with $B_0 = B \cap G_0$, $N_0 = N \cap G_0$, we wish to show that G_0 <u>is a (B,N)-pair with Weyl group</u> $W = W(T)$ <u>as before</u>.

<u>Lemma 14</u>. $\qquad G_0 = \bigcup_{w \in W} UH_0 \omega(w) U$

<u>Proof</u>. It suffices to show the right-hand side is stable under left multiplication by elements of U, H_0 and U^-. Since H_0 normalizes U, the first two are immediate. In view of the generation of $G' = \langle U, U^- \rangle$ as in ii) above, it suffices to show that the right side is stable under left multiplication by $\exp(\mathrm{ad} X_{-\alpha})$, $\alpha \in \Pi$. Now we have $\exp(\mathrm{ad} X_{-\alpha}) UH_0 \omega(w) U = \exp(\mathrm{ad} X_{-\alpha}) U_{(\alpha)} H_0 U_{P-\{\alpha, 2\alpha\}} \omega(w) U$, using the last Corollary to Lemma 6. By the Corollary to Lemma 1, $\exp(\mathrm{ad} X_{-\alpha}) \in U_{(\alpha)} H_0 \cup U_{(\alpha)} H_0 \omega(w_\alpha) U_{(\alpha)}$. The former is impossible unless $X_{-\alpha} = 0$, as one sees by applying to $T \in T$, $\alpha(T) \neq 0$; thus we are left to consider

(*) $U_{(\alpha)} H_0 \omega(w_\alpha) U_{(\alpha)} H_0 U_{P-\{\alpha, 2\alpha\}} \omega(w) U = UH_0 \omega(w_\alpha) U_{(\alpha)} \omega(w) U$, since $\omega(w_\alpha)$ normalizes H_0 and $U_{P-\{\alpha, 2\alpha\}}$ and H_0 normalizes $U_{(\alpha)}$, $U_{P-\{\alpha, 2\alpha\}}$. If $\alpha^w > 0$, then $U_{(\alpha)} \omega(w) U = \omega(w) U_{(\alpha^w)} U = \omega(w) U$, $\omega(w_\alpha) \omega(w)$ agrees on T with $\omega(w_\alpha w)$ and we have $UH_0 \omega(w_\alpha w) U$. If $\alpha^w < 0$, let $w' = w_\alpha w$, having $\omega(w_\alpha) U_{(\alpha)} \omega(w) = \omega(w_\alpha) U_{(\alpha)} \omega(w_\alpha w') \subseteq \omega(w_\alpha)^{-1} U_{(\alpha)} H_0 \omega(w_\alpha) \omega(w') = H_0 U_{(-\alpha)} \omega(w')$, $UH_0 \omega(w_\alpha) U_{(\alpha)} \omega(w) U \subseteq UH_0 U_{(-\alpha)} \omega(w') U \subseteq UH_0 U_{(\alpha)} H_0 \omega(w_\alpha) U_{(\alpha)} \omega(w') U \cup UH_0 U_{w_\alpha(\alpha)} \omega(w') U = UH_0 \omega(w_\alpha w') U_{(_\alpha w')} U \cup UH_0 \omega(w') U$. Since $\alpha^{w_\alpha w} = \alpha^{w'} = -\alpha^w > 0$, $U_{(\alpha w')} U = U$, and we are done.

Dropping the first two factors of (*) and following through the above proof, we see that we have proved:

<u>Lemma 15</u>. <u>Let</u> $\alpha \in \Pi$, $w \in W$. <u>Then</u> $\omega(w_\alpha) UH_0 \omega(w) U \subseteq UH_0 \omega(w) U \cup UH_0 \omega(w_\alpha w) U$.

<u>Lemma 16</u>. $B_0 = UH_0$; $N_0 = \bigcup_{w \in W} H_0 \omega(w)$; $B_0 \cap N_0 = H_0$.

<u>Proof</u>. Clearly UH_0 is a subgroup of B_0, and $\bigcup_w H_0 \omega(w)$ is a subgroup of N_0. Now let $\sigma \in B_0$, and write $\sigma = \rho \theta \omega(w) \tau$, ρ,

$\tau \in U$, $\theta \in H_0$, $w \in W$, $\sigma\tau^{-1} = \rho\theta\omega(w)$. Then $\omega(w) \in B_0$, $w \in W$ maps positive roots to positive roots, and is therefore the identity, $\omega(w) = 1$, $\sigma \in UH_0U = UH_0$, and $B_0 = UH_0$. Next suppose $\sigma \in N_0$, $\sigma = \rho\theta\omega(w)\tau$ as above, where we may assume by the last corollary to Lemma 30 that $\tau \in U_{S(w)}$. Then $\sigma\tau^{-1}\omega(w)^{-1} = \rho\theta = \theta\rho'$, $\rho' = \theta^{-1}\rho\theta \in U$. Arrange factors of τ^{-1} as given by Lemma 6 so that the roots of $S(w)$ are taken in ascending order; thus if $\tau \neq 1$, we have $\tau^{-1} = \exp(adX_\beta) \cdot \prod_{\substack{\gamma \in S(w) \\ \gamma > \beta}} \exp(adX_\gamma)$, where $0 \neq X_\beta$, $\beta \in S(w)$. Choose $T \in \mathcal{T}$ such that $\beta(T\sigma) \neq 0$. Then $T^{\sigma\tau^{-1}\omega(w)^{-1}}$

$$= (T^\sigma - \beta(T^\sigma)X_\beta + \sum_{\substack{\gamma \in S(w) \\ \gamma \neq \beta}} Y_\gamma)^{\omega(w)^{-1}}$$

$$= T^{\sigma\omega(w^{-1})} - \beta(T^\sigma)X_\beta^{\omega(w)^{-1}} + \sum_{\substack{\gamma \in S(w) \\ \gamma \neq \beta}} Y_\gamma^{\omega(w)^{-1}}$$

Now $X_\beta^{\omega(w^{-1})} \in L_{\beta^{w^{-1}}} \subseteq N^-$, $Y_\gamma^{\omega(w^{-1})} \in N^-$, $T^{\sigma\omega(w)^{-1}} \in \mathcal{T}$, so the above is in P^-, but not in L_0. On the other hand, $T^{\theta\rho'} = T^{\rho'}$ is in P, and $P \cap P^- = L_0$. This contradiction yields $\tau = 1$, hence $\rho \in N$. By the same argument, $\rho = 1$. The last assertion now follows, since we have seen that $\omega(w) \in B_0$ implies $w = 1$.

<u>Proposition 6</u>. (B_0, N_0) <u>is a</u> (B,N)-<u>pair for</u> G_0 <u>with Weyl group</u> W.

<u>Proof</u>.   We know there is an element w_0 of the Weyl group W mapping Π to $-\Pi$; hence $\omega(w_0) \in N_0$ conjugates U to U^-. Thus $\langle B_0, N_0 \rangle \supseteq \langle U, U^-, H_0 \rangle = G_0$. Since $B_0 \cap N_0 = H_0$, the second axiom is satisfied. From $N_0 = \bigcup H_0 \omega(w)$, we have $N_0/H_0 \cong W$. With $R = \{w_\alpha | \alpha \in \Pi\}$, 3)b is proved in Lemma 15. Finally we consider $w_\alpha B_0 w_\alpha$ $(\alpha \in \Pi) = \omega(w_\alpha)^{-1} U_{(\alpha)} U_{P-\{\alpha,2\alpha\}} H_0 \omega(w_\alpha) = \omega(w_\alpha)^{-1} U_{(\alpha)} \omega(w_\alpha) U_{P-\{\alpha,2\alpha\}} H_0 = U_{(-\alpha)} U_{P-\{\alpha,2\alpha\}} H_0 \supseteq U_{(-\alpha)}$. But any non-identity element of $U_{(-\alpha)}$ maps $\mathcal{T}$ outside of P, hence cannot be in $B \supseteq B_0$. Thus $w_\alpha B_0 w_\alpha \neq B_0$, and all axioms hold.

Now our group G' is normal in G and contained in G_0, hence is normal in G_0. Since G' contains U, U^- and the $\omega(w)$, we see by Proposition 10 that $G'B_0 = G'H_0 = G_0$, and $G_0/G' \cong H_0/G' \cap H$

is isomorphic to a quotient of H_0. (This quotient may be rather large, for instance if L has some compact simple summands.)

Lemma 17. $G' = (G',G')$.

Proof. It suffices to show that (G',G') contains every $\exp(\text{ad} X_\alpha)$, α primitive (Lemma 5). Thus let X_α be as indicated, $X_{-\alpha} \in L_{-\alpha}$, $[X_\alpha X_{-\alpha}] = H_\alpha$. By Lemma I.17, we have for $\xi \neq 0$ in F, $\omega_\alpha(\xi) = \exp(\text{ad } \xi X_{-\alpha}) \exp(\text{ad } \xi^{-1} X_\alpha) \exp(\text{ad} \xi X_{-\alpha}) \in G'$, and $\omega_\alpha(\xi)$ sends X_α to $(X_\alpha + \xi H_\alpha + \xi^2 X_{-\alpha}) \exp(\text{ad}\xi^{-1} X_\alpha) \exp(\text{ad}\xi X_{-\alpha}) = (X_\alpha + \xi H_\alpha + \xi^2 X_{-\alpha} - 2X_\alpha - \xi H_\alpha + X_\alpha) \exp(\text{ad}\xi X_{-\alpha}) = \xi^2 X_{-\alpha}$, and likewise $X_{-\alpha}$ to $X_{-\alpha} \exp(\text{ad}\xi^{-1} X_\alpha) \cdot \exp(\text{ad}\xi X_{-\alpha}) = (X_{-\alpha} - \xi^{-1} H_\alpha + \xi^{-2} X_\alpha) \exp(\text{ad}\xi X_{-\alpha}) = \xi^{-2} X_\alpha$. Thus $\omega_\alpha(\xi)\omega_\alpha(1)$ sends X_α to $\xi^2 X_\alpha$, from which $(\omega_\alpha(\xi)\omega_\alpha(1), \exp(\text{ad} X_\alpha)) = \exp(\text{ad}\xi^2 X_\alpha) \exp(\text{ad} X_\alpha) = \exp(\text{ad}(\xi^2 + 1)X_\alpha) \in (G',G')$. Now if A is the additive subgroup of F consisting of those $\lambda \in F$ with $\exp(\text{ad}\lambda X_\alpha) \in (G',G')$, we see that $1 + \xi^2 \in A$ for every $\xi \neq 0$, hence $\xi^2 - \eta^2 \in A$ for $\xi, \eta \neq 0$. Take $\xi = \frac{5}{4}$, $\eta = \frac{3}{4}$ to complete the proof.

Lemma 18. If $[LL]$ is simple, then Π does not decompose into non-empty orthogonal subsets, and $R = \{w_\alpha | \alpha \in \Pi\}$ does not decompose into non-empty commuting subsets.

Proof. Suppose $\Pi = \Pi_1 \cup \Pi_2$, non-empty orthogonal subsets with respect to the metric of §I.2. Then every root is a combination either of Π_1 or of Π_2; for if β is a minimal positive counterexample, $\beta = \Sigma m_i \alpha_i$, $m_i \geq 0$, $\alpha_i \in \Pi$, we see from $(\beta,\beta) > 0$ that $(\beta,\alpha_i) > 0$ for some i, hence $A_{\beta,\alpha_i} > 0$, and $\beta - \alpha_i$ is a root. We may assume $\alpha_i \in \Pi_1$. Since $m_j > 0$ for some $\alpha_j \in \Pi_2$ by assumption, $\beta - \alpha_i > 0$, so $\beta - \alpha_i$ is a combination of Π_2 by minimality: $\beta = \alpha_i + \sum_{\alpha_j \in \Pi_2} m_j \alpha_j$. But then $A_{\beta,\alpha_i} = A_{\alpha_i,\alpha_i} = 2, \beta - 2\alpha_i = -\alpha_i + \sum_{\alpha_j \in \Pi_2} m_j \alpha_j$ is a root, and we have contradicted a fundamental property of the simple roots. Now let Σ_i be the set of roots which are combinations of Π_i, and let $L_1 = \sum_{\alpha \in \Sigma_1}(L_\alpha + [L_\alpha L_{-\alpha}]) \subseteq [LL]$. Clearly $[L_0 L_1] \subseteq L_1$, $[L_\beta L_1] \subseteq L_1$, if $\beta \in \Sigma_1$, while if $\beta \in \Sigma - \Sigma_1$, $\beta \in \Sigma_2$, $\beta + \alpha \notin \Sigma$ for any $\alpha \in \Sigma_1$ by the above and $[L_\beta L_1] = 0 \subseteq L_1$. Hence

51

L_1 is an ideal of L contained in $[LL]$ properly (since any L_β, $\beta \in \Pi_2$, is in $[LL]$ but not in L_1). This proves the first assertion. For the second we have, for $\alpha \neq \beta \in \Pi$, that if w_α, w_β commute then $-\alpha + 2(\alpha,\beta)(\beta,\beta)^{-1}\beta = \alpha w_\alpha w_\beta = \alpha w_\beta w_\alpha = (\alpha - 2(\alpha,\beta)(\beta,\beta)^{-1}\beta)w_\alpha = -\alpha - 2(\alpha,\beta)(\beta,\beta)^{-1}\beta + 4(\alpha,\beta)^2(\beta,\beta)^{-1}(\alpha,\alpha)^{-1}\alpha$, and α, β are linearly independent, so that $(\alpha,\beta) = 0$. That is, if R decomposes, so does Π, and the proof is complete.

In view of i) we may describe G' as the subgroup of G_0 generated by all conjugates of U. Since the matrices of U may be taken simultaneously triangular (with 1 on the diagonal), U is solvable, and we have seen that $B_0 = H_0 U$, $G' = (G',G')$. Thus we may apply Lemma 18, Propositions 5 and 6, to obtain

Proposition 7. *If $[LL]$ is simple, then every subgroup of G_0 normalized by G' either contains G' or is contained in B_0.*

We investigate further the second alternative in Proposition 7. If K is a subgroup of B_0 normalized by G', then K is contained in every G'-conjugate of B_0, in particular in $\omega(w_0)^{-1} B_0 \, \omega(w_0) = H_0 U^-$, where $w_0 \in W$ maps N onto N^-. But we have seen before that $U^- \cap H_0 U$ is trivial, from which we have $K \subseteq H_0$, $(K,U) \subseteq K \cap U = \{1\}$, from which K centralizes U and fixes T. Likewise K centralizes U^-, hence G'. For $\sigma \in K$, we thus have $\exp(adX_\alpha) = \sigma^{-1} \exp(adX_\alpha) \sigma = \exp(adX_\alpha^\sigma)$, $X_\alpha, X_\alpha^\sigma \in L_\alpha$, from which σ fixes every L_α, hence N, hence every G'-conjugate of N, hence every nilpotent element of L.

Lemma 19. *If L is simple and contains non-zero nilpotents, then L is generated by nilpotents.*

Proof. Under the hypotheses, $T \neq 0$, and we may let $X_\alpha, X_{-\alpha}, H_\alpha$ be as usual. Considering the adjoint representation of $\{X_\alpha, X_{-\alpha}, H_\alpha\}$ on L, we see that L is a direct sum of irreducible modules as described in §I.1, and that all elements belonging to non-zero characteristic roots of adH_α are nilpotent; in particular, any irreducible summand of <u>even</u> dimension has a basis of nilpotent elements. For $i \in \mathbb{Z}$, let L_i be the space corresponding to the characteristic root i of adH_α, and let M be the sum of the irreducible summands not annihilated by $\{X_\alpha, X_{-\alpha}, H_\alpha\}$. Then $L = M + \sum_{i > 0}[L_i L_{-i}]$; for the right-hand side contains all L_j, $j \neq 0$, as well as all $[L_j L_{-j}]$, hence is stable under ad L_j, $j \neq 0$. Now ad L_0 stabilizes $[L_i L_{-i}]$ by the Jacobi identity, as well as all L_j, $j \neq 0$, and by definition of M, $M \cap L_0 = [L_{-2} X_\alpha] \subseteq [L_2 L_{-2}]$. Hence $M + \sum_{i > 0}[L_i L_{-i}]$ is stable under $\text{ad} L_j$ for all j, so is an ideal in L, therefore is L, as asserted. The above shows also that $L_0 = \sum_{i > 0}[L_i L_{-i}]$, and this completes the proof.

<u>Theorem 1.</u> <u>If L is simple, then every non-trivial subgroup of G_0 normalized by G' contains G'.</u>

<u>Proof.</u> If $N = 0$, then $G' = \{1\}$ and the result is trivial. If $N \neq 0$, and if K is such a subgroup, we may assume by Proposition 7 that $K \subseteq B_0$, which implies that K fixes every nilpotent element of L, hence is trivial by Lemma 18.

<u>Corollary.</u> <u>If L is a simple Lie algebra of characteristic zero, and if L contains elements $X \neq 0$ such that adX is nilpotent, then the group G' of automorphisms of L generated by all exp(adX) for such X is a simple group.</u>

<u>Corollary.</u> <u>Let L be a completely reducible Lie algebra of endomorphisms of a finite-dimensional vector space V of characteristic zero. Assume that:</u>

(1) <u>L contains non-zero nilpotents,</u>
(2) <u>$[LL]$ is simple.</u>
<u>Let G be the group of automorphisms of V generated by all exp(N), $N \in L$, N nilpotent. Then the quotient of G by its center is a simple group.</u>

<u>Proof.</u> The hypotheses assure that the group G' of automorphisms of $[LL]$ induced by G is a simple group. Thus it is only necessary to show that the center of G is the kernel of the obvious homomorphism $G \to G'$. Since G' is centerless, the center of G is in the kernel, and since all nilpotent elements $N \in L$ are in $[LL]$, we have $A^{-1} N A = N$ for all such N when A is in the kernel. But this shows that the kernel centralizes all exp(N), hence is the center of G.

Interpretations of this result for classical and other linear groups will be discussed in Chapter V.

CHAPTER III

CONSTRUCTIONS FOR SIMPLE LIE ALGEBRAS

§1. Decompositions relative to split subalgebras.

Let L, T, Σ, Π, F be as in earlier chapters. In case Π is of type BC_r:

$$\underset{\alpha_1}{\bullet}\!\!\rule[0.5ex]{2cm}{0.4pt}\!\!\underset{\alpha_2}{\bullet} \quad \cdots \quad \underset{\alpha_{r-1}}{\bullet}\!\!\Rightarrow\!\!\underset{\alpha_r}{\bullet} \qquad (r \geq 1),$$

let $\Pi' = \{\alpha_1, \alpha_2, \ldots, \alpha_{r-1}, 2\alpha_r\}$. Set $\Pi' = \Pi$ in all other cases, always denoting the members of Π by $\alpha_1, \ldots, \alpha_r$. If $r = 0$, so that $T = 0$, the considerations of this chapter will be vacuous; that is, we assume throughout that $T \neq 0$. We further assume that L is simple, therefore that Π' is indecomposable (cf. Lemma II.18).

For each $\alpha \in \Pi'$, we choose an arbitrary $e_\alpha \neq 0$ in L_α. As in Lemma I.3, we take $0 \neq e_{-\alpha} \in L_{-\alpha}$ such that $[e_\alpha e_{-\alpha}] \in T$, $\alpha([e_\alpha e_{-\alpha}]) = 2$. Let S be the subalgebra generated by these $e_{\pm\alpha}$, $\alpha \in \Pi'$. Clearly $T \subseteq S$, by Lemma I.6 and its corollary (since of necessity $H_\alpha = 2H_{2\alpha}$).

Now it follows from a result of Serre ([SR], Appendice, pp. VI - 19 through VI - 25; see also [HU], pp. 99 ff.) that S is a split simple Lie algebra with T as splitting Cartan subalgebra and Π' as fundamental system of roots. Namely, let $\Pi' = \{\beta_1, \ldots, \beta_r\}$; let $e_\alpha = e_{\beta_i} = X_i$ be as above, $e_{-\alpha} = e_{-\beta_i} = -Y_i$, and $H_i = [X_i, Y_i] = -H_{\beta_i}$, $1 \leq i \leq r$. The Cartan integer $n(j,i)$ is $\beta_i(H_{\beta_j})$.. Then X_i, Y_i, H_i satisfy the relations:

(W.1): $[H_i H_j] = 0$;

(W.2): $[X_i Y_i] = H_i$, $[X_i Y_j] = 0$ if $i \neq j$;

(W.3): $[H_i X_j] = n(i,j) X_j$, $[H_i Y_j] = -n(i,j) Y_j$,

$$X_j(\text{ad}(X_i)^{-n(i,j)+1}) = 0, \quad i \neq j,$$

$$Y_j(\text{ad}(Y_i)^{-n(i,j)+1}) = 0, \quad i \neq j.$$

For (W.1) is clear, as is the first line of (W.3) and the first part of (W.2). The second part of (W.2) holds since $\beta_i - \beta_j$ is not a root. Moreover, it follows from this observation that the string of roots $\beta_j, \beta_j \pm \beta_i, \ldots$ is in fact $\beta_j, \beta_j + \beta_i, \ldots, \beta_j + (-\beta_j(H_{\beta_i}))\beta_i = \beta_j - n(i,j)\beta_i$, so that $\beta_j + (-n(i,j) + 1)\beta_i) \notin \Sigma$, and this gives the rest of (W.3). Serre's theorem says that the generators X_i, Y_i, H_i and relations (W.1) – (W.3) define a Lie algebra $L^{(0)}$ which is semisimple, having the span H of the H_i as Cartan subalgebra and as system of roots the system Σ'. Since Π' is connected, $L^{(0)}$ is a simple split Lie algebra with H as maximal split torus, and S is a homomorphic image of $L^{(0)}$, hence must be isomorphic to $L^{(0)}$. Since H maps onto T, this establishes our assertion.

The totality Σ' of roots of S relative to T are determined by Π', and coincide with Σ if $\Pi = \Pi'$; if $\Pi' \neq \Pi$, so that Π is of type BC_r, we have

$$\Pi' : \quad \underset{\alpha_1}{\bullet} \underline{\qquad} \underset{\alpha_2}{\bullet} \quad \cdots \quad \underset{\alpha_{r-2}}{\bullet} \underline{\qquad} \underset{\alpha_{r-1}}{\bullet} \Longleftarrow \underset{2\alpha_r}{\bullet} \quad ,$$

of type C_r, and Σ' consists of the following and their negatives: $\alpha_i + \alpha_{i+1} + \cdots + \alpha_j, i \leq j < r;\ \alpha_i + \alpha_{i+1} + \cdots + 2\alpha_j + \cdots + 2\alpha_r,\ i < j \leq r;\ 2\alpha_i + \cdots + 2\alpha_r,\ i \leq r$ (cf.[J], Chapter IV, or [Bo], Chapter VI, p.222).

Now consider the adjoint representation of S on L. Since S is a split simple Lie algebra, this representation is completely reducible and the irreducible S-modules are all absolutely irreducible; i.e., if M is an irreducible S-module, $\text{Hom}_S(M,M)$ consists only of scalar multiplications by elements of F ([J], Chapter 7, p. 223). Moreover, each such M has a basis of elements v such that $vt = \lambda(t)v$ for all $t \in T$, where $\lambda(t) \in F$. The functions λ are called the <u>weights</u> of M. There is a unique <u>highest weight</u> $\hat{\lambda}$ of M, in the sense that no $\hat{\lambda} + \alpha_i, \alpha_i \in \Pi'$, is a weight, and M is determined, as S-module, by $\hat{\lambda}$ ([J], Chapter 7). The set of weights of any S-module is independent of the choice of basis consisting of eigenvectors for T; thus, in particular, when M is an irreducible S-summand of L, the set of weights of M is contained in the set of weights of L, i.e., in $\Sigma \cup \{0\}$.

First suppose all roots in Σ are of the same length, which also means that they all are conjugate under the Weyl group of Σ ([HU], p. 53), or that Π has a diagram of type A_r, D_r, E_6, E_7 or E_8 in the classification. Then, since the Weyl group permutes the weights of any representation, any irreducible summand M involving a weight $\neq 0$ (therefore an element of Σ) involves all elements of Σ. Since $\Sigma = \Sigma'$ and S is simple, there is a unique highest root in Σ, and this must be the highest weight of M. But then M is isomorphic to S as S-module. The other irreducible S-summands of L must have 0 as the only weight. Since a one-dimensional space on which S acts trivially is an irreducible S-module with highest weight 0, these summands must all be of this type. Their sum is precisely the centralizer D of S in L. Thus, as S-module,

$$L = D \oplus M_1 \oplus M_2 \oplus \ldots \oplus M_k, \text{ where}$$

D is the centralizer of S in L, and the S-modules M_i are all isomorphic to S.

Next, suppose $\Pi = \Pi'$, so $\Sigma = \Sigma'$, and that not all roots in Σ are of the same length. Here the orbits of the Weyl group in Σ are two in number, and it follows as above that any non-trivial S-irreducible summand M of L must have as highest weight a root in such an orbit. Now the weights of the representation with highest weight $\hat{\lambda}$ are all of the form $\hat{\lambda} - \Sigma m_i \alpha_i$, m_i non-negative integers ([J], Chapter 7). Hence the roots which can serve as highest weights must have the property that their orbit under the Weyl group consists of elements of this form. Using the classification of diagrams, we are able to give explicitly those roots in each orbit having this property. They are as follows:

$B_r (r \geq 2)$ $\underset{\alpha_1}{\bullet}\ \underset{\alpha_2}{\bullet}\ \cdots\ \underset{\alpha_{r-1}}{\bullet}\!\Rightarrow\!\underset{\alpha_r}{\bullet}$: $\alpha_1 + 2\alpha_2 + \ldots + 2\alpha_r$,

$\alpha_1 + \alpha_2 + \ldots + \alpha_r$;

$C_r (r \geq 2)$ $\underset{\alpha_1}{\bullet}\ \underset{\alpha_2}{\bullet}\ \cdots\ \underset{\alpha_{r-1}}{\bullet}\!\Leftarrow\!\underset{\alpha_r}{\bullet}$: $2\alpha_1 + 2\alpha_2 + \ldots + 2\alpha_{r-1} + \alpha_r$,

$\alpha_1 + 2\alpha_2 + \ldots + 2\alpha_{r-1} + \alpha_r$;

$F_4 (r = 4)$ $\underset{\alpha_1}{\bullet}\ \underset{\alpha_2}{\bullet}\!\Rightarrow\!\underset{\alpha_3}{\bullet}\ \underset{\alpha_4}{\bullet}$: $2\alpha_1 + 3\alpha_2 + 4\alpha_3 + 2\alpha_4$,

$\alpha_1 + 2\alpha_2 + 3\alpha_3 + 2\alpha_4$;

$G_2 (r = 2)$ ⇛ : $2\alpha_1 + 3\alpha_2,$
$ \alpha_1\ \alpha_2 \alpha_1 + 2\alpha_2 .$

In each case, the first root listed is the highest root of S, so the highest weight of the adjoint representation, and the second root listed is the "highest short root" $\hat{\mu}$. The latter has the property of being <u>dominant</u>, i.e., $\hat{\mu}(H_{\alpha_i}) \geq 0$ for all $\alpha_i \in \Pi$, and hence there exists a unique irreducible S-module M' of highest weight $\hat{\mu}$. <u>We therefore have an</u> S-<u>module decomposition</u>

$$L = D \oplus M_1 + \ldots \oplus M_k \oplus N_1 \oplus \ldots \oplus N_m ,$$

<u>where</u> $k \geq 1$, $m \geq 0$, <u>and where all</u> M_i <u>are isomorphic to</u> S <u>and all</u> N_j <u>isomorphic to</u> M'.

Finally, suppose $\Pi \neq \Pi'$, so $\Sigma \neq \Sigma'$. Then the roots in Σ form <u>three</u> orbits under the Weyl group (<u>two</u> if we are dealing with BC_1), represented in our labeling of the diagram BC_r by:

$BC_r (r \geq 2)$: •—• … •⇒• : $2\alpha_1 + 2\alpha_2 + \ldots + 2\alpha_r ,$
$ \alpha_1\ \alpha_2\alpha_{r-1}\ \alpha_r \alpha_1 + \alpha_2 + \ldots + \alpha_r ,$
$ \alpha_1 + 2\alpha_2 + \ldots + 2\alpha_r ;$

BC_1 : • : $2\alpha_1, \alpha_1 .$
$ \alpha_1$

Again these are dominant, both with respect to Π and to Π'. The first is the highest weight of the adjoint representation of S. If we let M'' be as above for the second of these dominant functions, and M' (if necessary) for the third, we note that the highest weight of M'' is <u>not</u> in Σ', hence is not a root of S; nor can it occur as a weight of M' as S-module, since these must have a coefficient of α_r which is <u>even</u> (the corresponding root of S is $2\alpha_r$, not α_r). Since this dominant weight of T actually does occur, some summands isomorphic to M'' <u>must actually be present</u>. Accordingly, <u>the decomposition of</u> L <u>as</u> S-<u>module is</u>

$$L = D \oplus M_1 \oplus \ldots \oplus M_k \oplus N_1 \oplus \ldots \oplus N_m \oplus O_1 \oplus \ldots \oplus O_p ,$$

where $k \geq 1$, $m \geq 0$, $p \geq 1$, and where $M_i \cong S$, $N_i \cong M'$, $O_i \cong M''$.

Now let A, B, C be trivial S-modules of respective dimensions k, m, p. Then, as S-modules,

$$M_1 \oplus \ldots \oplus M_k \cong S \otimes A,$$

$$N_1 \oplus \ldots \oplus N_m \cong M' \otimes B,$$

$$O_1 \oplus \ldots \oplus O_p \cong M'' \otimes C,$$

where an explicit isomorphism is given as soon as we choose a basis for A, B or C. Assuming such a choice, we have isomorphisms of S-modules as follows:

Types A, D, E: $\qquad L \cong D \oplus (S \otimes A)$

B, C, F, G: $\qquad L \cong D \oplus (S \otimes A) \oplus (M' \otimes B)$

BC_1: $\qquad L \cong D \oplus (S \otimes A) \oplus (M'' \otimes C)$

$BC_r (r > 1)$: $\qquad L \cong D \oplus (S \otimes A) \oplus (M' \otimes B) \oplus (M'' \otimes C)$,

where A, C are nonzero trivial S-modules, B and D are trivial S-modules, and M', M'' are as above. Here the notation has been chosen to conform approximately to that of Tits' 1962 paper in Proc. Neder. Akad. Series A 65 (= Indag. Math. 24), which, from our point of view, treats the case where $\Pi = \Pi'$ is of type A_1. We use this isomorphism to "transport the structure" of L to the right-hand side above, and thus write L = (right-hand side). Following the idea of the cited paper of Tits, we shall expect to infer from the properties of the S-modules S, M', M'' the existence of certain compositions among the trivial modules A, B, C, D, resulting from the bracket-operation in L. The anticommutative and Jacobi identities in L will force certain identities to be satisfied by these compositions. The simplicity of L will be reflected in a property suggestive of simplicity, and the fact that T is a maximal split torus in L will be reflected in a property of "anisotropy" appropriate to the particular systems.

The "properties of the S-modules S, M', M''", to which we shall appeal concern the dimension of the spaces $\text{Hom}_S(M_1 \otimes_F M_2, M_3)$, where M_1, M_2 are any two of these (not necessarily distinct), and where M_3 is either one of these or the irreducible trivial S-module F. Since all irreducible S-modules are absolutely irreducible, this dimension is the same as the multiplicity of M_3 as an irreducible summand in the tensor product module $M_1 \otimes_F M_2$. The necessary multiplicities are listed below, with their determination being deferred to an appendix (Appendix A.):

a) Multiplicity of F in $M_1 \otimes M_2$ is <u>one</u> if $M_1 = M_2$, <u>zero</u> otherwise.

b) Multiplicity of S in $S \otimes S$ is <u>one</u> except for type $A_r (r \geq 2)$, in which case it is <u>two</u>.

c) Multiplicity of S in $S \otimes M'$ is <u>zero</u> except for types C_r and $BC_r (r \geq 2)$, in which case it is <u>one</u>.

d) Multiplicity of S in $M' \otimes M'$ is <u>one</u> in all cases (where M' comes into consideration); likewise for multiplicity of S in $M'' \otimes M''$.

e) Multiplicity of S in $S \otimes M''$, or in $M' \otimes M''$, is <u>zero</u>.

f) Multiplicity of M' in $S \otimes S$ is <u>zero</u> in all cases except for types C_r and $BC_r (r \geq 2)$, in which cases it is <u>one</u>.

g) Multiplicity of M' in $S \otimes M'$ is <u>one</u> in all cases; likewise for multiplicity of M'' in $S \otimes M''$.

h) Multiplicity of M' in $S \otimes M''$, $M' \otimes M''$ is always <u>zero</u>.

i) Multiplicity of M' in $M' \otimes M'$ is <u>zero</u> except for types C_r and BC_r, with $r \geq 3$, F_4, and G_2; then it is <u>one</u>.

j) Multiplicity of M' in $M'' \otimes M''$ is <u>one</u>.

k) Multiplicity of M'' is <u>zero</u> in all tensor products except $S \otimes M''$, $M' \otimes M''$, in which it is <u>one</u>.

Now we consider explicit realizations of the spaces $\mathrm{Hom}_S(M_1 \otimes M_2, M_3)$ in the various cases. For $M_1 = M_2 = S$, $M_3 = F$, the <u>Killing form</u>: $x \otimes y \mapsto \mathrm{Tr}(\mathrm{ad}\, x\, \mathrm{ad}\, y) = B(x,y)$ of S is nonzero, and $B([xz], y) + B(x, [yz]) = 0 = B(x,y) \cdot z$ shows that B gives a homomorphism of S-modules. Thus we have a generator for $\mathrm{Hom}_S(S \otimes S, F)$, by a).

By b), the ordinary bracket: $x \otimes y \mapsto [xy]$ provides a basis for $\mathrm{Hom}_S(S \otimes S, S)$ unless S is of type A_n with $n \geq 2$. In that case ([J], Chap. IV), S may be realized as the $(n+1)$ by $(n+1)$ F-matrices of trace zero, with $[xy] = xy - yx$. The composition $x \otimes y \mapsto xy + yx - \frac{2}{n+1} \mathrm{Tr}(xy) I$ is seen in this case to define a second element of $\mathrm{Hom}_S(S \otimes S, S)$, linearly independent of $x \otimes y \mapsto [xy]$. Thus <u>these two form a basis for</u> $\mathrm{Hom}_S(S \otimes S, S)$ in the exceptional case.

For the remainder, we must distinguish by types of S, giving specific realizations of S, M', M''.

<u>Type</u> $B_r (r \geq 2)$: M' is to be a vector space over F of odd dimension $2r + 1$, carrying a nondegenerate symmetric bilinear form (u,v) of Witt index r and discriminant $(-1)^r$. S is to be the Lie algebra of all linear transformations x of M' satisfying $(ux,v) + (u,vx) = 0$ for all u, v, with $[xy] = xy - yx$. Here $x \otimes u \mapsto -ux$ is a basis for $\text{Hom}_S(S \otimes M', M')$, $u \otimes v \mapsto (u,v)$ for $\text{Hom}_S(M' \otimes M', F)$, and $u \otimes v \mapsto [uv]: w \mapsto (w,u)v - (w,v)u$ is a basis for $\text{Hom}_S(M' \otimes M', S)$.

<u>Types</u> C_r, $BC_r (r \geq 1)$: M'' is to be a vector space of dimension $2r$ over F carrying a nondegenerate alternate bilinear form (u,v). S is defined as for type B_r, using M'' in place of M'. The space M' is defined to be the space of all linear transformations s of M'', of trace zero, satisfying $(us,v) = (u,vs)$ for all u,v. That these are S-modules is included in the verifications of the following:

$x \otimes y \mapsto x \circ y = xy + yx - \frac{1}{r} \text{Tr}(xy)I$ is a basis for
 $\quad \text{Hom}_S(S \otimes S, M')$ $(r \geq 2)$.

$x \otimes s \mapsto x \circ s = xs + sx$ is a basis for $\text{Hom}_S(S \otimes M', S)$ $(r \geq 2)$.

$s \otimes t \mapsto [st] = st - ts$ is a basis for $\text{Hom}_S(M' \otimes M', S)$ $(r \geq 2)$.

$u \otimes v \mapsto u \circ v: w \mapsto (w,u)v + (w,v)u$ is a basis for
 $\quad \text{Hom}_S(M'' \otimes M'', S)$.

$x \otimes u \mapsto ux$ is a basis for $\text{Hom}_S(S \otimes M'', M'')$.

$x \otimes s \mapsto [xs] = xs - sx$ is a basis for $\text{Hom}_S(S \otimes M', M')$ $(r \geq 2)$.

$s \otimes t \mapsto s \circ t = st + ts - \frac{1}{r} \text{Tr}(st)I$ is a basis for
 $\quad \text{Hom}_S(M' \otimes M', M')$ $(r > 2)$ - (it is zero for $r = 2$).

$u \otimes v \mapsto [uv]: w \mapsto (w,u)v - (w,v)u + \frac{1}{r}(u,v)w$ is a basis for
 $\quad \text{Hom}_S(M'' \otimes M'', M')$ $(r \geq 2)$.

$s \otimes u \mapsto us$ is a basis for $\text{Hom}_S(M' \otimes M'', M'')$ $(r \geq 2)$.

$s \otimes t \mapsto \text{Tr}(st)$ is a basis for $\text{Hom}_S(M' \otimes M', F)$ $(r \geq 2)$.

$u \otimes v \mapsto (u,v)$ is a basis for $\text{Hom}_S(M'' \otimes M'', F)$.

To see that these are bases, it suffices in view of a)-k) to show that they define non-zero S-homomorphisms.

<u>Type</u> G_2: $M' = C_0$, the elements of trace zero in a split Cayley-Dickson algebra C over F (see [J], pp. 234-5), the multiplication in C being denoted uv and the standard involution as $u \mapsto u^*$; thus M' is the set of u in C with $u^* = -u$. S is the Lie algebra of derivations of C. These necessarily commute with the involution so that $M' = C_0$ is an S-module under the action of S on C. Then

we have:

$x \otimes u \mapsto ux$ is a basis for $\text{Hom}_S(S \otimes M', M')$.

$u \otimes v \mapsto (u,v) = (uv^* + vu^*) = -(uv + vu)$ is a basis for
$\text{Hom}_S(M' \otimes M', F)$.

$u \otimes v \mapsto [uv] = uv - vu$ is a basis for $\text{Hom}_S(M' \otimes M', M')$.

$u \otimes v \mapsto D_{u,v}: w \mapsto v(uw) - u(vw) + (uw)v - u(wv) + (wu)v - (wv)u$
is a basis for $\text{Hom}_S(M' \otimes M', S)$ ([J], p. 143).

<u>Type</u> F_4: $M' = J_0$, the elements of trace zero in the 3 by 3
Hermitian matrices J over C as above, with *-transpose as involution; J is a Jordan algebra under $uv + vu$, uv being the ordinary
product of C-matrices. S is the Lie algebra of derivations of this
Jordan algebra, with $M' = J_0$ as irreducible submodule of J. Then:

$x \otimes u \mapsto ux$ is a basis for $\text{Hom}_S(S \otimes M', M')$

$u \otimes v \mapsto (u,v) = \text{Trace}(uv + vu)$ is a basis for $\text{Hom}_S(M' \otimes M', F)$.

$u \otimes v \mapsto u \circ v = uv + vu - \frac{1}{3}(u,v)I$ is a basis for
$\text{Hom}_S(M' \otimes M', M')$

$u \otimes v \mapsto D_{u,v}: w \mapsto (uw)v - u(wv)$ is a basis for $\text{Hom}_S(M' \otimes M', S)$.

§2. <u>Algebras</u> <u>of</u> <u>Types</u> A_1, D, E.

First let Π' (= Π) be of type $D_r(r \geq 4)$, E_6, E_7, E_8, or A_1.
Let $L = D \oplus (S \otimes A)$ as in §1, and fix a basis $\{d_i\}$ for D and
a basis $\{a_j\}$ for A, as well as an element $d \in D$ and an $a \in A$.
Consider $[x \otimes a, d]$ for $x \in S$: Since d centralizes the action of
S on L, and since $x \otimes a$ is in the sum of all S-submodules of
L isomorphic to S, so is $[x \otimes a, d]$. That is, one has
$[x \otimes a, d] = \Sigma f_j(x) \otimes a_j$, where $f_j(x)$ are well-determined elements
of S, and the f_j are linear mappings of S to S. Now for
$y \in S$, we have $[[x \otimes a, d]y] = [[x \otimes a, y]d] = [[xy] \otimes a, d]$, or

$$\sum_j [f_j(x), y] \otimes a_j = \sum_j f_j([xy]) \otimes a_j .$$

That is, the linear mappings f_j commute with all $\text{ad } y$, $y \in S$, so
are in the <u>centroid</u> ([J], Chap. X) of S. But S is absolutely
simple, since T_K and Π' give an indecomposable diagram over any
field extension K of F, and this means $f_j(x) = \lambda_j x$ for all x,
where $\lambda_j \in F$. But then

$[x \otimes a, d] = x \otimes \sum_j \lambda_j a_j$. We write ad for the element $\sum \lambda_j a_j$ of A, and see that $(a,d) \to ad$ is a bilinear pairing of A and D to A. From
$$[x \otimes a, [dd']] = [[x \otimes a, d]d'] - [[x \otimes a, d']d],$$ we have
$$a[dd'] = (ad)d' - (ad')d,$$
so that A <u>is a</u> D-<u>module</u>.

Fixing $a, a' \in A$, we have for $x, y \in S$,
$$[x \otimes a, y \otimes a'] = \sum_i g_i(x,y) d_i + \sum_j h_j(x,y) \otimes a_j,$$
where $g_i(x,y) \in F$, $h_j(x,y) \in S$. Applying $ad\, z$, $z \in S$, to this relation gives

$[[xz] \otimes a, y \otimes a'] + [x \otimes a, [yz] \otimes a'] =$

$$\sum_j [h_j(x,y), z] \otimes a_j,$$

or $\qquad g_i([xz], y) + g_i(x, [yz]) = 0$,

$\qquad h_j([xz], y) + h_j(x, [yz]) = [h_j(x,y), z]$,

for all $x, y, z \in S$. By a) of §1 and the later remarks of that section, $g_i(x,y) = \mu_i B(x,y)$, where B is the Killing form and $\mu_i \in F$. By b) of §1 and the remarks following, $h_j(x,y) = \nu_j [xy]$, $\nu_j \in F$. Thus we have
$$[x \otimes a, y \otimes a'] = B(x,y) \cdot \sum \mu_i d_i + [xy] \otimes \sum \nu_j a_j, \text{ or}$$
$$[x \otimes a, y \otimes a'] = B(x,y) <a, a'> + [xy] \otimes aa',$$
where the bilinear mappings $(a, a') \to <a, a'>$, aa' of $A \otimes A$ into D, resp. A, are determined from the above. By the anticommutativity of L and of S, and by the symmetry of B, we see that $<a', a> = - <a, a'>$, $a'a = aa'$. For $d \in D$, the Jacobi identity applied to the triple $(x \otimes a, y \otimes a', d)$ yields

(1) $\quad [<a, a'>, d] = <ad, a'> + <a, a'd>$, and
(2) $\quad (aa')d = (ad)a' + a(a'd)$.

The second of these says that D is represented on A by <u>derivations</u> with respect to the product aa'; the first implies that the space $<A, A>$ is an ideal in D. From this we see that $<A, A> + S \otimes A$ is a non-zero ideal in L, hence is L, so that $D = <A, A>$.

Finally, we consider the Jacobi identity as applied to triples $x \otimes a, y \otimes a', z \otimes a''$. Except in the case of A_1, we may choose

$x, y, z \in S$ such that $B(x,y) = 0 = B(y,z) = B(x,z)$, and such that $[x[yz]]$ and $[y[zx]]$ are linearly independent. (Take α_1, α_2 to be adjacent roots in Π; take $x = e_{\alpha_1}$, $y = e_{\alpha_2}$, $z = [e_{-\alpha_1} e_{-\alpha_2}]$.) Then $[x \otimes a[y \otimes a', z \otimes a'']] = [x \otimes a, [yz] \otimes a'a''] = B(x,[yz]) <a,a'a''> + [x[yz]] \otimes a(a'a'')$. The sum of the three terms resulting from this by cyclic permutation of the factors must be zero. Using that $B(x,[yz]) = B(y,[zx]) = B(z,[xy])$ and that $[z[xy]] = - [x[yz]] - [y[zx]]$, together with the linear independence of these last two terms, we have

$B(x,[yz]) \{ <a,a'a''> + <a',a''a> + <a'',aa'> \} = 0$ and
$[x[yz]] \otimes \{ a(a'a'') - a''(aa') \} = 0.$

Since we may choose x, y, z to satisfy our conditions with $B(x,[yz]) \neq 0$ (<u>cf</u>. the indicated choice above), it follows that

(3) $\quad <a,a'a''> + <a',a''a> + <a'',aa'> = 0,$
(4) $\quad a(a'a'') = a''(aa') \quad$ for all $a,a',a'' \in A$.

Since A is commutative, (4) shows that A is <u>associative</u> as well.

Next we consider the Jacobi identity as above, but with $x = y = H_{\alpha_1}$, $z = e_{\alpha_1}$, a choice which may be invoked for A_1 as well: We have

$$[x \otimes a, y \otimes a'] = B(x,x) <a,a'>,$$
$$[z \otimes a''[x \otimes a, y \otimes a']] = B(x,x) z \otimes a'' <a,a'>;$$
$$[z \otimes a'', x \otimes a] = 2z \otimes a''a,$$
$$[y \otimes a'[z \otimes a'', x \otimes a]] = -4z \otimes a'(a''a);$$
$$[y \otimes a', z \otimes a''] = -2z \otimes a'a'',$$
$$[x \otimes a[y \otimes a', z \otimes a'']] = 4z \otimes a(a'a''), \text{ yielding}$$

(5) $\quad B(H_{\alpha_1},H_{\alpha_1}) a'' <a,a'> = 4(a a'')a' - 4 a(a''a').$

But the right-hand side is zero by the associativity of A, and $B(H_{\alpha_1},H_{\alpha_1}) \neq 0$. Thus $A <A,A> = 0$, or $AD = 0$ since $D = <A,A>$. But then $[S \otimes A, D] = 0$, and D is an ideal in L, clearly distinct from L since it centralizes S. Therefore $D = 0$, and $L = S \otimes A$, where A is a commutative associative algebra over F. Now it is clear that if I is an ideal in A, then $S \otimes I$ is an ideal in L, from which A is <u>simple</u>, hence a field extension of F, and L is just the split A-algebra $S \otimes_F A$, regarded as Lie algebra over F. We have thus proved:

Theorem III.1. Let L be a simple Lie algebra over F of relative type $D_r (r \geq 4)$ or $E_r (r = 6,7,8)$. Then there is a field extension A of F such that L is the split simple Lie algebra of the same type over A, regarded as Lie algebra over F.

It should be remarked that the field A is uniquely determined by L (it is the "centroid" of L). Thus the simple Lie algebras of Theorem III.1 and any type given there are in one-one correspondence with the (finite) field extensions of F. One may also notice that, once the decomposition $L = \mathcal{D} \oplus (S \otimes A)$ is assumed, no further use is apparently made of the finite-dimensionality of L over F.

When the relative type is A_1, we are in the situation treated by Tits. Since $S = [SS]$ and B is non-degenerate, we have $x, y, z \in S$ with $B(x,[yz]) \neq 0$, and this was really all that was used in deriving (3). Moreover, (5) has been seen to hold for A_1. Combined with (2), this says that the map $a'' \mapsto (aa'')a' - a(a''a')$ is a derivation of A for each fixed choice of a and a' in A; (3) with $a = a' = a''$ gives $< a, aa > = 0$, which with (5) yields $(aa'')(aa) = a(a''(aa))$ for all a, a'', thus that A is a Jordan algebra over F, and $< A, A > = \mathcal{D}$ induces the Lie algebra of inner derivations of A. Since the annihilator of A in $\mathcal{D}$ is central in L (see (1)), this is zero, and we may identify $\mathcal{D} = < A, A >$ with the Lie algebra Inder(A) of inner derivations of A. Now if I is an ideal in the Jordan algebra A, we have
$A < A, I > \subseteq I$, $I\mathcal{D} = I < A, A > \subseteq I$, and
$[< A, I >, \mathcal{D}] \subseteq < A\mathcal{D}, I > + < A, I\mathcal{D} > \subseteq < A, I >$, from which $< A, I > + S \otimes I$ is an ideal in L. It follows that $I = A$ or $I = 0$, so that A is a simple Jordan algebra. It follows that $\mathcal{D} = <A,A> = $ Der(A) ([JJ], Chap. 7).

In fact, A is a Jordan division algebra; that is, for every $a \neq 0$ in A there is an element $b \in A$ such that $ab = 1$ (unit element of A) and $a^2 b = a$. For clearly our $T \subseteq L$ is identified with $T \otimes 1$, h_α with $h_\alpha \otimes 1$, and $L_{\pm \alpha}$ with $e_{\pm \alpha} \otimes A$. If $0 \neq a \in A$, then by Lemma I.3 we have $e_{-\alpha} \otimes b \in e_{-\alpha} \otimes A = L_{-\alpha}$ such that $[e_\alpha \otimes a, e_{-\alpha} \otimes b] = h_\alpha \otimes 1$. It follows that $< a, b > = 0$, $ab = 1$. But then $a^2 b = a< a, b > + a(ab) = a$, as required.

Theorem III.2. Let L be a simple Lie algebra over F of relative type A_1. Then there is a Jordan division algebra A over F such that

$$L \cong \text{Der}(A) \oplus S \otimes A, \text{ with}$$

$[x \otimes a, d] = x \otimes ad,$

$[x \otimes a, y \otimes a'] = B(x,y) <a,a'> + [xy] \otimes aa'$;

$x, y \in S$; $a, a' \in A$, $d \in \text{Der}(A)$, where

$B(H_{\alpha_1}, H_{\alpha_1}) a'' <a,a'> = 4(a''a)a' - 4 a(a''a'),$

for all $a'' \in A$. Here S is the split algebra of type A_1 over F.

That such a construction always results in a (simple) Lie algebra has been shown by Tits [TA] and also by Koecher [K]. (See also ([JE],§9). Verification that the relative type is A_1 may be done by retracing our steps above to identify the centralizer of $H_{\alpha_1} \otimes 1$ with $\text{Der}(A) + R_A$, and then observing that this is without nilpotents if A is a division algebra. Since $B(H_{\alpha_1}, H_{\alpha_1}) = 8$, we have $2 a'' <a,a'> = (a''a)a' - a(a''a')$.

§3. Algebras of Relative Type $A_r (r \geq 2)$

In this case, we have $L = \mathcal{D} + S \otimes A$ as in the previous case, but here we must have, for fixed $a, a' \in A$,

$[x \otimes a, y \otimes a'] = B(x,y) <a,a'> + \sum_j \lambda_j [xy] \otimes a_j + \sum_j \mu_j (x \circ y) \otimes a_j,$

where $x \circ y = xy + yx - \frac{2}{n+1} \text{Tr}(xy) I$ when S is identified with the $(n+1)$ by $(n+1)$ F-matrices of trace zero, and $[xy]$ with $xy - yx$ (see §1). Thus we have, as before,

(6) $[x \otimes a, y \otimes a'] = B(x,y) <a,a'> + [xy] \otimes f_1(a,a') +$

$(x \circ y) \otimes f_2(a,a')$, where f_1 and f_2 are uniquely determined bilinear multiplications on A, and $<a,a'> \in \mathcal{D}$ is bilinear in a, a'. It follows that f_1 is symmetric, while f_2 and $<a,a'>$ are skew.

Application of the Jacobi identity to the triple $x \otimes a, y \otimes a', d$ yields (1) as before, as well as

(7) $\quad f_i(a,a')d = f_i(ad,a') + f_i(a,a'd), \quad i = 1,2.$

As before, we deduce $\mathcal{D} = <A,A>$.

Now we concentrate on the component in $S \otimes A$ of the Jacobi identity applied to the triple $x \otimes a, y \otimes a', z \otimes a''$. From (6) we see that this is the sum of the cyclic permutations of

$$B(x,y)z \otimes a'' \quad <a,a'> \quad + [z[xy]] \otimes f_1(a'', f_1(a,a'))$$
$$+ [z, x \circ y] \otimes f_1(a'', f_2(a,a')) + (z \circ [xy]) \otimes f_2(a'', f_1(a,a'))$$
$$+ z \circ (x \circ y) \otimes f_2(a,a')).$$

Let us specialize x, y, z in our standard realization of S, used in defining $x \circ y$. We write e_{ij} for the usual $(n+1)$ by $(n+1)$ matrix unit. Setting $x = y = e_{11} - e_{22}$, $z = e_{22} - e_{33}$, we see that any term involving a commutator is zero, so that

$$B(x,x)z \otimes a'' \quad <a,a'> \quad + z \circ (x \circ x) \otimes f_2(a'', f_2(a,a'))$$
$$+ B(x,z)x \otimes a \quad <a',a''> \quad + x \circ (x \circ z) \otimes f_2(a, f_2(a',a''))$$
$$+ B(z,x)x \otimes a' \quad <a'',a> \quad + x \circ (z \circ x) \otimes f_2(a', f_2(a'',a)) = 0.$$

Now z and x are linearly independent, $x \circ x = 2(e_{11} + e_{22}) - \frac{4}{n+1} I$,
$z \circ x = -2e_{22} + \frac{2}{n+1} I$, $z \circ (x \circ x) = 4e_{22} - \frac{8}{n+1} e_{22} + \frac{8}{n+1} e_{33} - \frac{4}{n+1} I =$
$-\frac{4}{n+1}(e_{11} - e_{22}) + \frac{4(n-3)}{n+1}(e_{22} - e_{33})$
$\quad - \frac{4}{n+1}((2-n)e_{33} + e_{44} + \ldots + e_{n+1,n+1})$,
$x \circ (z \circ x) = 4e_{22} + \frac{4}{n+1} e_{11} - \frac{4}{n+1} e_{22} - \frac{4}{n+1} I$
$= \frac{4(n-1)}{n+1}(e_{22} - e_{33}) - \frac{4}{n+1}((2-n)e_{33} + e_{44} + \ldots + e_{n+1,n+1}).$

If $n > 2$, then x, z and $(2-n)e_{33} + e_{44} + \ldots + e_{n+1,n+1}$ are linearly independent; the coefficient of the last of these gives

(8) $\quad f_2(a'', f_2(a,a')) + f_2(a, f_2(a'a'')) + f_2(a', f_2(a'',a)) = 0$, if $n > 2$.

The coefficient of $z = e_{22} - e_{33}$ gives
$B(x,x) \, a'' \, <a,a'> \, + \frac{4(n-3)}{n+1} f_2(a'', f_2(a,a')) +$
$\frac{4(n-1)}{n+1} f_2(a, f_2(a',a'')) + \frac{4(n-1)}{n+1} f_2(a', f_2(a'',a)) = 0;$

using $B(x,x) = 4(n+1)$, and substituting for the sum of the last two terms by (8) if $n > 2$, we have

(9) $\quad a'' \, <a,a'> \, = -\frac{2}{(n+1)^2} f_2(a'', f_2(a,a'))$, if $n > 2$.

For $n = 2$, we can conclude only that

(9') $\quad a'' \, <a,a'> \, = \frac{1}{9}(f_2(a'', f_2(a,a')) - f_2(a, f_2(a',a''))$
$\quad - f_2(a', f_2(a'',a))).$

(It will be noted that (8) says that (A, f_2) is a Lie algebra if $n > 2$, and (9) says that $<a,a'>$ acts, except for a scalar factor, as the inner derivation by $f_2(a,a')$; together with the derivation property for $<a,a'>$, (9') says that (A, f_2) is a <u>Malcev algebra</u> [Ma], [SAG] for $n = 2$. However, we shall do better than this below.)

Next we take $x = e_{12}$, $y = e_{21}$, $z = e_{13}$, so that $[zx] = 0 = z \circ x$, $B(x,z) = 0 = B(y,z)$, $B(x,y) = 2(n+1)$, $[z[xy]] = -e_{13}$, $[x[yz]] = e_{13}$, $z \circ [xy] = e_{13}$, $[z, x \circ y] = -e_{13}$, $[x, y \circ z] = e_{13}$, $x \circ [yz] = e_{13}$, $x \circ (y \circ z) = e_{13}$, $z \circ (x \circ y) = \frac{n-3}{n+1} e_{13}$, and our Jacobi identity gives

$$
\begin{aligned}
(10) \quad & 2(n+1)a'' <a,a'> - f_1(a'', f_1(a,a')) - f_1(a'', f_2(a,a')) \\
& + f_2(a'', f_1(a,a')) + \frac{n-3}{n+1} f_2(a'', f_2(a,a')) + f_1(a, f_1(a',a'')) \\
& + f_1(a, f_2(a',a'')) + f_2(a, f_1(a',a'')) + f_2(a, f_2(a',a'')) = 0 .
\end{aligned}
$$

Substituting for $a'' <a,a'>$ from (9) if $n > 2$, we see that (10) yields that $f_1 + f_2$ <u>is an associative bilinear product on</u> A. Denoting this product simply by aa', we have $aa' = f_1(a,a') + f_2(a,a')$, $a'a = f_1(a,a') - f_2(a,a')$, so that

$$f_1(a,a') = \frac{1}{2}(aa' + a'a), \quad f_2(a,a') = \frac{1}{2}(aa' - a'a) .$$

When $a = a'$, (10) gives

$$f_1(f_1(a,a), a'') + f_2(f_1(a,a), a'') =$$
$$f_1(a, f_1(a,a'')) + f_1(a, f_2(a,a'')) + f_2(a, f_1(a,a'')) + f_2(a, f_2(a,a'')),$$

or $(f_1+f_2)((f_1+f_2)(a,a),a'') = (f_1+f_2)(a,(f_1+f_2)(a,a''))$,

or that f_1+f_2 <u>is a left-alternative product on</u> A; <u>for</u> $n > 2$, this (as well as right-alternativity) follows from the associativity. For $n = 2$, setting $a'' = a'$ in (9') and substituting in (10) gives the right-alternative law

$$(f_1+f_2)(a,(f_1+f_2)(a',a')) = (f_1+f_2)((f_1+f_2)(a,a'), a') .$$

Thus (A, f_1+f_2) <u>is an alternative algebra for all</u> $n \geq 2$. We again write aa' for $f_1(a,a') + f_2(a,a')$.

From (7) we see that $\mathcal{D} = <A,A>$ acts by derivations on (A, f_1+f_2), and the simplicity of L implies as in §2 that $\mathcal{D}$ may be identified with a subalgebra of $\mathrm{Der}(A)$.

For $n \geq 3$, we have $a''(aa' - a'a) - (aa' - a'a)a''$
$= 4 f_2(a'', f_2(a,a')) = 2(n+1)^2 a'' <a,a'>$ by (9), so that

$\mathcal{D} = \langle A, A \rangle$ identifies with the inner derivations of A determined by elements of the derived algebra $[AA]$.

The above is really valid whenever A is associative; namely for $n = 2$, we have for $a, a', a'' \in A$,

$$a'(aa'') - a(a'a'') + (aa'')a' - a(a''a') + (a''a)a'$$
$$- (a''a')a \text{ as the element } a''D_{a,a'}, \text{ where } D_{a,a'}$$

is the inner derivation of the alternative algebra A determined by a and a' (see "Type G_2" of §5). Now in any alternative algebra, one verifies directly from the identities that

$$a''D_{a,a'} = \frac{1}{2} \{a'(aa'' - a''a) - (aa'' - a''a)a'$$
$$- a(a'a'' - a''a') + (a'a'' - a''a')a$$
$$+ a''(aa' - a'a) - (aa' - a'a)a'' \},$$

which in our case is given by

$$2\{f_2(a', f_2(a, a'')) - f_2(a, f_2(a', a'')) + f_2(a'', f_2(a, a'))\}$$
$= 18 \, a'' \langle a, a' \rangle$. That is, $D_{a,a'} = 18 \langle a, a' \rangle$, and if one knows that all the derivations of A are inner, we have $\mathcal{D} = \langle A, A \rangle = \text{Der}(A)$. When A is associative, $a''D_{a,a'} = [a'', [a, a']]$, and we have the same situation as for $n > 2$.

Now let I be an ideal in A. Since $\mathcal{D}$ acts on A by inner derivations, $I\mathcal{D} \subseteq I$. Consider the subspace

$$\langle A, I \rangle + S \otimes I \quad \text{of} \quad L.$$

By (1), it follows that $\langle A, I \rangle$ is an ideal in $\mathcal{D}$, so the above is stable under brackets with $\mathcal{D}$. By the form of $a'' \langle a, a' \rangle$ we have $A \langle A, I \rangle \subseteq I$, from which our space is stable under brackets with $S \otimes A$, hence is an ideal in L. By simplicity, $I = 0$ or $I = A$, so that A is simple as associative (alternative) algebra. (From this one knows [cf. Appendix B] that $\mathcal{D} = \text{Der}(A)$; however, we shall not use this fact in any essential way.) Now T identifies with the subalgebra $T \otimes 1$, where 1 is the identity element of A, and the centralizer of T with $\mathcal{D} + T \otimes A$. Also $\mathcal{D}$ identifies with inner derivations of A, and contains nilpotent inner derivations unless A is a division algebra; for otherwise A is either a full matrix algebra over a division algebra, containing off-diagonal matrix units e_{ij} (which are in $[A, A]$), or is a split Caley-Dickson algebra over its center, in which case

68

$\mathcal{D} = \text{Der}(A) = \text{Inder}(A)$ is a split Lie algebra G_2, and so contains nilpotent elements. Since $\mathcal{D}$ is completely reducible in its action on L, as the centralizer of S in L, the theorem of Jacobson-Morozov implies that T is not a maximal split torus in this case. Hence A <u>is a division algebra</u>. Thus:

<u>Theorem III.3</u>. Let L <u>be a simple Lie algebra over</u> F <u>of relative type</u> $A_n (n > 2)$. <u>Then there is an associative division algebra</u> A <u>over</u> F <u>such that</u>

$$L \cong \text{Der}(A) \oplus S \otimes A, \text{ with}$$

$[x \otimes a, d] = x \otimes ad$,

$[x \otimes a, y \otimes a'] = B(x,y) <a,a'>$
$\quad + [xy] \otimes \frac{aa' + a'a}{2} + (x \circ y) \otimes \frac{aa' - a'a}{2}$

<u>where</u> $<a,a'>$ <u>is the inner derivation of</u> A <u>given by</u> $a'' \mapsto (\frac{1}{2(n+1)^2}) [a''[aa']]$; S <u>is the</u> $(n+1)$ <u>by</u> $(n+1)$ <u>F-matrices of trace zero with</u> $[xy] = xy - yx$, $x \circ y = xy + yx - \frac{2}{n+1} \text{Tr}(xy)I$, <u>and</u> $B(x,y)$ <u>is the Killing form</u>. For $n = 2$, L <u>is either as above, or there is an alternative, not associative, division algebra</u> A <u>over</u> F (<u>hence a Cayley-Dickson division algebra over its center</u>) <u>such that</u>

$$L \cong \text{Der}(A) \oplus S \otimes A$$

<u>as above, with</u> $<a,a'>$ <u>the inner derivation of</u> A <u>given by</u> $<a,a'> = \frac{1}{18} D_{a,a'}$, <u>where</u> $D_{a,a'}$ <u>is the inner derivation of</u> A <u>determined by</u> a <u>and</u> a'.

When A is associative, the algebra L above is isomorphic to the derived Lie algebra of the $(n+1)$ by $(n+1)$ A-matrices; we defer this and other explicit realizations to Chapter V.

§4. <u>Algebras of Relative Type</u> $B_r (r > 2)$:

Here M' is a $(2r+1)$-dimensional F-space with a nondegenerate quadratic form (u,u) of maximal Witt index and discriminant $(-1)^r$; thus there is a basis $e_1,\ldots,e_{2r+1}$ such that the associated bilinear form has $(e_i, e_{2r+2-i}) = 1$, $1 \le i \le 2r+1$, all other (e_i, e_j) being 0. S is the Lie algebra of skew transformations of M' with respect to this form, and
$L = \mathcal{D} \oplus (S \otimes A) \oplus (M' \otimes B)$, where $\mathcal{D}$, A, B are trivial S-modules. By the results of §1, we may apply reasoning analogous to that of the

69

previous cases to conclude:

$\mathcal{D}$ is a subalgebra of L ;

$[x \otimes a, d] = x \otimes ad$, where $(a,d) \mapsto ad$ is a bilinear pairing of A and $\mathcal{D}$ to A, giving A the structure of $\mathcal{D}$-module;

$[u \otimes b, d] = u \otimes bd$, where $(b,d) \mapsto bd$ gives B the structure of $\mathcal{D}$-module;

$[u \otimes b, x \otimes a] = ux \otimes g(b,a)$, where $g: B \otimes A \to B$ is a bilinear pairing;

$[x \otimes a, y \otimes a'] = B(x,y) <a,a'> + [xy] \otimes aa'$, as in previous cases;

$[u \otimes b, v \otimes b'] = (u,v)<b,b'> + [uv] \otimes f(b,b')$, where $<b,b'>$ is a skew bilinear pairing: $B \times B \to \mathcal{D}$, $f(b,b')$ is a symmetric bilinear pairing $B \times B \to A$, (u,v) is the given form on M', and $[uv] \in S$ is as in §1.

The Jacobi identity applied to triples $x \otimes a, y \otimes a', d$; $x \otimes a, u \otimes b, d$; $u \otimes b, v \otimes b', d$ yields respectively:

(1) $[<a,a'>, d] = <ad,a'> + <a,a'd>$;
(2) $(aa')d = (ad)a' + a(a'd)$.
(11) $g(b,a)d = g(bd,a) + g(b,ad)$.
(12) $[<b,b'>, d] = <bd,b'> + <b,b'd>$;
(13) $f(b,b')d = f(bd,b') + f(b,b'd)$.

As in the cases of types $D_n (n \geq 4)$, E_6, E_7, E_8, we may choose $x,y,z \in S$ with $B(x,y) = 0 = B(y,z) = B(x,z)$, and with $[x[yz]]$ and $[y[zx]]$ linearly independent, and conclude thereby that:

(4) <u>the product</u> aa' <u>is associative</u> (<u>and</u> <u>commutative</u>).

The considerations that led to (5) now give us

(14) $A <A, A> = 0$.

Applying the Jacobi identity to $x \otimes a, y \otimes a', u \otimes b$ gives
$0 = [[u \otimes b, x \otimes a] y \otimes a'] - [[u \otimes b, y \otimes a'] x \otimes a] - [u \otimes b [x \otimes a, y \otimes a']]$
$= (ux)y \otimes g(g(b,a),a') - (uy)x \otimes g(g(b,a'),a)$
$- u[xy] \otimes g(b,aa') - B(x,y) u \otimes b <a,a'>$.

With $u = e_i \in M'$, $x = e_{1,2} - e_{2r,2r+1} \in S$, $y = e_{2,3} - e_{2r-1,2r} \in S (r \geq 3)$, we have $(ux)y = u[xy] \neq 0$, $(uy)x = 0$, $B(x,y) = 0$, hence

(15) $\qquad g(g(b,a),a') = g(b,aa')$.

That is, the mapping g endows B with a structure of A-module. We write ba (or ab) for g(b,a).

Since A is commutative, we also have (ba')a = b(a'a) = b(aa') = (ba)a', so for all x, y, u,

u[xy] ⊗ b(aa') = (ux)y ⊗ b(aa') - (uy)x ⊗ b(a'a)   reduces our Jacobi relation to $B(x,y)u \otimes b <a,a'> = 0$, and this implies

(16) $B <A,A> = 0$.

Now the intersection of the annihilators in D of the D-modules A and B is an ideal in L, so must be zero. From (14) and (16), we therefore have

(17) $<A,A> = 0$.

Next we apply the Jacobi identity to the triple u ⊗ b, v ⊗ b', x ⊗ a:

[u ⊗ b[v ⊗ b', x ⊗ a]] - [v ⊗ b'[u ⊗ b, x ⊗ a]] + [x ⊗ a[u ⊗ b, v ⊗ b']] = 0
= (u,vx) <b,b'a> - (v,ux) <b',ba> + (u,v)x ⊗ a <b,b'>
+ [u,vx] ⊗ f(b,b'a) - [v,ux] ⊗ f(b',ba) + [x,[uv]] ⊗ af(b,b').

From (u,vx) = -(ux,v) = -(v,ux) and <b',ba> = -<ba,b'>, we have

(18) $<b, b'a> = <ba, b'>$.

Now [x,[uv]] = -[u,vx] + [v,ux], giving
0 = [u,vx] ⊗ (f(b,b'a) - af(b,b')) - [v,ux] ⊗ (f(b',ba) - af(b,b'))
   + (u,v) x ⊗ a <b,b'>.

With $u = e_1$, $v = e_2$, $x = e_{2,3} - e_{2r-1,2r}$ we have (u,v) = 0, ux = 0, [u,vx] = [e_1,e_3] = $e_{2r+1,3} - e_{2r-1,1}$ ≠ 0, from which

(19) $f(b,b'a) = af(b,b')$,

or f is a symmetric A-bilinear function from $B \otimes B$ to A. Also, the condition (19), together with the symmetry of f, forces (u,v)x ⊗ a <b,b'> = 0 for all u,v,x,a,b,b', hence

(20) $A <B,B> = 0$.

Finally, we have
[u ⊗ b[v ⊗ b', w ⊗ b'']] = (v,w)u ⊗ b <b',b''>
                          + u[vw] ⊗ bf(b',b''),

and the Jacobi identity here gives a cyclic sum equal to zero.

Choosing v and w in M' with $(v,w) = 1$, v and w linearly independent, and $u \neq 0$ with $(u,v) = 0 = (u,w)$, this cyclic sum becomes

$$u \otimes b <b',b''> - u \otimes b''f(b,b') + u \otimes b'f(b'',b) = 0,$$

or

(21) $\qquad b <b',b''> = f(b,b')b'' - f(b,b'')b'$.

It is now true that the numbered relations we have listed make the space $D + S \otimes A + M' \otimes B$ into a Lie algebra. Since our interest is in simple algebras, we note from (12), (17) and the multiplication that $<B,B> + S \otimes A + M' \otimes B$ is an ideal in L, so is equal to L, and $D = <B,B>$. Next let I be an ideal in A, and consider $K = <IB,B> + S \otimes I + M' \otimes IB$. Since $AD = A<B,B> = 0$, $AI \subseteq I$, $A(IB) \subseteq (AI) B \subseteq IB$, K is closed under brackets with $S \otimes A$. Since
$[<IB,B>,D] \subseteq <(ID) B,B> + <I(BD),B> + <IB,BD>$, the first term being zero and the last two in $<IB,B>$; since $ID \subseteq AD = 0$; and since $(IB)D \subseteq (ID) B + I(BD) = I(BD) \subseteq IB$, K is closed under brackets with D. Finally, $f(B,IB) \subseteq If(B,B) \subseteq IA \subseteq I$ and $B <IB,B> \subseteq (IB)f(B,B) + Bf(IB,B)$ (by (21))
$\subseteq (IB) A + B (IA) \subseteq IB$
show that K is closed under brackets with $M' \otimes B$. Thus K is an ideal in L, so $I = A$ or 0 and A <u>is a simple commutative associative algebra, hence a field extension of</u> F.

(We have not explicitly eliminated the case where A is a one-dimensional space with trivial multiplication. This is easily seen to be impossible, for instance by the fact that $<BA,B> + S \otimes A + M' \otimes BA$ would then be an abelian ideal in L. However, it is advantageous to have on hand for reference some deductions from the existence of the injection $x \mapsto x'$ of S into L. If $\{a_j\}$ is our basis for A as before, we must have $x' = \Sigma f_j(x) \otimes a_j, f_j(x) \in S$, with all the f_j being in $\text{Hom}_S(S,S)$ hence scalars. Thus: $x' = x \otimes a_o$, where a_o is a fixed element of A. But then from $Xx = [Xx']$ for all $X \in L$, $x \in S$, we have $[yx] \otimes a = [y \otimes a, x'] = [y \otimes a, x \otimes a_o] = [yx] \otimes aa_o$ for all $x,y \in S$, $a \in A$, so that a_o <u>is an identity element for</u> A. We write 1 for a_o. Now for $u \in M'$, $b \in B$, we have $ux \otimes b = [u \otimes b, x'] = [u \otimes b, x \otimes 1] = ux \otimes b1$, from which B is a "unital", or "unitary" A-module. Similar considerations will be

72

used implicitly in other cases.

Since B is an A-module, hence a vector space over the field A, f is now a symmetric bilinear form on B over A, and (21) says that the effect of $<b',b''>$ on B is exactly analogous to that of the mapping of M' that we have designated $[uv]$. It follows that if R is the radical of the form f, we have $<B,R> = 0$, and $M' \otimes R$ is an ideal in L. Hence $R = 0$, and f <u>is non-degenerate on</u> B. The annihilator of B in D will be an ideal in L, so must be zero, and D identifies with a Lie algebra of $(A-)$ endomorphisms of B, all realized from $<B,B>$. From (21), one sees that these are all skew with respect to f, and choice of an orthogonal basis for B over A enables one quickly to obtain a basis of the form $<b',b''>$ for all skew transformations of B over A. That is, $D = <B,B>$ <u>identifies with the Lie algebra of all f-skew A-endomorphisms of</u> B.

Our maximal torus T of L identifies with the set of elements $t \otimes 1$ of $S \otimes A$, where t is in the maximal torus T_0 of S, which consists of diagonal matrices relative to the basis $e_1, \ldots, e_{2r+1}$ for M'. The subspace of M' annihilated by T_0 is Fe_{r+1} and T_0 is its own centralizer in S. It follows that the centralizer $Z_L(T)$ of T in L is $D + T_0 \otimes A + e_{r+1} \otimes B$, with derived algebra $D + e_{r+1} \otimes BD = D + e_{r+1} \otimes B$ unless B has dimension zero or one over A, in which cases $D + e_{r+1} \otimes BD = 0$.

Now consider the A-module $V = A \otimes B$, with symmetric A-bilinear form g: $g(a+b, a'+b') = aa' - f(b,b')$. Clearly, D may be identified with the g-skew transformations of V mapping B into B (hence A into A, and therefore necessarily annihilating A). If T is any g-skew transformation of V, with $1T = a+b$, we have $a = 0$ from $0 = (1T,1) = a$, so that $1T = b \in B$. If $b' \in B$, we have $0 = (1T,b') + (1,b'T)$, or $b'T = f(b,b') + b''$, $b'' \in B$. Thus $b \in B$ defines the skew mapping $a + b' \to f(b,b') + ab$, and T is the sum of this mapping and an element d of D. It follows that $T \leftrightarrow d + e_{r+1} \otimes b$ is an isomorphism between the A-Lie algebra of g-skew transformations of V and the subalgebra $D + e_{r+1} \otimes B$ of $Z_L(T)$, a subalgebra containing the derived algebra of $Z_L(T)$ and equal to this derived algebra if B has A-dimension at least <u>two</u>. Hence V can admit no nilpotent g-skew transformations if B has dimension at least 2 over A, which means that g is anisotropic on V, or that <u>the</u>

form f on B does not represent 1.

If $B = 0$, we have $L = S \otimes A$; if B has A-dimension one, we have $D = 0$ and $Z_L(T) = T_0 \otimes A + e_{r+1} \otimes B$, $L = S \otimes A + M' \otimes B$. In this case, suppose there is $b' \in B$, with $f(b',b') = 1$. L has an A-basis consisting of elements of the form $y_i \otimes 1$, where $y_i \in S$ are a basis for the annihilator of e_{r+1}, together with the elements $[e_i, e_{r+1}] \otimes 1 \pm e_i \otimes b'$, $1 \le i \le r+1$. Evidently, $[y_i \otimes 1, e_{r+1} \otimes b'] = 0 = [e_{r+1} \otimes b', e_{r+1} \otimes b']$, while for $i \ne r+1$,
$[[e_i e_{r+1}] \otimes 1 \pm e_i \otimes b', e_{r+1} \otimes b'] =$
$\pm [e_i e_{r+1}] \otimes f(b',b') - e_{r+1}[e_i e_{r+1}] \otimes b'$
$= \pm [e_i e_{r+1}] \otimes 1 + e_i \otimes b' = \pm([e_i e_{r+1}] \otimes 1 \pm e_i \otimes b')$.
In this case $e_{r+1} \otimes b' \in L$ is not in T, centralizes T, and $\mathrm{ad}(e_{r+1} \otimes b')$ has characteristic roots $0, \pm 1 \in F$, contradicting the maximality of T. That is, f cannot represent 1 in any case. We now have the

Theorem III.4. Let L be a simple Lie algebra over F of relative type $B_r (r \ge 3)$. Then there is a field extension A of F, finite-dimensional vector space B over A (possibly $B = 0$), and a nondegenerate symmetric A-bilinear form f on B, not representing 1, such that
$L \cong D \oplus S \otimes A \oplus M' \otimes B$, where D is the Lie algebra of f-skew A-linear transformations of B, S and M' are as in §1, and
$$[x \otimes a, d] = 0, [u \otimes b, x \otimes a] = ux \otimes ab,$$
$[u \otimes b, d] = u \otimes bd, [u \otimes b, v \otimes b'] = (u,v) <b,b'> + [uv] \otimes f(b,b')$, where (u,v) and $[uv]$ are as above, and where $<b,b'> \in D$ is given by $b'' <b,b'> = f(b'',b)b' - f(b'',b')b$.

(For a more conventional realization, see Chapter V.)

§5. Algebras of Types G_2 and F_4.

With S and M' as in §1, one has here
$L \cong D \oplus (S \otimes A) \oplus (M' \otimes B)$, with compositions between D and each of the summands as before, likewise between $S \otimes A$ and each of the summands, but with
$[u \otimes b, v \otimes b'] = (u,v) <b,b'> + D_{u,v} \otimes f(b,b') + \{u,v\} \otimes h(b,b')$,
where $D_{u,v}$ is the inner derivation (of C or J) determined by

u and v, and $\{u,v\}$ is either $[uv] = uv - vu$ (if $M' = C_0$) or $u \cdot v - \frac{1}{3} \text{Tr}(u \cdot v)I$ (if $M' = J_0$). $f: B \times B \to A$ is skew, while $h: B \times B \to B$ is symmetric for G_2, skew for F_4. By considering the subalgebra $D + S \otimes A$, we find, as for B_r, that

A <u>is commutative and associative, and</u> $A <A,A> = 0$.

By considering the Jacobi identity on triples $x \otimes a$, $y \otimes a'$, $u \otimes b$, we see as before that

B <u>is an</u> A-<u>module and</u> $B < A,A> = 0$ <u>hence</u> $< A,A >= 0$; <u>moreover,</u> A <u>and</u> B <u>are modules for the subalgebra</u> D (<u>which we now know is equal to</u> $< B,B >$), <u>and</u> D <u>acts as derivations with regard to all compositions among them.</u>

The situation is a little more complicated when we consider the Jacobi identity for a triple $u \otimes b$, $v \otimes b'$, $x \otimes a$:

$0 = [u \otimes b[v \otimes b', x \otimes a]] - [v \otimes b'[u \otimes b, x \otimes a]]$
$\quad + [x \otimes a[u \otimes b, v \otimes b']]$
$= (u,vx) < b,b'a > - (v,ux) < b',ba > + (u,v)x \otimes a < b,b'>$
$+ D_{u,vx} \otimes f(b,b'a) - D_{v,ux} \otimes f(b',ba) + [x, D_{u,v}] \otimes af(b,b')$
$+ \{u,vx\} \otimes h(b,b'a) - \{v,ux\} \otimes h(b',ba) - \{u,v\} x \otimes h(b,b')a$.

From the skewness of $x \in S$ with respect to the form (u,v) on M', we have $(u,vx) = -(ux,v) = -(v,ux)$, giving by the skewness of $< b,b' >$:

(22) $\qquad\qquad < b,b'a > = < ba,b' >$.

From the definition of the inner derivation $D_{u,v}$ in each case, we have for $x \in S$, $[D_{u,v}, x] = D_{ux,v} + D_{u,vx} = D_{u,vx} - D_{v,ux}$, so that $D_{u,vx} \otimes (f(b,b'a) - af(b,b')) - D_{v,ux} \otimes (f(b',ba) - af(b,b'))$
$+ (u,v)x \otimes a < b,b' > = 0$.

In each case we may take u and v to be linearly independent in a maximal totally isotropic subspace of M' (where the form has respective Witt indices 3 and 12 or 13, the latter depending on the arithmetic of F), and x to be a derivation of C or J annihilating v and sending u to a third linearly independent element w of this subspace for which $D_{w,v} \neq 0$. (The possibility of such a choice may be verified directly from the explicit computations for C and $\text{Der}(C)$ in [SLA II], or for J and $\text{Der}(J)$ in [SLA III].) For purposes below, it should also be remarked that this choice can be made in such a way that $\{v,w\} \neq 0$. From the above, we then have

$$D_{w,v} \otimes (f(b',ba) - af(b,b')) = 0, \text{ so}$$

$$f: B \otimes B \to A \quad \underline{\text{is}} \quad A\text{-}\underline{\text{bilinear}},$$

and then, as before,

(20') $$A < B, B > = 0.$$

The S-invariance of $\{u,v\}$ yields $\{u,v\}x = \{ux,v\} + \{u,vx\}$, from which the component of our Jacobi identity in $M' \otimes B$ gives for u, v, x, w as above,

$$- \{v,w\} \otimes h(b',ba) - \{w,v\} \otimes h(b,b')a = 0.$$

In each case, $\{v,w\} \otimes h(b',ba) = -\{w,v\} \otimes h(ba,b')$, from which, using its symmetry or anti-symmetry,

(23) $$h: B \otimes B \to B \quad \underline{\text{is}} \quad A\text{-}\underline{\text{bilinear}}.$$

Now we consider the Jacobi identity for triples from $M' \otimes B$:

$$[u \otimes b[v \otimes b', w \otimes b'']] = (u, \{v,w\}) < b, h(b',b'') >$$
$$+ D_{u, \{v,w\}} \otimes f(b,h(b',b'')) + (v,w)u \otimes b \quad < b',b'' >$$
$$+ u D_{v,w} \otimes f(b',b'')b + \{u, \{v,w\}\} \otimes h(b,h(b',b'')),$$

and the cyclic sum is to be zero. In each of our two cases, one has $(u, \{v,w\}) = (v, \{w,u\}) = (w, \{u,v\})$, not identically zero, from which

(24) $$< b,h(b',b'') > + < b',h(b'',b) > + < b'',h(b,b') > = 0.$$

In each of our two cases, we have

$$D_{u, \{v,w\}} + D_{v, \{w,u\}} + D_{w, \{u,v\}} = 0 \text{ for all } u, v, w;$$

in the case of G_2, this results from [SCH], p. 78, and, for F_4, from [SCH], p. 92. Moreover, in both cases it is possible to choose $u, v, w \in M'$ so that $D_{u, \{v,w\}}$ and $D_{v, \{w,u\}}$ are linearly independent in S; for G_2, we may take u, v, w to be a basis for a totally isotropic subspace M_0 of M' (with respect to the form (u,v)), so that $\{v,w\} = vw - wv$ becomes a non-zero scalar multiple of an element u' of a dual maximal totally isotropic subspace M_1 with $(u,u') = 1$, $(v,u') = 0 = (w,u')$, and likewise such that $\{w,u\}$ is a scalar multiple of an analogous $v' \in M_1$. Then $D_{u, \{v,w\}}$ annihilates u', but not v', while $D_{v, \{w,u\}}$ annihilates v', but not u'. For F_4, suitable choices of u,v,w corresponding to the (1,2) -, (1,3) -, (2,3) - matrix positions in [SLA III] or in [JJ], Chapter IX are readily seen to display the linear independence. From these observations and from

$$D_{u,\{v,w\}} \otimes f(b,h(b',b'')) + D_{v,\{w,u\}} \otimes f(b',h(b'',b))$$
$$+ D_{w,\{u,v\}} \otimes f(b'',h(b,b')) = 0$$

we conclude:

(25) $\qquad f(b,h(b',b'')) - f(b'',h(b,b')) = 0,$

or, using the skewness resp. symmetry of h, that $b \mapsto h(b',b)$ __is an__ f-__symmetric transformation of__ B __in the case__ G_2, __an__ f-__skew transformation in the case__ F_4.

Finally, we consider the effect of our Jacobi identity on the terms in $M' \otimes B$; here it seems unavoidable to separate the cases G_2 and F_4. The Jacobi identity forces the cyclic sum derived from

$(v,w)u \otimes b <b',b''> + uD_{v,w} \otimes f(b',b'')b + \{u,\{v,w\}\} \otimes h(b,h(b',b''))$

to be identically zero. In the case of G_2, we have

$(v,w)u = \frac{1}{2}(v\bar{w} + w\bar{v})u = -\frac{1}{2}(vw + wv)u = -\frac{1}{2}u(vw + wv),$

$uD_{v,w} = w(vu) - v(wu) + (uw)v + (vu)w - v(uw), \{u,\{v,w\}\}$

$= u(vw - wv) - (vw - wv)u$

for all $u,v,w \in M'$, the multiplication being that of the split Cayley algebra C. Now

$2uD_{v,w} + \{v,\{w,u\}\} = 2uD_{v,w} + v(wu - uw) - (wu - uw)v$

$= 2w(vu) - v(wu) + 2(uv)w - (uw)v + 2(vu)w - 3v(uw) - (wu)v$

$= 2w(vu) - v(wu + uw) - 2v(uw) + 2(uv + vu)w - (uw + wu)v$

$= 2w(vu) - 2v(uw) + 4(u,w)v - 4(u,v)w.$

The alternative law gives $2w(vu) = 2(vw)u - 2v(wu) + 2(wv)u,$ from which the above is equal to

$8(u,w)v - 4(u,v)w - 4(v,w)u.$ Thus we can eliminate $\{u,\{v,w\}\}, \{v,\{w,u\}\}, \{w,\{u,v\}\}$ from our relation to obtain

$(v,w)u \otimes ((b <b',b''> + 8h(b,h(b',b'')) - 4h(b',h(b'',b)) -$
$4h(b'',h(b,b'))) + (w,u)v \otimes (\qquad) + (u,v)w \otimes (\qquad)$

(26)
$$+ uD_{v,w} \otimes (f(b',b'')b - 2h(b',h(b'',b))) + vD_{w,u} \otimes (\quad)$$
$$+ wD_{u,v} \otimes (\quad) = 0,$$

where the empty parentheses indicate terms obtained from the preceding by cyclic permutation of b,b',b''. Now we also have

$$uD_{v,w} = 4(u,w)v - 4(u,v)w + (u,v,w),$$ where (u,v,w) is the <u>associator</u> $(uv)w - u(vw)$ (cf. [SCH], p. 125).

(For the definitions give $uD_{v,w} = 2(w,u)v - 2(u,v)w + w(vu) - (uw)v$. Writing $w(vu) = -2(u,v)w - w(uv)$ and $-(uw)v = 2(w,u)v + (wu)v$, we have

$$uD_{v,w} = 4(u,w)v - 4(u,v)w + (w,u,v),$$

which is our assertion.) We now eliminate $uD_{v,w}$, etc., in favor of $(u,v)w, (v,w)u, (w,u)v$ and (u,v,w) to get

$$(v,w)u \otimes (b<b',b''> - 4f(b'',b)b' + 4f(b,b')b'' - 4h(b',h(b'',b))$$
(27) $$+ 4h(b'',h(b,b'))) + (w,u)v \otimes (\quad) + (u,v)w \otimes (\quad)$$
$$+ (u,v,w) \otimes (f(b',b'')b + f(b'',b)b' + f(b,b')b'' - 2h(b',h(b'',b))$$
$$- 2h(b'',h(b,b')) - 2h(b,h(b',b''))) = 0,$$

the contents of the empty parentheses being evident as before. Now taking u,v,w to be a basis of a maximal totally isotropic subspace of M', we find $(u,v,w) \neq 0$ (<u>e.g.</u>, use u_1, u_2, u_3 from [SLA III], p. 287), from which

(28) $$f(b,b')b'' + f(b',b'')b + f(b'',b)b'$$
$$= 2(h(b',h(b'',b)) + h(b'',h(b,b')) + h(b,h(b',b'')))$$

On the other hand, if u and v are chosen to be isotropic, but not orthogonal, and $w = uv - vu$, then $(u,v,w) = 0$, $(u,w) = (v,w) = 0$ but $(u,v)w \neq 0$ (<u>e.g.</u>, use $u = u_1$, $v = u_5$ in [SLA III], <u>loc. cit.</u>), from which

(29) $$b<b',b''> = 4f(b'',b)b' - 4f(b,b')b''$$
$$+ 4h(b',h(b'',b)) - 4h(b'',h(b,b')).$$

Next we consider the case F_4, where
$\{w, \{u,v\}\} = w \cdot (u \cdot v) - \frac{1}{3}(u,v)w - \frac{1}{3}(u \cdot v, w)I$, (with $u \cdot v = \frac{1}{2}(uv + vu)$), from which, since $(u \cdot v, w) = (w \cdot u, v)$, we find
$\{w, \{u,v\}\} - \{v, \{w,u\}\} + \frac{1}{3}(u,v)w - \frac{1}{3}(u,w)v = (v \cdot u) \cdot w - v \cdot (u \cdot w)$

that is, $uD_{v,w} = \{w\ \{u,v\}\} - \{v\ \{w,u\}\} + \frac{1}{3}(u,v)w - \frac{1}{3}(u,w)v$.

Thus we may eliminate the terms $uD_{v,w}$, $vD_{w,u}$, $wD_{u,v}$ from our Jacobi relation to obtain

$(v,w)u \otimes (b < b',b''> + \frac{1}{3}f(b'',b)b' - \frac{1}{3}f(b,b')b'')$

$+ (w,u)v \otimes (\quad) + (u,v)w \otimes (\quad)$

(30)

$+ \{u,\ \{v,w\}\} \otimes (h(b,h(b',b'')) + f(b'',b)b' - f(b,b')b'')$

$+ \{v,\ \{w,u\}\} \otimes (\quad) + \{w,\{u,v\}\} \otimes (\quad) = 0$,

with conventions regarding empty parentheses as before.

From the fact that elements of the exceptional Jordan algebra satisfy a cubic equation, we have ([SCH], p. 126)

$$\{u\ \{v,w\}\} + \{v\ \{w,u\}\} + \{w\ \{u,v\}\}$$
$$= \frac{1}{6}((u,v)w + (w,u)v + (v,w)u),$$

which we may use to eliminate $\{u\ \{v,w\}\}$ from (30), obtaining

(31) $(v,w)u \otimes (b < b',b''> + \frac{1}{2}f(b'',b)b' - \frac{1}{2}f(b,b')b'' +$

$+ \frac{1}{6}h(b,h(b',b''))) + (w,u)v \otimes (b' < b'',b> + \frac{1}{6}f(b,b'')b'' -$

$- \frac{1}{3}f(b',b'')b + \frac{1}{6}f(b'',b)b' + \frac{1}{6}h(b,h(b',b''))) +$

$+ (u,v)w \otimes (b'' < b,b' > + \frac{1}{3}f(b',b'')b - \frac{1}{6}f(b,b)b' -$

$- \frac{1}{6}f(b,b')b'' + \frac{1}{6}h(b,h(b',b''))) + \{v\ \{w,u\}\} \otimes (h(b',h(b'',b))$

$- h(b,h(b',b'')) + 2f(b,b')b'' - f(b',b'')b - f(b'',b)b') +$

$+ \{w,\ \{u,v\}\} \otimes (\quad) = 0$

In the notation of [SLA III], p. 288, we take $u = u_1(1,2)$, $v = u_2(2,3)$, $w = u_3(1,3)$, so that $(u,v) = 0 = (w,u) = (v,w)$, $\{w,u\} = w \cdot u = \frac{1}{4} u_6(2,3)$, $\{u,v\} = \frac{1}{4} u_7(1,3)$, $v \cdot \{w,u\} = \frac{1}{2}$ diag $\{0,1,1\}$, $\{v\ \{w,u\}\} =$ diag $\{-\frac{1}{3}, +\frac{1}{6}, \frac{1}{6}\}$, $w \cdot \{u,v\} = \frac{1}{2}$ diag $\{1,0,1\}$, $\{w\ \{u,v\}\} =$ diag $\{\frac{1}{6}, -\frac{1}{3}, \frac{1}{6}\}$, the latter two elements being linearly independent. We thus conclude

$$h(b',h(b'',b)) - h(b,h(b',b''))$$

(32)

$$= f(b',b'')b + f(b'',b)b' - 2f(b,b')b''.$$

Finally, taking $u = v = \text{diag}\{1,-1,0\}$, $w = \text{diag}\{1,1,-2\}$, we have $(u,w) = 0 = (v,w)$, $(u,v) = 2$, $u.v = \text{diag}\{1,1,0\}$, $w.u = v.w = u = \{w,u\} = \{v,w\}$, $\{u,v\} = \{u,u\} = \frac{1}{3}w$, $\{u,\{v,w\}\} = \frac{1}{3}w = \{v,\{w,u\}\}$, and $\{w\{u,v\}\} = \frac{1}{3}\{w,w\} = \frac{1}{3}\text{diag}\{-1,-1,2\} = -\frac{1}{3}w$. From this we find, as coefficient of w,

$2b'' < b,b' > + \frac{2}{3} f(b',b'')b - \frac{1}{3} f(b'',b)b' - \frac{1}{3} f(b,b')b''$
$+ \frac{1}{3} h(b,h(b',b'')) + \frac{1}{3} h(b',h(b'',b)) - \frac{1}{3} h(b,h(b',b''))$
$+ \frac{2}{3} f(b,b')b'' - \frac{1}{3} f(b',b'')b - \frac{1}{3} f(b'',b)b' - \frac{1}{3} h(b'',h(b,b'))$
$+ \frac{1}{3} h(b',h(b'',b)) - \frac{2}{3} f(b',b'')b + \frac{1}{3} f(b'',b)b' + \frac{1}{3} f(b,b')b''$,

so that

$2 b'' < b,b' > - \frac{1}{3} f(b',b'')b - \frac{1}{3} f(b'',b)b' + \frac{2}{3} f(b,b')b''$
$\quad + \frac{2}{3} h(b',h(b'',b)) - \frac{1}{3} h(b'',h(b,b')) = 0$,

which combines with (32) to yield

$6b'' < b,b' > = 2h(b',h(b'',b)) - 2h(b'',h(b,b')) + 2h(b,h(b',b''))$, or
(33) $\quad b'' < b,b' > = -\frac{1}{3}(h(b,h(b',b'')) + h(b',h(b'',b)) - h(b'',h(b,b')))$.

Now we draw some inferences from the identities deduced above. The intersection of the annihilators of A and of B in D is again an ideal in L, so must be zero, and we saw at the beginning of our discussion that $< A,A >$ is contained in this annihilator. Then

$$< B,B > + S \otimes A + M' \otimes B$$

is an ideal in L, so is L, and $< B,B > = D$. Since $A < B,B > = 0$ by (20'), $D = < B,B >$ is __faithfully__ represented on B, and every ideal in the commutative, associative algebra A is D-stable. If I is such an ideal, then consider

$$K = < B,BI > + S \otimes I + M' \otimes BI.$$

We claim this is an ideal in L. Clearly $[S \otimes I, L] \subseteq K$. Now $[M' \otimes BI, D] \subseteq M' \otimes (BI) D \subseteq M' \otimes (BD) I + M' \otimes B(ID) \subseteq M' \otimes BI$, $[M' \otimes BI, S \otimes A] \subseteq M' \otimes BIA \subseteq M' \otimes BI$, and $[M' \otimes BI, M' \otimes B] \subseteq <BI, B> + S \otimes f(BI,B) \subseteq < B,BI > + S \otimes If(B,B) \subseteq K$, by the A-bilinearity of f. Thus $[M' \otimes BI, L] \subseteq K$. Finally, $[< B,BI >, D] \subseteq < BD, (BI) > + < B, (BI) D > \subseteq < B,BI > + < B,(BD)I > \subseteq < B,BI >$, $A < B,BI > = 0$, and

$B<B,BI>$ is seen, for the case G_2 by (29), to be contained in
$$f(BI,B)B + f(B,B)BI + h(B,h(BI,B)) + h(BI,h(B,B)) \subseteq BI \quad \text{by}$$
(20) and (23).
For the case F_4, (33) and (23) likewise yield that $B<B,BI> \subseteq BI$, which suffices to complete the proof that K is an ideal in L. Hence, as before, we have that A <u>is a field extension of</u> F.

Again B is a vector space over A, f is a symmetric A-bilinear form on B, and h is an A-bilinear composition on B, symmetric in the case of G_2 and skew-symmetric in the case of F_4. The condition (13) listed under B_n, combined with the remarks above, shows that $D = <B,B>$ is identified with a Lie subalgebra of the Lie algebra of f-skew A-linear transformations of B.

Now let R be the radical of f on B. Then R is an A-subspace of B and $RD \subseteq R$. To show that $K = <B,R> + M' \otimes R$ is an ideal of L, it therefore suffices to show that $[<B,R>,L] \subseteq K$. That $[<B,R>,D] \subseteq <B,R>$ follows as above since $RD \subseteq R$. Since $A<B,R> = 0$, it suffices to show $B<B,R> \subseteq R$, for which, by (29) and (33), it suffices to show $h(B,R) \subseteq R$; but this is immediate from (25). It follows that $K = 0$, so $R = 0$, and f is nondegenerate.

Let us concentrate for now on the case F_4. We define a quadratic form q on the A-vector space $A \oplus B$ by $q(a+b) = a^2 - f(b,b)$, and a bilinear product in the same space by
$$(a+b)(a' + b') = (aa' + f(b,b')) + (ab' + a'b + h(b,b')).$$
We claim that q <u>admits</u> <u>this</u> <u>product</u> <u>as</u> <u>composition</u>, i.e.,
$$q((a+b)(a' + b')) = q(a+b)q(a' + b').$$
The right-hand side is $a^2a'^2 - a^2f(b',b') - a'^2f(b,b) + f(b,b)f(b',b')$ while the left side is

$a^2a'^2 + 2\,aa'f(b,b') + f(b,b')^2$

$\quad - a^2f(b',b') - 2\,aa'f(b,b') - a'^2f(b,b)$

$\quad - 2\,af(b',h(b,b')) - 2\,a'f(b,h(b,b')) - f(h(b,b'),h(b,b')).$

In view of (25), all we need to show is that
$$f(b,b)f(b',b') = f(b,b')^2 - f(h(b,b'),h(b,b')).$$

Now replacing b by b', b' by b", b" by b in (32) and subtracting the resulting relation from (32) gives

$$f(b',b")b - f(b,b')b" = \frac{1}{3}(2h(b',h(b",b))$$
$$- h(b,h(b',b")) - h(b",h(b,b'))),$$

which for b' = b" gives
$f(b',b')b - f(b,b')b' = h(b',h(b,b'))$, hence
$f(b,b)f(b',b') - f(b,b')^2 = f(b,h(b',h(b,b')))$, which finally is equal to $f(h(b,b'), h(b,b'))$, by (25).

It follows (e.g., from [JC]) that the A-algebra $A + B$ has (A-)dimension 1,2,4 or 8, and is either A (for degree 1); $A \oplus A$, or a quadratic extension field of A (degree 2); a (generalized) quaternion division algebra over A, or the 2 by A-matrices (degree 4); or a Cayley-Dickson division algebra of octonions over A, or else a split Cayley-Dickson algebra (degree 8). It remains to identify the operators $< b,b' >$ on B, and thereby the subalgebra D. We have seen $A <B,B> = 0$, and that
$3b" <b,b'> = h(b",h(b,b')) - h(b,h(b',b")) - h(b',h(b",b))$.

But our definition of multiplication in $A + B$ and the skewness of h, together with the symmetry of f, give $2h(b,b') = bb' - b'b$, so that

$12 b" < b,b' > = b"(bb' - b'b) - (bb' - b'b)b" - b(b b" - b"b')$

(33)
$\qquad + (b'b" - b"b')b - b'(b"b - bb") + (b"b - bb")b'.$

Where the degree is 1 or 2, $A + B$ is commutative and associative, and this is zero; where the degree is 4, our algebra is associative, and we find

$12 b" < b,b' > = 2 b"(bb' - b'b) - 2(bb' - b'b)b",$

or $6 < b,b' >$ is the inner derivation of $A + B$ determined by $bb' - b'b$. From the split case we see that D thereby identifies with $\mathrm{Der}_A(A + B)$, or simply with B, as Lie subalgebra of $A + B$, and is three-dimensional. If $A + B$ is split, then B is the 2 by 2 matrices of trace zero and contains nilpotent elements, so that D would also. Hence $A + B$ must be a quaternionic division algebra in this case, and q its norm form.

Where the degree is 1, we have $B = 0 =< B,B > = D$, and

$L = S \otimes A$ is a split algebra of type F_4 over the field extension A. Where the degree is 2, we have B of dimension 1 over A, $<B,B> = 0$, and $L = S \otimes A + M' \otimes B$. The centralizer in L of $T = T \otimes 1$ is $T \otimes A + M'_0 \otimes B$, where M'_0 is the 2-dimensional subspace of M' which may be identified with the diagonal elements in our realization of M' (See [SLA III], §2). If $A + B$ is split, the form q represents 0, so there is $b \in B$ with $f(b,b) = 4$. Let $u_0 = \text{diag}\{1,-1,0\} \in M'_0$, and consider the element $u_0 \otimes b \in M'_0 \otimes B$; we claim $\text{ad}(u_0 \otimes b)$ is diagonal, with integral eigenvalues $0, \pm 1, \pm 2$. For by [SLA III], if $x \in S$, we have $(\text{diag}\{1,0,0\})x = a_{12}(1,2) + a_{13}(1,3)$, $(\text{diag}\{0,1,0\})x = -a_{12}(1,2) - a_{23}(2,3)$, $(\text{diag}\{0,0,1\})x = -a_{13}(1,3) + a_{23}(2,3)$, where $a(i,j)$ has a in the (i,j)-position, $\bar{a}$ in the (j,i)-position, and 0 elsewhere. Direct calculation shows that such an x agrees on diagonal elements of J with the inner derivation $D_{u_0, 2a_{12}(1,2)} + D_{u_0, 4a_{13}(1,3)} + D_{u_0, 4a_{23}(2,3)}$. It follows that S has a basis consisting of elements x annihilating u_0, together with elements of the form $D_{u_0, a(i,j)}$ as above, and that if we let a run over a basis for the split octonions O these 24 elements $D_{u_0, a(i,j)}$, together with a basis for the annihilator S_0 of u_0, form an F-basis for S. An A-basis for $L = S \otimes A + M' \otimes B$ is thus comprised of tensor products with $1 \in A$ of a basis for S_0, tensor products with b of a basis for M', together with the elements $D_{u_0, a(i,j)} \otimes 1$ and $a_{(i,j)} \otimes b$, a running over a basis for O. Elements of the first two types listed are annihilated by $\text{ad}(u_0 \otimes b)$; from the above, we have $[D_{u_0, a(i,j)} \otimes 1, u_0 \otimes b] = -u_0 D_{u_0, a(i,j)} \otimes b = -\frac{1}{4} a(i,j) \otimes b$ if $(i,j) \neq (1,2)$, and $-a(1,2) \otimes b$ if $(i,j) = (1,2)$.

Also $[a(i,j) \otimes b, u_0 \otimes b] = D_{a(i,j), u_0} \otimes f(b,b)$

$= -4 D_{u_0, a(i,j)} \otimes 1$. It follows that, for $(i,j) \neq (1,2)$, $4 D_{u_0, a(i,j)} \otimes 1 \pm a(i,j) \otimes b$ belongs to the eigenvalue ∓ 1 of $\text{ad}(u_0 \otimes b)$, while $2 D_{u_0, a(1,2)} \otimes 1 \pm a(1,2) \otimes b$ belongs to the eigenvalue ∓ 2. This completes our assertion, and shows that g cannot represent zero, hence that $A + B$ <u>must be a quadratic field extension</u>.

Where the degree is 8, we see by [SCH], p. 125, that the right-hand side of (33) is $2b'' D_{b,b'}$, where $D_{b,b'}$ is the inner derivation of $A + B$ determined by b and b', so that

$$< b,b' > = \tfrac{1}{6} D_{b,b'}$$

in this case. By [SCH], p. 87, (cf. also Appendix B), every derivation of $A + B$ is inner, so that $\mathcal{D} = \mathrm{Der}_A(A + B)$ (all these annihilate A, of necessity, as does $< B,B >$). Now if $A + B$ is a split Cayley-Dickson algebra, $\mathcal{D}$ contains nilpotent elements ([SLA II], p. 476); it follows that $A + B$ is a division algebra. We can now formulate our conclusions for F_4 as follows:

Theorem III.5. Let L be a simple Lie algebra over F of relative type F_4. Then there is a field extension A of F and a composition division algebra C over A such that

$$L \cong \mathcal{D} + S \otimes A + M' \otimes B ,$$

where S and M' and their compositions are as in §1, where $\mathcal{D}$ is the Lie algebra of all A-derivations of C, where B is the subspace of C orthogonal to the unit element with respect to the norm form, and where $[x \otimes a, d] = 0$, $[u \otimes b, d] = u \otimes bd$, $[u \otimes b, x \otimes a] = ux \otimes ab$, $[u \otimes b, v \otimes b'] = (u,v) < b,b'> + D_{u,v} \otimes f(b,b') + (u \circ v) \otimes \tfrac{1}{2}[b,b']$, where f is the norm form of C, and $< b,b'> = \tfrac{1}{6} D_{b,b'}$, $D_{b,b'}$ being the inner derivation of C determined by b and b' (defined to be the inner derivation $\mathrm{ad}(bb' - b'b)$ if C is associative). In particular, the dimension of L over A is one of 52, 78, 133, 248.

Noting that $S \otimes A$ is the derivation algebra of the split exceptional Jordan algebra $J \otimes A$ over A and that $M' \otimes_F B$ is the A-space $(J_0 \otimes A) \otimes_A C_0$, one sees that our algebra is exactly what results from the "second construction" of Tits (explained in detail in [SCH], pp. 120ff., or [JE], §10) using a split exceptional Jordan algebra and a composition division algebra.

We return to the case G_2. As in the case of F_4, we define a product in $A \otimes B$ by

$$(a+b)(a'+b') = (aa' + f(b,b')) + (ab' + a'b + h(b,b')).$$

Since h and f are symmetric, this is a commutative product, and A-bilinear since all the compositions involved are. We claim that it

84

is a <u>Jordan</u> composition on $\mathcal{A} + \mathcal{B}$, i.e., that
$(a+b)^2((a' + b')(a+b)) = ((a+b)^2(a' + b'))(a+b)$. Using the bilinearity
and symmetry of the forms involved, one sees upon expansion that this
will follow at once if we know

(34) $\qquad f(h(b,b),h(b',b)) = f(h(h(b,b),b'),b)$ $\qquad\qquad$ and

(35) $\qquad f(h(b,b),b')b + h(h(h(b,b),b'),b)$
$\qquad\qquad = f(b,b')h(b,b) + h(h(b,b),h(b',b))$.

Now (34) is immediate from (25). From (24), we have $< b,h(b,b)> = 0$.
The formula (29) applied to $b' < b,h(b,b) > = 0$ then gives

$\qquad f(h(b,b),b')b + h(b,h(h(b,b),b'))$
$\qquad\quad = f(b,b')h(b,b) + h(h(b,b),h(b,b'))$, or (35).

$\qquad$ Moreover, we have $(a+b)^3 = (a+b)^2(a+b)$
$= \{a^3 + 3af(b,b) + f(h(b,b),b)\}$
$\qquad + \{3a^2 b + 3ah(b,b) + f(b,b)b + h(h(b,b),b)\}$.
Thus $(a+b)^3 - 3a(a+b)^2 + (3a^2 - \frac{3}{2} f(b,b))(a+b)$ $\qquad$ is equal to
$\{a^3 - \frac{3}{2} af(b,b) + f(h(b,b),b)\} + \{ - \frac{1}{2} f(b,b)b + h(h(b,b),b)\}$, and
the second bracketed term is zero by (28). That is, <u>each element</u>
$c = a + b$ <u>of the Jordan algebra</u> $\mathcal{J} = \mathcal{A} + \mathcal{B}$ $\quad$ <u>satisfies the cubic equation</u>
$X^3 - 3a X^2 + (3a^2 - \frac{3}{2} f(b,b))X - (a^3 - \frac{3}{2} af(b,b) + f(h(b,b),b)) = 0$ over
$\mathcal{A}$.

$\qquad$ We know from the above that $\mathcal{D} = \; < \mathcal{B},\mathcal{B} >$ $\;$ is identified with a
sub-Lie algebra of the Lie algebra of derivations of $\mathcal{J}$ over $\mathcal{A}$.
These necessarily annihilate $\mathcal{A}$, as do all inner derivations.
Computing $((a+b)b'')(a' + b')$, we find $(f(b,b'') + ab'' + h(b,b''))$
$(a' + b')$, which is equal to
$\qquad \{ a'f(b,b'') + af(b'',b') + f(h(b,b''),b') \}$ $\qquad +$
$\qquad\quad \{aa'b'' + a'h(b,b'') + f(b,b'')b' + ah(b'',b') + h(h(b,b''),b')\}$.
From this we see that $b''D_{a+b,a'+b'} = ((a+b)b'')(a'+b')-((a'+b')b'')(a+b)$
is equal to $\{f(h(b,b''),b') - f(h(b',b''),b)\} + \{f(b,b'')b' +$
$h(h(b,b''),b') - f(b',b'')b - h(h(b',b''),b)\} = -\frac{1}{4} b'' < b,b' >$, by
(25) and (29). Also $D_{a+b,a'+b'} = D_{b,b'}$, so we have $< b,b' > = -$
$4 D_{b,b'}$, and $\mathcal{D} = \; < \mathcal{B},\mathcal{B} >$ $\;$ is the set of all <u>inner derivations</u> of $\mathcal{J}$
(hence all derivations of $\mathcal{J}$, as soon as we know that $\mathcal{J}$ is

simple, a fact which will appear below.)

In terms of the vector-matrices of [SLA II], with

$\underline{e}_i = \begin{pmatrix} 0 & e_i \\ 0 & 0 \end{pmatrix}$, $\underline{f}_i = \begin{pmatrix} 0 & 0 \\ e_i & 0 \end{pmatrix}$, $i = 1, 2, 3$, the root-space of S corresponding to the short root $\alpha = -\alpha_1 - 2\alpha_2$ has as basis the derivation $e_\alpha = 2(e_3, 0, 0)$, sending the vector-matrix $\begin{pmatrix} 1 & 0 \\ 0 & -1 \end{pmatrix}$ to $\frac{1}{2}\begin{pmatrix} 0 & 4e_3 \\ 0 & 0 \end{pmatrix} = 2\underline{e}_3$, $\begin{pmatrix} 0 & a \\ b & 0 \end{pmatrix}$ to $\begin{pmatrix} -b \cdot e_3 & 0 \\ e_3 \wedge a & b \cdot e_3 \end{pmatrix}$, where the dot and wedge denote the usual scalar and vector products in F^3. Likewise, we have $e_{-\alpha} = 2(0, e_3, 0)$, sending $\begin{pmatrix} 1 & 0 \\ 0 & -1 \end{pmatrix}$ to $2\underline{f}_3$, $\begin{pmatrix} 0 & a \\ b & 0 \end{pmatrix}$ to $\begin{pmatrix} -a \cdot e_3 & e_3 \wedge b \\ 0 & a \cdot e_3 \end{pmatrix}$.

The weight-space of M' corresponding to α has basis $v_\alpha = \underline{e}_3$ and $M'_{-\alpha}$ has basis $v_{-\alpha} = \underline{f}_3$. Thus an arbitrary element of L_α has the form $X_\alpha = e_\alpha \otimes a + v_\alpha \otimes b$ ($a \in A$, $b \in B$), and, if $X_\alpha \neq 0$, there is a unique $Y_{-\alpha} \in L_{-\alpha}$ such that $[X_\alpha Y_{-\alpha}] = H_\alpha \in T$, i.e., $[X_\alpha Y_{-\alpha}] = h_\alpha \otimes 1$, where $h_\alpha = [e_\alpha e_{-\alpha}]$.

Now $Y_{-\alpha} = e_{-\alpha} \otimes a' - v_{-\alpha} \otimes b'$ for unique $a' \in A$, $b' \in B$. Our formulas give

$$h_\alpha \otimes 1 = [X_\alpha Y_{-\alpha}] = -(v_\alpha, v_{-\alpha}) <b, b'>$$
$$+ h_\alpha \otimes aa' - D_{v_\alpha, v_{-\alpha}} \otimes f(b, b')$$
$$+ v_{-\alpha} e_\alpha \otimes ab' + v_\alpha e_{-\alpha} \otimes a'b - [v_\alpha v_{-\alpha}] \otimes h(b, b').$$

Now one sees from the definitions that $D_{v_\alpha, v_{-\alpha}} = -h_\alpha$, $v_{-\alpha} e_\alpha = -\begin{pmatrix} 1 & 0 \\ 0 & -1 \end{pmatrix} = v_\alpha e_{-\alpha} = -[v_\alpha v_{-\alpha}]$. From the components in the above we find $<b, b'> = 0$, $aa' + f(b, b') = 1$, $ab' + a'b + h(b, b') = 0$.

The last two say $(a+b)(a'+b') = 1$, and this combines with the first as in the case of type A_1 to show that $A \oplus B$ <u>is a Jordan division algebra</u>.

We now have our conclusions for type G_2:

<u>Theorem III.6.</u>   <u>Let L be a simple Lie algebra over F of relative type G_2. Then there is a field extension A of F and a Jordan division algebra J over A, satisfying a generic cubic polynomial, such that</u>

$$L \cong D + S \otimes A + M' \otimes B,$$

where S and M' and their compositions are as in §1, where $\mathcal{D}$ is the Lie algebra of all A-derivations of J, where B is the subspace of J orthogonal to the unit element with respect to the generic trace form 3f, and where

$$[x \otimes a, d] = 0, \quad [u \otimes b, d] = u \otimes bd, \quad [u \otimes b, x \otimes a] = ux \otimes ab,$$
$$[u \otimes b, v \otimes b'] = (u,v) <b,b'> + D_{u,v} \otimes f(b,b') + [uv] \otimes (bb' - f(b,b')1),$$

where $<b,b'> = -4 D_{b,b'}$, $D_{b,b'}$ being the inner derivation of J determined by b and b'.

§6. Algebras of Type $C_r (r \geq 2)$.

Here we have a symplectic space V over F of dimension $2r$, S is the endomorphisms x of V satisfying $(ux,v) + (u,vx) = 0$ for all $u,v \in V$, and M' is the endomorphisms s of trace zero satisfying $(us,v) = (u,vs)$ for all u and v. As before, we may identify L with

$$\mathcal{D} \oplus (S \otimes A) \oplus (M' \otimes B),$$

where $\mathcal{D}, A, B$ are trivial S-modules, and where $\mathcal{D}$ is a subalgebra having A and B as $\mathcal{D}$-modules, and $[x \otimes a, d] = x \otimes ad$, $[s \otimes b, d] = s \otimes bd$ $(x \in S, s \in M')$ as before. Further, we have $[x \otimes a, x' \otimes a'] = B(x,x') <a,a'>' + [xx'] \otimes f_1(a,a') + (x \circ x') \otimes f_2(a,a')$, where B is the Killing form, $x \circ x'$ is as in §1, and $<a,a'>'$, $f_i(a,a')$ are bilinear from $A \times A$ to $\mathcal{D}, A, B$, respectively, f_1 being symmetric and the others skew. Also,

$$[s \otimes b, x \otimes a] = (s \circ x) \otimes g_1(b,a) + [sx] \otimes g_2(b,a),$$

where g_1, g_2 are F-bilinear from $B \times A$ to A resp. B, and $[s \otimes b, s' \otimes b'] = \text{Tr}(ss') <b,b'> + [ss'] \otimes h_1(b,b') + (s \circ s') \otimes h_2(b,b')$, where $<b,b'>$, h_1, h_2 are bilinear from $B \times B$ to $\mathcal{D}, A, B$, respectively, h_1 being symmetric and the others skew. (if $r = 2$, we have seen in §1 that $s \circ s'$ is zero; therefore there is no h_2 in this case.) In the interest of uniformity, we replace $B(x,x')$ by $\text{Tr}(xx')$, a fixed (and easily determined) F-multiple, and $<a,a'>'$ by $<a,a'>$ such that $B(x,x') <a,a'>' = \text{Tr}(xx') <a,a'>$. Then we write (p,p') for $\text{Tr}(pp')$ in either the case of S or that of M'.

The Jacobi identity involving one $d \in \mathcal{D}$ yields that the action of $\mathcal{D}$ on $\mathcal{D}, A, B$ is one by derivations with respect to all the operations in question; for example,

87

$$[\langle a,a'\rangle ,d] = \langle ad,a'\rangle + \langle a,a'd\rangle \quad \text{and}$$
$$g_1(b,a)d = g_1(bd,a) + g_1(b,ad).$$

For three factors from $S \otimes A$, we have
$$[x \otimes a[y \otimes a', z \otimes a'']] = (x,[yz])\langle a, f_1(a',a'')\rangle$$
$$+ (y,z) \, x \otimes a \langle a',a''\rangle + [x[yz]] \otimes f_1(a, f_1(a',a''))$$
$$- x \circ (y \circ z) \otimes g_1(f_2(a',a''),a) + x \circ [yz] \otimes f_2(a, f_1(a',a''))$$
$$+ [x, y \circ z] \otimes g_2(f_2(a',a''),a),$$

and the cyclic sum of three such quantities must vanish. It will be convenient to fix a symplectic basis $e_1,\ldots,e_{2r}$ for the space V, with $(e_i, e_j) = \mathrm{sgn}(i-j)\,\delta_{i+j, 2r+1}$, and to make choices of elements of S, or of M', in terms of the matrix units E_{ij} relative to this basis. Thus if we take $x = E_{31} - E_{2r, 2r-2}$, $y = E_{12} - E_{2r-1, 2r}$, $z = E_{23} - E_{2r-2, 2r-1} \in S$ in the above, these three are pairwise orthogonal with respect to the trace form (indeed, they are three root-vectors of the type that yielded the associativity of A in several earlier cases). From the universal relation $(x,[yz]) = (y,[zx]) = (z,[xy])$, we find

(36) $\quad \langle a, f_1(a',a'')\rangle + \langle a', f_1(a'',a)\rangle + \langle a'', f_1(a,a')\rangle = 0.$

For our special choices of x, y, z, we have

$$[x[yz]] = E_{33} - E_{11} - E_{2r-2, 2r-2} + E_{2r, 2r} = H_3 - H_1 \quad \text{where}$$
$$H_i = E_{ii} - E_{2r+1-i, 2r+1-i},$$
$$[y[zx]] = H_1 - H_2,$$
$$[z[xy]] = -[x[yz]] - [y[zx]],$$
$$x \circ (y \circ z) = H_1 + H_3,$$
$$y \circ (z \circ x) = H_1 + H_2,$$
$$z \circ (x \circ y) = H_2 + H_3,$$

all of them combinations of the linearly independent elements H_i, $1 \le i \le 3$, of S. Collecting the coefficients in A of H_1 in the terms of $S \otimes A$ from the Jacobi relation gives, for $r \ge 3$,

$$0 = -f_1(a, f_1(a',a'')) + f_1(a', f_1(a'',a))$$
$$- g_1(f_2(a',a''),a) - g_1(f_2(a'',a),a'), \quad \text{or}$$

(37) $\quad f_1(a, f_1(a',a'')) - f_1(a', f_1(a'',a))$
$$= g_1(f_2(a'',a'),a) - g_1(f_2(a'',a),a') \quad (r \ge 3).$$

Setting $U_i = E_{ii} + E_{2r-1-i, 2r+1-i} - \frac{1}{r} I \in M'$, $1 \le i \le r$, we find that the U_i, $1 \le i \le 3$, are linearly independent if $r > 3$, and that $U_1 + U_2 + U_3 = 0$ for $r = 3$ is the only relation of dependence in this case. Now, with x, y, z as above,

$x \circ [yz] = U_1 + U_3$, $y \circ [zx] = U_1 + U_2$, $z \circ [xy] = U_2 + U_3$,

$[x, y \circ z] = U_3 - U_1$, $[y, z \circ x] = U_1 - U_2$, $[z, x \circ y] = U_2 - U_3$.

Collecting the coefficients of U_1 in $\mathcal{B}$ from the terms in $M' \otimes \mathcal{B}$ of our Jacobi relation, we find

(38) $\quad \begin{aligned} & f_2(a, f_1(a', a'')) + f_2(a', f_1(a'', a)) \\ & = g_2(f_2(a', a''), a) - g_2(f_2(a'', a), a') \end{aligned} \quad (r \ge 4)$,

(38') $\quad \begin{aligned} & f_2(a', f_1(a'', a)) - f_2(a'', f_1(a, a')) - 2 g_2(f_2(a', a''), a) \\ & + g_2(f_2(a'', a), a') + g_2(f_2(a, a'), a'') = 0 \end{aligned} \quad (r = 3)$.

For $r = 2$, we take $x = E_{12} - E_{34}$, $y = E_{23}$, $z = E_{31} + E_{42}$, again pairwise orthogonal with respect to the trace form. Here $U_1 + U_2 = 0$, and we have $x \circ [yz] = 0$, $y \circ [xz] = U_2 - U_1 = -2U_1$, $z \circ [xy] = 0$, $[x, y \circ z] = U_1 - U_2 = 2U_1$, $[y, z \circ x] = 0$, $[z, x \circ y] = U_2 - U_1 = -2U_1$. The terms in $M' \otimes \mathcal{B}$ of our Jacobi relation now give

(38'') $\quad f_2(a', f_1(a'', a)) = g_2(f_2(a', a''), a) - g_2(f_2(a, a'), a'') \quad (r = 2)$.

Also, for $r = 2$, x, y, z as above give $[x[yz]] = H_1 - H_2$, $[y[zx]] = 2H_2$, $[z[xy]] = -H_1 - H_2$, $x \circ (y \circ z) = H_1 + H_2$, $y \circ (z \circ x) = 0$, $z \circ (x \circ y) = H_1 - H_2$ and, as with (37) above, for $r = 2$,

(37') $\quad f_1(a, f_1(a', a'')) - f_1(a'', f_1(a, a')) = g_1(f_2(a', a''), a) + g_1(f_2(a, a'), a'')$.

The choice $y = z = H_1$ with $x = H_2$, results in $(y, z)x = 2x$, $y \circ z = U_1$, $x \circ (y \circ z) = -\frac{4}{r} x$, and

(39) $\quad \begin{aligned} & 2a <a', a''> + \frac{4}{r} g_1(f_2(a', a''), a) = 0 \quad \text{or} \\ & a <a', a''> = -\frac{2}{r} g_1(f_2(a', a''), a). \end{aligned}$

With the same choice of $y = z = H_1$, and with $x = E_{2r,1}$, we find

$a <a', a''> = 2 \left\{ \frac{r-1}{r} g_1(f_2(a', a''), a) - f_1(a', f_1(a'', a)) + f_1(a'', f_1(a, a')) \right\}$

which combines with (39) to yield

(39') $\quad g_1(f_2(a',a''),a) = f_1(a',f_1(a'',a)) - f_1(a'',f_1(a,a'))$

and

(39") $\quad a<a',a''> = \frac{2}{r}(f_1(f_1(a,a'),a'') - f_1(a',f_1(a,a'')))$.

Now consider the Jacobi identity for a triple $x \otimes a$, $y \otimes a'$, $v \otimes b$, with $v \in M'$, $b \in B$:

$[x \otimes a, [y \otimes a', v \otimes b]] = [x \otimes a, -y \circ v \otimes g_1(b,a') + [yv] \otimes g_2(b,a')]$
$= (x,y \circ v) < g_1(b,a'), a > - [x, y \circ v] \otimes f_1(a, g_1(b,a'))$
$- x \circ [yv] \otimes g_1(g_2(b,a'),a) - x \circ (y \circ v) \otimes f_2(a, g_1(b,a'))$
$+ [x[yv]] \otimes g_2(g_2(b,a'),a)$, and hence
$[y \otimes a'[v \otimes b, x \otimes a]] = - (y, x \circ v) < g_1(b,a),a' >$
$+ [y, x \circ v] \otimes f_1(a', g_1(b,a)) + y \circ [xv] \otimes g_1(g_2(b,a),a')$
$+ y \circ (x \circ v) \otimes f_2(a', g_1(b,a)) - [y[xv]] \otimes g_2(g_2(b,a),a')$;
$[v \otimes b, [x \otimes a, y \otimes a']] = (v, x \circ y) < b, f_2(a,a') >$
$+ v \circ [xy] \otimes g_1(b, f_1(a,a')) + [v, x \circ y] \otimes h_1(b, f_2(a,a'))$
$+ (x,y) v \otimes b < a, a' > + [v[x \circ y]] \otimes g_2(b, f_1(a,a')) + v \circ (x \circ y') \otimes h_2(b, f_2(a,a'))$,

the term involving h_2 not present if $r = 2$. Since $(x, y \circ v) = (y, x \circ v) = (v, x \circ y)$, we conclude

(40) $\quad < b, f_2(a,a') > = < g_1(b,a), a' > - < g_1(b,a'), a >$.

For $r \geq 3$, we specialize x to $E_{31} - E_{2r,2r-2}$, y to $E_{12} - E_{2r-1,2r}$, v to $E_{23} + E_{2r-2,2r-1}$, finding

$[x, y \circ v] = H_3 - H_1$,
$x \circ [yv] = H_1 + H_3$,
$[y, x \circ v] = H_1 - H_2$,
$z \circ [xv] = - H_1 - H_2$,
$[v, x \circ y] = H_2 - H_3$,
$v \circ [xy] = H_2 + H_3$.

Setting equal to zero the coefficient of $H_1 \in S$ in the Jacobi relation gives:

(41) $\quad f_1(a, g_1(b,a')) + f_1(a', g_1(b,a)) = g_1(g_2(b,a'),a) + g_1(g_2(b,a),a')$.

Likewise, the coefficient of H_2 gives:

(42) $\quad f_1(a, g_1(b,a')) + g_1(g_2(b,a'),a)$
$= g_1(b, f_1(a,a')) - h_1(b, f_2(a,a'))$.

Taking $x = y = H_1$, $v = U_2 - U_3$, the only non-zero terms in $M' \otimes B$ in our relation are those involving $(x,y)v = 2v$ and

$v \circ (x \circ y) = v \circ (2E_{11} + 2E_{2r,2r} - \frac{2}{r} I) = -\frac{4}{r} v.$ Thus, for $r \geq 3$,

(43) $\qquad b <a,a'> = \frac{2}{r} h_2(b, f_2(a,a')).$

Taking $x = E_{2r,1}$, $y = E_{12} - E_{2r-1,2r}$, $v = E_{1,2r-1} - E_{2,2r}$, we have $x \circ (y \circ v) = -2U_1$, $[x[yv]] = 0$, $y \circ (x \circ v) = -U_1 - U_2$, $[y[xv]] = U_1 - U_2$, $(x,y) = 0$, $[v[xy]] = U_1 - U_2$, $v \circ (x \circ y) = -U_1 - U_2$; the coefficients of U_1 and U_2 give, respectively, for $r \geq 3$:

(44) $\qquad 2f_2(a, g_1(b,a')) - f_2(a', g_1(b,a))$
$\qquad\qquad = h_2(b, f_2(a,a')) + g_2(g_2(b,a), a') - g_2(b, f_1(a,a')),$

(45) $\qquad f_2(a', g_1(b,a)) - g_2(g_2(b,a), a') + g_2(b, f_1(a,a'))$
$\qquad\qquad + h_2(b, f_2(a,a')) = 0.$

For $r = 2$, $U_2 = -U_1$ results in:

(44') $\qquad f_2(a, g_1(b,a')) = g_2(g_2(b,a), a') - g_2(b, f_1(a,a')).$

Also for $r = 2$, choice of $x = E_{12} - E_{34}$, $y = E_{23}$, $v = E_{31} - E_{42}$ gives $v \circ [xy] = H_1 - H_2$, $[v, x \circ y] = -H_1 - H_2$, $[x, v \circ y] = H_1 - H_2$, $x \circ [yv] = H_1 + H_2$, $[y, x \circ v] = 2H_2$, $y \circ [xv] = 0$, hence from the component in $S \otimes A$ of our Jacobi relation,

(42') $\qquad g_1(g_2(b,a'), a) + f_1(a, g_1(b,a'))$
$\qquad\qquad = g_1(b, f_1(a,a')) - h_1(b, f_2(a,a')),$

(41') $\qquad 2f_1(a', g_1(b,a)) = -f_1(a, g_1(b,a'))$
$\qquad\qquad + g_1(g_2(b,a'), a) + g_2(b, f_1(a,a')) + h_1(b, f_2(a,a')).$

Addition and subtraction of (41') and (42') yields, for $r = 2$:

(41'') $\qquad g_1(b, f_1(a,a')) = f_1(a, g_1(b,a')) + f_1(a', g_1(b,a)),$

(42'') $\qquad f_1(a', g_1(b,a)) = h_1(b, f_2(a,a')) + g_1(g_2(b,a'), a).$

Finally, taking $x = y = H_1$, $v = 2U_1$, all brackets are zero, $(x,y) = 2$, and $x \circ (y \circ v) = y \circ (x \circ v) = 4U_1 = 2v$, giving, for $r = 2$,

(43') $\qquad b <a,a'> = f_2(a, g_1(b,a')) - f_2(a', g_1(b,a)).$

Now consider the Jacobi identity for a triple $u \otimes b$, $v \otimes b'$, $x \otimes a$, where we have
$[x \otimes a[u \otimes b, v \otimes b']] = (x, [uv]) <a, h_1(b,b')> + (u,v) x \otimes a <b, b'>$
$\qquad + [x[uv]] \otimes f_1(a, h_1(b,b')) - x \circ (u \circ v) \otimes g_1(h_2(b,b'), a)$
$\qquad + x \circ [uv] \otimes f_2(a, h_1(b,b')) + [x, u \circ v] \otimes g_2(h_2(b,b'), a),$
$[u \otimes b[v \otimes b', x \otimes a]] = (u, [vx]) <b, g_2(b',a)>$

$$+ u \circ (v \circ x) \otimes g_1(b,g_1(b',a)) + [u[vx]] \otimes h_1(b,g_2(b',a))$$
$$+ [u, v \circ x] \otimes g_2(b,g_1(b',a)) + u \circ [vx] \otimes h_2(b,g_2(b',a)),$$
$$[v \otimes b'[x \otimes a, u \otimes b]] = -(v,[ux]) < b', g_2(b,a) >$$
$$- v \circ (u \circ x) \otimes g_1(b',g_1(b,a)) - [v[ux]] \otimes h_1(b',g_2(b,a))$$
$$- [v, u \circ x] \otimes g_2(b',g_1(b,a)) - v \circ [ux] \otimes h_2(b',g_2(b,a)),$$

with $h_2 = 0$ if $r = 2$.

From $(x,[uv]) = (u,[vx]) = (v,[xu])$, we have

(46) $\quad <a, h_1(b,b') > + < b, g_2(b',a) > + < b', g_2(b,a) > = 0$.

For $r \geq 3$, we first choose $x = E_{31} - E_{2r, 2r-2}$, $u = E_{12} + E_{2r-1, 2r}$, $v = E_{23} + E_{2r-2, 2r-1}$, and have

$$[x[uv]] = H_3 - H_1, \quad (u,v)x = 0,$$
$$x \circ (u \circ v) = H_1 + H_3, \quad u \circ (v \circ x) = H_1 + H_2.$$
$$[u[vx]] = H_1 - H_2, \quad v \circ (u \circ x) = H_2 + H_3,$$
$$[v[ux]] = H_3 - H_2.$$

From the coefficient of H_1, we have, for $r \geq 3$,

(47) $\quad f_1(a, h_1(b,b')) + g_1(h_2(b,b'),a)$
$$= g_1(b, g_1(b',a)) + h_1(b, g_2(b',a)),$$

and from that of H_2,

(48) $\quad h_1(b, g_2(b',a)) - h_1(b', g_2(b,a))$
$$= g_1(b, g_1(b',a)) - g_1(b', g_1(b,a)).$$

With the above choices, we have $x \circ [uv] = U_1 + U_3$, $[x, u \circ v] = U_3 - U_1$, $[u, v \circ x] = U_1 - U_2$, $u \circ [vx] = U_1 + U_2$, $[v, u \circ x] = U_2 - U_3$, $v \circ [ux] = -U_2 - U_3$. For $r > 3$, the coefficients of U_1 and U_2 yield, respectively,

(49) $\quad g_2(h_2(b,b'),a)$
$$= f_2(a, h_1(b,b')) + g_2(b, g_1(b',a)) + h_2(b, g_2(b',a)),$$

(50) $\quad g_2(b, g_1(b',a)) + g_2(b', g_1(b,a))$
$$= h_2(b, g_2(b',a)) + h_2(b', g_2(b,a)).$$

When $r = 3$, we have $U_1 + U_3 = -U_2$, $U_3 - U_1 = -U_2 - 2U_1$, $U_2 - U_3 = 2U_2 + U_1$, $-U_2 - U_3 = U_1$ above, giving, from the coefficients of U_2, U_1, respectively,

(50') $f_2(a,h_1(b,b')) + g_2(h_2(b,b'),a) + g_2(b,g_1(b',a))$

$\qquad = h_2(b,g_2(b',a)) - 2g_2(b',g_1(b,a))$, and

(49') $2g_2(h_2(b,b'),a) + g_2(b',g_1(b,a)) + h_2(b',g_2(b,a))$

$\qquad = g_2(b,g_1(b',a)) + h_2(b,g_2(b',a))$, <u>for</u> $r = 3$.

For $r = 3$, taking $x = E_{16}$, $u = E_{62} - E_{51}$, $v = E_{21} + E_{65}$, we have $x \circ [uv] = 2U_1$, $[x,u \circ v] = 0$, $[u,v \circ x] = U_1 - U_2$, $u \circ [vx] = U_1 + U_2$, $[v,u \circ x] = U_2 - U_1$, $v \circ [ux] = -U_1 - U_2$.

Collecting coefficients of U_2, we see that

(50) <u>is valid for all</u> $r \geq 3$.

Replacing one term $g_2(h_2(b,b'),a)$ in (49') by its value from (50') gives

$g_2(h_2(b,b'),a) - g_2(b',g_1(b,a)) + h_2(b',g_2(b,a))$

$\qquad = f_2(a,h_1(b,b')) + 2g_2(b,g_1(b',a))$, and

substitution from (50) for $h_2(b',g_2(b,a)) - g_2(b',g_1(b,a))$ yields that

(49) <u>is valid for all</u> $r \geq 3$.

Finally, if we take $u = v = U_1 - U_2$, $x = E_{1,2r}$, we have $(u,v) = 4$, u, v, and x commute, $u \circ v = 2(U_1+U_2)$, $x \circ (u \circ v) = \frac{4(r-2)}{r} E_{1,2r}$, $u \circ (v \circ x) = v \circ (u \circ x) = 4E_{1,2r}$, and

(51) $a < b,b' > = g_1(b',g_1(b,a)) - g_1(b,g_1(b',a))$

$\qquad\qquad + \frac{r-2}{r} g_1(h_2(b,b'),a) \qquad (r \geq 2)$.

For $r = 2$, taking $u = E_{12} + E_{34}$, $v = E_{13} - E_{24}$, $x = E_{41}$, we have $(u,v) x = 0$, $[x[uv]] = 2H_1$, $[u[v,x]] = H_2 - H_1$, $[v[ux]] = H_1 + H_2$, $u \circ (v \circ x) = -H_1 - H_2$, $v \circ (u \circ x) = H_1 - H_2$, from which the coefficient of H_2 gives

(51') $h_1(b,g_2(b',a)) - h_1(b',g_2(b,a)) - g_1(b,g_1(b',a))$

$\qquad + g_1(b',g_1(b,a)) = 0$;

combining with (51) and the fact that $h_2 = 0$, we have

(51") $a < b,b' > = h_1(b',g_2(b,a)) - h_1(b,g_2(b',a)) \qquad (r = 2)$.

The coefficient of H_1 in the above gives

(52)
$$2f_1(a,h_1(b,b')) = h_1(b,g_2(b',a)) + h_1(b',g_2(b,a))$$
$$+ g_1(b,g_1(b',a)) + g_1(b',g_1(b,a)) \quad (r = 2) .$$

Finally, with u, v, x as above, we have $x \circ [uv] = -2U_1$, $[u,v \circ x] = -2U_1$, $[v,u \circ x] = 2U_1$, so that

(53) $\quad f_2(a,h_1(b,b')) + g_2(b,g_1(b',a)) + g_2(b',g_1(b,a)) = 0 \quad (r = 2)$.

Our last set of identities results from the Jacobi identity applied to triples $u \otimes b$, $v \otimes b'$, $w \otimes b''$:

$[w \otimes b''[u \otimes b, v \otimes b']] = (w, u \circ v) < b'', h_2(b,b') >$
$+ w \circ [uv] \otimes g_1(b'', h_1(b,b')) + [w, u \circ v] \otimes h_1(b'', h_2(b,b'))$
$+ (u,v)w \otimes b'' < b, b' > + [w[uv]] \otimes g_2(b'', h_1(b,b')) + w \circ (u \circ v) \otimes$

$h_2(b'', h_2(b,b'))$, and the cyclic sum of three such is zero.

From the fact that $(w, u \circ v) = (u, v \circ w) = (v, w \circ u)$ is not identically zero, we have

(54) $\quad < b, h_2(b',b'') > + < b', h_2(b'',b) > + < b'', h_2(b,b') > + 0$.

When $r \geq 4$, we choose $u = v = U_2 - U_3$, $w = U_1 - U_4$; all three commute, $w \circ u = 0 = v \circ w$, $(w,u) = 0 = (u,w)$, and $(u,v) = 4$, $u \circ v = 2(U_2 + U_3)$, $w \circ (u \circ v) = (E_{11} - E_{44} - E_{2r-3,2r-3} + E_{2r,2r}) \circ$

$(2E_{22} + 2E_{33} + 2E_{2r-2,2r-2} + 2E_{2r-1,2r-1} - \frac{4}{r} I) = -\frac{8}{r}(E_{11} - E_{44}$

$- E_{2r-3,2r-3} + E_{2r,2r}) = -\frac{8}{r} w$. Our Jacobi relation thus yields

(55) $\quad b'' < b,b' > = \frac{2}{r} h_2(b'', h_2(b,b')) \quad (r \geq 4)$.

For $r = 3$, we take $u = v = U_2 - U_3$, $w = U_1 = E_{11} + E_{66} - \frac{1}{3} I$. Here $u \circ v = -\frac{4}{3} U_1 + \frac{2}{3} U_2 + \frac{2}{3} U_3$, $u \circ w = -\frac{2}{3}(U_2 - U_3) = -\frac{2}{3} u$, $v \circ (w \circ u) = u \circ (v \circ w) = \frac{4}{3} U_1$ (since $U_1 + U_2 + U_3 = 0$), and $w \circ (u \circ v) = -2(w \circ w) = -\frac{4}{3} U_1$, yielding

(55') $\quad b'' < b,b' > = \frac{1}{3} (h_2(b'', h_2(b,b')) - h_2(b, h_2(b',b'')) -$

$h_2(b', h_2(b'',b))) \quad (r = 3)$.

For $r \geq 3$, we may take $u = E_{31} + E_{2r,2r-2}$, $v = E_{12} + E_{2r-1,2r}$,

$w = E_{23} + E_{2r-2,2r-1}$, obtaining $w \circ [uv] = E_{22} + E_{33} - E_{2r-2,2r-2} - E_{2r-1,2r-1} = H_2 + H_3$, $[w, u \circ v] = H_2 - H_3$, $u \circ [vw] = H_1 + H_3$, $[u, v \circ w] = H_3 - H_1$, $v \circ [wu] = H_1 + H_2$, $[v, w \circ u] = H_1 - H_2$, yielding from the coefficient in A of H_1,

(56) $\qquad g_1(b, h_1(b', b'')) + g_1(b', h_1(b'', b))$
$$= h_1(b, h_2(b', b'')) - h_1(b', h_2(b'', b)) \qquad (r \geq 3).$$

The same choice of u, v, w results in $[u[vw]] = U_3 - U_1$, $[v[wu]] = U_1 - U_2$, $[w[uv]] = U_2 - U_3$, $u \circ (v \circ w) = U_1 + U_3$, $v \circ (w \circ u) = U_1 + U_2$, $w \circ (u \circ v) + U_2 + U_3$. If $r > 3$, the coefficient in B of U_1 gives (since $(u,v) = 0 = (v,w) = (u,w)$):

(57) $\qquad g_2(b, h_1(b', b'')) - g_2(b', h_1(b'', b)) \qquad (r > 3)$
$$= h_2(b, h_2(b', b'')) + h_2(b', h_2(b'', b)).$$

Where $r = 3$, $U_3 - U_1 = -U_2 - 2U_1$, $U_2 - U_3 = U_1 + 2U_2$, $U_1 + U_3 = -U_2$, $U_2 + U_3 = -U_1$, and the coefficient of U_1 gives:

(57') $\qquad 2g_2(b, h_1(b', b'')) - g_2(b', h_1(b'', b)) - g_2(b'', h_1(b, b'))$
$$= h_2(b', h_2(b'', b)) - h_2(b'', h_2(b, b')) \qquad (r = 3).$$

When $r = 2$, we have $h_2 = 0$; taking $u = U_1 - U_2 = 2U_1$, $v = E_{12} + E_{34}$, $w = E_{21} + E_{43}$ gives $w \circ [uv] = 2(H_1 + H_2) = u \circ [vw] = v \circ [wu]$, hence, from the Jacobi-identity,

(56') $\qquad g_1(b, h_1(b', b'')) + g_1(b', h_1(b'', b)) + g_1(b'', h_1(b, b')) = 0$.

(From (56), we see that this holds for $r \geq 2$.)

Finally, for u, v, w as above, we have $(u,v) = 0 = (w,u)$, $(v,w) = 2$, so $(v,w)u = 4U_1$. Moreover, $[u[vw]] = 0$, while $[v[wu]] = 2[E_{12} + E_{34}, E_{21} - E_{43}] = 2(U_1 - U_2) = 4U_1$ and likewise $[w[uv]] = -4U_1$. The Jacobi identity gives

(55'') $\qquad b < b', b'' > \; = g_2(b'', h_1(b, b')) - g_2(b', h_1(b'', b)) \qquad (r = 2)$.

For $r > 2$, we define a product on $A \oplus B$ by $(a+b)(a'+b') =$
$\{f_1(a, a') + g_1(b, a') - g_1(b', a) + h_1(b, b')\} + \{ f_2(a, a') + g_2(b, a')$
$$+ g_2(b', a) + h_2(b, b')\}.$$

We test this product for associativity; it will be associative if and only if each of the pairs a) - p) of entries in the table below coincide:

	$((a+b)(a'+b'))(a''+b'')$	$(a+b)((a'+b')(a''+b''))$
a)	$f_1(f_1(a,a'),a'') + g_1(f_2(a,a'),a'')$	$f_1(a,f_1(a',a'')) - g_1(f_2(a',a''),a)$
b)	$f_1(g_1(b,a'),a'') + g_1(g_2(b,a'),a'')$	$g_1(b,f_1(a',a'')) + h_1(b,f_2(a',a''))$
c)	$-f_1(g_1(b',a),a'') + g_1(g_2(b',a),a'')$	$f_1(a,g_1(b',a'')) - g_1(g_2(b',a''),a)$
d)	$-g_1(b'',f_1(a,a')) + h_1(f_2(a,a'),b'')$	$-f_1(a,g_1(b'',a')) - g_1(g_2(b'',a'),a)$
e)	$f_1(h_1(b,b'),a'') + g_1(h_2(b,b'),a'')$	$g_1(b,g_1(b',a'')) + h_1(b,g_2(b',a''))$
f)	$-g_1(b'',g_1(b,a')) + h_1(g_2(b,a'),b'')$	$-g_1(b,g_1(b'',a')) + h_1(b,g_2(b'',a'))$
g)	$g_1(b'',g_1(b',a)) + h_1(g_2(b',a),b'')$	$f_1(a,h_1(b',b'')) - g_1(h_2(b',b''),a)$
h)	$-g_1(b'',h_1(b,b')) + h_1(h_2(b,b'),b'')$	$g_1(b,h_1(b',b'')) + h_1(b,h_2(b',b''))$
i)	$f_2(f_1(a,a'),a'') + g_2(f_2(a,a'),a'')$	$f_2(a,f_1(a',a'')) + g_2(f_2(a',a''),a)$
j)	$g_2(b'',f_1(a,a')) + h_2(f_2(a,a'),b'')$	$-f_2(a,g_1(b'',a')) + g_2(g_2(b'',a'),a)$
k)	$-f_2(g_1(b',a),a'') + g_2(g_2(b',a),a'')$	$f_2(a,g_1(b',a'')) + g_2(g_2(b',a''),a)$
l)	$f_2(g_1(b,a'),a'') + g_2(g_2(b,a'),a'')$	$g_2(b,f_1(a',a'')) + h_2(b,f_2(a',a''))$
m)	$f_2(h_1(b,b'),a'') + g_2(h_2(b,b'),a'')$	$g_2(b,g_1(b',a'')) + h_2(b,g_2(b',a''))$
n)	$g_2(b'',g_1(b,a')) + h_2(g_2(b,a'),b'')$	$-g_2(b,g_1(b'',a')) + h_2(b,g_2(b'',a'))$
o)	$-g_2(b'',g_1(b',a)) + h_2(g_2(b',a),b'')$	$f_2(a,h_1(b',b'')) + g_2(h_2(b',b''),a)$
p)	$g_2(b'',h_1(b,b')) + h_2(h_2(b,b'),b'')$	$g_2(b,h_1(b',b'')) + h_2(b,h_2(b',b''))$

Now a) is a consequence of (37), the symmetry of f_1 and the skewness of f_2; c) follows from (41), and b) and d) from (42). From (48), we have f), and (47) yields e) and g). We get h) directly from (56). From (38) we have i) and j), and l) follows from (45). Using (45) to replace one summand $f_2(a,g_1(b,a'))$ in the left-hand side of (44), we find

$$f_2(a,g_1(b,a')) + g_2(g_2(b,a'),a) - g_2(b,f_1(a',a))$$
$$- h_2(b,f_2(a',a)) = f_2(a',g_1(b,a)) + h_2(b,f_2(a,a'))$$
$$+ g_2(g_2(b,a),a') - g_2(b,f_1(a,a')), \text{ or}$$
$$f_2(a,g_1(b,a')) + g_2(g_2(b,a'),a) = f_2(a',g_1(b,a)) + g_2(g_2(b,a),a'),$$

which is k). From (49), we have m) and o), and from (50), we get n). Finally, (57) yields p) if $r > 3$.

We conclude that a) - o) hold for all $r \geq 3$, while p) holds for $r > 3$. Thus $A + B$ is associative for $r > 3$.

With $b = b'$, p) becomes a comparison of $g_2(b'',h_1(b,b))$ and $g_2(b,h_1(b,b'')) + h_2(b,h_2(b,b''))$, and these two are equal for $r = 3$ by setting $b = b'$ in (57'). With $b' = b''$, p) compares

$g_2(b',h_1(b,b')) + h_2(h_2(b,b'),b')$ and $g_2(b,h_1(b',b'))$, and these are again equal for $r = 3$ by (57'). That is, $A + B$ <u>is an alternative algebra for</u> $r \geq 3$, <u>associative for</u> $r > 3$.

Next we note from the definition of our multiplication in $A + B$, from the symmetry of f_1 and h_1, and from the skewness of f_2 and h_2, that the mapping $*$ sending $a + b$ to $a - b$ is an <u>involution</u> in $A + B$, i.e., an anti-automorphism of period two, with A as symmetric elements ($a* = a$) and B as skew elements ($b* = -b$), and that, relative to our product in $A + B$, we have

$$f_1(a,a') = \tfrac{1}{2}(aa' + a'a), \quad f_2(a,a') = \tfrac{1}{2}(aa' - a'a),$$
$$g_1(b,a) = \tfrac{1}{2}(ba=ab), \quad g_2(b,a) = \tfrac{1}{2}(ba+ab),$$
$$h_1(b,b') = \tfrac{1}{2}(bb'+b'b), \quad h_2(b,b') = \tfrac{1}{2}(bb'-b'b),$$

all for $r \geq 3$. Since the action of D on $A + B$ stabilizes A and B, <u>all the derivations of</u> $A + B$ <u>induced from</u> D <u>commute with the involution</u>.

Now let I be an ideal in $A + B$, with $I^* = I$. Then $I = I^+ + I^-$, where $I^+ = I \cap A$, $I^- = I \cap B$. From the fact that $D = <AA> + <B,B>$ as before, and from (39), (43), (51), (55) and (55'), we see that I is D-stable. Now consider

$$K = <I^+, A> + <I^-, B> + S \otimes I^+ + M' \otimes I^-.$$

we claim K is an ideal in L. From the formulas (39) - (55') for $<A,A>$ and $<B,B>$ cited above, K is stable under brackets with D, and the first two summands carry $S \otimes A$, $M' \otimes B$, respectively, by bracketing, into $S \otimes I^+$, $M' \otimes I^-$. The assertion follows easily.

As a simple involutorial algebra in this sense, $A + B$ is either simple or the direct sum of two ideals $R_1 \oplus R_2$, the R_i anti-isomorphic simple algebras, the involution being exchange of these summands. We show that the latter case is impossible by the maximality of T in L. For writing $1 = e_1 + e_2$, where e_i is the unit element of R_i, we have $e = e_1 - e_2 \in B$, $e^2 = 1$, and e commutes and associates in every product from $A + B$ in which it is present. Letting $U_1 \in M'$ be as before, $U_1 \otimes e$ centralizes $T = T \otimes 1$. If $d \in D$, we have $e^2 d = 0 = 2e(ed)$; but if $ed = a_1 + a_2$, $a_i \in R_i$, we find $0 = e(a_1+a_2) = a_1 - a_2$, so that $ed = 0$. Thus $U_1 \otimes e$

centralizes D. For $a \in A$, $x \in S$, $[x \otimes a, U_1 \otimes e] = [x, U_1] \otimes ae$, and for $b \in B$, $v \in M'$, $[v \otimes b, U_1 \otimes e] = [v, U_1] \otimes be$ (since $<b,e> = 0$ by (39) - (55') as above.) Now $B = Ae$, $A = Be$, and $\text{ad}(U_1)$ acts diagonally on $S + M'$, the 2r-rowed square F-matrices of trace zero, with $(\text{ad } U_1)^2$ acting diagonally on both S and M', all characteristic roots lying in F^2. If $[[x\ U_1]U_1] = \lambda^2 x, \lambda \in F$, then $\lambda x \otimes a \pm [x\ U_1] \otimes ae$ belongs to the characteristic root $\pm \lambda$ of $\text{ad}(U_1 \otimes e)$ for all $a \in A$; if $\lambda = 0$ above, then $[x\ U_1] = 0$, and $[x \otimes a, U_1 \otimes e] = 0$. Likewise, if $[[v\ U_1]U_1] = \lambda^2 v$, $v \in M'$, then $\lambda v \otimes b \pm [v\ U_1] \otimes be$ belongs to $\pm \lambda$ for $\text{ad}(U_1 \otimes e)$, with all $h \in B$, and $[[v\ U_1]U_1] = 0$ implies $[v\ U_1] = 0$, $[v \otimes b, U_1 \otimes e] = 0$. This shows that $\text{ad}(U_1 \otimes e)$ diagonalizes in F, contradicting the maximality of T. Therefore $A + B$ is a simple algebra.

If $A + B$ is not associative (which occurs only for $r = 3$), $A + B$ must be a Cayley-Dickson algebra over its center, which is identifiable with the set of elements z satisfying $A(c,z,d) = 0$ for all $c,d \in A + B$, where $A(c,z,d) = (cz)d - c(zd)$ ([SCH], p. 48). From our relations a) - o), we see that A satisfies these conditions, thus lies in the center, which also may be viewed in the usual way as the set of z such that $[z,c] = 0$ for all c. On the other hand, if $z \in B$, $[z,c] = 0$ for all c means that $h_2(b,z) = 0$ for all $b \in B$. Letting Z be the set of such $z \in B$, we have $h_2(b,az) = h_2(b,g_2(z,a)) = g_2(z,g_1(b,a)) + g_2(b,g_1(z,a))$, by n). But since A is central, $g_1 = 0$ here, and $AZ = ZA = Z$. Also, by h), $h_1(h_2(B,B),Z) = 0$, and by (57'),

$$h_1(b,b')z = \frac{1}{2}(h_1(b,z)b' + h_1(b',z)b) \in Z$$

for all $b,b' \in B$, $z \in Z$. Since A is a subalgebra over F of the center, which is a field, A is a field. If $h_1(b,z) \neq 0$ for some $b \in B$, $z \in Z$, we see from the latter that $b' \in Z$ whenever $h_1(b',z) = 0$, thus that Z contains an A-subspace of B of A-codimension at most one. Moreover, B has A-dimension at least seven. With $h_1(b,z) \neq 0$ as above, we can therefore find $0 \neq z' \in Z$ with $h_1(b,z') = 0$. If $h_1(B,z') \neq 0$, then $b \in Z$ as above and $Z = B$, $h_2 = 0$, so that our algebra $A + B$ would be commutative. Since this is not the case, we have elements $0 \neq z' \in Z$ with $h_1(B,z') = 0$. But then $z'B = Bz' = 0$, and these elements form an ideal in $A + B$, a contradiction. Therefore A is the center of $A + B$. As in §5, we see that $<b,b'> = \frac{1}{12}D_{b,b'}$ and that the annihilator of $A + B$ (=annihilator of B) in D is zero, so that D = inner derivations of $A + B$ = $\text{Der}(A + B)$. Now $A + B$ is either a split Cayley-Dickson algebra over A, in which case D is a split G_2 and thus contains nilpotent elements centralizing T, or $A + B$ is an octonion division algebra. By our assumption of maximality for T, the latter must be the case. This fully describes L in terms of A and the octonion division algebra $A + B$ over A in the case $r = 3$, $A + B$ not associative.

Now suppose $A + B$ is associative, in which case $A + B$ identifies with the full algebra $\text{Hom}_E(U,U)$, U a finite-dimensional left vector space over the division algebra E over F, and the involution $*$ in $A + B$ is identified as follows: E has an involution σ over F, and, if σ is not the identity, U carries a non-degenerate σ-hermitian form; the mapping $*$ is the adjoint with respect to this form. If σ is the identity, so that E is commutative, then U carries either a non-degenerate symmetric E-bilinear form, or a non-degenerate alternate E-bilinear form and $*$ is again the adjoint. When $[U:E] = 1$, U identifies with E and $A + B$ with E, the involution being σ. When $[U:E] > 1$, and we are not in the alternate case with σ=identity, U has an E-basis $u_1,\ldots,u_n$ with $(u_i,u_j) = 0$, $i \neq j$. The matrix-unit $E_{11} = e \in A + B$ satisfies $e* = e$, $e^2 = e$ in this case, so is in A, but not in $F \cdot 1 \subseteq A$. With $H_1 = E_{11} - E_{2r,2r} \in S$ as before, $H_1 \otimes e \in T \otimes A$, so is in the centralizer in L of $T = T \otimes 1 \subseteq T \otimes A$, but is not in $T \otimes 1$. We claim $\text{ad}(H_1 \otimes e)$ is diagonalizable in F.

It is a standard fact in Jordan theory (and is easily verified directly) that the map $a \mapsto f_1(a,e)$ of A into A satisfies the polynomial $X(X-1)(X-\frac{1}{2})$, so that A is the direct sum of the three eigenspaces $A = A(0) + A(\frac{1}{2}) + A(1)$. If $a \in A(0)$, we have $0 = ae + ea = eae + ea = ae + eae$; from the latter, $ae = ea$, hence $ae = ea = 0$. Thus $f_2(a,e) = 0 = <a,e>$, and we have $[x \otimes a, H_1 \otimes e] = 0$ for all $x \in S$, all $a \in A(0)$. If $a \in A(1)$, we have $ae + ea = 2a$, $eae + ea = 2ea$, $ae + eae = 2ae$, from which $eae = ae = ea = a$. Again we have $[a,e] = 0 = <a,e>$, and if x runs over a basis for S relative to which $\text{ad } H_1$ acts diagonally with characteristic roots $\lambda \in F$, a over a basis for $A(1)$, $[x \otimes a, H_1 \otimes e] = \lambda(x \otimes a)$ diagonalizes $\text{ad}(H_1 \otimes e)$ on $S \otimes A(1)$. Likewise, the map

$b \mapsto g_2(b,e)$ of B satisfies the same polynomial, so $B = B(0) + B(\frac{1}{2}) + B(1)$ as before; $\text{ad}(H_1 \otimes e)$ annihilates $M' \otimes B(0)$ and acts diagonally on $M' \otimes B(1)$, just as before. Finally, if $\mathcal{D}_0 = \{d \in \mathcal{D} | ed = 0\}$, $\text{ad}(H_1 \otimes e)$ annihilates $\mathcal{D}_0$, and if $d \in \mathcal{D}$ is arbitrary, $e^2 = e$ implies $(e^2d) = ed = e(ed) + (ed)e$, or $ed \in A(\frac{1}{2})$, and $a \in A(\frac{1}{2})$ implies $ae + ae = a$, $eae + ae = ae$, so that $eae = 0$. Thus $e < a, e > = \frac{1}{2r}[e[ae]] = -\frac{1}{2r}(ae+ea) = -\frac{1}{2r}a$. It follows that, to prove our claim, it suffices to show $\text{ad}(H_1 \otimes e)$ is diagonalizable in $< A(\frac{1}{2}), e > + S \otimes A(\frac{1}{2}) + M' \otimes B(\frac{1}{2})$.

Now $(H_1, H_1) = \text{Tr}(H_1^2) = 2$, and $\text{ad } H_1$ stabilizes the subspace S_0 of S orthogonal to H_1 with respect to the trace form, so acts diagonally there. For $x \in S_0$, $a \in A(\frac{1}{2})$, we have $[x \otimes a, H_1 \otimes e] = \frac{1}{2}[x H_1] \otimes a + \frac{1}{2}(x \circ H_1) \otimes [ae]$. Now a basis for S_0 may be taken as follows:

i) combinations of matrix units involving neither of the indices 1 nor $2r$; for these x, $xH_1 = H_1 x = 0$, so $[x \otimes a, H_1 \otimes e] = 0$;

ii) $E_{j1} - E_{2r,2r-j+1}$, $1 < j \leq r$; $E_{2r-j+1,1} + E_{2r,j}$, $1 < j \leq r$; for these x, $[xH_1] = x$, $x \circ H_1 = E_{j1} + E_{2r,2r-j+1}$ resp. $E_{2r-j+1,1} - E_{2r,j}$, $[x \circ H_1, H_1] = x \circ H_1$, $(x \circ H_1) \circ H_1 = x$;

iii) $E_{1j} - E_{2r-j+1,2r}$, $E_{1,2r-j+1} + E_{j,2r}$, $1 < j \leq r$; for these x, $[xH_1] = -x$, $x \circ H_1 = E_{1j} + E_{2r-j+1,2r}$ resp. $E_{1,2r-j+1} - E_{j,2r}$, $[x \circ H_1, H_1] = -x \circ H_1$, $(x \circ H_1) \circ H_1 = x$;

iv) $E_{1,2r}$: $[xH_1] = -2x$, $x \circ H_1 = 0$;

v) $E_{2r,1}$: $[xH_1] = 2x$, $x \circ H_1 = 0$.

In addition to the $E_{j,1} + E_{2r,2r-j+1}$, $E_{2r-j+1,1} - E_{2r,j}$, $E_{1j} + E_{2r-j+1,2r}$, $E_{1,2r-j+1} - E_{j,2r}$ above, we can obtain a basis for M' by taking U_1 and certain combinations u of matrix units involving neither index 1 nor $2r$, hence satisfying $[u \otimes b, H_1 \otimes e] = 0$ for all $b \in B$.

With $x \in S$ as in ii) or iii) above, we have $[xH_1] = \pm x$, $[x \circ H_1, H_1] = \pm x \circ H_1$, $(x \circ H_1) \circ H_1 = x$. Thus since $[[ae]e] = a$ for $a \in A(\frac{1}{2})$ (and likewise for $b \in B(\frac{1}{2})$), we have

$[x \otimes a \mp (x \circ H_1) \otimes [ae], H_1 \otimes e] = 0$, $[x \otimes a \pm (x \circ H_1) \otimes [ae], H_1 \otimes e]$
$= \pm (x \otimes a \pm (x \circ H_1) \otimes [ae])$ for all $a \in A(\frac{1}{2})$, and $[ae]$ runs over
$B(\frac{1}{2})$ as a runs over $A(\frac{1}{2})$. Thus it suffices now to show $\operatorname{ad}(H_1 \otimes e)$
is diagonalizable in

$$< A(\tfrac{1}{2}), e > + H_1 \otimes A(\tfrac{1}{2}) + U_1 \otimes B(\tfrac{1}{2}).$$

But now for fixed $a \in A(\frac{1}{2})$, we have

$$[< a,e >, H_1 \otimes e] = - H_1 \otimes e < a,e > = \frac{1}{2r} H_1 \otimes a,$$

$$[H_1 \otimes a, H_1 \otimes e] = 2 < a,e > + U_1 \otimes [ae],$$

$$[U_1 \otimes [ae], H_1 \otimes e] = U_1 \circ H_1 \otimes \tfrac{1}{2}[[ae]e] = (1 - \tfrac{1}{r}) H_1 \otimes a.$$

With $a \neq 0$, it follows that in the three-dimensional subspace of L
spanned by $< a,e >$, $H_1 \otimes a$, $U_1 \otimes [ae]$, $\operatorname{ad}(H_1 \otimes e)$ has the three eigenvalues $0, \pm 1$, so is diagonalizable. Taking account of the remarks
above, this shows that $\operatorname{ad}(H_1 \otimes e)$ is diagonalizable on L.

There remains the case where $A + B = \operatorname{Hom}_E(V,V)$, E is a field,
V a symplectic space over E, with the resulting involution in
$\operatorname{Hom}_E(V,V)$. Here $[V:E] \geq 2$, and B contains $n \neq 0$ with $n^2 = 0$
(compare $E_{1,2r} \in S$ for a completely analogous situation); in fact,
$n = [n,b]$ for some $b \in B$; then $< n,b > \in D$ has the effect on A or
B of sending an element c into $[c,n]$ (to within scalar multiples), so
acts nilpotently on $A + B$, and is not zero (since $b < n,b > \neq 0$).
In this case $< n,b >$ is a nilpotent element of the centralizer of T in
L, and this is a contradiction. Thus we have: $A + B$ <u>is a division</u>
<u>algebra</u>.

We have now proved the structure theorem for Lie algebras of
relative type $C_r (r \geq 3)$:

<u>Theorem III.7.</u> <u>Let</u> L <u>be a simple Lie algebra over</u> F <u>of relative</u>
<u>type</u> $C_r (r \geq 3)$. <u>If</u> $r > 3$, <u>there is an involutorial associative division</u>
<u>algebra</u> $(E,*)$ <u>over</u> F <u>such that</u>

$$L \cong D \oplus S \otimes A \oplus M' \otimes B,$$

<u>where</u> S <u>and</u> M' <u>and their compositions are as in §1, where</u> A <u>and</u> B
<u>are, respectively, the</u> *-<u>fixed and</u> *- <u>skew elements of</u> E, <u>where</u> D
<u>is the set of derivations of</u> E <u>which are linear combinations of maps</u>
$D_{b,b'}: c \mapsto [c[bb']]$, $D_{a,a'}: c \mapsto [c[aa']]$ $(a,a' \in A ; b,b' \in B)$ <u>and where</u>

101

$$[x \otimes a, d] = x \otimes ad, \quad [u \otimes b, d] = u \otimes bd,$$

$$[x \otimes a, y \otimes a'] = \text{Tr}(xy) <a,a'> + [xy] \otimes \frac{aa'+a'a}{2} + (x \circ y) \otimes \frac{aa'-a'a}{2},$$

$$[u \otimes b, x \otimes a] = u \circ x \otimes \frac{ba-ab}{2} + [ux] \otimes \frac{ba+ab}{2},$$

$$[u \otimes b, v \otimes b'] = \text{Tr}(uv) <b,b'> + [uv] \otimes \frac{bb'+b'b}{2} + (u \circ v) \otimes \frac{bb'-b'b}{2},$$

with $<a,a'> = \frac{1}{2r} D_{a,a'}$, $<b,b'> = \frac{1}{2r} D_{b,b'}$.

If $r = 3$, L <u>is either as given above, or there is an alternative, non-associative, division algebra</u> O <u>over</u> F, <u>with center</u> A <u>and skew elements</u> B <u>relative to the standard involution, such that</u> L <u>has a decomposition as above,</u> D <u>being the set of (inner) derivations spanned by the</u> $D_{b,b'}$: $c \mapsto [c[bb']] - [b[b'c]] - [b'[cb]]$ <u>and with</u>

$$[x \otimes a, d] = 0, \quad [u \otimes b, d] = u \otimes bd,$$

$$[x \otimes a, y \otimes a'] = [xy] \otimes aa', \quad [u \otimes b, x \otimes a] = [ux] \otimes ab,$$

$$[u \otimes b, v \otimes b'] = \text{Tr}(u,v) <b,b'> + [uv] \otimes \frac{bb'+b'b}{2} + (u \circ v) \otimes \frac{bb'-b'b}{2},$$

$$<b,b'> = \frac{1}{12} D_{b,b'}.$$

More traditional realizations will be found in Chapter V.

We now consider the case $r = 2$. Here, as will be seen below, it is natural to distinguish two cases, according as (A,f_1) is or is not associative.

<u>Case</u> a): (A,f_1) <u>is associative</u>.

Here f_1 is a commutative, associative product on A. (37') yields

$$g_1(f_2(a',a''),a) + g_1(f_2(a,a'),a'') = 0,$$

and, from (39),

$$a <a',a''> = g_1(f_2(a',a''),a).$$

Other relevant identities are (38"), (40), (44'), (41"), (42"), (43'), (46), (51'), (52), (53), (56'), (55").

In particular, it will be noted that (41") says that, for each $b \in B$, the map $a \mapsto g_1(b,a)$ is a <u>derivation</u> of (A,f_1). Thus if I is an ideal of (A,f_1), stable under all derivations, we have $ID \subseteq I$, $g_1(B,I) \subseteq I$. Let

$$K = <A,I> + <B, g_2(B,I)> + S \otimes I$$
$$+ M' \otimes (f_2(A,I) + g_2(B,I)).$$

From (37') above, $g_1(f_2(A,I),A) \subseteq g_1(B,I) \subseteq I$, so that (39) as above yields $A < A, I > \subseteq I$. Now $g_1(g_2(B,I),A)$ is seen from (42") to lie in $I + h_1(B, f_2(A,I))$; in fact, we have

$$g_1(g_2(B,I),A) + I = h_1(B,f_2(A,I)) + I .$$

Now (42"), applied for $a \in I$, gives $h_1(B,f_2(I,A)) \subseteq I$, hence $h_1(B,f_2(A,I)) \subseteq I$, $g_1(g_2(B,I),A) \subseteq I$, and

$$A < B, g_2(B,I) > \subseteq h_1(B,g_2(g_2(B,I),A))$$
$$+ h_1(g_2(B,I),B).$$

By (44'), $g_2(g_2(B,I),A) \subseteq f_2(A,I) + g_2(B,I)$, so $A < B, g_2(B,I) > \subseteq h_1(g_2(B,I),B) + h_1(B,f_2(A,I))$. From (38") we see that $g_2(f_2(A,I),A) \subseteq g_2(B,I) + f_2(A,I)$.
By the above, $h_1(B,f_2(A,I)) \subseteq I$. By (51') and (52") we have, for $a \in I$, $b,b' \in B$, both of $h_1(b',g_2(b,a)) \pm h_1(b,g_2(b',a)) \in I$, from which $h_1(g_2(B,I),B) \subseteq I$. It now follows that $[K, S \otimes A] \subseteq K$. Since $ID \subseteq I$, it is clear that $[K,D] \subseteq K$, from the derivation-properties of the action of D. Finally, we show $[K,M' \otimes B] \subseteq K$. Clearly, $[S \otimes I, M' \otimes B] \subseteq S \otimes g_1(B,I) + M' \otimes g_2(B,I) \subseteq K$; now $[M' \otimes g_2(B,I), M' \otimes B] \subseteq <B, g_2(B,I)> + S \otimes h_1(g_2(B,I),B) \subseteq K$, by the above, and $[M' \otimes f_2(A,I), M' \otimes B] \subseteq <B, f_2(A,I)> + S \otimes h_1(f_2(A,I),B)$. By (40), $<B,f_2(A,I)> \subseteq <A,I>$, and $h_1(f_2(A,S),B) \subseteq I$ by the above. Finally, we have $B<A,I> \subseteq f_2(A,I)$ by (43') and $B<B,g_2(B,I)> \subseteq g_2(g_2(B,I),A) + g_2(B,h_1(g_2(B,I),B))$, by (55"). The first of these terms has been seen above to lie in $f_2(A,I) + g_2(B,I)$, and the second is in $g_2(B,I)$ since $h_1(g_2(B,I),B) \subseteq I$. This completes the proof that K is an ideal in L, and thus that $I = 0$ or A. Since the radical of a commutative, associative algebra of characteristic zero is easily seen to be stable under all derivations, we conclude that (A,f_1) is <u>semi-simple</u>, hence a direct sum of fields, and therefore <u>all</u> ideals are stable, since the only derivation is zero. It follows that (A,f_1) <u>is a field</u>, and, by the derivation-property for g_1, that $g_1(B,A) = 0$, as well as $AD = 0$.

Now (43'), together with the fact that the intersection in D of the annihilators of A and B is zero, shows that $<A,A> = 0$. By (44'), g_2 gives B the structure of vector space over (A,f_1). From (51'), (52), $AD = 0$, $g_1(B,A) = 0$, we have

$h_1(b,g_2(b',a)) = h_1(b',g_2(b,a)) = f_1(a,h_1(b,b'))$ for all a,b,b', so that h_1 is a symmetric bilinear form on B (with respect to f_1 and g_2).

Let R be the radical of h_1 in B. Then $g_2(R,A) \subseteq R$. Let $K = <B,R> + M' \otimes R$ which we claim is an ideal in L. We have seen that it is stable under brackets with $S \otimes A$; also $RD \subseteq R$, since $h_1(RD,B) \subseteq h_1(R,B)D + h_1(R,BD) = 0$, and hence $[K,D] \subseteq K$. Clearly $[M' \otimes R, M' \otimes B] \subseteq <R,B> \subseteq K$. Finally, $B<B,R>$ is seen from (55") to lie in $g_2(R,A) \subseteq R$, and K is an ideal unequal to L, hence zero, and h_1 is <u>non-degenerate</u>.

From (53), $f_2(A,h_1(B,B)) = 0$; but if $B \neq 0$, the above shows $h_1(B,B) = A$, so that $f_2 = 0$ in either case. Thus $f_2 = 0$. We now write aa' for the product f_1 on A, ab for $g_2(b,a)$, and h for h_1, and have, from (55"),

$$b <b',b"> = h(b,b')b" - h(b,b")b'.$$

The situation is now exactly analogous to that of $B_r (r \geq 3)$; S acts as skew transformations of the 5-dimensional space M' with respect to the form (u,v), which is symmetric and nondegenerate, of Witt index 2, dual maximal totally isotropic subspaces being $\{E_{12} + E_{34}, E_{13} - E_{24}\}$ and $\{E_{21} + E_{43}, E_{31} - E_{42}\}$, with $(U_1, U_1) = 1$. Thus M' plays the part of our V in those considerations, and A and B are exactly as there. A precise summary of this case will be given in the statement of Theorem 3.8 below.

<u>Case b)</u>: (A,f_1) <u>is not associative</u>.

From the commutativity of (A,f_1), the formula (39") for $<a',a">$, and the fact that $<a,f_1(a,a)> = 0$ by (36), we see that (A,f_1) <u>is a Jordan algebra</u>. As in case a), if I is an ideal stable under all derivations of (A,f_1), we have $g_1(B,I) \subseteq I$, and $K = <A,I> + <B, g_2(B,I)> + S \otimes I + M' \otimes (f_2(A,I) + g_2(B,I))$ is an ideal of L. Hence $I = 0$ or $I = A$, and (A,f_1) <u>has no proper characteristic ideals</u>. Since the radical of a Jordan algebra is a derivation-stable ideal (in characteristic zero - cf. Appendix B) (A,f_1) is semisimple, and hence all derivations are inner (Appendix B). Therefore (A,f_1) <u>is a simple Jordan algebra</u>.

Now let $B_0 = \{b \in B \mid g_1(b,A) = 0\}$, the kernel of the

F-linear mapping $b \to g_1(b, \ )$ of B into the derivations of (A, f_1). Since all derivations are inner, i.e., linear combinations of elements $<a, a'>$, by (39")), we see by (39') that the image of $f_2(A,A)$ under this map is all of Der(A), and that $B = f_2(A,A) + B_0$.

We claim this is a direct sum; to show this, consider the Killing form of L, written (X,Y). Fixing $d \in \mathcal{D}$, $a \in A$, $b \in B$, we have $(d, [xy] \otimes a) = -([dy], x \otimes a) = 0$ for all $x, y \in S$, $(d, [uy] \otimes b) = -([dy], u \otimes b) = 0$ for all $u \in M'$, and from this, $(\mathcal{D}, S \otimes A) = 0 = (\mathcal{D}, M' \otimes B)$. Also, $(x \otimes a, y \otimes b) = g(x,u)$ is an S-invariant bilinear pairing of S and M' to F, so must be zero: $(S \otimes A, M' \otimes B) = 0$. Finally, fixing $b, b' \in B$, $f(u,v) = (u \otimes b, v \otimes b')$ is an S-invariant pairing of M' and M' to F, so has the form $\lambda(b,b')$ Tr(uv), and one easily sees that $\lambda(b,b')$ defines a symmetric bilinear form on B. If $\lambda(b, B) = 0$, then $(M' \otimes b, L) = 0$ by the above, so λ is non-degenerate. Now it suffices to show $\lambda(f_2(A,A), B_0) = 0$. By the same line of reasoning, we have a non-degenerate symmetric bilinear form μ on A such that $(x \otimes a, y \otimes a') =$ Tr$(xy)\mu(a,a')$ for all x, y, a, a'. Now for $u \in M'$, $b \in B$, we have

$$([u \otimes b, x \otimes a], y' \otimes a') = (u \otimes b, [x \otimes a, y' \otimes a']),$$

whence

$$\text{Tr}((u \circ x)y')\mu(g_1(b,a),a') = \text{Tr}(u(x \circ y'))\lambda(b, f_2(a,a'))$$

Since Tr$((u \circ x)y') = $ Tr$(u(x \circ y'))$ is not identically zero, we have $\mu(g_1(b,a),a') = \lambda(b, f_2(a,a'))$, from which $\lambda(f_2(A,A), B_0) = 0$ follows, and

$$B = f_2(A,A) \oplus B_0 .$$

The map $b \to g_1(b,.)$ induces an isomorphism of the subspace $f_2(A,A)$ of B onto the derivations of (A, f_1). We therefore define a map of $B \times B$ onto $f_2(A,A) \subseteq B$, calling it h_2, by: $h_2(b',b'')$ shall be that unique element of $f_2(A,A)$ such that, for all $a \in A$,

(58) $\qquad\qquad a <b'', b'> = g_1(h_2(b',b''), a)$.

Having defined h_2, we define a product in $A \oplus B$ exactly as for $r > 2$ (p.), and proceed to verify a) - p) as in that case. Now a) is (37'), b) and d) are (42'); interchanging a and a' in (42') and adding gives

$$g_1(g_2(b,a),a') + g_1(g_2(b,a'),a) + f_1(a,g_1(b,a'))$$
$$+ f_1(a',g_1(b,a)) = 2g_1(b,f_1(a,a')).$$

By (41''), we have c). The relation f) for $r = 2$ is (51'). Our definition of h_2 gives

$$g_1(h_2(b,b'),a) = a < b'b > = h_1(b,g_2(b',a))$$

$- h_1(b',g_2(b,a))$, by (51''). Verifying e) thus amounts to showing that

$$f_1(h_1(b,b'),a) - h_1(b',g_2(b,a)) = g_1(b,g_1(b',a)),$$

which is clear from (51') and (52). This relation also has g) as a consequence. Now, leaving h) aside for the moment, i) follows from (38''). Interchanging a and a' in (44') and subtracting the two relations gives k). To see j) and l), note that $f_2(g_2(b,a),a') + g_2(g_2(b,a),a') - g_2(b,f_1(a,a')) = f_2(g_1(b,a),a') - f_2(g_1(b,a'),a)$, by (44'), and thus lies in $f_2(A,A)$. For all $a'' \in A$, we have $g_1(f_2(g_1(b,a),a'),a'') - g_1(f_2(g_1(b,a'),a),a'') = a'' < g_1(b,a'),a > - a'' < g_1(b,a),a' >$, by (39). By (40), this is $a'' < b,f_2(a',a) >$, which, by our definition, is $g_1(h_2(f_2(a,a'),b),a'')$. Thus we have

(59) $h_2(f_2(a,a'),b) = f_2(g_1(b,a),a') - f_2(g_1(b,a'),a),$

and this yields j) and l) as indicated above. In view of (53), verification of n) reduces to showing that $h_2(g_2(b,a'),b'') = f_2(a',h_1(b,b'')) + h_2(b,g_2(b'',a'))$, or since all terms are in $f_2(A,A)$, that for all $a \in A$,

$$g_1(h_2(g_2(b,a'),b''),a) + g_1(h_2(g_2(b'',a'),b),a)$$
$$= g_1(f_2(a',h_1(b,b'')),a).$$

By definition, the left-hand side is $a < b'',g_2(b,a') > + a < b,g_2(b'',a') >$, which by (46) is $a < h_1(b,b''),a' >$, and the relation now follows from (39).

For the remaining relations h), m), o), p), we require some preparation, consisting mainly of analysis of $< B,B > = < f_2(A,A) + B_0, f_2(A,A) + B_0 >$. By (40), $< f_2(A,A), B_0 > = 0$, so the above is

$$< f_2(A,A), f_2(A,A) > + < B_0, B_0 > \subseteq < A,A > + < B_0, B_0 >,$$

again by (40). By (51) and (51''), $A < B_0, B_0 > = 0$, so $AD = A < A,A >$ is contained in the subspace of A generated by all

$f_1(f_1(a,a'),a'') - f_1(a,f_1(a',a''))$. It also follows that $f_2(A,A) < B_0,B_0 > = 0$, so that $B < B_0 B_0 > = B_0 < B_0,B_0 > \subseteq B_0$. Since $< B,B_0 > = < B_0,B_0 >$ by the above, we have $B < B,B_0 > = B_0 < B_0,B_0 >$; by (43'), $B_0 < A,A > = 0$, so $B_0 < B,B > = B_0 < B_0,B_0 >$ as well. Now let

$$K = < B,B_0 < B_0,B_0 >> + S \otimes h_1(B,B_0 < B_0,B_0 >)$$
$$+ M' \otimes B_0 < B_0,B_0 > .$$

We claim K is an ideal in L. Since $B_0 \mathcal{D} \subseteq B_0$, $[K\mathcal{D}] \subseteq K$. The construction of K forces $[M' \otimes B_0 < B_0,B_0 >, M' \otimes B] \subseteq K$, and the first component of K lies in $< B_0,B_0 >$ by the above, so that $B < B_0,B_0 > = B_0 < B_0,B_0 >$ assures that $[< B,B_0 < B_0,B_0 >>, M' \otimes B] \subseteq K$. We now consider $[S \otimes h_1(B,B_0 < B_0,B_0 >), M' \otimes B]$; Now $g_1(B,h_1(B,B_0 < B_0,B_0 >)) = g_1(f_2(A,A), h_1(B,B_0 < B_0,B_0 >))$ and

$$h_1(B,B_0 < B_0,B_0 >) \subseteq h_1(B,B_0) < B_0,B_0 >$$
$$+ h_1(B < B_0,B_0 >, B_0) \subseteq A< B_0,B_0 > + h_1(B_0 < B_0,B_0 >, B_0)$$

$\subseteq h_1(B_0 < B_0 B_0 >, B_0)$. Thus we are considering

$$g_1(f_2(A,A), h_1(B_0 < B_0,B_0 >, B_0)); \text{ by } (39),$$

this is contained in

$$h_1(B_0 < B_0,B_0 >, B_0) < A,A > = 0, \text{ since}$$

$B_0 < A,A > = 0$. That is, we have

$$g_1(B,h_1(B,B_0 < B_0,B_0 >)) = 0. \text{ Next,}$$
$$g_2(B,h_1(B,B_0 < B_0,B_0 >))$$
$$= g_2(B,h_1(B_0,B_0 < B_0,B_0 >)), \text{ which by } (55'')$$

is contained in $g_2(B_0,h_1(B,B_0 < B_0,B_0 >))$
$$+ (B_0 < B_0,B_0 >) < B,B_0 > \subseteq g_2(B_0,h_1(B_0,B_0 < B_0 B_0 >))$$
$$+ B_0 < B_0,B_0 >; \text{ another application of } (55'')$$
shows this to lie in $g_2(B_0 < B_0,B_0 >, h_1(B_0,B_0)) + B_0 < B_0,B_0 >$. But $g_2(B_0 < B_0,B_0 >, A) \subseteq g_2(B_0,A) < B_0,B_0 >$
$+ g_2(B_0, A < B_0,B_0 >) = g_2(B_0,A) < B_0,B_0 >$
$\subseteq B < B_0,B_0 > = B_0 < B_0,B_0 >$, so the above shows
$$g_2(B,h_1(B,B_0 < B_0,B_0 >)) \subseteq B_0 < B_0,B_0 >, \text{ and we conclude that}$$

$[K, M' \otimes B] \subseteq K$. Finally, we consider $[K, S \otimes A]$. Since $g_2(B_0 <B_0, B_0>, A) \subseteq B_0 <B_0, B_0>$ by the above; $g_1(B_0 <B_0, B_0>, A) = 0$; $A <B, B_0 <B_0, B_0 >> = A <B_0, B_0> = 0$, it suffices to show $f_1(A, h_1(B, B_0 <B_0, B_0>)) \subseteq h_1(B, B_0 <B_0, B_0>)$ and $f_2(A, h_1(B, B_0 <B_0, B_0>)) \subseteq B_0 <B_0, B_0>$. We use $h_1(B, B_0 <B_0, B_0>) = h_1(B_0, B_0 <B_0, B_0>)$; then the former inclusion follows from (52) and the observation above that $g_2(B_0 <B_0, B_0>, A) \subseteq B_0 <B_0, B_0>$; the latter inclusion follows since the left-hand side is zero by (53). Thus K is an ideal in L. If $K = L$, then $g_1 = 0$ and (A, f_1) is associative, by (39'); therefore $K = 0$, and we conclude:

(60) $\quad B_0 <B_0, B_0> = 0 = B <B_0, B_0>$, **and therefore**

$<B_0, B_0> = 0 = <B, B_0>$, $B_0 <B, B> = 0$,

and $\quad <B, B> = <f_2(A, A), f_2(A, A)> \subseteq <A, A>$.

Thus $<A, A> = D$, and $BD \subseteq B <A, A> \subseteq f_2(A, A) <A, A> \subseteq f_2(A, A)$.

The property p) now follows at once; for $g_2(b'', h_1(b, b')) - g_2(b, h_1(b', b'')) = b' <b, b''>$ by (55"); by (60), this is in $f_2(A, A)$. Thus to show p), it suffices to show that, for all $a \in A$,

$$g_1(b' <b, b''>, a) + g_1(h_2(h_2(b, b'), b''), a)$$
$$= g_1(h_2(b, h_2(b', b'')), a), \text{ or that}$$
$$g_1(b', a) <b, b''> - g_1(b', a <b, b''>)$$
$$+ a <b'', h_2(b, b')> = a <h_2(b', b''), b>,$$

which amounts to showing

$$g_1(b', a) <b, b''> - g_1(b', a <b, b''>) + g_1(h_2(b, b'), g_1(b'', a))$$
$$- g_1(b'', g_1(h_2(b, b'), a)) = g_1(b, g_1(h_2(b', b''), a))$$
$$- g_1(h_2(b', b''), g_1(b, a)), \text{ by (51') and (51''), or}$$

that
$$g_1(b', a) <b, b''> - g_1(b', a <b, b''>) + g_1(b'', a) <b'b>$$
$$- g_1(b'', a <b', b>) = g_1(b, a <b'', b'>) - g_1(b, a) <b'', b'>,$$

or that

$$g_1(b' <b, b''> + b'' <b', b> + b <b'', b'>, a) = 0. \text{ But}$$
we see from (55") that $b' <b, b''> + b'' <b', b> + b <b'', b'> = 0$,

and p) is established.

To establish m) (and o), by symmetry under our involution $a + b \mapsto a - b$), it will suffice by our decomposition of B to treat the two cases: i) $b' = b_0 \in B_0$; ii) $b' = f_2(a,a'), a,a' \in A$. When $b' = b_0$, we have $h_2(b,b') = 0$ (since $<B,B_0> = 0$), and our relation reduces to showing

$$f_2(h_1(b,b_0),a'') = h_2(b,g_2(b_0,a'')).$$

Since both members are in $f_2(A,A)$ and $f_2(A,A) \cap B_0 = 0$, it suffices to show that, for $a \in A$,

$$g_1(f_2(h_1(b,b_0),a''),a) = g_1(h_2(b,g_2(b_0,a'')),a).$$

By (39), the left-hand side is $a < a'', h_1(b,b_0) >$, and the right-hand side is, by definition of h_2, $a < g_2(b_0,a''),b >$. Now (46) shows that the difference of the two is $a < b_0, g_2(b,a'') > = 0$ since $< B_0, B > = 0$. Thus m) holds if $b' \in B_0$.

ii) Now suppose $b' = f_2(a,a')$, where $a,a' \in A$. Here m) reads:

(61)
$$f_2(h_1(b,f_2(a,a')),a'') + g_2(h_2(b,f_2(a,a')),a'')$$
$$= g_2(b,g_1(f_2(a,a'),a'')) + h_2(b,g_2(f_2(a,a'),a'')).$$

By j), $h_2(b,f_2(a,a')) = g_2(b,f_1(a,a')) + f_2(a,g_1(b,a'))$
$- g_2(g_2(b,a'),a)$. Substitution shows the second term on the left-hand side of (61) to be:

$$g_2(g_2(b,f_1(a,a')),a'') + g_2(f_2(a,g_1(b,a')),a'') - g_2(g_2(g_2(b,a'),a)a'').$$

Since $g_1(f_2(a,a'),a'') = a'' < a',a >$ by (39), the first term on the right in (61) is

$$g_2(b,a'' < a',a >) = g_2(b,a'') < a',a > + g_2(b < a,a' >, a'')$$

by the derivation-property. This is equal to

$$f_2(a',g_1(g_2(b,a''),a)) - f_2(a,g_1(g_2(b,a''),a'))$$
$$+ g_2(f_2(a,g_1(b,a')),a'') - g_2(f_2(a',g_1(b,a)),a''),$$

by (43'). Subtracting the first term on the right from the second term on the left in (61) gives

$$f_2(a,g_1(g_2(b,a''),a')) - f_2(a',g_1(g_2(b,a''),a))$$
$$+ g_2([g_2(b,f_1(a,a')) - g_2(g_2(b,a'),a) + f_2(a',g_1(b,a))],a''),$$

and the member in square brackets is zero by (44'). Thus m) reduces to showing

(62)
$$f_2(h_1(b,f_2(a,a')),a'') + f_2(a',g_1(g_2(b,a''),a)) \\ = f_2(a,g_1(g_2(b,a''),a')) + h_2(b,g_2(f_2(a,a'),a'')).$$

Now all terms lie in $f_2(A,A)$, so it suffices to show $g_1(b_1, \cdot) = g_1(b_2, \cdot)$, where these are the derivations of (A,f_1) given by the two sides, b_1 and b_2, of (62). By (39) and the definition of h_2, $g_1(b_1, \cdot)$ and $g_1(b_2, \cdot)$ agree, respectively, with

$$< a'',h_1(b,f_2(a,a')) > + < g_1(g_2(b,a''),a),a' > , \text{ and}$$
$$< g_1(g_2(b,a''),a'),a > - < b,g_2(f_2(a,a'),a'' > .$$

By (46), the former is

$$< g_2(f_2(a,a'),a''),b > + < g_2(b,a''),f_2(a,a') > \\ + < g_1(g_2(b,a''),a'),a' > \quad \text{leaving to be shown:}$$

$$< g_2(b,a''),f_2(a,a') > + < g_1(g_2(b,a''),a),a' > \\ = < g_1(g_2(b,a''),a'),a > ,$$

and this is an immediate consequence of (40).

Thus only (h) remains of our list a) - p). In our case i) as above, where $b' \in B_0$, we must show

$$g_1(b,h_1(b',b'')) + g_1(b'',h_1(b,b')) = 0 \text{ for all } b,b'' \in B, b' \in B_0.$$

However, this is immediate from (56'). Finally, with $b' = f_2(a,a')$ (the case ii) above), we use j) to substitute for $h_2(b,f_2(a,a'))$ and $h_2(f_2(a,a'),b'')$, then apply (44') to obtain

$$h_2(b,b') = f_2(a,g_1(b,a')) - f_2(a',g_2(b,a)),$$
$$h_2(b',b'') = f_2(a',g_1(b'',a)) - f_2(a,g_1(b'',a')).$$

Now showing (h) reduces, by (56'), to showing

$$g_1(b',h_1(b'',b)) + h_1(h_2((b,b'),b'') + h_1(h_2(b',b''),b) = 0$$

for $b' = f_2(a,a')$, or that

(63)
$$g_1(f_2(a,a'), h_1(b'',b)) + h_1(f_2(a,g_1(b,a')),b'') \\ - h_1(f_2(a',g_1(b,a)),b'') - h_1(f_2(a',g_1(b'',a)),b) \\ + h_1(f_2(a,g_1(b'',a')),b) = 0.$$

Now $g_1(f_2(a,a'), h_1(b'',b)) = h_1(b'',b) <a',a>$ by (39), and this is $h_1(b'' <a',a>, b) + h_1(b'', b <a',a>)$

$= h_1(f_2(a', g_1(b'',a)), b) - h_1(f_2(a, g_1(b'',a')), b)$
$+ h_1(f_2(a', g_1(b,a)), b'') - h_1(f_2(a, g_1(b,a')), b'')$

by (43'). The relation (63) follows, and h) is proved. We thus have the associativity of our product on $A \oplus B$.

The associative algebra $A \oplus B$ is involutorial as before, and, as before, is a simple involutorial algebra. If $A + B$ is not a simple algebra, we have $1 = e_1 + e_2$ as before, $e = e_1 - e_2 \in B$, $e^2 = 1$, $ed = 0$ for all derivations d, and $U_1 \otimes e$ centralizes $T(= T \otimes 1)$ and has adjoint action diagonalizable in F, just as before. Thus the associative algebra $A + B$ is simple. To see that $A + B$ is a division algebra, we follow exactly the same reasoning as in the cases $r \geq 3$, and now have our structure theorem:

Theorem III.8. <u>Let</u> L <u>be a simple Lie algebra over</u> F <u>of relative type</u> C_2. <u>Then, either</u>: i) <u>there is a field extension</u> A <u>of</u> F <u>and a vector space</u> B <u>over</u> A <u>with a non-degenerate symmetric bilinear form</u> h, <u>not representing</u> 1, <u>such that</u> $L \cong D \oplus S \otimes A \oplus M' \otimes B$, <u>where</u> D <u>is the Lie algebra of</u> h-<u>skew</u> A-<u>linear transformations of</u> B, S <u>and</u> M' <u>are as in</u> §1, <u>and</u>

$[x \otimes a, d] = 0$, $[u \otimes b, d] = u \otimes bd$, $[u \otimes b, x \otimes a] = [ux] \otimes ab$,
$[x \otimes a, y \otimes a'] = [xy] \otimes aa'$, $[u \otimes b, v \otimes b'] = (u,v) <b,b'>$
$+ [uv] \otimes h(b,b')$, <u>where</u> $(u,v) = Tr(uv)$ <u>and where</u> $<b,b'>$ <u>is given by</u> $b'' <b,b'> = h(b'',b)b' - h(b'',b')b$; <u>or</u>:
ii) <u>there is an involutorial associative division algebra</u> $E = A \oplus B$ <u>over</u> F, <u>with symmetric elements</u> A <u>and skew elements</u> B, <u>such that</u>

$$L \cong D \oplus S \otimes A \oplus M' \otimes B .$$

<u>where</u> D <u>is the Lie algebra of all derivations of</u> E <u>commuting with the involution, and where</u>

$[x \otimes a, d] = x \otimes ad$, $[u \otimes b, d] = u \otimes bd$,
$[u \otimes b, x \otimes a] = (u \circ x) \otimes \frac{ba-ab}{2} + [ux] \otimes \frac{ba+ab}{2}$,
$[x \otimes a, y \otimes a'] = Tr(xy) <a,a'> + [xy] \otimes \frac{aa'+a'a}{2} + (x \circ y) \otimes \frac{aa'-a'a}{2}$.
$[u \otimes b, v \otimes b'] = Tr(uv) <b,b'> + [uv] \otimes \frac{bb'+b'b}{2}$.

<u>Here</u> $< c,c' >$ (c <u>and</u> c' <u>both in</u> A <u>or both in</u> B) <u>sends</u> $c'' \in E$ <u>to</u>

$$c'' < c,c' > = \frac{1}{4} [c''[cc']].$$

(That $\mathcal{D} = < B,B > + < A,A >$ is the Lie algebra of derivations claimed in ii) is clear if E is commutative, since then E is a field and $\mathcal{D} = 0$; otherwise, from the fact that all F-derivations of E are inner it follows that those of the form $c \mapsto [cf]$, $f + f* = \lambda$ central in E, are the derivations centralizing the involution *. Replacing f by $f - \frac{1}{2}\lambda$, we may assume $f \in B$. Now $B = [BB]$ unless either the involution is of second kind, or E is a quaternion algebra and B is one-dimensional over the center of E ([J], Chapter X; [H]). In the former of these cases, $B = [BB] \oplus (B \cap Z)$, Z the center of E, and in the latter, $B = [AA]$. (Both cases i) and ii) are "classical", as we shall see in Chapter V.)

With theorems III.1 - III.8, we have thus given constructions for all simple Lie algebras of positive relative rank, and for which the relative root-system is <u>reduced</u> (i.e., only $\pm\alpha$ are roots among the multiples of the root α). We next consider the non-reduced case, where only the type $BC_r (r \geq 1)$ is encountered.

§7.   **Algebras of type** $BC_r (r \geq 3)$.

Here we have
$$L \cong \mathcal{D} \oplus S \otimes A \oplus M' \otimes B \oplus M'' \otimes C \ ,$$
where S is as in the last section, as is M', and where M'' is the 2r-dimensional symplectic space defining S. As we have seen in §1, there are no S-invariant compositions $S \times S \to M''$, $S \times M' \to M''$, $M' \times M' \to M''$, so $\mathcal{D} + S \otimes A + M'' \otimes B$ is a subalgebra, containing, as before, the ideal $R = <A,A> + <B,B> + S \otimes A + M' \otimes B$. Restricting our attention to this subalgebra, we have established all the identities (36) - (57') in §6, and those calculations apply here. This is the case for all $r \geq 2$; for $r = 1$, M' is a trivial module, so $M' \otimes B$ is absorbed in $\mathcal{D}, S$ is of type A_1, and the considerations of §2 apply to the ideal $R = <A,A> + S \otimes A$ in $\mathcal{D} + S \otimes A$, a subalgebra, at least insofar as the derivation of identities (viz. (1), (2), (3), (5)) is concerned.

Thus we may conclude that, for $r > 3$, $R = A \oplus B$ is an involutorial associative algebra with A as symmetric elements and B as skew elements, and the compositions in R are as in Theorem III.7; for $r = 3$, one needs only substitute "alternative" for "associative" in the last statement; for $r = 2$, we know that (A, f_1) is a Jordan algebra (and somewhat more, to be discussed later); for $r = 1$, we know that A is a Jordan algebra, and that R is generated from A as in §2. We have **not** yet shown that R is as in the conclusions of those theorems, since we have not even shown the simplicity of the algebras (associative, alternative, or Jordan) involved. In particular, our information in case $r = 2$ is well short of Theorem III.8. However, all that will be shown to follow.

Using the remarks of §1, it follows as in previous cases that
$$[u \otimes c, d] = u \otimes cd,$$
$$[u \otimes c, x \otimes a] = ux \otimes p_1(c,a),$$
$$[u \otimes c, s \otimes b] = us \otimes p_2(c,b),$$
$$[u \otimes c, v \otimes c'] = (u,v) <c,c'> + (u \circ v) \otimes q_1(c,c') + [uv] \otimes q_2(c,c'),$$
where $u, v \in M''$; $x \in S$; $s \in M'$; $d \in \mathcal{D}$; $c, c' \in C$; $b \in B$; $a \in A$; cd, $p_1(c,a)$, $p_2(c,b) \in C$; $<c,c'> \in \mathcal{D}$; $q_1(c,c') \in A$; $q_2(c,c') \in B$; and where ux, us, (u,v), $u \circ v$, $[uv]$ are as in §1.

113

The compositions $<c,c'>$ and $q_2(c,c')$ are symmetric, while q_1 is skew. As before, $\mathcal{D}$ is a subalgebra and the action of $\mathcal{D}$ on each of $\mathcal{A}$, $\mathcal{B}$, $\mathcal{C}$ is a $\mathcal{D}$-module action by derivations with respect to all compositions.

We restrict attention to the case $r \geq 3$, where we have the alternative product $(a+b)(a'+b')$ in $\mathcal{A} \oplus \mathcal{B}$ as in §6, in terms of which the composition in $\mathcal{R}$ is defined. First consider the Jacobi identity for triples $u \otimes c$, $x \otimes a$, $y \otimes a'$, where we have

$[x \otimes a, [y \otimes a', u \otimes c]] = [[u \otimes c, y \otimes a'] x \otimes a] = (uy)x \otimes p_1(p_1(c,a'),a)$,

$[y \otimes a'[u \otimes c, x \otimes a]] = -[[u \otimes c, x \otimes a], y \otimes a'] = -(ux)y \otimes p_1(p_1(c,a),a')$,

$[u \otimes c, [x \otimes a, y \otimes a']] = (x,y) u \otimes c <a,a'>$

$+ u[xy] \otimes p_1(c, \frac{aa' + a'a}{2}) + u(xy+yx - \frac{1}{r}(x,y)1) \otimes p_2(c, \frac{aa-a'a}{2})$.

If we specialize u to our first basis vector e_1, x to $E_{12} - E_{2r-1,2r}$, y to $E_{23} - E_{2r-2,2r-1}$, we have $(x,y) = 0$, $uyx = 0$, $uxy = e_3$, and the Jacobi identity yields

(64) $\quad p_1(p_1(c,a),a') = p_1(c, \frac{aa' + a'a}{2}) + p_2(c, \frac{aa'-a'a}{2})$.

Specializing x and y to $E_{11} - E_{2r,2r}$, u to e_2, we have $(x,y) = 2$, $ux = uy = 0$ and find from the Jacobi identity that

(65) $\quad c <a,a'> = \frac{1}{r} p_2(c, \frac{aa' - a'a}{2})$.

For triples $u \otimes c$, $x \otimes a$, $s \otimes b$, we have

$[x \otimes a, [u \otimes c, s \otimes b]] = - usx \otimes p_1(p_2(c,b),a)$,

$[s \otimes b, [x \otimes a, u \otimes c]] = uxs \otimes p_2(p_1(c,a),b)$,

$[u \otimes c[s \otimes b, x \otimes a]] = u(sx + xs) \otimes p_1(c, \frac{ba - ab}{2})$

$+ u(sx-xs) \otimes p_2(c, \frac{ba + ab}{2})$.

Specializing u to e_1, x to $E_{12} - E_{2r-1,2r}$, s to $E_{23} + E_{2r-2,2r-1}$ we find as before

(66) $\quad p_2(p_1(c,a),b) = p_1(c, \frac{ab - ba}{2}) + p_2(c, \frac{ab + ba}{2})$.

With $u = e_1$, $x = E_{23} - E_{2r-2,2r-1}$, $s = E_{12} + E_{2r-1,2r}$, we have

(67) $\quad p_1(p_2(c,b),a) = p_1(c, \frac{ba - ab}{2}) + p_2(c, \frac{ba + ab}{2})$.

For $s,s' \in \mathcal{M}'$, $b,b' \in \mathcal{B}$, we have

$$[s \otimes b[s' \otimes b', u \otimes c]] = us's \otimes p_2(p_2(c,b'),b),$$
$$[s' \otimes b'[u \otimes c, s \otimes b]] = - uss' \otimes p_2(p_2(c,b),b'),$$
$$[u \otimes c[s \otimes b, s' \otimes b']] = (s,s') u \otimes c <b,b'>$$
$$+ u[ss'] \otimes p_1(c, \frac{bb'+b'b}{2}) + u(ss'+s's - \frac{1}{r}(s,s')I) \otimes p_2(c, \frac{bb'-b'b}{2}).$$

The specializations $u = e_1$, $s = E_{12} + E_{2r-1,2r}$, $s' = E_{23} + E_{2r-2,2r-1}$ and $u = e_3$, $s = E_{11} - E_{22} - E_{2r-1,2r-1} + E_{2r,2r} = s'$, and the Jacobi identity then yield the relations:

(68) $$p_2(p_2(c,b),b') = p_1(c, \frac{bb'+b'b}{2}) + p_2(c, \frac{bb'-b'b}{2}),$$

(69) $$c<b,b'> = \frac{1}{r} p_2(c, \frac{bb'-b'b}{2}).$$

It follows from (64), (67), (68), that, if we define
$$c(a+b) = p_1(c,a) + p_2(c,b), \text{ then}$$
$(c(a+b))(a'+b') = c((a+b)(a'+b'))$; in other words, C is an associative right $A + B$-module; that $1 \in A$ acts as the identity on C follows by our remarks at the end of §4. Accordingly, we have
$$ca = p_1(c,a), \quad cb = p_2(c,b), \quad \text{and (65) and (69) become}$$

(70) $$c<a,a'> = \frac{1}{2r} c[a,a'], \quad c<b,b'> = \frac{1}{2r} c[b,b'].$$

We turn to the Jacobi identity for $u \otimes c$, $v \otimes c'$, $x \otimes a$;
$$[x \otimes a[u \otimes c, v \otimes c']] = (u,v) x \otimes a <c,c'> + [x, u \circ v] \otimes \frac{aq_1(c,c') + q_1(c,c')a}{2}$$
$$+ x \circ [uv] \otimes \frac{aq_2(c,c') - q_2(c,c')a}{2} + (x, u \circ v) <a, q_1(c,c')>$$
$$+ x \circ (u \circ v) \otimes \frac{aq_1(c,c') - q_1(c,c')a}{2} + [x[uv]] \otimes \frac{aq_2(c,c') + q_2(c,c')a}{2},$$
$$[u \otimes c[v \otimes c', x \otimes a]] = (u, vx) <c, c'a> + (u \circ vx) \otimes q_1(c, c'a)$$
$$+ [u, vx] \otimes q_2(c, c'a),$$
$$[v \otimes c'[x \otimes a, u \otimes c]] = - (v, ux) <c', ca> - (v \circ ux) \otimes q_1(c', ca)$$
$$- [v, ux] \otimes q_2(c', ca).$$

Now $(x, u \circ v) = \text{Tr}(x(u \circ v)) = 2(ux, v) = - 2(v, ux) = - 2(u, vx)$, and this is not identically zero. The component in $\mathcal{D}$ of our Jacobi relation thus gives
$$- 2 <a, q_1(c,c')> - <c', ca> + <c, c'a> = 0 \text{ or}$$

(71) $$2 <q_1(c,c'), a> = <ca, c'> - <c, c'a>.$$

115

Specializing u and v to e_1 in the above, we have $(u,v) = 0$, $[uv] = 0$, and $u \circ v = -2E_{2r,1} \in S$. Specializing x to $E_{11} - E_{2r,2r}$, we have $ux = vx = e_1$, $x \circ (u \circ v) = 0$, $[x, u \circ v] = 4E_{2r,1}$, $u \circ vx = v \circ ux = -2E_{2r,1}$, and from the Jacobi identity,

(72) $\qquad aq_1(c,c') + q_1(c,c')a = q_1(c,c'a) - q_1(c',ca).$

Next take $u = e_1$, $v = e_2$, so $u \circ v = -E_{2r,2} - E_{2r-1,1}$, $(u,v) = 0$, $[uv] = E_{2r-1,1} - E_{2r,2}$. With $x = E_{1,2} - E_{2r-1,2r}$, we have $vx = 0$, $ux = e_2$,

$[x, u \circ v] = 2E_{2r-1,2} = x \circ [uv]$, $v \circ (ux) = -2E_{2r-1,2}$,

yielding, from the component in $S \otimes A$ in the Jacobi identity,

$$aq_1(c,c') + q_1(c,c')a + aq_2(c,c') - q_2(c,c')a$$
$$= -2q_1(c',ca), \text{ or, upon substitution from (72)},$$

(73) $\qquad aq_2(c,c') - q_2(c,c')a = -q_1(c,c'a) - q_1(c',ca).$

With $u = e_1 = v$, $x = E_{1,2r}$, our component of the Jacobi relation in $M' \otimes B$ gives

(74) $\qquad aq_1(c,c') - q_1(c,c')a = q_2(c',ca) - q_2(c,c'a)$

With $u = e_1$, $v = e_2$, $x = E_{1,2r}$, we have $vx = 0$, $x \circ (u \circ v) = E_{1,2r} \circ (-E_{2r,2} - E_{2r-1,1}) = -E_{1,2} - E_{2r-1,2r}$ $= [x[uv]] = [v,ux]$, and the component in $M' \otimes B$ of the Jacobi relation gives

$2q_2(c',ca) = aq_1(c,c') - q_1(c,c')a + aq_2(c,c') + q_2(c,c')a$,
and, upon substitution from (74),

(75) $\qquad q_2(c',ca) + q_2(c,c'a) = aq_2(c,c') + q_2(c,c')a.$

We now define $q(c,c') = q_1(c,c') + q_2(c,c') \in A \oplus B$. Then $q(c',c) = -q_1(c,c') + q_2(c,c') = -q(c,c')*$, where $*$ is the involution in $A + B$. Subtraction of (73) from (72) gives

$$2q_1(c,c'a) = a(q_1(c,c') - q_2(c,c')) + q(c,c')a,$$

and subtraction of (74) from (75) gives

$$2q_2(c,c'a) = a(q_2(c,a') - q_1(c,c')) + q(c,c')a.$$

Addition of these relations yields

(76) $\qquad q(c,c'a) = q(c,c')a \qquad (c,c' \in C, a \in A).$

It will soon be shown that (76) holds with a replaced by an arbitrary $b \in B$, from which we shall know that q is an antihermitian form (with respect to $*$) on the right $A + B$-module C.

Finally, let $u = e_1$, $v = e_{2r}$, $x = E_{22} - E_{2r-1,2r-1}$. Then $(u,v) = 1$, $u \circ v = E_{11} - E_{2r,2r} = H_1$, $[uv] = -U_1$, and the Jacobi identity gives

(77) $$a<c,c'> = \frac{1}{r}(q_2(c,c')a - aq_2(c,c')).$$

The Jacobi identity for $u \otimes c$, $v \otimes c'$, $s \otimes b$ yields for $u = v = e_1$, $s = U_1 - U_2$,

$0 = [s \otimes b, [u \otimes c, v \otimes c']] + [u \otimes c [v \otimes c', s \otimes b]] + [v \otimes c'[s \otimes b, u \otimes c]]$

$= [s \otimes b, -2E_{2r,1} \otimes q_1(c,c')] - 2E_{2r,1} \otimes q_1(c,c'b)$

$+ 2E_{2r,1} \otimes q_1(c',cb)$, or

(78) $$bq_1(c,c') - q_1(c,c')b = q_1(c',cb) - q_1(c,c'b).$$

Generally, we have $(s,[uv]) = 2(vs,u) = 2(v,us)$, yielding

(79) $$2<b,q_2(c,c')> = <c,c'b> + <c',cb>.$$

With $u = e_1$, $v = e_2$, $s = U_1 - U_3$, we have

$0 = [s \otimes b, [e_1 \otimes c, e_2 \otimes c']] - [e_2 \otimes c', e_1 \otimes cb]$

$= [(U_1 - U_3) \otimes b, (-E_{2r,2} - E_{2r-1,1}) \otimes q_-(c,c') + (E_{2r-1,1} - E_{2r,2}) \otimes q_2(c,c')]$

$+ (E_{2r,2} + E_{2r-1,1}) \otimes q_1(c',cb) + (E_{2r,2} - E_{2r-1,1}) \otimes q_2(c',cb)$

$= -(E_{2r-1,1} + E_{2r,2}) \otimes \left\{ \frac{bq_1(c,c') - q_1(c,c')b}{2} + \frac{bq_2(c,c') + q_2(c,c')b}{2} \right\}$

$+ (E_{2r-1,1} - E_{2r,2}) \otimes \left\{ \frac{bq_1(c,c') + q_1(c,c')b}{2} + \frac{bq_2(c,c') - q_2(c,c')b}{2} \right\}$

$+ (E_{2r-1,1} + E_{2r,2}) \otimes q_1(c',cb) + (E_{2r-1,1} - E_{2r,2}) \otimes q_2(c',cb)$,

yielding

$bq_1(c,c') - q_1(c,c')b + bq_2(c,c') + q_2(c,c')b = 2q_1(c',cb)$,

and, with (78),

(80) $$bq_2(c,c') + q_2(c,c')b = q_1(c',cb) + q_1(c,c'b),$$

as well as

(81) $bq_1(c,c') + q_1(c,c')b + bq_2(c,c') - q_2(c,c')b = -2q_2(c',cb)$.

The same choice of u and v, with $s = U_1 - U_2$ gives, as component in $M' \otimes B$ of the Jacobi relation,

$$-(E_{2r-1,1} - E_{2r,2}) \otimes bq_1(c,c') + q_1(c,c')b - q_2(c,c'b) + q_2(c',cb)$$

or

(82) $\quad bq_1(c,c') + q_1(c,c')b = q_2(c,c'b) - q_2(c',cb),$

which combines with (81) to give

(83) $\quad -bq_2(c,c') + q_2(c,c')b = q_2(c',cb) + q_2(c,c'b).$

As before, adding (82) and (83) gives

$$2q_2(c,c'b) = b(q_1(c,c') - q_2(c,c')) + (q_1(c,c') + q_2(c,c'))b,$$

and subtracting (78) from (80) gives

$$2q_1(c,c'b) = b(q_2(c,c') - q_1(c,c')) + (q_1(c,c') + q_2(c,c'))b,$$

from which $q(c,c'b) = q(c,c')b$ follows.

Thus q <u>is an anti-hermitian sesquilinear form on the right</u> $A \oplus B$ <u>-module</u> C <u>with respect to the involution</u> $*$, <u>and</u> $q_1 = \frac{q+q^*}{2}$, $q_2 = \frac{q-q^*}{2}$.

Finally, taking $u = e_1$, $v = e_{2r}$, $s = U_2 - U_3$, we have $[v \otimes c', s \otimes b] = 0 = [u \otimes c, s \otimes b]$, $u \circ v = H_1$, so $[u \circ v, s] = 0 = (u \circ v) \circ s$, $[u,v] = -U_1$, so $[s[u,v]] = 0$, $s \circ [uv] = \frac{2}{r} s$, and $(u,v) = 1$. The component in $M' \otimes B$ of our Jacobi identity gives

(84) $\quad b < c,c' > = -\frac{1}{r} [b, q_2(c,c')]$

Thus, combining (84) with (77), we see that the action of $< c,c' >$ on $A \oplus B$ is by the derivation $e \mapsto -\frac{1}{r}[e, q_2(c,c')]$, which evidently commutes with the involution.

As a final relation, we consider the Jacobi identity for $u \otimes c, v \otimes c', w \otimes c''$, where we have

$$[u \otimes c[v \otimes c', w \otimes c'']] = (v,w) u \otimes c < c',c'' >$$
$$+ u(v \circ w) \otimes cq_1(c',c'') + u[vw] \otimes cq_2(c',c''),$$

and the cyclic sum of three such terms is zero.

With $u = e_1$, $v = e_2$, $w = e_{2r}$, we have $v \circ w = E_{1,2} - E_{2r-1,2r}$, $[vw] = -(E_{12} + E_{2r-1,2r})$, so $u(v \circ w) = e_2$, $u[vw] = -e_2$; likewise $[uv] = E_{2r-1,1} - E_{2r,2}$, $u \circ v = -(E_{2r-1,1} + E_{2r,2})$, $w[uv] = -e_2 = w(u \circ v)$; $w \circ u = H_1$, $[wu] = U_1$, and $v(w \circ u) = 0$, $v[wu] = -\frac{1}{r} e_2$. Also $(u,v) = 0 = (v,w)$, $(w,u) = -1$, and the Jacobi identity gives

118

(85)
$$c' <c'',c> = cq_1(c',c'') - cq_2(c',c'') - c''q_1(c,c')$$
$$- c''q_2(c,c') - \frac{1}{r} c'q_2(c'',c)$$
$$= cq(c',c'')* - c''q(c,c') - \frac{1}{2r} c'(q(c'',c) - q(c'',c)*)$$

It will be noted that the map of C into C sending c' to $cq(c',c'')* - c''q(c,c')$ is an $A \oplus B$-module endomorphism ϕ satisfying $q(c'\phi,c''') + q(c',c'''\phi) = 0$ for all $c',c''' \in C$, and from this that $<c'',c>$ satisfies the derivation condition (using (77) and (84)),

(86) $q(c' <c'',c>,c''') + q(c',c''' <c'',c>) = q(c',c''') <c'',c>$.

Now we show $A \oplus B$ is an associative division algebra. Since it is easily seen that

$<A,A> + <B,B> + <C,C> + S \otimes A + M' \otimes B + M'' \otimes C$

is an ideal in L, D is the sum of the first three terms, and the formulas giving their action on $A \oplus B$ show that every two-sided ideal of $A \oplus B$ is stable under the action of D. Now one sees as before that if I is a *-stable two sided ideal in $A \oplus B$, then

$K = <A,I^+> + <B,I^-> + <C,CI> + S \otimes I^+ + M' \otimes I^- + M'' \otimes CI$

is an ideal in L, and it follows that $A \oplus B$ is *-simple. If $A + B$ is not simple, we have $e = e_1 - e_2 \in B$, commuting and associating with all elements of $A \oplus B$ as in §6, and with $e^2 = 1$, $ed = 0$ for all $d \in D$. Our maximal split torus T of L identifies with $T \otimes 1$, and centralizes $U_1 \otimes e$, with $ad(U_1 \otimes e)$ acting semi-simply on $D + S \otimes A + M' \otimes B$, with characteristic roots in F (as in §6.) For $c \in C$, we have $(ce)e = ce^2 = c$, so the action of e on C is semisimple with eigenvalues ± 1, while that of U_1 on M'' is evidently semisimple with rational eigenvalues. Thus the adjoint action of $U_1 \otimes e$ on $M'' \otimes C$ is semisimple with rational eigenvalues, contradicting the maximality of T, and we have proved that $A \oplus B$ is a simple algebra. Now the annihilator of C in $A \oplus B$ is an ideal of $A \oplus B$, not containing 1, so must be zero; since C is an associative $A \oplus B$-module, $A \oplus B$ is a simple associative algebra.

If $A \oplus B$ is not a division algebra, we can produce, as in §6, either a matrix unit $E_{11} \neq 1 \in A$, or an $n \neq 0$ in B, $n^2 = 0$. Review of that argument shows that it suffices to note that $H_1 \otimes E_{11}$

resp. $U_1 \otimes n$ acts F-diagonally resp. nilpotently in $M'' \otimes C$ in order to obtain a contradiction. These are obvious, so we conclude that $A \oplus B$ <u>is an associative division algebra</u> E.

Thus C is a right vector space over E, and q an antihermitian form with respect to our involution in E. If N is the radical of q on C, N is an E-subspace, and

$$K = <C,N> + M'' \otimes N$$

is an ideal in L; in view of (77) and (84), it suffices to note that $C<C,N> \subseteq N$, and this follows by (85) and the fact that N is an E-subspace of C. Thus $N = 0$, and q <u>is nondegenerate</u>. If $c \neq 0$, $q(c,c) = 0$, we have $c' \in C$, $q(c,c') = 1$, and may replace c' by $c' + c\lambda$ for suitable $\lambda \in E$ to assume $q(c',c') = 0$. Then $cE + c'E$ is a 2-dimensional nondegenerate E-subspace of C. We consider $<c,c> \in \mathcal{D}$. From (77) and (84), $<c,c>$ annihilates $S \otimes A + M' \otimes B$. From (85) and the remarks following that relation, $<c,c>$ acts on C as a q-skew transformation (E-linear), annihilating c and the subspace q-orthogonal to $cE + c'E$, and mapping c' to $cq(c',c)* - cq(c,c') = -2c$, thus having square zero in C. Now the representation of $\mathcal{D}$ in $A \oplus B \oplus C$ is faithful, since the kernel would be an ideal in L, and $<c,c> \neq 0$ is represented by a map of square zero; it follows that $(\text{ad} <c,c>)^3 = 0$ in L, and thus that the centralizer of T contains non-zero nilpotents. Thus q <u>is anisotropic</u> on C.

Our conclusions for type $BC_r (r \geq 3)$ now may be formulated as follows:

<u>Theorem III.9.</u> <u>Let</u> L <u>be a simple Lie algebra of relative type</u> $BC_r (r \geq 3)$. <u>Then there is an associative involutorial division algebra</u> $E = A \oplus B$ <u>over</u> F, <u>with symmetric elements</u> A <u>and skew elements</u> B, <u>and a right</u> E-<u>vector space</u> C <u>bearing an anisotropic skew-hermitian form</u> q, <u>such that</u>

$$L \cong \mathcal{D} \oplus (S \otimes A) \oplus (M' \otimes B) \oplus (M'' \otimes C),$$

<u>where, if the involution in</u> E <u>is of second kind,</u> $\mathcal{D} = [BB] \oplus (Z(E) \cap B) \oplus K'(C,q)$ <u>may be identified with the F-linear mappings</u> d <u>of</u> C <u>of the form</u> $c \to cd = cb + cT$, $b \in B$, $T \in K'(C,q)$, <u>the derived algebra of the q-skew F-linear mappings</u> $K(C,q)$ <u>of</u> C. <u>If the involution in</u> E <u>is of first kind,</u> $\mathcal{D} = B \oplus K(C,q)$, <u>similarly identified. With this identification,</u> $\mathcal{D}$ <u>is a subalgebra, its summands</u>

are ideals in $\mathcal{D}$, and

$[u \otimes c, d] = u \otimes cd$, $[x \otimes a, d] = x \otimes [a,b]$.

$[s \otimes b', d] = s \otimes [b',b]$, $[x \otimes a, s \otimes b'] = (x \circ s) \otimes \frac{[ab']}{2} + [xs] \otimes \frac{ab'+b'a}{2}$,

$[u \otimes c, x \otimes a] = ux \otimes ca$, $[u \otimes c, s \otimes b'] = us \otimes cb'$,

$[x \otimes a, y \otimes a'] = \mathrm{Tr}(xy) <a,a'> + [xy] \otimes \frac{aa'+a'a}{2} + (x \circ y) \otimes \frac{aa'-a'a}{2}$,

$[s \otimes b', t \otimes b''] = \mathrm{Tr}(st) <b',b''> + [st] \otimes \frac{b'b''+b''b'}{2} + (s \circ t) \otimes \frac{b'b''-b''b'}{2}$,

$[u \otimes c, v \otimes c'] = (u,v) <c,c'> + (u \circ v) \otimes \frac{q(c,c')-q(c',c)}{2} + [uv] \otimes \frac{q(c,c')+q(c',c)}{2}$,

with $<a,a'> \in [\mathcal{B}\mathcal{B}]$, $<b',b''> \in [\mathcal{B},\mathcal{B}]$ given by

$$c<a,a'> = \frac{1}{2r} c[a,a'], \quad c<b',b''> = \frac{1}{2r} c[b',b''],$$

and $<c,c'> \in \mathcal{D}$ given by

$c''<c,c'> = -c'q(c,c'') - cq(c',c'') - \frac{1}{2r} c''(q(c,c') + q(c',c))$.

Everything in the theorem has been established, with the exception of our assertions about $\mathcal{D}$. Now since
(87) $q(c,c'')d = q(cd,c') + q(c,c'd)$ for all $c, c' \in \mathcal{C}$, and since $q(c,c')$ can be made to take on all values in E, it follows that the annihilator of $\mathcal{C}$ in $\mathcal{D}$ annihilates $\mathcal{A}$ and $\mathcal{B}$ as well, so must be zero. Thus it is legitimate to regard $\mathcal{D}$ as contained in $\mathrm{Hom}_F(\mathcal{C},\mathcal{C})$. Since $\mathcal{D} = <\mathcal{A},\mathcal{A}> + <\mathcal{B},\mathcal{B}> + <\mathcal{C},\mathcal{C}>$, our observations above show that $\mathcal{D}$ is contained in the set of all mappings $c \to cb + cT$, $b \in \mathcal{B}$, $T \in K(\mathcal{C},q)$, the full Lie algebra of q-skew E-endomorphisms of $\mathcal{C}$. But, in this sense, $\mathcal{B} \cap K(\mathcal{C},q) = Z(\mathcal{B}) = Z(E) \cap \mathcal{B}$, and $K(\mathcal{C},q) = K'(\mathcal{C},q) \oplus (Z(E) \cap \mathcal{B})$ whenever the involution is of second kind (and even for first kind except when $[\mathcal{C}:E] = 1$ and E is a quaternion algebra over its center $Z(E)$ - cf. [H]; [J], Chapter X). Likewise, $\mathcal{B} = [\mathcal{B}\mathcal{B}] \oplus (Z(E) \cap \mathcal{B})$ under the same conditions. Thus $\mathcal{D}$ is contained in the indicated algebra, and one sees from (77), (84), and Theorem III.7, the consistency of our action of this algebra on $\mathcal{A}$ and $\mathcal{B}$. It remains only to show that $\mathcal{D}$ contains the indicated algebra. Now $[\mathcal{B}\mathcal{B}] \subseteq \mathcal{D}$ by (69), and $K'(\mathcal{C},q)$ is contained in every noncentral ideal of $K(\mathcal{C},q)$ ([H]; [J], Chapter X). If $0 \neq e \in \mathcal{C}$, we choose $0 \neq \lambda \in E$ such that $q(e,e\lambda) = q(e,e)\lambda \in \mathcal{A}$. Then $<e,e\lambda> \in K(\mathcal{C},q)$ (since $q(e,e\lambda) - q(e,e\lambda)^* = 0$), and $e<e,e\lambda> = -e\lambda q(e,e) - eq(e\lambda,e) = -e(\lambda q(e,e) + \lambda^* q(e,e))$, while $f<e,e\lambda> = 0$ if $q(e,f) = 0$.

Now $q(e,e)\lambda \in A$ implies $q(e,e)\lambda = \lambda^* q(e,e)^* = -\lambda^* q(e,e)$. Thus $<e,e\lambda> = 0$ for all λ implies that λ centralizes $q(e,e)$, hence that $q(e,e)$ centralizes A. Thus we have displayed a non-central element $<c,c'> \in K(C,q)$, $q(c,c') \in A$, provided the dimension $[C:E] > 1$ and there is an element $e \in C$ such that $q(e,e)$ does not centralize A. In the former case, we may also let $q(e_1,e_2) = 0$, $0 \neq e_i \in C$, and form $<e_1,e_2> \in K(C,q)$, sending e_1 to $-e_2 q(e_1,e_1)$ and e_2 to $-e_1 q(e_2,e_2)$, so certainly not central.

If we let $<C,C>_0 \subseteq \mathcal{D}$ be the set of elements representable as $\Sigma <c_i,c_i'>$, where $\Sigma q(c_i,c_i') \in A$, these are all in $K(C,q)$, and not all central in $K(C,q)$ if $[C:E] > 1$. If $T \in K(C,q)$, we have $[<c_i,c_i'>,T] = <c_i T,c_i'> + <c_i,c_i'T>$, and $q(c_i T,c_i') + q(c_i,c_i'T) = 0$, so that $<C,C>_0$ is an ideal in $K(C,q)$, hence contains $K'(C,q)$ if $[C:E] > 1$. If $[C:E] = 1$, $0 \neq e \in C$, $q(e,e) = \beta \in B$, $T \in K(C,q)$ sends $e\lambda$ to $e\gamma\lambda$, where $0 = q(e\gamma,e) + q(e,e\gamma) = \gamma^*\beta + \beta\gamma$, or $\gamma^* = -\beta\gamma\beta^{-1}$, or $(\beta\gamma)^* = \beta\gamma$. Thus $K(C,q)$ is isomorphic as F-module to the set of elements of E skew with respect to the involution $\lambda \to \beta^{-1}\lambda^*\beta$, and is anti-isomorphic to the latter as Lie algebra. Here $<C,C>_0$ contains all $<e,e\mu>$ with $\mu^*\beta = -\beta\mu$, and this sends e to $e(\beta\mu - \mu\beta)$. If all such are central in $K(C,q)$, we have $[\beta\mu - \mu\beta,\beta] = 0$ whenever $\mu^*\beta = -\beta\mu$, or whenever $\beta\mu \in A$. But then $[[\beta\mu,\beta]\beta] = 0$, or $[[A\beta]\beta] = 0$. It follows ([J], Lemma II.4) that β centralizes A, or else $K'(C,q) \subseteq <C,C>_0$, as before.

If β centralizes A, then A is (one-dimensional over) the center of E, over which E is a quaternion algebra, or β is central in E [H]. In the former case, $K(C,q)$ is one-dimensional over the center of E, and $0 = K'(C,q) \subseteq <C,C>_0$. In the latter case $K(C,q)$ is anti-isomorphic as Lie algebra to the skew elements B of E, and $K'(C,q)$ to $[B,B]$. The typical element of $<C,C>$ here has the form $\Sigma <e\gamma_i, e\delta_i>$, and we have
$(e\lambda) <e\gamma, e\delta> = -e\beta\gamma\delta^*\lambda - e\beta\delta\gamma^*\lambda$
$- \frac{1}{2r} e(\beta\lambda\gamma^*\delta + \beta\lambda\delta^*\gamma)$, from which $<e\gamma, e\delta> \in K(C,q)$ if and only if $\gamma^*\delta + \delta^*\gamma$ is central in E. Taking $\gamma = 1, \delta \in Z(E)$, we have $<e,e\delta> \in K(C,q)$, $\delta + \delta^*$ runs over $Z(E) \cap A$, and $(\delta + \delta^*)\beta$ runs over $Z(E) \cap B$, $(e\lambda) <e,e\delta> = -(1 + \frac{1}{2r})e(\delta + \delta^*)\beta\lambda$, and $<e,e\delta>$ runs over the center of $K(C,q)$, which thus is

122

contained in $<C,C>$.

Letting $\rho \in Z(E) \cap A$, $\mu \in B$, $\delta \neq 0$ in E, we let $\gamma^*\delta = \rho + \mu$ (thereby defining γ^*), and have $\gamma^*\delta + \delta^*\gamma \in Z(E)$, hence $<e\gamma, e\delta> \in K(C,q)$, sending e to $-e\beta(\gamma\delta^* + \delta\gamma^*) - \frac{1}{2r}e\beta(\gamma^*\delta + \delta^*\gamma)$. Now $<C,C> \cap K(C,q)$ is seen, as was $<C,C>_0$, to be an ideal in $K(C,q)$, and we know it contains the center. Thus it will contain K' if for some such choice of ρ, μ, δ, γ, $\gamma\delta^* + \delta\gamma^*$ is not central in E; and then $K(C,q) \subseteq <C,C>$ follows by remarks above. Now $\gamma\delta^* + \delta\gamma^* = \delta^{*-1}\rho\delta^* - \delta^{*-1}\mu\delta^* + \delta\rho\delta^{-1} + \delta\mu\delta^{-1} = 2\rho + \delta\mu\delta^{-1} - \delta^{*-1}\mu\delta^*$. To say this is central for all $\mu \in B$, $\delta \in E$ is to say $\delta\mu\delta^{-1} - \delta^{*-1}\mu\delta^*$ is, or that $(\delta^*\delta)\mu - \mu\delta^*\delta = \zeta\delta^*\delta$ for all μ, δ, where $\zeta = \zeta(\mu,\delta)$ is central. But then $[[\mu,\delta^*\delta]\delta^*\delta] = 0$, from which $[\mu,\delta^*\delta] = 0$, by a remark used previously. It follows that B centralizes all elements $\delta^*\delta$, $\delta \neq 0$ in E, hence that either B is central in E, and $K'(C,q) = 0$, or E is a quaternion algebra over its center, the fixed elements [H], in which case $K'(C,q) = 0$ as well. Thus we have shown $K'(C,q) \subseteq <C,C>$ in all cases, and that $K(C,q) \subseteq <C,C>$ whenever $[C:E] = 1$ and a basis e for C over C has $q(e,e)$ central in E. We now show $K(C,q) \subseteq <C,C>$ in the remaining cases as well, for which it suffices in the case $Z(E) \cap B \neq 0$ to show that all maps $c \to c\beta$, $0 \neq \beta \subset Z(E) \cap B$, are in $<C,C>$. But if $e_1,\ldots,e_m$ is a basis for C over E with $q(e_i,e_j) = 0$ if $i \neq j$, and $q(e_i,e_i) = \beta_i \neq 0$ in B, we choose $\gamma_i \in E$ such that
$-(1 + \frac{m}{2r})(\gamma_i + \gamma_i^*)\beta_i = \beta$, where $\beta \in Z(E) \cap B$ is fixed.
(This is possible, since $\beta\beta_i^{-1} \in A$.) Then $\Sigma <e_i, e_i\gamma_i>$ sends each $e_i\lambda_i$ to $e_i\lambda_i\beta$, so is our map $c \mapsto c\beta$ as desired, and $K(C,q) \subseteq <C,C>$, so that $\mathcal{D}$ is as claimed. (Indeed, $\mathcal{D} = <C,C>$, since it follows from (79) that $<B,B> \subseteq <C,C>$, which means $[BB] \subseteq \mathcal{D}$ in our interpretation.)

Finally, we consider the cases with involution of first kind where it is possible that $[BB] \neq B$ or that $K'(C,q) \neq K(C,q)$. The former occurs only if E is a quaternion algebra over $Z = Z(E)$, with $[B:Z] = 1$. In this case, $B = [AA] \subseteq \mathcal{D}$, so that $B \subseteq \mathcal{D}$ in all cases where the involution is of first kind. To have $K'(C,q) \neq K(C,q)$ when the involution is of first kind, we must have $[C:E] = 1$, E a quaternion algebra over Z. If e is a basis for C over E, $q(e,e) = \beta \in B$, we have two cases: i) $B = Z\beta$; then β has (generic) trace zero, hence square in Z and E has Z-basis 1, β, κ, λ, where $\kappa, \lambda \in \beta^{-1}A$ have trace zero. An element of $K(C,q)$ is given by $e\nu \mapsto e\tilde{s}\nu$, where $\beta\tilde{s} \in A$. Thus all of the values $\tilde{s} = \beta, \kappa, \lambda$ are admissible, $K(C,q)$ is anti-isomorphic to E', and is equal to its derived algebra. The case ii) is that where $A = Z$, in which $K(C,q)$ is abelian, with Z-basis $e\nu \mapsto e\beta\nu$ and where $e\nu <e,e> = -2e\beta\nu - \frac{1}{r}e\nu\beta$. We have seen above that the map $e\nu \mapsto e\nu\beta$ is in $\mathcal{D}$, from which $K(C,q) \in \mathcal{D}$ in this case as well.

§8. Algebras of type BC_2.

Here the situation is as at the beginning of §7, with $r = 2$. As at that stage, we have for the subalgebra $<A,A> + <B,B> + S \otimes A + M' \otimes B$ all the identities of §6 for the case of type C_2. For easy reference, these are:

(36) $\qquad <a, f_1(a',a'')> + <a', f_1(a'',a)> + <a'', f_1(a,a')> = 0$;

(38") $\qquad f_2(a', f_1(a'',a)) = g_2(f_2(a',a''),a) - g_2(f_2(a,a'),a'')$;

(39') $\qquad g_1(f_2(a',a''),a) = f_1(f_1(a'',a),a') - f_1(f_1(a,a'),a'')$;

(39") $\qquad a <a',a''> = g_1(f_2(a'',a'),a)$;

(40) $\qquad <b, f_2(a,a')> = <g_1(b,a),a'> - <g_1(b,a'),a>$;

(41') $\qquad g_1(b, f_1(a,a')) = f_1(a, g_1(b,a')) + f_1(a', g_1(b,a))$;

(42") $\qquad f_1(b', g_1(b,a)) = h_1(b, f_2(a,a')) + g_1(g_2(b,a'),a)$;

(43') $\qquad b <a,a'> = f_2(a, g_1(b,a')) - f_2(a', g_1(b,a))$;

(44') $\qquad f_2(a, g_1(b,a')) = g_2(g_2(b,a),a') - g_2(b, f_1(a,a'))$;

(46) $\qquad <a, h_1(b,b')> + <b, g_2(b',a)> + <b', g_2(b,a)> = 0$;

(51") $\qquad a <b,b'> = h_1(b', g_2(b,a)) - h_1(b, g_2(b',a))$;

simple algebraic combinations of (51'), (51"), (52) show that these are equivalent to (51"), combined with the next two identities:

(51''') $\qquad a <b,b'> = g_1(b', g_1(b,a)) - g_1(b, g_1(b',a))$;

(52') $\qquad a <b,b'> + f_1(a, h_1(b,b')) = h_1(b', g_2(b,a)) + g_1(b', g_1(b,a))$;

finally, we have

(53) $\qquad f_2(a, h_1(b,b')) + g_2(b, g_1(b',a)) + g_2(b', g_1(b,a)) = 0$;

(55") $\quad b <b',b"> \;=\; g_2(b",h_1(b,b')) - g_2(b',h_1(b",b))\,;$

(56') $\quad g_1(b,h_1(b',b")) + g_1(b',h_1(b",b)) + g_1(b",h_1(b,b')) = 0.$

To obtain analogues to the identities (64) - (86) of §7, involving elements of C, we must make somewhat more careful choices for our specializations. For an analogue to (64), we specialize u to e_1, x to $E_{12} - E_{34}$, y to E_{23}, obtaining

(64') $\quad p_1(p_1(c,a),a') = p_1(c,f_1(a,a')) + p_2(c,f_2(a,a'))\,.$

The same specialization as before gives

(65') $\quad 2c <a,a'> \;=\; p_2(c,f_2(a,a'))$

For an analogue of (66), use $u = e_1$, $x = E_{12} - E_{34}$, $s = E_{21} + E_{43}$, to get

(66') $\quad p_2(p_1(c,a),b) = -p_1(c,g_1(b,a)) + p_2(c,g_2(b,a));$

for (67), $u = e_1$, $x = E_{23}$, $s = E_{12} + E_{34}$ gives

(67') $\quad p_1(p_2(c,b),a) = p_1(c,g_1(b,a)) + p_2(c,g_2(b,a))\,.$

For (68), let $u = e_1$, $s = E_{13} - E_{24}$, $s' = E_{12} + E_{34}$, to get

(68') $\quad 2p_1(c,h_1(b,b')) = p_2(p_2(c,b),b') + p_2(p_2(c,b'),b)\,.$

For (69), we have, specializing as before,

(69') $\quad 4c <b,b'> \;=\; p_2(p_2(c,b),b') - p_2(p_2(c,b'),b)\,.$

From (71) - (77), all specializations made in §7 are acceptable when $r = 2$. In our notation, the results are:

(71') $\quad 2 <q_1(c,c'),a> \;=\; <p_1(c,a),c'> - <c,p_1(c',a)>\,;$

(72') $\quad 2f_1(a,q_1(c,c')) = q_1(c,p_1(c',a)) - q_1(c',p_1(c,a));$

(73') $\quad 2g_1(q_2(c,c'),a) = q_1(c',p_1(c,a)) + q_1(c,p_1(c',a));$

(74') $\quad 2f_2(a,q_1(c,c')) = q_2(c',p_1(c,a)) - q_2(c,p_1(c',a));$

(75') $\quad 2g_2(q_2(c,c'),a) = q_2(c',p_1(c,a)) + q_2(c,p_1(c',a));$

(77') $\quad a <c,c'> \;=\; g_2(q_2(c,c'),a)\,.$

Using $s = U_1$ instead of $s = U_1 - U_2$ in the derivation of (78) gives

(78') $\quad 2g_1(b,q_1(c,c')) = q_1(c',p_2(c,b)) - q_1(c,p_2(c',b));$

as before, we have

(79') $\quad 2 <b,q_2(c,c')> = <c,p_2(c',b)> + <c',p_2(c,b)>$.

Changing the specialization in (80) to $s = U_1$ yields, as before,

(80') $\quad 2h_1(b,q_2(c,c')) = q_1(c',p_2(c,b)) + q_1(c,p_2(c',b))$.

This choice, and the choice $u = e_1$, $v = e_4$, $s = U_1$, give

(82') $\quad 2g_2(b,q_1(c,c')) = q_2(c,p_2(c',b)) - q_2(c',p_2(c,b))$;

(84') $\quad 2b <c,c'> = q_2(c,p_2(c',b)) + q_2(c',p_2(c,b))$.

Finally, we have as before

(85') $\quad c' <c'',c> = p_1(c,q_1(c',c'')) - p_2(c,q_2(c',c''))$
$\quad\quad\quad\quad\quad\quad - p_1(c'',q_1(c,c')) - p_2(c'',q_2(c,c'))$
$\quad\quad\quad\quad\quad\quad - \frac{1}{2} p_2(c',q_2(c'',c))$.

As before, (A,f_1) is a Jordan algebra. Once again, we can show this Jordan algebra is <u>simple</u>, for which it suffices to show (A,f_1) contains no non-trivial ideal stable under all derivations. If I is such an ideal, it follows from (41") that $g_1(B,I) \subseteq I$. Consider

$$K = <A,I> + <B,g_2(B,I)> + <C,p_1(C,I)>$$
$$+ S \otimes I + M' \otimes (g_2(B,I) + f_2(A,I))$$
$$+ M'' \otimes (p_1(C,I) + p_2(C,f_2(A,I)) + p_2(C,g_2(B,I))) .$$

We claim K is an ideal (necessarily proper) in L . By the properties of composition with D , $[K,D] \subseteq K$. Next, we consider $[K, S \otimes A]$; by combining (39') and (39"), with $a'' \in A$, we have $A <A,I> \subseteq I$. As in §6, we have $[S \otimes I, S \otimes A] \subseteq <A,I> + S \otimes I + M' \otimes f_2(A,I)$ $\subseteq K$, $[M' \otimes f_2(A,I), S \otimes A] \subseteq S \otimes g_1(f_2(A,I),A)$
$+ M' \otimes g_2(f_2(A,I),A) \subseteq K$, since (39') gives $g_1(f_2(A,I),A) \subseteq I$, and (38") gives $g_2(f_2(A,I),A) \subseteq g_2(f_2(A,A),I) + f_2(A,f_1(A,I))$
$\subseteq g_2(B,I) + f_2(A,I)$. Then $[M' \otimes f_2(A,I), S \otimes A] \subseteq S \otimes g_1(f_2(A,I),A)$
$+ M' \otimes g_2(f_2(A,I),A) \subseteq K$ by the last sentence, and
$[M' \otimes g_2(B,I), S \otimes A] \subseteq S \otimes g_1(g_2(B,I),A) + M' \otimes g_2(g_2(B,I),A)$.
From (41"), $h_1(B,f_2(A,I)) \subseteq f_1(A,g_1(B,I)) + g_1(g_2(B,A),I) \subseteq I$, and also by (41"), $g_1(g_2(B,I),A) \subseteq f_1(I,g_1(B,A)) + h_1(B,f_2(A,I)) \subseteq I$.
From (44'), $g_2(g_2(B,I),A) \subseteq g_2(B,I) + f_2(A,I)$, so

$[M' \otimes g_2(B,I), S \otimes A] \subseteq K$. Also, $p_1(p_1(C,I),A) \subseteq p_1(C,I) + p_2(C,f_2(A,I))$, by (64'), so $[M'' \otimes p_1(C,I), S \otimes A] \subseteq K$. Next we have $p_1(p_2(C,f_2(A,I)),A) \subseteq p_1(C,g_1(f_2(A,I),A)) + p_2(C,g_2(f_2(A,I),A)$, by (67'); the first term is in $p_1(C,I)$, and we have seen that $g_2(f_2(A,I),A) \subseteq g_2(B,I) + f_2(A,I)$. Thus $[M'' \otimes p_2(C,f_2(A,I)), S \otimes A] \subseteq K$. That $[M'' \otimes p_2(C,g_2(B,I)), S \otimes A] \subseteq K$ follows likewise from $p_1(p_2(C,g_2(B,I)),A) \subseteq p_1(C,g_1(g_2(B,I),A)) + p_2(C,g_2(g_2(B,I),A)) \subseteq p_1(C,I) + p_2(C,g_2(B,I)) + p_2(C,f_2(A,I))$ by the above. Finally we have $A < B, g_2(B,I) > \subseteq g_1(g_2(B,I),A) + g_1(B,g_1(g_2(B,I),A))$ by (51'''), and $g_1(g_2(B,I),A) \subseteq I$ has been seen above; thus $[< B,g_2(B,I) >, S \otimes A] \subseteq S \otimes I \subseteq K$, and likewise $A < C,p_1(C,I) > \subseteq g_1(q_2(C,p_1(C,I)),A)$, by (77'), while addition of (74') and (75') gives $q_2(C,p_1(C,I)) \subseteq f_2(A,I) + g_2(B,I)$, yielding $[< C,p_1(C,I) >, S \otimes A] \subseteq S \otimes I \subseteq K$, and thus that $[K, S \otimes A] \subseteq K$. Now $B < A,I > \subseteq f_2(A,I)$ by (43'); $[S \otimes I, M' \otimes B] \subseteq S \otimes g_1(B,I) + M' \otimes g_2(B,I) \subseteq K$; $< B,f_2(A,I) > \subseteq < A,I >$ (by (40)); $h_1(g_2(B,I),B) \subseteq I$ by (52'); $h_1(f_2(A,I),B) \subseteq I$ by (42'''); $p_2(p_1(C,I),B) \subseteq p_1(C,I) + p_2(C,g_2(B,I))$, by (66'); $p_2(p_2(C,f_2(A,I)),B) \subseteq p_1(C,h_1(f_2(A,I),B)) + C < f_2(A,I),B >$, by adding (68') and (69'), and we know $h_1(f_2(A,I),B) \subseteq I$, while $< f_2(A,I),B > \subseteq < A,I >$ by (40), and $C < A,I > \subseteq p_2(C,f_2(A,I))$ by (65'), so $p_2(p_2(C,f_2(A,I)),B) \subseteq p_1(C,I) + p_2(C,f_2(A,I))$; $p_2(p_2(C,g_2(B,I)),B) \subseteq p_1(C,h_1(g_2(B,I),B)) + p_2(p_2(C,B),g_2(B,I))$ (by (68')) $\subseteq p_1(C,I) + p_2(C,g_2(B,I))$; $B < B,g_2(B,I) > \subseteq g_2(g_2(B,I),A) + g_2(B,h_1(g_2(B,I),B))$ (by (55'')) $\subseteq g_2(B,I) + f_2(A,I)$, by the above; $B < C,p_1(C,I) > \subseteq q_2(C,p_2(p_1(C,I),B)) + q_2(p_1(C,I),C)$ (by (84')) $\subseteq q_2(C,p_1(C,I)) + q_2(C,p_2(C,g_2(B,I))) + q_2(C,p_2(C,f_2(A,I)))$, by the above; now adding (74') and (75') gives $q_2(C,p_1(C,I)) \subseteq f_2(A,I) + g_2(B,I)$, while adding (82') and (84') gives $q_2(C,p_2(C,g_2(B,I))) \subseteq g_2(g_2(B,I),A) + g_2(B,I) < C,C > \subseteq g_2(B,I) + f_2(A,I)$, and $q_2(C,p_2(C,f_2(A,I))) \subseteq g_2(f_2(A,I),A) + f_2(A,I) < C,C > \subseteq g_2(B,I) + f_2(A,I)$. The steps separated by semicolons in this paragraph combine to show $[K, M' \otimes B] \subseteq K$.

Finally, we have $C < A,I > \subseteq p_2(C,f_2(A,I))$, by (65'); $q_1(C,p_1(C,I)) \subseteq I$, by adding (72') and (73'); $q_2(C,p_1(C,I)) \subseteq f_2(A,I) + g_2(B,I)$, as above; $< C,p_2(C,g_2(B,I)) > \subseteq < C,p_1(C,I) >$, by (67'); $< C,p_2(C,f_2(A,I)) > \subseteq < C,p_1(C,I) >$, by (64');

$q_1(C, p_2(C, g_2(B,I))) \subseteq h_1(g_2(B,I), B) + g_1(g_2(B,I), A)$, by adding (78') and (80'), and this contained in I, as is $q_1(C, p_2(C, f_2(A,I))) \subseteq h_1(f_2(A,I), B) + g_1(f_2(A,I), A)$; by (82') and (84'), $q_2(C, p_2(C, g_2(B,I))) \subseteq g_2(B,I) < C,C > + g_2(g_2(B,I), A) \subseteq g_2(B,I) + f_2(A,I)$, and likewise for $q_2(C, p_2(C, f_2(A,I)))$; $C < B, g_2(B,I) > \subseteq p_2(C, g_2(B,I)) + p_2(p_2(C, g_2(B,I)), B)$, by (69'), and this is already known to be contained in $p_1(C,I) + p_2(C, g_2(B,I)) + p_2(C, f_2(A,I))$; finally, a similar assertion holds for $C < C, p_1(C,I) >$, which by (85') is contained in $p_1(C, q_1(C, p_1(C,I))) + p_2(C, q_2(C, p_1(C,I))) + p_1(p_1(C,I), A) + p_2(p_1(C,I), B)$. This sequence of inclusions shows that $[K, M' \otimes C] \subseteq K$, thus that K is an ideal, and therefore that <u>the Jordan algebra</u> (A, f_1) <u>is simple</u>.

As in §6, it is convenient to split the situation into the cases where (A, f_1) is or is not associative. We take up the latter first:

<u>Case</u> a): (A, f_1) <u>is</u> <u>not</u> <u>associative</u>.

Letting $B_0 \subseteq B$ be defined, as in §6, by $g_1(B_0, A) = 0$, we find as there that $B = f_2(A,A) \oplus B_0$, with g_1 inducing an isomorphism of vector spaces between $f_2(A,A)$ and $\text{Der}(A, f_1)$. This enables us to define $h_2: B \times B \to f_2(A,A) \subseteq B$ as in (58) of §6. As in §6, we show $B_0 < B_0, B_0 > = 0$ by showing that

$K = \ < B, B_0 < B_0, B_0 >> + < C, p_2(C, B_0 < B_0, B_0 >) >$
$\quad + < C, p_1(C, h_1(B, B_0 < B_0, B_0 >)) > + S \otimes h_1(B, B_0 < B_0, B_0 >)$
$\quad + M' \otimes B_0 < B_0, B_0 > + M'' \otimes p_2(C, B_0 < B_0, B_0 >)$
$\quad + M''' \otimes p_1(C, h_1(B, B_0 < B_0, B_0 >))$

is an ideal in L, necessarily $\neq L$, since $B_0 < B_0, B_0 > \subseteq B_0 \neq B$.

The first, fourth and fifth terms above were already shown in §6 to yield elements of K upon bracketing with $S \otimes A$ and $M' \otimes B$, and all terms are stable under brackets with D. Bracketing of $M'' \otimes C$ with the fourth and fifth term leads to the seventh resp. sixth term. Moreover, by (69'),

$C < B, B_0 < B_0, B_0 >> \subseteq p_2(C, B_0 < B_0, B_0 >)$
$\qquad + p_2(p_2(C, B_0 < B_0, B_0 >), B)$, and this is, by

(68'), contained in

$p_2(C, B_0 < B_0, B_0 >) + p_1(C, h_1(C, B_0 < B_0, B_0 >))$.

Thus all brackets involving the first term again fall in K, and it is only brackets involving the second, third, sixth and seventh terms that we must check. First, we check the brackets of the sixth and seventh with $S \otimes A$ and with $M' \otimes B$: By (67') and work of §6,

$$p_1(p_2(C,B_0 < B_0,B_0 >),A) \subseteq p_2(C,B_0 < B_0,B_0 >).$$

By (64'), $p_1(p_1(C,h_1(B,B_0 < B_0,B_0 >)),A)$
$$\subseteq p_1(C,f_1(A,h_1(B,B_0 < B_0,B_0 >))) + p_2(C,f_2(A,h_1(B,B_0 < B_0,B_0 >))),$$
and this is contained in $p_2(C,B_0 < B_0,B_0 >) + p_1(C,h_1(B,B_0 < B_0,B_0 >))$ by §6. We have seen above that $p_2(p_2(C,B_0 < B_0,B_0 >),B)$
$$\subseteq p_2(C,B_0 < B_0,B_0 >) + p_1(C,h_1(B,B_0 < B_0,B_0 >)), \text{ using (68'), and}$$
from (66') we find

$$p_2(p_1(C,h_1(B,B_0 < B_0,B_0 >)),B)$$
$$\subseteq p_1(C,g_1(B,h_1(B,B_0 < B_0,B_0 >)))$$
$$+ p_2(C,g_2(B,h_1(B,B_0 < B_0,B_0 >)));$$

these terms are contained in

$p_2(C,B_0 < B_0,B_0 >) + p_1(C,h_1(B,B_0 < B_0,B_0 >))$ by §6. Thus K contains the brackets with $S \otimes A$ and $M' \otimes B$ of its sixth and seventh terms. Clearly, K contains the D-parts of the brackets with $M'' \otimes C$ of these terms. The remaining parts are in K, for the following reasons:

Addition of (78') and (80') shows

$q_1(C,p_2(C,B_0 < B_0,B_0 >))$
$$\subseteq g_1(B_0 < B_0,B_0 >,A) + h_1(B_0 < B_0,B_0 >,B);$$

addition of (82') and (84') shows

$q_2(C,p_2(C,B_0 < B_0,B_0 >))$
$$\subseteq g_2(B_0 < B_0,B_0 >,A) + (B_0 < B_0,B_0 >) < C,C >,$$
and this is contained in $B_0 < B_0,B_0 >$ by §6. Next, adding (72') and (73') shows that

$$q_1(C,p_1(B,h_1(B,B_0 < B_0,B_0 >)))$$
$$\subseteq f_1(h_1(B,B_0 < B_0,B_0 >),A) + g_1(B,h_1(B,B_0 < B_0,B_0 >)),$$
and these terms are in $h_1(B,B_0 < B_0,B_0 >)$, by §6; moreover,

129

adding (74') and (75') gives

$$q_2(C, p_1(C, h_1(B, B_0 < B_0, B_0 >)))$$
$$\subseteq f_2(h_1(B, B_0 < B_0, B_0 >), A) + g_2(B, h_1(B, B_0 < B_0, B_0 >))$$
$$\subseteq B_0 < B_0, B_0 >, \text{ by } \S 6. \text{ Thus brackets with } M'' \text{ \& } C \text{ of}$$

the sixth and seventh terms fall in K, and it remains only to check the effects on A, B, C of the second and third terms.

The last inclusion established above combines with (77') to show

$$A < C, p_1(C, h_1(B, B_0 < B_0, B_0 >)) > = 0;$$

also, by (77'),
$$A < C, p_2(C, B_0 < B_0, B_0 >) >$$
$$\subseteq g_1(q_2(C, p_2(C, B_0 < B_0, B_0 >)), A),$$ and this is zero since we have seen above that

$$q_2(C, p_2(C, B_0 < B_0, B_0 >)) \subseteq B_0 < B_0, B_0 > .$$

By (84'),
$$B < C, p_2(C, B_0 < B_0, B_0 >) >$$
$$\subseteq q_2(C, p_2(p_2(C, B_0 < B_0, B_0 >), B)) + q_2(p_2(C, B_0 < B_0, B_0 >), C),$$

and these terms are in $B_0 < B_0, B_0 >$ by remarks above; also, by (84'),

$$B < C, p_1(C, h_1(B, B_0 < B_0, B_0 >)) >$$
$$\subseteq q_2(C, p_2(p_1(C, h_1(B, B_0 < B_0, B_0 >)), B))$$
$$+ q_2(p_1(C, h_1(B, B_0 < B_0, B_0 >)), C)$$ and, by use of observations already made, this is in $B_0 < B_0, B_0 >$. Finally, from (85'),

$$C < C, p_2(C, B_0 < B_0, B_0 >) > \subseteq p_1(p_2(C, B_0 < B_0, B_0 >), A)$$
$$+ p_1(C, q_1(C, p_2(C, B_0 < B_0, B_0 >)))$$
$$+ p_2(p_2(C, B_0 < B_0, B_0 >), B) + p_2(C, q_2(C, p_2(C, B_0 < B_0, B_0 >))),$$

and
$$C < C, p_1(C, h_1(B, B_0 < B_0, B_0 >)) >$$
$$\subseteq p_1(p_1(C, h_1(B, B_0 < B_0, B_0 >)), A)$$
$$+ p_1(C, q_1(C, p_1(C, h_1(B, B_0 < B_0, B_0 >))))$$
$$+ p_2(p_1(C, h_1(B, B_0 < B_0, B_0 >)), B)$$
$$+ p_2(C, q_2(C, p_1(C, h_1(B, B_0 < B_0, B_0 >))));$$

we have already seen that each of the displayed eight terms is in

$p_2(C, B_0 <B_0, B_0>) + p_1(C, h_1(B, B_0 <B_0, B_0>))$, so that K is an ideal and $B_0 <B_0, B_0> = 0$.

Thus, as in §6, $B<B_0, B_0> = 0 = A<B_0, B_0>$, $<f_2(A,A), B_0> = 0$, and $<B, B_0> = <B_0, B_0>$.

From (46), $<A, h_1(B_0, B_0)> \subseteq <B, B_0> = <B_0, B_0>$, and therefore annihilates A. Combined with (39') and (39"), this implies that every f_1-product of three factors from A, of which one is in $h_1(B_0, B_0)$, is associative, and thus that $h_1(B_0, B_0) \neq A$. Even though we have not yet shown that $<B, B_0> = 0$, as was the case in §6, we have shown that $<B, B_0>$ annihilates A and B, and this is sufficient to show that, with the product

$$(a+b)(a'+b') = f_1(a,a') + g_1(b,a') - g_1(b',a) + h_1(b,b')$$
$$+ \{f_2(a,a') + g_2(b,a') + g_2(b',a) + h_2(b,b')\}$$

and the involution $(a+b)* = a - b$, $E = A \oplus B$ becomes an involutorial associative algebra. Writing $c(a+b)$ for $p_1(c,a) + p_2(c,b)$, we see from (64') that $(ca)a' = c(aa')$, from (66') that $(ca)b = c(ab)$, from (67') that $(cb)a = c(ba)$. To show that C is a right E-module, it thus suffices to show that

$$p_2(p_2(c,b), b') = p_1(c, h_1(b,b')) + p_2(c, h_2(b,b')),$$

or, by virtue of (68') and (69'), that

(87) $\qquad p_2(c, h_2(b,b')) = 2c <b, b'>$.

From the above, we have seen that $h_2(B, B_0) = 0$, so that, since $<B, B_0> = <B_0, B_0>$, to prove (87) when $b' \in B_0$ it suffices to show $C <B_0, B_0> = 0$. This we shall do below. It remains to prove (87) when $b' = f_2(a,a')$. Now, from (40),

$$2c <b, f_2(a,a')> = 2c <g_1(b,a), a'> - 2c <g_1(b,a'), a>,$$

which is equal by (65') to

$$p_2(c, f_2(g_1(b,a), a')) - p_2(c, f_2(g_1(b,a'), a)),$$

which, by (44'), is

(88) $\qquad p_2(c, f_2(g_1(b,a), a')) + p_2(c, g_2(g_2(b,a), a'))$
$\qquad - p_2(c, g_2(b, f_1(a,a')))$.

By relation 1) from §6, $p_2(c,h_2(b,f_2(a,a')))$
$= p_2(c,f_2(g_1(b,a),a')) + p_2(c,g_2(g_2(b,a),a')) - p_2(c,g_2(b,f_1(a,a')))$,
and this is (88), as required.

To show $C<B_0,B_0> = 0$, we first show $B_0<C,C> = 0$, and this by showing that B_0 lies in a commutative subalgebra $A_0 + B_0$, without nilpotent elements, of the associative algebra $A+B$, such that all derivations of $A+B$ stabilize $A_0 + B_0$; it then follows that all such derivations must annihilate $A_0 + B_0$, so that, in particular $B_0<C,C> = 0$.

We let A_0 be the centralizer of A in A, i.e., the set of those $a \in A$ with $f_2(a,A) = 0$ in our notation. For fixed $a' \in A$, the mapping $D: a \mapsto f_2(a',a)$ of A into B is seen (by associative computations) to satisfy $f_1(a_1,a_2)D = g_2(a_1D,a_2) + g_2(a_2D,a_1)$ (cf.(38")), i.e., D <u>is a derivation of the Jordan algebra</u> (A,f_1) <u>into its Jordan bimodule</u> B, <u>with action of</u> A <u>on</u> B <u>given by</u> g_2. (See [JJ],p.292; the definition of Jordan bimodule is to be found in <u>op.cit.</u>,p. 82, and is checked by verifying, using associative calculations, that

$$g_2(g_2(b,f_1(a,a)),a) = g_2(g_2(b,a),f_1(a,a)),$$
$$2g_2(g_2(g_2(b,a),a'),a) + g_2(b,f_1(f_1(a,a),a'))$$
$$= 2g_2(g_2(b,a),f_1(a,a')) + g_2(g_2(b,a'),f_1(a,a))$$

for all b,a,a', where $g_2(b,a) = \frac{1}{2}(ab+ba), f_1(a,a') = \frac{1}{2}(aa'+a'a)$.)
Now Jacobson has proved (see [JJ], p. 293) that each such derivation D is <u>inner</u>, in the sense that there exist elements $b_i \in B$, $a_i \in A$ with

$$aD = \Sigma a[R_{b_i} R_{a_i}] = \Sigma g_2(g_2(b_i,a),a_i) - g_2(b_i,f_1(a,a_i))$$
$$= f_2(a, \Sigma g_1(b_i,a_i)), \text{ by (44')}.$$

It follows that the mapping $a' \mapsto f_2(a', \cdot)$ carries A <u>onto</u> the derivations of A into B, and in fact carries $g_1(B,A)$ onto these derivations. The kernel is A_0, so we have $A = A_0 + g_1(B,A)$. By considering, as in §6, the bilinear forms λ, μ on B resp. A induced from the Killing form of L, we have that μ is non-degenerate on A, and $a_0 \in A_0$, $b \in B$, $a \in A$ imply $\mu(a_0,g_1(b,a)) = \lambda(f_2(a,a_0),b) = 0$, so that A_0 and $g_1(B,A)$ are μ-orthogonal. Thus their dimensions are complementary in A, and $A = A_0 \oplus g_1(B,A)$. Also $\mu(a,g_1(b,a_0)) = \lambda(f_2(a_0,a),b) = 0$, so that $g_1(B,A_0) = 0$.

To show that $A_0 + B_0$ is a subalgebra of $A+B$ it is enough to show: $f_1(A_0,A_0) \subseteq A_0$; $h_1(B_0,B_0) \subseteq A_0$; $g_2(B_0,A_0) \subseteq B_0$. These follow at once from (38"), (53) and (42"), respectively. Since g_1, f_2 and h_2 vanish here, $A_0 + B_0$ is commutative. Now suppose $a_0 + b_0 \in A_0 + B_0$ is non-zero, with square zero; then $f_1(a_0,a_0) + h_1(b_0,b_0) = 0$, $g_2(b_0,a_0) = 0$. If $b_0 = 0$, we have $a_0^2 = f_1(a_0,a_0) = 0$; if $b_0 \neq 0$, $g_2(b_0,h_1(b_0,b_0)) = -g_2(b_0,f_1(a_0,a_0)) = -g_2(g_2(b_0,a_0),a_0) = 0$, by (44') and the above, so that $b_0^3 = 0$. Thus either $b_0^2 = 0$ or $0 \neq b_0^2 \in A_0$ has square 0. Hence we may assume our element of square zero is in $A_0 \cup B_0$. As remarked above, we shall have proved that $B_0 <C,C> = 0$ as soon as we have shown there is no element of square zero in $A_0 \cup B_0$, except zero.

First let $0 \neq a_0 \in A_0$, $a_0^2 = 0$, and let $0 \neq h \in T$. Then $h \otimes a_0$ centralizes T, and $\mathrm{ad}(h \otimes a_0)$ is nilpotent. The latter is seen from the following:

$[v \otimes c, h \otimes a_0] = vh \otimes p_1(c,a_0)$,

$[[v \otimes c, h \otimes a_0], h \otimes a_0] = vh^2 \otimes p_1(p_1(c,a_0),a_0) = 0$, by (64');

$[s \otimes b, h \otimes a_0] = (s \circ h) \otimes g_1(b,a_0) + [sh] \otimes g_2(b,a_0)$

$= [sh] \otimes g_2(b,a_0)$,

$[[s \otimes b, h \otimes a_0] h \otimes a_0] = [[sh]h] \otimes g_2(g_2(b,a_0),a_0) = 0$, by (44');

$[x \otimes a, h \otimes a_0] = (x,h) <a,a_0> + [xh] \otimes f_1(a,a_0)$; now $A<a,a_0> = 0$ by (39"), so

$[[x \otimes a, h \otimes a_0] h \otimes a_0] = ([xh],h) < f_1(a,a_0),a_0 > +$

$[[xh]h] \otimes f_1(f_1(a,a_0),a_0) = 0$ since $([xh],h) = 0$ and $f_1(f_1(a,a_0),a_0)$ is equal by (39') to $f_1(a,f_1(a_0,a_0)) = 0$. Finally, for $d \in D$,

$[d, h \otimes a_0] = -h \otimes a_0 d$,

$[[d, h \otimes a_0] h \otimes a_0] = -(h,h) <a_0,a_0 d>$,

$d(\mathrm{ad}(h \otimes a_0))^3 = (h,h)h \otimes a_0 < a_0, a_0 d > = 0$, by (39").

Thus $(\mathrm{ad}(h \otimes a_0))^3 = 0$, which is impossible. Likewise, if $b_0 \neq 0$ in B_0, $b_0^2 = 0 = h_1(b_0,b_0)$, we let $0 \neq s \in M'$, $[s,T] = 0$, so that $s \otimes b_0$ centralizes T; we show this element is nilpotent:

133

$$[v \otimes c, s \otimes b_0] = vs \otimes p_2(c,b_0),$$

$$[[v \otimes c, s \otimes b_0]s \otimes b_0] = vs^2 \otimes p_2(p_2(c,b_0),b_0) = 0, \text{ by (68')};$$

$$[r \otimes b, s \otimes b_0] = (r,s) < b,b_0 > + [rs] \otimes h_1(b,b_0),$$

$$[[r \otimes b, s \otimes b_0]s \otimes b_0] = -(r,s)s \otimes b_0 < b,b_0 >$$

$$+ [[rs]s] \otimes g_2(b_0, h_1(b,b_0)) = 0 \text{ by } B_0 < B, B_0 > = 0 \text{ and}$$

(55");

$$[x \otimes a, s \otimes b_0] = [xs] \otimes g_2(b_0, a),$$

$$[[x \otimes a, s \otimes b_0]s \otimes b_0] = ([xs], s) < g_2(b_0, a), b_0 >$$

$$+ [[xs]s] \otimes h_1(g_2(b_0,a)b_0),$$

of which the first term is zero since $([xs],s) = 0$ and the second by (52');

$$[d, s \otimes b_0] = -s \otimes b_0 d, \ d(ad(s \otimes b_0))^2 = -(s,s) < b_0 d, b_0 >,$$

and $d(ad(s \otimes b_0))^3 = 0$. Thus $(ad(s \otimes b_0))^3 = 0$, again a contradiction. Our proof shows that $B_0 D = 0 = A_0 D$, in particular that $B_0 < C, C > = 0$.

To see that $C < B_0, B_0 > = 0$, we show that

$$U = < C, C < B_0, B_0 >> + M" \otimes C < B_0, B_0 >$$

is an ideal in L, hence necessarily is zero. Each term is stable under brackets with D, and the derivation property for $< B_0, B_0 >$, together with $A < B_0, B_0 > = 0$, $B < B_0, B_0 > = 0$, gives $p_1(C < B_0, B_0 >, A) \subseteq p_1(C,A) < B_0, B_0 > = C < B_0, B_0 >$, $p_2(C < B_0, B_0 >, B) \subseteq p_2(C,B) < B_0, B_0 > \subseteq C < B_0, B_0 >$.

For $c, c' \in C$ and $b, b' \in B_0$, we have, by (69'),

$$4q_1(c, c' < b, b' >)$$

$$= q_1(c, p_2(p_2(c',b),b')) - q_1(c, p_2(p_2(c',b'),b));$$

by (78'), (80') and the definition of B_0, this is

$$h_1(b', q_2(p_2(c',b),c)) - h_1(b, q_2(p_2(c',b'),c)),$$

which by (82'), (84') and $B_0 < C, C > = 0$, is

$$h_1(b', g_2(b, q_1(c,c'))) - h_1(b, g_2(b', q_1(c,c'))),$$

which (51") shows is $q_1(c,c') < b,b' > = 0$. Thus $q_1(C, C < B_0, B_0 >) = 0$. Analogous steps based on, in this order: (82'),

(84') and $B_0 < C,C > = 0$; (78'), (80') and definition of B_0; (55") and $B < B_0, B_0 > = 0$, show that $q_2(C, C < B_0, B_0 >) = 0$. Thus bracketing any element of L with the second term of U gives a result in U. The above and (77') show that $A < C, C < B_0, B_0 >> = 0$, and likewise (82') and the above yield $B < C, C < B_0, B_0 >> = 0$. Finally, we have, from (85') and the above,

$$C < C, C < B_0, B_0 >> \subseteq p_1(C < B_0, B_0 >, A) + p_2(C < B_0, B_0 >, B)$$

$$\subseteq C < B_0, B_0 > ,$$

completing the proof that U is an ideal, so that $C < B_0, B_0 > = 0$, and with it that C <u>is a</u> (<u>right</u>) <u>module for the associative algebra</u> $A \oplus B$.

We now write the product in $E = A \oplus B$ by juxtaposition, so that, e.g., $g_1(b,a) = \frac{1}{2}(ba-ab)$, $h_2(b,b') = \frac{1}{2}(bb'-b'b)$. We denote the involution in E fixing A and mapping b to $-b$ by $*$, and the module structure map $C \times E \mapsto C$ by juxtaposition. We set $q = q_1 + q_2 : C \times C \mapsto E$; then evidently $q(c,c') = -q(c',c)*$. Moreover, we have for $a \in A$, $q(c,c'a) = q_1(c,c'a) + q_2(c,c'a)$; which by addition of (72') and (73'), and subtraction of (74') and (75'), is equal to

$$\frac{1}{2}(aq_1(c,c') + q_1(c,c')a) + \frac{1}{2}(q_2(c,c')a - aq_2(c,c'))$$
$$+ \frac{1}{2}(aq_2(c,c') + q_2(c,c')a) - \frac{1}{2}(aq_1(c,c') - q_1(c,c')a)$$
$$= q(c,c')a .$$

Likewise, for $b \in B$, $q(c,c'b)$ is seen, by subtracting (78') and (80'), and adding (82') and (84'), to be equal to

$$\frac{1}{2}(bq_2(c,c') + q_2(c,c')b) - \frac{1}{2}(bq_1(c,c') - q_1(c,c')b)$$
$$+ \frac{1}{2}(bq_1(c,c') + q_1(c,c')b) + b < c,c' >$$
$$= q_1(c,c')b + b < c,c' > + \frac{1}{2}(bq_2(c,c') + q_2(c,c')b) .$$

We wish to show this is equal to $q(c,c')b$, for which it suffices to show $b < c,c' > = \frac{1}{2}(q_2(c,c')b - bq_2(c,c'))$. When $b \in B_0$, we have seen that both members are zero; thus it suffices to show $g_1(b_1,a) = g_1(b_2,a)$ for all $a \in A$, where b_1, b_2 are the two members. By the derivation property of $< c,c' >$ and by (77'),

$$g_1(b < c,c' >, a) = -g_1(b, g_1(q_2(c,c'),a)) + g_1(q_2(c,c'), g_1(b,a)),$$

which is equal to a $< b,q_2(c,c') >$, i.e., to $g_1(h_2(q_2(c,c'),b),a)$, by (51") and the definition of h_2. Thus we have, for $e \in E$, $q(c,c'e) = q(c,c')e$, and q is a sesquilinear antihermitian form on the right E module C .

As in §7, we may show that E is a division algebra by showing first that E is *-simple, then by displaying in the non-simple case a non-zero element $e \in B_0$, $e^2 = 1$, so that $U_1 \otimes e$ centralizes T and acts rational-diagonally in the adjoint representation. Once E is known to be simple, our argument of §7 (and §6) is repeated to show that E is a division algebra.

As in §7, we now show the E-form q is nondegenerate on C, and is in fact anisotropic. Once this is shown, the conclusions of Theorem III.9 hold, with r = 2, with the following change:

$$[s \otimes b', t \otimes b''] = \text{Tr}(st) < b',b'' > + [st] \otimes \frac{b'b''+b''b'}{2} .$$

We turn to the case: (A,f_1) is associative; being simple, it is thus a field extension of F, so admits no F-derivations, and $g_1 = 0$, as in §6. From (44'), $g_2: B \times A \to B$ gives B the structure of vector space over (A,f_1), and from (38") we see that for fixed $a' \in A$, the mapping $a \to f_2(a',a)$ is a derivation D of A into B, in the sense that $(aa'')D = (aD)a'' + (a''D)a$. Thus $a^nD = n(aD)a^{n-1}$, from which we see upon applying D to a polynomial in A of minimal degree over F vanishing at a that aD = 0, hence that $f_2 = 0$. It follows from (65') that $C < A,A > = 0$, and $B < A,A > = 0$ by (43'), so that $< A,A > = 0$. Now (52') shows that h_1 is a (symmetric) A-bilinear form on B, and (55") gives the action of $< b',b'' >$ on B in terms of h_1, as in §6. From (64') we see that, relative to $p_1: C \times A \to C$, C is a vector space over A , and (66') and (67') say that $p_2: C \times B \to C$ is A-bilinear.

Now $A \oplus B$ carries the structure of the Jordan algebra associated with the form h_1 if a product is defined by $ab = ba = g_2(b,a)$, $aa' = f_1(a,a')$, $bb' = h_1(b,b')$. With this definition, (68') completes the demonstration that $p_1 + p_2: C \times (A + B) \to C$ defines a structure of special unital $A \oplus B$- bimodule on C (from [JJ]; for our purposes, this notion may be regarded as defined by the mentioned relations). From (74') and (75'), we see that these are equivalent to saying that $q_2: C \times C \to B$ is A-bilinear, and then we see from (55"), (69'), (84'), (85') that $\mathcal{D} = < B,B > + < C,C >$ acts

A-linearly on $B \oplus C$, and that the maps $\langle\,,\,\rangle : B \times B \to D$ and $C \times C \to D \subseteq \mathrm{End}_A(B \oplus C)$ are A-bilinear.

The mapping $a + b \mapsto a - b$ defines an involution $*$ in the Jordan algebra $A \oplus B$, and if $q = q_1 + q_2 : C \times C \to A \oplus B$, we have $q(c,c')^* = -q(c',c)$. Moreover, it follows from (72'), (73'), (74'), (75'), (78'), (80'), (82'), (84') that, for $c, c' \in C$, $e \in A \oplus B$,

(89) $\qquad q(c, c' \cdot e) = q(c, c') \cdot e + e \langle c, c' \rangle$,

where $c' \cdot e$ is the module-action of $E = A \oplus B$ on C, and where $q(c, c') \cdot e$ is the Jordan composition in E. Thus our "form" q fails to be skew-hermitian by a term associated with $\langle c, c' \rangle$. By breaking (89) into components, all of the identities indicated just before (89) are recovered. The only identities from the list at the beginning of §8 which have not yet been translated into the new context are:

(69') $\qquad 4c \langle b, b' \rangle = (c \cdot b) b' - (c \cdot b') \cdot b$,

(79') $\qquad \langle b, q(c,c') + q(c',c) \rangle = \langle c, c' \cdot b \rangle + \langle c', c \cdot b \rangle$,

(85') $\qquad c' \langle c'', c \rangle = -c \cdot q(c'', c') - c'' \cdot q(c, c')$

$\qquad\qquad\qquad - \frac{1}{4} c' \cdot (q(c'', c) + q(c, c''))$.

As in §6, we show that the form $h_1 : B \times B \to A$ is nondegenerate, and that $h_1(b,b) \neq 1$ for all $b \in B$; it follows that E is a Jordan division algebra. For the former, let R be the radical of h_1. Then R is clearly an A-subspace of B; we show that

$$K = \langle B, R \rangle + \langle C, C \cdot R \rangle + M' \otimes R + M'' \otimes C \cdot R$$

is an ideal of L, necessarily zero by simplicity. All summands are evidently stable under brackets with D and with $S \otimes A$; we have $[M' \otimes B, M' \otimes R] \subseteq \langle B, R \rangle$; $[M' \otimes B, \langle B, R \rangle] \subseteq M' \otimes B \langle B, R \rangle \subseteq M' \otimes R$, from (55"); $[M' \otimes B, M'' \otimes C \cdot R] \subseteq M'' \otimes (C \cdot R) \cdot B \subseteq M'' \otimes (C \cdot B) \cdot R$, by (68'), so in $M'' \otimes C \cdot R$; $[M' \otimes B, \langle C, C \cdot R \rangle] \subseteq M' \otimes B \langle C, C \cdot R \rangle \subseteq M' \otimes q_2(C, C \cdot R)$, by (84') and the above. Now by adding (82') and (84') we find $q_2(C, C \cdot R) \subseteq R \langle C, C \rangle + R \cdot q_1(C, C) \subseteq R + R \cdot A \subseteq R$, so bracketing with $M' \otimes B$ preserves K. Finally, to show that bracketing with $M'' \otimes C$ preserves K, it remains to check that $C \langle B, R \rangle \subseteq C \cdot R$, $C \langle C, C \cdot R \rangle \subseteq C \cdot R$, $q_1(C, C \cdot R) = 0$. The first of these follows from (69') and (68'), the third from (78') and (80'); by (85') and the above, we have

$C < C, C \cdot R > \subseteq (C \cdot R) \cdot A + (C \cdot R) \cdot B + C \cdot q_1(C, C \cdot R) + C \cdot q_2(C, C \cdot R) \subseteq C \cdot R$, thus completing the proof that K is an ideal, and therefore that $R = 0$. Thus E is a <u>simple</u> Jordan algebra.

If $h_1(b,b) = 1$, and if $U_1 \in M'$ is as in §6, we show that $\text{ad}(U_1 \otimes b)$ is diagonalizable in F. First, $\text{ad}(U_1 \otimes b)$ annihilates $\{D \in \mathcal{D} \mid bD = 0\}$, and if $D \in \mathcal{D}$, $1D = 0 = h_1(b,b)D$ means $h_1(b,bD) = 0$. Then $b < b, bD > = bD$, by (55"), so that every element of $\mathcal{D}$ can be written as the sum of an element in the annihilator of b and an element $< b, b' >$ with $h_1(b, b') = 0$, with $[U_1 \otimes b, < b, b' >] = U_1 \otimes b'$, and with $[U_1 \otimes b, U_1 \otimes b'] = (U_1, U_1) < b, b' > = < b, b' >$. Thus $< b, b' >$ and $U_1 \otimes b'$ span a stable subspace, where $\text{ad}(U_1 \otimes b)$ has matrix $\begin{pmatrix} 0 & -1 \\ -1 & 0 \end{pmatrix}$, of square 1, so diagonalizable with eigenvalues ± 1. Now arguments as in §6 complete the demonstration that $\text{ad}(U_1 \otimes b)$ acts F-diagonally on $\mathcal{D} + S \otimes A + M' \otimes B$. Finally, let $v \in M''$, $c \in C$; then $[v \otimes c, U_1 \otimes b] = vU_1 \otimes c \cdot b$, $[[v \otimes c, U_1 \otimes b]U_1 \otimes b] = vU_1^2 \otimes (c \cdot b) \cdot b = vU_1^2 \otimes c$, by (68'). We may take generators with $vU_1 = \lambda v$, $\lambda \in F$, so that $vU_1^2 = \lambda^2 v$, $(v \otimes c)(\text{ad}(U_1 \otimes b))^2 = \lambda^2 v \otimes c$, again yielding the diagonalizability of $\text{ad}(U_1 \otimes b)$ on $M'' \otimes C$, and therefore on L. Since $U_1 \otimes b$ centralizes T, but is not in T, we have a contradiction.

Next we note that <u>the alternate form</u> $q_1 : C \times C \mapsto A$ <u>is nondegenerate</u>; for if N is the radical of q_1, then $K = < C, N > + M'' \otimes N$ is an ideal in L. For K is evidently stable under brackets with $\mathcal{D}$ and with $S \otimes A$; from (78'), $M'' \otimes N$ is stable under brackets with $M' \otimes B$, and then from (80') we see that $h_1(B, q_2(C,N)) = 0$, so that $q_2(C,N) = 0$. Thus $B < C, N > = 0$, by (84'), and $[K, M' \otimes B] \subseteq K$. Finally, by (85'), $C < C, N > \subseteq p_1(N, A) + p_2(N, B) \subseteq N$, from which $[K, M'' \otimes C] \subseteq K$, and K is an ideal, therefore zero, and $N = 0$.

As in §6, we see from (55") that the restrictions to B of elements of $< B, B >$ exhaust the A-linear skew transformations of B with respect to h_1. Since the latter have dimension $\binom{[B:A]}{2}$, and since this is an upper bound for the A-dimension of $< B, B >$, we see that <u>restriction to</u> B <u>identifies</u> $< B, B >$ <u>with the</u> A-<u>linear</u>, h_1-<u>skew transformations of</u> B.

If $B = 0$, then $q_2 = 0$, and C is simply a symplectic A-space relative to q_1, hence of even A-dimension. If $c, c' \in C$, $q_1(c, c') = 1$, we see by (85') that $< c, c' >$ acts diagonally on C with eigenvalues $0, \pm 1$, hence on $D = \langle C,C \rangle$ diagonally with integral eigenvalues in the adjoint representation. Since $\langle c,c' \rangle$ annihilates A, this would violate the assumption that T is a maximal F-split torus. Thus <u>if</u> $B = 0$, <u>then</u> $C = 0$, <u>and</u> $L = S \oplus A$, <u>a case already treated in</u> §6.

Next let $[B:A] = 1$, so that, $B = Ab$, $0 \neq b \in B$. Here again $\langle B,B \rangle = 0$; $h_1(b,b)$ is not a square in A; (79') yields $\langle c,p_2(c',b) \rangle + \langle p_2(c,b),c' \rangle = 0$. $E = A \oplus B$ is a quadratic field extension of A, so is annihilated by $\langle C,C \rangle$. From (84') and the above, we have $q_2(c,c'b) = q_1(c,c')b$, while $q_1(c,c'b) = h_1(b,q_2(c,c')) = q_2(c,c')b$ in E. It follows that $q = q_1 + q_2$ is a sesquilinear E-form on the E-vector space C, necessarily nondegenerate since q_1 is, and anti-hermitian with respect to our involution in E. The formula (85') amounts to $c' \langle c'',c \rangle = -q(c'',c')c - q(c,c')c'' - \frac{1}{4}(q(c,c'') + q(c'',c))c'$. As in the argument just preceding the statement of Theorem III.9 in §7, we conclude that q <u>is anisotropic</u>. As in that theorem, $D = \langle C,C \rangle$ is identified with $K(C,q) \cong B \oplus K'(C,q)$, and the structure of L again follows the pattern of the case where (A,f_1) is not associative, but here with $E = A \oplus B$ being commutative.

A similar situation is found when $[B:A] = 3$, where our conditions result in an identification of $A \oplus B$ with a quaternionic division algebra E over A (the <u>even Clifford algebra</u> of (B,h_1)) and (C,q) with an anisotropic antihermitian E-space. The rather detailed considerations on Clifford algebras used to give an exhaustive treatment of the case b) are deferred to Chapter V. It is convenient, however, to remark here that the verification that $p_1 + p_2$ gives C the structure of special unital bimodule for the Jordan algebra $A \oplus B$ also amounts to showing that this composition extends to a structure on C of right module for the Clifford algebra (over A) of the form (B,h_1). We denote this Clifford algebra by $H(B)$. Since $g_1 = 0$, we see from (78') that $q_1(cb,c') + q_1(c,c'b) = 0$ for all $b \in B$, $c,c' \in C$, or that <u>the elements of</u> B <u>act on</u> C <u>as</u> q_1<u>-skew</u> A<u>-linear transformations</u>. From (80') and this, we have $h_1(b,q_2(c,c')) = q_1(c,c'b)$; in view of the nondegeneracy of h_1, this shows that q_2 is defined in terms of h_1 and q_1. Then (84') and (85') ultimately lead the definition of $\langle c,c' \rangle$ back to these forms as well. The present state of our conclusions concerning algebras of type BC_2 is as follows:

Theorem III.10. Let L be a simple Lie algebra of relative type BC_2. Then either:

a) the conclusions of Theorem III.9 hold with the following modifications: i) $[s \otimes b', t \otimes b''] = \text{Tr}(st) <b',b''> + [st] \otimes \frac{b'b''+b''b'}{2}$; ii) $r = 2$; or b) there is a field extension A of F, a nonzero vector space B over A admitting a nondegenerate symmetric A-bilinear form h not representing 1, and a right module C for the Clifford algebra $H(B)$ of (B,h) over A, carrying moreover a nondegenerate alternate A-bilinear form q_1, such that $B \subseteq H(B)$ acts on C by q_1-skew transformations, and:

$L = D \oplus (S \otimes A) \oplus (M' \otimes B) \oplus (M'' \otimes C)$, with
$D = <B,B> + <C,C>$; $[x \otimes a, d] = 0$; $[s \otimes b, d] = s \otimes bd$;
$[u \otimes c, d] = u \otimes cd$; $[x \otimes a, y \otimes a'] = [xy] \otimes aa'$; $[s \otimes b, t \otimes b'] = \text{Tr}(st) <b,b'> + [st] \otimes h(b,b')$; $[s \otimes b, x \otimes a] = [sx] \otimes ab$;
$[u \otimes c, x \otimes a] = ux \otimes ac$; $[u \otimes c, s \otimes b] = us \otimes cb$ (action of Clifford algebra $H(B)$); $[u \otimes c, v \otimes c'] = (u,v) <c,c'> + (u \circ v) \otimes q_1(c,c') + [uv] \otimes q_2(c,c')$. Here $q_2(c,c') \in B$ satisfies $h(b, q_2(c,c')) = q_1(c,c'b)$ for all $b \in B$, and we have

$$b'' <b,b'> = h(b,b'')b' - h(b',b'')b;$$
$$c\ <b,b'> = \frac{1}{4} ((cb)b' - (cb')b);$$
$$b\ <c,c'> = \frac{1}{2} (q_2(cb,c') + q_2(c,c'b));$$
$$c''<c,c'> = -cq_1(c',c'') - cq_2(c',c'') - c'q_1(c,c'')$$
$$- c'q_2(c,c'') - \frac{1}{2} c''q_2(c,c').$$

9.   **Algebras of type BC_1.**

Here S may be identified with the 2 by 2 F-matrices of trace zero, M'' with the space F^2, with the canonical action of S on M''. We have

$$L = D \oplus (S \otimes A) \oplus (M'' \otimes C)$$

as S-modules as before, with D , the centralizer of S, represented

140

on both A and C, and with

$$[x \otimes a, y \otimes a'] = \text{Tr}(xy) <a,a'> + [xy] \otimes f(a,a'),$$

$$[u \otimes c, x \otimes a] = ux \otimes p(c,a),$$

$$[u \otimes c, v \otimes c'] = (u,v) <c,c'> + (u \circ v) \otimes q(c,c'),$$

where $<a,a'> \in \mathcal{D}$, $f(a,a') \in A$, $p(c,a) \in C$, $<c,c'> \in \mathcal{D}$, $q(c,c') \in A$, and where (u,v) is an alternate non-degenerate bilinear form on F^2 relative which S is the set of skew transformations, and $u \circ v \in S$ is given in its effect on $w \in M''$ by $w(u \circ v) = (w,u)v + (w,v)u$. Thus $<a,a'>$ is skew, as is q while f and $<c,c'>$ are symmetric. Hence $<A,A> + S \otimes A$ is a subalgebra, and one sees as in the case of A_1 that (A,f) is a Jordan algebra, and $<a,a'>$ acts on A as an inner derivation. More precisely,

(5') $\qquad a'' <a,a'> = 2f(f(a,a''),a') - 2f(a,f(a'',a')).$

Applying the Jacobi identity to a triple $u \otimes c$, $x \otimes a$, $y \otimes a'$ results in an identity

$$\text{Tr}(xy)u \otimes c <a,a'> + u[xy] \otimes p(c,f(a,a'))$$

$$= (ux)y \otimes p(p(c,a),a') - (uy)x \otimes p(p(c,a'),a);$$

with $x = E_{11} - E_{22}$, $y = E_{12}$, $u = e_1$ (in customary notation) this gives

(90) $\qquad 2p(c,f(a,a')) = p(p(c,a),a') + p(p(c,a'),a).$

Combined with the fact that the unit element 1 of A has $p(c,1) = c$ for all $c \in C$, this says that (C,p) is a structure of special unital A-module on C.

With $x = y = E_{11} - E_{22}$, $u = e_1$, we find

(91) $\qquad 2c <a,a'> = p(p(c,a),a') - p(p(c,a'),a).$

Next we apply the Jacobi identity to a triple $x \otimes a$, $u \otimes c$, $v \otimes c'$, resulting in an identity

$$(u,v) \; x \otimes a \; <c,c'> + [x, u \circ v] \otimes f(a,q(c,c'))$$

$$+ \text{Tr}((u \circ v)x) <a,q(c,c')>$$

$$= - (ux,v) <p(c,a),c'> - ((ux) \circ v) \otimes q(p(c,a),c')$$

$$- (u,vx) <c,p(c',a)> - (u \circ (vx)) \otimes q(c,p(c',a)).$$

With $(e_1,e_2) = 1 = -(e_2,e_1)$, setting $u = v = e_2$, $x = E_{21}$ gives $(ux,v) = 1$, $(u,vx) = -1$, $(u \circ v) = 2E_{12}$, $\text{Tr}((u \circ v)x) = 2$, hence

(92) $\qquad 2 <a,q(c,c')> \; = \; <c,p(c',a)> \; - \; <p(c,a),c'>$.

Also $[x, u \circ v] = 2(E_{22} - E_{11})$, $(ux) \circ v = e_1 \circ e_2 = E_{11} - E_{22} = u \circ (vx)$, giving

(93) $\qquad 2f(a,q(c,c')) = q(p(c,a),c') + q(c,p(c',a))$.

Likewise, $u = e_1$, $v = e_2$, $x = E_{11} - E_{22}$ gives

(94) $\qquad a <c,c'> \; = q(c,p(c',a)) - q(p(c,a),c')$.

Finally, we have $[u \otimes c [v \otimes c', w \otimes c'']] = (v,w)u \otimes c <c',c''> + u(v \circ w) \otimes p(c,q(c',c''))$, and the Jacobi identity requires that the cyclic sum of three such terms be zero. With $v = w = e_1$, $u = e_2$ we obtain

(95) $\qquad c' <c'',c> \; - \; c'' <c,c'>$

$= 2p(c,q(c',c'')) - p(c',q(c'',c)) - p(c'',q(c,c'))$.

Now if I is an ideal in (A,f) stable under all derivations of A, it follows by adding (93) and (94) that $q(C,p(C,I)) \subseteq I$ (again D acts as derivations with respect to all compositions). From (5'), $A <A,I> \; \subseteq I$, and from (90), $p(p(C,I),A) \subseteq p(C,I)$. Then, from (91), $C <A,I> \; \subseteq p(C,I)$; from (94), $A <C,p(C,I)> \; \subseteq q(C,p(p(C,I),A)) + q(p(C,A),p(C,I)) \subseteq q(C,p(C,I)) \subseteq I$ by the above, and from (95) with $c' \in p(C,I)$, $C <C,p(C,I)> \subseteq p(C,I) <C,C> + p(C,q(p(C,I),C)) + p(p(C,I),A) \subseteq p(C,I)$ by the above. Thus

$K = \; <A,I> \; + \; <C,p(C,I)> \; + (S \otimes I) + (M'' \otimes p(C,I))$

is an ideal in L. It follows as in previous cases that A is a <u>simple</u> Jordan <u>algebra</u>.

Then (90) shows that the annihilator of C in A is an ideal, unequal to A by the action of $1 \in A$, and hence is zero. Thus (A,f) <u>may be regarded as a</u> (special) <u>Jordan subalgebra of the F-endomorphisms of</u> C, <u>and we may use associative computations to compute the Jordan product</u> f <u>in</u> A. (That $C \neq 0$ is part of the context here, distinguishing the case BC_1 from the case A_1.)

As in the case of type A_1, it follows, by embedding a non-zero element of $L_{2\alpha}$ ($= E_{21} \otimes A$, in our realization) in a split three-dimensional subalgebra containing $(E_{11} - E_{22}) \otimes 1$ and an element of $L_{-2\alpha}$, that A <u>is a Jordan division algebra</u>.

If R is the set of $c \in C$ such that $q(c,C) = 0$, it follows from the derivation-property for D that $RD \subseteq R$, and then, rather easily, that

$$< R,C > + M'' \otimes R$$

is an ideal (note that $p(R,A) \subseteq R$ by (93), that $A < R,C > = 0$ by this and (94), and that $C < R,C > \subseteq R$ by the above and (95) with $c'' \in R$). Thus $R = 0$, and $\cdot q$ <u>is a nondegenerate pairing of</u> C <u>and</u> C <u>to</u> A.

The special case where $A = F$ has been investigated in greater detail by Faulkner [FT]. The general case has been carried out, in somewhat greater generality than here, by Allison [AJ] and, independently, by Benkart (unpublished). Allison has some results sharpening the implications of the situation. We defer such considerations to Chapter V, summarizing our information so far in the following form:

<u>Theorem III.11</u>. <u>Let</u> L <u>be a simple Lie algebra over</u> F <u>of type</u> BC_1. <u>Then there is a non-zero vector space</u> C <u>over</u> F, <u>and a Jordan division subalgebra</u> A <u>of</u> $End_F(C)$, <u>such that</u>

$$L \cong D \oplus (S \otimes A) \oplus (M'' \otimes C),$$

<u>where</u> M'' <u>is a</u> 2-<u>dimensional nondegenerate symplectic F-space with form</u> (u,v), S <u>is the skew transformations of</u> M'' <u>relative to</u> (u,v), <u>and the subalgebra</u> $D = < C,C > + < A,A >$ <u>is a Lie algebra of F-endomorphisms of</u> $A \oplus C$, <u>stabilizing each summand. We have a nondegenerate antisymmetric F-bilinear pairing</u> $q: C \times C \mapsto A$, <u>and a symmetric F-bilinear pairing</u> $< , > : C \times C \mapsto D$ <u>such that</u>:
$[x \otimes a, d] = x \otimes ad$: $[u \otimes c, d] = u \otimes cd$;
$[x \otimes a, y \otimes a'] = Tr(xy) < a,a' > + [xy] \otimes \frac{aa' + a'a}{2}$;
$[u \otimes c, x \otimes a] = ux \otimes ca$;
$[u \otimes c, v \otimes c'] = (u,v) < c,c' > + (u \circ v) \otimes q(c,c')$.
<u>Here</u> $< a, a' > \in D$ <u>sends</u> a'' <u>to</u> $f(f(a,a''),a') - f(a,f(a'',a'))$, <u>where</u> $2f(a,a') = aa' + a'a$, <u>and</u> $c \in C$ <u>to</u> $\frac{1}{2}((ca)a' - (ca')a)$. <u>Also</u>, $< c,c' > \in D$ <u>sends</u> a <u>to</u> $q(c,c'a) - q(ca,c')$, <u>and satisfies</u> (95), <u>where</u> $p(c,a) = ca$. <u>The elements of</u> D <u>act as derivations with respect to all compositions and moreover</u> (92) <u>and</u> (93) <u>hold</u>.

Chapter IV. **Data for Isomorphism.**

1. **Introduction; necessary conditions.**

Let L and M be simple Lie algebras over the field F of characteristic zero, and let Θ be an isomorphism of L onto M. Let T be a maximal split torus in L, so T^Θ is a maximal split torus in M. If L_0 is the centralizer of T in L, then L_0^Θ is the centralizer M_0 of T^Θ in M, and if Σ is the set of roots of L relative to T, then each $L_\gamma (\gamma \in \Sigma)$ is mapped by Θ onto a root-space M_γ^Θ of M relative to T^Θ. This yields a one-one mapping $\gamma \to \gamma^\Theta$ of Σ onto Σ', the set of roots of M relative to T^Θ.

From $[L_\gamma L_\delta]^\Theta = [M_\gamma^\Theta M_\delta^\Theta]$ and the fact that $\gamma + \delta (\neq 0)$ is in Σ if and only if $[L_\gamma L_\delta] \neq 0$, it follows that our map of Σ onto Σ' is an isomorphism of root-systems. That is, if $\gamma, \delta \in \Sigma$, then $\gamma + \delta \in \Sigma$ if and only if $\gamma^\Theta + \delta^\Theta \in \Sigma'$, and then $(\gamma + \delta)^\Theta = \gamma^\Theta + \delta^\Theta$; also $(-\gamma)^\Theta = -\gamma^\Theta$. In particular, if Π is a fundamental system for Σ, Π^Θ is one for Σ'. Moreover, for $\alpha_i \neq \alpha_j$ in Π, $A_{ij} = A_{\alpha_i, \alpha_j}$ is determined by the maximal string of roots $\alpha_i, \alpha_i + \alpha_j, \ldots, \alpha_i + (-A_{ij})\alpha_j$, so is equal to $A_{\alpha_i^\Theta, \alpha_j^\Theta}$, and $2\alpha_i$ is a root in Σ if and only if $2\alpha_i^\Theta$ is a root in Σ'. Thus the map of Π to $\Pi^\Theta = \Pi'$ induced by Θ gives an isomorphism of Dynkin diagrams, in the sense of §I.2.

As has been remarked in §I.3, we cannot hope to recover an isomorphism Θ as above from an isomorphism Θ_0 of L_0 onto M_0, mapping T onto a maximal split torus in M and inducing an isomorphism of root-systems. Further rational conditions necessary and sufficient for isomorphism have been given by Allison [A], and what is presented here is a modification of his arguments.

With the notations as in the first two paragraphs, let Θ_0 be the restriction of Θ to L_0. For each $\gamma \in \Sigma$, $L_\gamma^\Theta = M_\gamma^\Theta$ becomes an (irreducible) L_0-module by setting

$$X_\gamma^\Theta \cdot Z = [X_\gamma Z]^\Theta = [X_\gamma^\Theta Z^\Theta] = [X_\gamma^\Theta, Z^{\Theta_0}],$$

for $X_\gamma \in L_\gamma$, $Z \in L_0$, and the relations written show that the restriction of Θ to L_γ is an isomorphism of L_0-modules $L_\gamma \to M_\gamma^\Theta$.

If $\gamma \in \Sigma$, $0 \neq X_\gamma \in L_\gamma$, $0 \neq X_{-\gamma} \in L_{-\gamma}$, $H_\gamma = [X_\gamma X_{-\gamma}] \in T, \gamma(H_\gamma)=2$, then $\omega(w_\gamma) = \exp(\text{ad } X_\gamma) \exp(\text{ad } X_{-\gamma}) \exp(\text{ad } X_\gamma)$ is an automorphism of L,

stabilizing T and inducing in T the mapping fixing each T with $\gamma(T) = 0$ and sending H_γ to $-H_\gamma$ (§1.3). The contragredient (inverse-transpose) in T^* maps Σ to Σ and induces there the element w_γ of the Weyl group:
$\alpha^{w_\gamma} = \alpha - A_{\alpha,\gamma}\gamma, \alpha \in \Sigma$ (i.e., $<t^{\omega(w_\gamma)}, \alpha^{w_\gamma}> \; = \; <t,\alpha>$ for all $t \in T$, $\alpha \in \Sigma$). Our choice of $\omega(w_\gamma)$ evidently depends not only on γ, but also on X_γ, $X_{-\gamma}$, but we shall express properties used and/or assumed in such a way as to be independent of this choice. As an example, any element of L centralizing $L_\gamma + L_{-\gamma}$ is fixed by $\omega(w_\gamma)$; L_0 is stabilized by $\omega(w_\gamma)$, and if $\beta = \alpha^{w}\gamma$, we have $L_\alpha^{\omega(w_\gamma)} = L_\beta$ (see §I.3).

With the same notations, we have $X_\beta^\Theta \in M_\beta^\Theta$, $\beta = \pm\gamma$, $[X_\gamma^\Theta X_{-\gamma}^\Theta] = H_\gamma^\Theta \in T^\Theta$, $\gamma^\Theta(H_\gamma^\Theta) = 2$, so that $H_\gamma^\Theta = H_{\gamma^\Theta}$ and we may take
$$\omega(w_\gamma^\Theta) = \exp(\text{ad} X_\gamma^\Theta)\exp(\text{ad} X_{-\gamma}^\Theta)\exp(\text{ad} X_\gamma^\Theta),$$
an automorphism of M satisfying the same conditions relative to $T^\Theta, \Sigma', \gamma^\Theta$ as $\omega(w_\gamma)$ does relative to T, Σ, γ. Indeed, $\omega(w_\gamma^\Theta)$ as so chosen is equal to $\Theta^{-1}\omega(w_\gamma)\Theta$, so that in particular by restricting all mappings to L_0 resp. M_0, we have that $\Theta_0 \omega(w_\gamma^\Theta)\Theta_0^{-1}\omega(w_\gamma)^{-1}$ is the identity automorphism of L_0, when $\omega(w_\gamma^\Theta)$ and $\omega(w_\gamma)$ are related as indicated. In the next sections, we develop properties of these mappings, without the underlined restriction.

§2. On certain canonical subalgebras.

We let P be a subset of Π, $<P>$ the set of roots which are linear combinations of P, $<P>^+$ (resp. $<P>^-$) the positive (resp. negative) elements of $<P>$. Let $G_+ = \Sigma_{\gamma \in <P>^+} L_\gamma$; G_- is defined analogously, and $G_0 = L_0$. We set $G = G_- + G_0 + G_+$, a direct sum of vector spaces. Let Z be the centralizer in L_0 of all $L_{\pm\alpha_i}$ ($\alpha_i \in P$), and set $L_0^P = \Sigma_{\gamma \in <P>}[L_\gamma L_{-\gamma}]$. Clearly Z^P and L_0^P are ideals in L_0. Moreover, G_+ (resp. G_-) is generated by those $L_{\alpha_i}, \alpha_i \in P$ (resp. by those $L_{-\alpha_i}, \alpha_i \in P$). [One needs only consider a minimal $\gamma \in <P>^+$ for which L_γ is not in this subalgebra; since $\gamma \notin \Pi$, $\gamma = \beta + \alpha_i$ for some positive β, some α_i. By linear independence of Π, $\beta \in <P>^+$ and $\alpha_i \in P$; thus L_β is in the subalgebra by induction, and so is $L_\gamma = [L_\beta L_{\alpha_i}]$, this last equality

145

holding by §I.1 whenever $A_{\gamma,\alpha_i} > 0$; but we may assume $A_{\gamma,\alpha_i} > 0$ since otherwise $A_{\gamma,\alpha_i} \leq 0$ for all $\alpha_i \in P$, from which $(\gamma,\gamma) \leq 0$, contrary to the positive-definiteness of the scalar product of §I.2 on X_Σ.] Using the same kind of argument, since $[L_\gamma L_{-\gamma}] \subseteq [L_{\alpha_i}, L_{-\alpha_i}] + [L_\beta L_{-\beta}]$ with γ, α_i, β as in the bracketed reasoning, we have $L_0^P = \Sigma_{\alpha_i \in P} [L_{\alpha_i} L_{-\alpha_i}]$. Thus Z^P centralizes all L_γ, $\gamma \in <P>$, and hence L_0^P as well. If $B(X,Y)$ is the Killing form of L, we have for

$Z \in L_0$, $X_\gamma \in L_\gamma$, $X_{-\gamma} \in L_{-\gamma}$, $B(Z,[X_\gamma X_{-\gamma}]) = B([ZX_\gamma], X_{-\gamma})$,

and this is identically zero in X_γ and $X_{-\gamma}$ if and only if Z centralizes L_γ; it follows that Z^P <u>is the orthogonal complement in</u> L_0 <u>of</u> L_0^P with respect to the Killing form of L.

If $Z \in Z^P \cap L_0^P$, Z can be written as a sum $\Sigma [X_\gamma X_{-\gamma}]$, Z centralizing all X_γ, $X_{-\gamma}$. It follows from [J] (Lemma 2.4) that $\text{ad } Z$ is a nilpotent transformation of L. But we have seen at the beginning of §I.3 that $\text{ad } Z$ is semisimple for each $Z \in L_0$; from this $\text{ad } Z = 0$, and $Z = 0$ since L is simple. Thus $L_0 = Z_0^P \oplus L_0^P$, a direct sum of ideals. Since L_0 acts reductively in L, we have $L_0 = Z(L_0) \oplus [L_0 L_0]$, with $Z(L_0)$ the center of L_0 and $[L_0 L_0]$ semisimple ([J], Chapter 2). If K is an ideal in L_0, the projection π of L_0 onto $[L_0 L_0]$ in the above decomposition has $[K^\pi, L_0] = [K, L_0]^\pi \subseteq K^\pi$, so K^π is an ideal of L_0 contained in $[L_0 L_0]$, so must be semisimple. Thus $K^\pi = [K^\pi, K^\pi] = [KK]$, since the $Z(L_0)$-components centralize L_0. Thus $K^\pi \subseteq K$, so K is <u>homogeneous</u> with respect to the decomposition:

$$K = (K \cap Z(L_0)) \oplus (K \cap [L_0 L_0]),$$

with $Z(K) = K \cap Z(L_0)$, $[KK] = K \cap [L_0 L_0]$. In particular, we may apply this to $K = Z^P$ or L_0^P.

Now let $L^P = L_0^P + G_+ + G_-$, a subalgebra of L. Since $L_0^P \cap T$ contains all $H_{\alpha_i}, \alpha_i \in P$, but meets $Z^P \cap T$, a subspace of T of dimension $|\Pi| - |P|$, only in zero, we see that $L_0^P \cap T$ has as basis the $H_{\alpha_i}, \alpha_i \in P$, and that $T = (L_0^P \cap T) \oplus (Z^P \cap T)$. Any ideal A in L^P is normalized by L_0^P and by Z^P, hence by L_0, and is therefore a sum of certain L_γ and of $A \cap L_0^P$. If $L_\gamma \subseteq A$, then $H_\gamma \in A$, and $L_\gamma = [L_\gamma H_\gamma] \subseteq [AA]$; if $\underline{no}\ L_\gamma \subseteq A$, $A \subseteq L_0^P$ and $[L_\gamma A] \subseteq A \cap L_\gamma = 0$ for all $\gamma \in \langle P \rangle$; but then $A \subseteq Z^P \cap L_0^P = 0$. It follows that L^P $\underline{\text{contains no abelian ideals}}$ $A \neq 0$, i.e., L^P $\underline{\text{is semi-simple}}$.

If $S \in L^P$ is semisimple (in its action on L^P), centralizing $L_0^P \cap T$, and with the characteristic roots of $\text{ad}_{L^P}(S)$ all in F, then, as in Chapters I, II, $\text{ad}_L(S)$ is also diagonalizable in F, and S centralizes $Z^P \cap T$, therefore all of T. But then $S \in L^P \cap T = L_0^P \cap T$, so $L_0^P \cap T$ $\underline{\text{is a maximal split torus in}}$ L^P. Since each $\gamma \in \langle P \rangle$ annihilates $Z^P \cap T$, $\langle P \rangle$ identifies with its set of restrictions to $L_0^P \cap T$, so that $\underline{\text{the}}$ $L_\gamma(\gamma \in \langle P \rangle)$ $\underline{\text{are the}}$ $\underline{\text{root-spaces of}}$ L^P $\underline{\text{relative to its maximal split torus}}$ $L_0^P \cap T$, $\underline{\text{whose centralizer is}}$ L_0^P.

In particular, the above considerations apply when $P = \{\alpha\}$ consists of a single $\alpha \in \Pi$. Then L_0^α is the sum of $L_0^\alpha \cap Z(L_0)$ and $L_0^\alpha \cap [L_0 L_0]$, the former being the center of L_0^α, the latter its derived algebra, and likewise for Z^α, the centralizer of $L_{\pm\alpha}$ in L_0. From $L_0 = Z^\alpha \oplus L_0^\alpha$, we have that

$$[L_0 L_0] = (Z^\alpha \cap [L_0 L_0]) \oplus (L_0^\alpha \cap [L_0 L_0]).$$

The former of these ideals is the annihilator of $L_{\pm\alpha}$ in $[L_0 L_0]$, while the latter is the sum of those simple summands of $[L_0, L_0]$ which act non-trivially on L_α or $L_{-\alpha}$. If Θ and Θ_0 are as in §1, we see from these characterizations that Θ_0 maps

$$Z^\alpha \cap [L_0 L_0] \text{ to } Z^{\alpha'} \cap [M_0, M_0],$$
$$L_0^\alpha \cap [L_0 L_0] \text{ to } M_0^{\alpha'} \cap [M_0 M_0].$$

Likewise, $Z(L_0) = (Z(L_0) \cap Z^\alpha) \oplus (Z(L_0) \cap L_0^\alpha)$.

Let $\omega(w_\alpha)$ be chosen as in §1. This is an automorphism of L stabilizing L_0 and fixing the centralizer of $L_\alpha + L_{-\alpha}$; hence in particular, $\omega(w_\alpha)$ fixes $Z(L_0) \cap Z^\alpha$. It also necessarily stabilizes $Z(L_0)$ and, since it interchanges L_α and $L_{-\alpha}$, $\omega(w_\alpha)$ stabilizes

147

$L_0^\alpha = [L_\alpha L_{-\alpha}]$. Thus $Z(L_0) \cap L_0^\alpha$ is stable, and $H_\alpha \in T \cap L_0^\alpha \subseteq Z(L_0) \cap L_0^\alpha$ is sent to its negative by $\omega(w_\alpha)$.

Now let $Z \in Z(L_0)$, and let $X_\alpha \in L_\alpha$, $X_{-\alpha} \in L_{-\alpha}$, $[X_\alpha X_{-\alpha}] = H_\alpha$, $\omega(w_\alpha)$ defined in terms of X_α, $X_{-\alpha}$ as before. Then, if $2\alpha \notin \Sigma$,

$$Z^{\omega(w_\alpha)} = (Z + [ZX_\alpha])\exp(\text{ad}X_{-\alpha})\exp(\text{ad}X_\alpha)$$
$$= (Z + [ZX_\alpha] + [ZX_{-\alpha}] + [[ZX_\alpha]X_{-\alpha}] + \tfrac{1}{2}[[[ZX_\alpha]X_{-\alpha}]X_{-\alpha}]).$$
$$\exp(\text{ad}X_\alpha) \equiv ([ZX_{-\alpha}] + \tfrac{1}{2}[[[ZX_\alpha]X_{-\alpha}]X_{-\alpha}])$$

(mod $L_0 + L_\alpha$). But $Z^{\omega(w_\alpha)} \in L_0$, so we have

(1) $\quad [[[ZX_\alpha]X_{-\alpha}]X_{-\alpha}] = -2[ZX_{-\alpha}]$. Since $Z \in Z(L_0)$,

$0 = [Z[X_\alpha X_{-\alpha}]] = [[ZX_\alpha]X_{-\alpha}] + [X_\alpha[ZX_{-\alpha}]]$, so that

$[[ZX_{-\alpha}]X_\alpha] = [[ZX_\alpha]X_{-\alpha}]$. Reversing the roles of X_α and $X_{-\alpha}$ in the above gives $[[[ZX_{-\alpha}]X_\alpha]X_\alpha] = -2[ZX_\alpha]$, and this combines with our last remark to give

(2) $\quad [[[ZX_\alpha]X_{-\alpha}]X_\alpha] = -2[ZX_\alpha]$.

Now, by (1), $Z^{\omega(w_\alpha)} = (Z + [ZX_\alpha] + [[ZX_\alpha]X_{-\alpha}])\exp(\text{ad}X_\alpha)$
$= Z + 2[ZX_\alpha] + [[ZX_\alpha]X_{-\alpha}] + [[[ZX_\alpha]X_{-\alpha}]X_\alpha]$
$= Z + [[ZX_\alpha]X_{-\alpha}]$, by (2). Thus $[[ZX_\alpha]X_{-\alpha}]$

is in $Z(L_0)$ in this case.

Continuing to assume $2\alpha \notin \Sigma$, we form $Z^{\omega(w_\alpha)^2}$, which by the above is $Z + 2[[ZX_\alpha]X_{-\alpha}] + [[[[ZX_\alpha]X_{-\alpha}]X_\alpha]X_{-\alpha}] = Z$, by substitution from (2). Thus $\omega(w_\alpha)^2$ is the identity on $Z(L_0)$. If $Z \in Z(L_0)$ is fixed under $\omega(w_\alpha)$, we see by the above that

$$[[ZX_\alpha]X_{-\alpha}] = 0 = [[ZX_{-\alpha}]X_\alpha].$$

Since $\text{ad}X_\alpha$ is injective on $L_{-\alpha}$ and $\text{ad}X_{-\alpha}$ on L_α, we have $[ZX_\alpha] = 0 = [ZX_{-\alpha}]$. But $\text{ad}Z$ is in the centralizer of $\text{ad}L_0$ on $L_{\pm\alpha}$, where $\text{ad}L_0$ acts irreducibly, by Lemma 7 of Chapter I applied to $\alpha = \beta$. Then Schur's lemma implies that $Z \in Z^\alpha$. That is,

(3) The <u>fixed space of</u> $\omega(w_\alpha)$ <u>in</u> $Z(L_0)$ <u>is</u> $Z^\alpha \cap Z(L_0)$.

Since $\omega(w_\alpha)$ is diagonalizable with eigenvalues ± 1 in $Z(L_0)$ and stabilizes $L_0^\alpha \cap Z(L_0)$, we have:

(4) The (-1) - <u>space of</u> $\omega(w_\alpha)$ <u>in</u> $Z(L_0)$ <u>is</u> $L_0^\alpha \cap Z(L_0)$.

These assertions have only been proved here when 2α is not a root. When 2α is a root, considering the component in $L_{-2\alpha}$ of $Z^{\omega(w_\alpha)}$ gives

(5) $\quad \frac{1}{2} Z(adX_{-\alpha})^2 + \frac{1}{6}[ZX_\alpha](adX_{-\alpha})^3 + \frac{1}{24} Z(adX_\alpha)^2(adX_{-\alpha})^4 = 0.$

Eliminating this term before applying $\exp(adX_\alpha)$ for the second time gives, for the component in $L_{-\alpha}$,

(6) $\quad [ZX_{-\alpha}] + \frac{1}{2}[ZX_\alpha](adX_{-\alpha})^2 + \frac{1}{6} Z(adX_\alpha)^2(adX_{-\alpha})^3 = 0.$

Applying $adX_{-\alpha}$ to (6) and eliminating the last term against (5) gives

(7) $\quad\quad\quad\quad [ZX_\alpha](adX_{-\alpha})^3 = -6Z(adX_{-\alpha})^2.$

Eliminating the terms (5) and (6) before applying the final $\exp(adX_\alpha)$, and using that $Z^{\omega(w_\alpha)} \in L_0$ gives

(8) $\quad Z^{\omega(w_\alpha)} = Z + [[ZX_\alpha]X_{-\alpha}] + \frac{1}{2} Z(adX_\alpha)^2(adX_{-\alpha})^2.$

Now applying adX_α to the left-hand side of (7) and passing successively by the several $(adX_{-\alpha})$ gives (with some cancellation), $Z(adX_\alpha)^2(adX_{-\alpha})^3$, thus

$$Z(adX_\alpha)^2(adX_{-\alpha})^3 = -6Z(adX_{-\alpha})^2(adX_\alpha),$$

and "straightening" the right side yields

(9) $\quad Z(adX_\alpha)^2(adX_{-\alpha})^3 = -6Z(adX_\alpha)(adX_{-\alpha})^2 - 12[ZX_{-\alpha}].$

Combining (9) and (6), we have

(10) $\quad Z(adX_\alpha)(adX_{-\alpha})^2 = -2[ZX_{-\alpha}]$, which is the same as (1), and now yields

(11) $\quad Z^{\omega(w_\alpha)} = Z + [[ZX_\alpha]X_{-\alpha}]$ as before, and from this again that $\omega(w_\alpha)^2$ is the identity on $Z(L_0)$, so that (3) and (4) hold without restriction on α.

§3. The criteria for isomorphism.

We return to the setting of §1. First we deduce some further properties of the restrictions of the isomorphism Θ to L_0 and to certain root-spaces, and then turn these around to obtain sufficient conditions for extending an isomorphism $\Theta_0: L_0 \to M_0$ to an isomorphism of L onto M.

149

Let $\gamma \in \Sigma$, $\frac{1}{2}\gamma \notin \Sigma$, and with Θ as in §1 let $\gamma' = \gamma^\Theta \in \Sigma'$. Let $\exp(\operatorname{ad} Y_\gamma^\Theta) \exp(\operatorname{ad} Y_{-\gamma}^\Theta) \exp(\operatorname{ad} Y_\gamma^\Theta) = \omega(w_{\gamma'})$ as in §1, where $Y_\gamma \in L_\gamma$, $Y_{-\gamma} \in L_{-\gamma}$ canonically generate the three-dimensional algebra with basis Y_γ, $Y_{-\gamma}$, H_γ. Let X_δ, $X_{-\delta}$ ($\delta = \pm\gamma$) be in L_δ, $L_{-\delta}$, again canonically generating the 3-dimensional algebra with basis X_δ, $X_{-\delta}$, H_δ, hence with basis X_δ, $X_{-\delta}$, H_γ. Then $w_\delta = w_\gamma$. We let $\omega(w_\gamma) = \exp(\operatorname{ad} X_\delta) \exp(\operatorname{ad} X_{-\delta}) \exp(\operatorname{ad} X_\delta)$, and consider the restriction to L_0 of $\eta = \Theta_0\, \omega(w_{\gamma'}) \Theta_0^{-1} \omega(w_\gamma)^{-1}$. If we let

$$\omega_0 = \exp(\operatorname{ad} Y_\gamma) \exp(\operatorname{ad} Y_{-\gamma}) \exp(\operatorname{ad} Y_\gamma),$$ then

$\eta = [\Theta_0\, \omega(w_{\gamma'}) \Theta_0^{-1} \omega_0^{-1}] \omega_0\, \omega(w_\gamma)^{-1}$; we have seen in §1 that the factor in brackets is the restriction to L_0 of the identity automorphism $\Theta\omega(w_{\gamma'})\Theta^{-1} \omega_0^{-1}$ of L. As for the remaining factor, it evidently fixes the centralizer in L_0 of $L_{\pm\gamma}$ (which we have denoted in §2 by Z^γ, provided the fundamental system Π is chosen to contain γ - this is always possible since $\frac{1}{2}\gamma \notin \Sigma$; we shall use the notation of §2 freely with $\alpha = \gamma$; in this case it will be noted that the results on L^γ, etc., are really independent of the choice of Π with $\gamma \in \Pi$).

We have seen in §2 that each of ω_0, $\omega(w_\gamma)$ acts as -1 on $L_0^\gamma \cap Z(L_0)$; hence their quotient fixes $L_0^\gamma \cap Z(L_0)$, therefore fixes

$$Z(L_0) = (Z^\gamma \cap Z(L_0)) \oplus (L_0^\gamma \cap Z(L_0)),$$

and therefore η <u>fixes</u> $Z(L_0)$.

Moreover, $\omega_0 \omega(w_\gamma)^{-1}$ maps each L_β to itself, hence stabilizes each subspace of L_0 of the form

$$\Sigma_\beta [L_\beta L_{-\beta}],\text{ where }\beta\text{ runs over some subset of }\Sigma.$$

Therefore η does likewise; in particular, η <u>stabilizes</u> L_0^γ.

Composition of η with the adjoint representation of L_0 on L_β affords a second representation of L_0 on L_β, call it ρ, in which $X_\beta(X_0^\rho) = [X_\beta, X_0^\eta] = [X_\beta, X_0^{\omega_0 \omega(w_\gamma)^{-1}}]$. Thus we have $[X_\beta^{\omega_0 \omega(w_\gamma)^{-1}}, X_0^\eta] = [X_\beta X_0]^{\omega_0 \omega(w_\gamma)^{-1}}$, and the mapping $\omega_0 \omega(w_\gamma)^{-1}$ provides an equivalence <u>of the representation</u> ρ <u>of</u> L_0 <u>on</u> L_β <u>with the adjoint representation of</u> L_0 <u>on</u> L_β.

Now let Π be a fixed fundamental system in Σ. For each $w \in W(\Sigma)$, the Weyl group, we choose a fixed minimal representation as $w = w_{\alpha_1} \ldots w_{\alpha_m}$, $\alpha_i \in \Pi$ (not necessarily distinct). Our isomorphism $\Sigma \to \Sigma'$ of root-systems yields a minimal representation $w' = w_{\alpha_1'} \ldots w_{\alpha_m'}$ for the image w' of w in $W(\Sigma')$ under the induced isomorphism of Weyl groups, $\alpha_i' \in \Pi'$. We choose a canonical representative $\omega(w) = \omega(w_{\alpha_1}) \ldots \omega(w_{\alpha_m})$ as a product of canonical representatives, and a canonical representative $\omega(w') = \omega(w_{\alpha_1'}) \ldots \omega(w_{\alpha_m'})$. With Θ and Θ_0 as above, let $\Psi = \Theta_0 \omega(w') \Theta_0^{-1} \omega(w)^{-1}$, acting in L_0. Since Ψ is the restriction to L_0 of an automorphism of L stabilizing each L_β, it is clear that Ψ stabilizes each subspace of the form $\Sigma_\beta [L_\beta L_{-\beta}]$, running over a set of roots. We further show that two fundamental properties demonstrated above carry over from η to Ψ, namely that

i) $Z(L_0)$ is fixed by Ψ;

ii) The representations of L_0 on L_β ($\beta \in \Sigma$) in which X_0 is represented by $\mathrm{ad}\,X_0$ and by $\mathrm{ad}(X_0 \Psi)$ are equivalent.

Our proof of i) and ii) for η constitutes a proof of these assertions for Ψ when $m = 1$. For i), we may write
$$\Psi = \Theta_0 \omega(w') \Theta_0^{-1} (\omega(w')^{\Theta^{-1}})^{-1} \omega(w')^{\Theta^{-1}} \omega(w)^{-1},$$
where $\omega(w')^{\Theta^{-1}} = \Theta \omega(w') \Theta^{-1}$, so that $\Psi = \omega(w')^{\Theta^{-1}} \omega(w)^{-1}$

(12) $\qquad = \omega(w_{\alpha_1'})^{\Theta^{-1}} \ldots \omega(w_{\alpha_m'})^{\Theta^{-1}} \omega(w_{\alpha_m})^{-1} \ldots \omega(w_{\alpha_1})^{-1}.$

Now each $\omega(w_{\alpha_i'})^{\Theta^{-1}}$ maps L_0 to L_0, $Z(L_0)$ to $Z(L_0)$, and by our argument for $m = 1$,
$$\omega(w_{\alpha_m'})^{\Theta^{-1}} \omega(w_{\alpha_m})^{-1} \text{ fixes } Z(L_0). \text{ Hence } \Psi \text{ agrees on } Z(L_0) \text{ with}$$
$$\omega(w_{\alpha_1'})^{\Theta^{-1}} \ldots \omega(w_{\alpha_{m-1}'})^{\Theta^{-1}} \omega(w_{\alpha_{m-1}})^{-1} \ldots \omega(w_{\alpha_1})^{-1},$$
which, we may assume by induction, fixes $Z(L_0)$.

For ii), we may replace Ψ by (12) above. Setting $\mu_i = \omega(w_{\alpha_i'})^{\Theta^{-1}}$, $\nu_i = \omega(w_{\alpha_i})$, $1 \leq i \leq m$, we have that (12) is equal to $\mu_1 \ldots \mu_m \nu_m^{-1} \ldots \nu_1^{-1}$, and thus Ψ is equal to
$$\mu_1 \ldots \mu_{m-1}(\mu_m \nu_m^{-1})(\mu_1 \ldots \mu_{m-1})^{-1} \mu_1 \ldots \mu_{m-1} \nu_{m-1}^{-1} \ldots \nu_1^{-1}.$$

We set $\tau = \mu_1 \ldots \mu_{m-1}$, $L_\beta^\tau = L_\alpha$. If λ_1 is a linear automorphism of L_β satisfying

$$[X_\beta, X_0]^{\lambda_1} = [X_\beta^{\lambda_1}, X_0^{\mu_1 \ldots \mu_{m-1} \nu_{m-1}^{-1} \ldots \nu_1^{-1}}]$$

(the existence of such an equivalence being assumed by induction on m), and if λ_2 is a mapping $L_\alpha \to L_\alpha$, with

$$[X_\alpha, X_0]^{\lambda_2} = [X_\alpha^{\lambda_2}, X_0^{\mu_m \nu_m^{-1}}]$$

(the case $m = 1$, already done), then with $\tau = \mu_1 \ldots \mu_{m-1}$, we have

$$[X_\beta, X_0]^{\tau \lambda_2 \tau^{-1} \lambda_1}$$

$$= [X_\beta^\tau, X_0^\tau]^{\lambda_2 \tau^{-1} \lambda_1} = [X_\beta^{\tau \lambda_2}, X_0^{\tau \mu_m \nu_m^{-1} \tau^{-1} \lambda_1}]$$

$$= [X_\beta^{\tau \lambda_2 \tau^{-1}}, X_0^{\tau \mu_m \nu_m^{-1} \tau^{-1} \lambda_1}] = [X_\beta^{\tau \lambda_2 \tau^{-1} \lambda_1}, X_0^\Psi],$$

giving the desired equivalence ii).

The conditions i) and ii) are thus necessary conditions on the map $\Theta_0: L_0 \to M_0$ for Θ_0 to be extendible to an isomorphism Θ of L_0 to M_0. From §1 we also have the necessary conditions, implicit in the definition of the mappings η, Ψ, that Θ_0 map T to a maximal split torus U in M and induce an isomorphism $\alpha \to \alpha'$ of root-systems. Finally, for each α, there will necessarily be a linear isomorphism $\Theta_\alpha: L_\alpha \to M_{\alpha'}$ satisfying $[X_\alpha, X_0]^{\Theta_\alpha} = [X_\alpha^\Theta, X_0^{\Theta_0}]$ for all $X_\alpha \in L_\alpha$, $X_0 \in L_0$. Allison's theorem states that it is enough to assume the existence of such a Θ_α for one root α of each of the different root-lengths (at most three) present in Σ. Our version replaces the versions of i) and ii) utilized as hypotheses in Allison's work, namely that the maps Ψ (or some maps defined slightly more generally) are in the component of the identity of the group of automorphisms of L_0 fixing $Z(L_0)$. His hypothesis is less cumbersome, but necessitates an appeal to groups over the algebraic closure. I have tried to stress just those properties actually used in his argument.

The Isomorphism Theorem ([A], Theorem 9.1). Let L and M be simple Lie algebras over the field F of characteristic zero. Let T and U be maximal split tori with centralizers L_0 and M_0. Let $\Theta_0: L_0 \to M_0$

be an isomorphism mapping T onto U and inducing an isomorphism
$\alpha \to \alpha'$ of root-systems by contragredience. Suppose further that Θ_0 is
compatible with (a fixed minimal representation for) each element of the
Weyl group of L with respect to T, in the sense of i) and ii)
above, as well as that if $w = w_\gamma$, $\gamma \in \Sigma$, i) and ii) hold for a
canonical representation of $\omega(w_\gamma)$ and for one of $\omega(w_{\gamma'})$; moreover the
resulting Ψ shall fix $Z^\gamma \cap L_0$.

Let δ_i, $1 \le i \le t$, be roots representing the distinct root-
lengths present in Σ (thus $t \le 3$), and for each i, let $\Theta_i : L_{\delta_i} \to M_{\delta_i'}$
be an isomorphism of L_0-modules, using Θ_0 to make $M_{\delta_i'}$ into an
L_0-module.

Then there is an isomorphism $\Theta : L \to M$ extending Θ_0 and Θ_1,
and extending Θ_2 when $t = 3$, if the δ_i are indexed according to
increasing root-lengths.

A few remarks as to the conditions: That of compatibility assumes
i) and ii) when a particular minimal representation $w = w_{\alpha_1} \ldots w_{\alpha_m}$
has been chosen for $w \in W$, and then particular choices of $\omega(w_{\alpha_i})$,
$\omega(w_{\alpha_i'})$ in canonical form. In fact, it is independent of the choices of
these canonical representations for the $\omega(w_{\alpha_i})$, $\omega(w_{\alpha_i'})$. For if
$\Psi = \Theta_0 \, \omega(w_{\alpha_1'}) \ldots \omega(w_{\alpha_m'}) \Theta_0^{-1} (\omega(w_{\alpha_1}) \ldots (w_{\alpha_m}))^{-1}$ satisfies i)
and ii), and if $\{\omega^*(w_{\alpha_i}), \omega^{*\prime}(w_{\alpha_i'})\}$ are another set of canonical
representatives,
$\Psi^* = \Theta_0 \, \omega^*(w_{\alpha_1'}) \ldots \omega^*(w_{\alpha_m'}) \Theta_0^{-1} (\omega^*(w_{\alpha_1}) \ldots \omega^*(w_{\alpha_m}))^{-1}$ will likewise
satisfy them if we can show Ψ^* satisfies i) and ii) when:

a) $\omega^*(w_{\alpha_i}) = \omega(w_{\alpha_i})$, all i: b) $\omega^*(w_{\alpha_i'}) = \omega(w_{\alpha_i'})$, all i.
In case b), let $\mu = \omega(w_{\alpha_1'}) \ldots \omega(w_{\alpha_m'}) = \omega^*(w_{\alpha_1}) \ldots \omega^*(w_{\alpha_m})$,
$\nu = \omega(w_{\alpha_1}) \ldots \omega(w_{\alpha_m})$, $\nu^* = \omega^*(w_{\alpha_1}) \ldots \omega^*(w_{\alpha_m})$. Then $\Psi^* = \Psi \nu \nu^{*-1}$.
Since Ψ stabilizes L_0 and fixes $Z(L_0)$ and since we have seen above
that $\nu \nu^{*-1}$ fixes $Z(L_0)$, i) holds for Ψ^*. If $\beta \in \Sigma$, and if
$\eta : L_\beta \to L_\beta$ satisfies $[X_\beta^\eta, X_0^\Psi] = [X_\beta, X_0]^\eta$ for all $X_\beta \in L_\beta$, $X_0 \in L_0$,
then $\eta^* = \eta^{\nu \nu^{*-1}}$ satisfies $[X_\beta^{\eta^*}, X_0^{\Psi^*}] = [X_\beta, X_0]^{\eta^*}$ for all X_β, X_0.
Thus ii) holds for Ψ^*. The case a) is dealt with similarly, letting

$\mu^* = \omega^*(w_{\alpha_1'}) \ldots \omega^*(w_{\alpha_m'})$, so that $\Psi^* = \Theta_0 \mu^* \Theta_0^{-1} \nu^{-1} = \Theta_0 \mu^* \mu^{-1} \mu \Theta_0^{-1} \nu^{-1} = \Theta_0 \mu^* \mu^{-1} \Theta_0^{-1} \Psi$. Now Θ_0 maps L_0 to M_0, $Z(L_0)$ to $Z(M_0)$, and the latter is fixed by $\mu^* \mu^{-1}$, as above. It follows that i) holds for Ψ^*. We shall see below that:

(13) <u>If $\beta \in \Sigma$, there is a linear isomorphism</u> $\Theta_\beta : L_\beta \to M_\beta$ <u>such that for all</u> $X_\beta \in L_\beta$, $X_0 \in L_0$, $[X_\beta^{\Theta_\beta}, X_0^{\Theta_0}] = [X_\beta, X_0]^{\Theta_\beta}$.

With η as above $L_\beta \to L_\beta$, we set $\zeta = \Theta_\beta \mu^* \mu^{-1} \Theta_\beta^{-1} \eta$. Then for $X_\beta \in L_\beta$, $X_0 \in L_0$, we have

$$[X_\beta^\zeta, X_0^{\Psi^*}] = [X_\beta^{\Theta_\beta \mu^* \mu^{-1} \Theta_\beta^{-1} \eta}, X_0^{\Theta_0 \mu^* \mu^{-1} \Theta_0^{-1} \Psi}]$$

$$= [X_\beta, X_0]^{\Theta_\beta \mu^* \mu^{-1} \Theta_\beta^{-1} \eta} = [X_\beta X_0]^\zeta,$$

by repeated application of our equivalences and the fact that $\mu^* \mu^{-1}$ is an automorphism of M stabilizing L_0 and all L_β.

To verify (13), the simplicity of L implies (Lemma 2.18) that the system of roots Σ is indecomposable, hence that all roots of the same length are conjugate under the Weyl group ([BO],p. 151,Proposition 11). Therefore we have $w = w_{\alpha_1} \ldots w_{\alpha_m}$ with $\beta^w = \delta_j$, $1 \leq j \leq t$. Let $\Psi = \Theta_0 \omega(w_{\alpha_1'}) \ldots \omega(w_{\alpha_m'}) \Theta_0^{-1} (\omega(w_{\alpha_1}) \ldots \omega(w_{\alpha_m}))^{-1}$ satisfy our conditions of compatibility, and let η be the corresponding equivalence $L_\beta \to L_\beta$ of L_0-modules. Let

$\omega = \omega(w_{\alpha_1}) \ldots \omega(w_{\alpha_m}) \in \text{Aut}(L)$, $\omega' = \omega(w_{\alpha_1'}) \ldots \omega(w_{\alpha_m'}) \in \text{Aut}(M)$.

Then $\eta \omega \Theta_j \omega'^{-1} = \Theta_\beta$ maps L_β to M_β, and for $X_\beta \in L_\beta$, $X_0 \in L_0$,

$$\left[X_\beta^{\eta \omega \Theta_j \omega'^{-1}}, X_0^{\Theta_0}\right] = [X_\beta^{\eta \omega \Theta_j}, X_0^{\Theta_0 \omega'}]^{\omega'^{-1}}$$

$$= \left[X_\beta^{\eta \omega}, X_0^{\Theta_0 \omega' \Theta_0^{-1}}\right]^{\Theta_j \omega'^{-1}} = \left[X_\beta^\eta, X_0^\Psi\right]^{\omega \Theta_j \omega'^{-1}}$$

$$= \left[X_\beta, X_0\right]^{\eta \omega \Theta_j \omega'^{-1}},$$

giving that Θ_β is the desired equivalence of L_0-modules.

In particular, the existence of the equivalences Θ_β assures that Θ_0 maps the centralizer of L_β in L_0 to the centralizer of $M_{\beta'}$ in M_0, therefore maps Z^P to $Z^{P'}$ (definition at the beginning of §2). In §2, we also characterized $L_0^\alpha \cap [L_0 L_0]$ as the sum of those simple summands of $[L_0 L_0]$ which act nontrivially on L_α. From this characterization and the above, Θ_0 maps $L_0^\alpha \cap [L_0 L_0]$ to $M_0^{\alpha'} \cap [M_0, M_0]$. Finally $Z(L_0) \cap L_0^\alpha$ was characterized as the elements of $Z(L_0)$ mapped to their negatives by a canonical representative $\omega(w_\alpha)$. Then if $\omega(w_{\alpha'})$ is a canonical representative of $w_{\alpha'} \in W'$, $X_0 \in L_0^\alpha \cap Z(L_0)$, we have

$$X_0^{\Theta_0 \omega(w_{\alpha'})} = X_0^{\Psi \omega(w_\alpha) \Theta_0}, \quad \Psi = \Theta_0 \omega(w_{\alpha'}) \Theta_0^{-1} \omega(w_\alpha)^{-1}.$$

But Ψ fixes $Z(L_0)$, so $X_0^\Psi = X_0$, and the assertion follows. We summarize these consequences of the hypotheses of the isomorphism theorem (using that $L_0^\alpha = (Z(L_0) \cap L_0^\alpha) + ([L_0 L_0] \cap L_0^\alpha)$):

<u>Lemma 4.1.</u>   <u>Let the hypotheses and notations be as in the isomorphism theorem. Then</u>

   a) <u>For each</u> $\beta \in \Sigma$, <u>there is an equivalence of</u> L_0-<u>modules</u>
   $$L_\beta \to M_{\beta'}.$$

   b) <u>If</u> $P \subseteq \Pi$, Θ_0 <u>maps</u> Z^P <u>to</u> $Z^{P'}$.

   c) <u>If</u> $\alpha \in \Sigma$, $\frac{1}{2}\alpha \notin \Sigma$, Θ_0 <u>maps</u> L_0^α <u>to</u> $M_0^{\alpha'}$.

That the isomorphism $\Theta: L \to M$ cannot always be chosen to extend Θ_t as well is most easily seen in the case BC_1, where if Θ is to extend Θ_2, we must have $[X_\alpha Y_\alpha]^{\Theta_2} = [X_\alpha^{\Theta_1}, Y_\alpha^{\Theta_1}]$ if $\delta_1 = \alpha$, $\delta_2 = 2\alpha$. One might also ask for a criterion for isomorphism not specifying so much data to be extended. For algebraic groups, in terms involving data for Galois descent, other criteria have been given by Tits [TG], [TD] and Satake [SR], [SC]. What is there specified amounts to the isomorphism Θ_0, an action of the Galois group Γ of a common Galois splitting field K on the Dynkin diagrams of the algebras as split by K, and a Γ-isomorphism of these Dynkin diagrams compatible with one determined by Θ_0 in its restriction to the diagram of L_{0K}. It was the presentation of that approach in an earlier version of these lectures that suggested to Allison the desirability of "rational" criteria, and led to the approach here.

155

§4. Lemmas on the extension of isomorphisms.

Lemma 4.2. Let L and M be simple Lie algebras over F of characteristic zero. Assume that each is graded as a direct sum of subspaces $L_i(M_i)$, $-n \leq i \leq n$ ($n \geq 0$), and that L_i and L_j (M_i and M_j) are orthogonal with respect to the Killing form unless $i + j = 0$. Suppose that each $L_i(M_i)$ ($i \neq 0$) is a nontrivial $L_0 - (M_0-)$ module, and that $L(M)$ is generated by L_0 and $L_{\pm 1}$ (M_0 and $M_{\pm 1}$). Let Θ_0 be an isomorphism of L_0 onto M_0 (these are subalgebras), and for each i, $1 \leq i \leq n$, let Θ_i be an L_0-isomorphism of L_i onto M_i, the latter being considered as L_0-module by means of Θ_0. Then there is an isomorphism Θ of L onto M extending Θ_0 and Θ_1. If Θ is further required to map L_{-1} to M_{-1}, then Θ is unique.

Proof. We note that the graded structure of L (or M) involves the assumption $[L_i L_j] \subseteq L_{i+j}$, so that L_0 is a subalgebra, each L_i is an L_0-module and $[L_i L_j] = 0$ if $|i+j| > n$. We may assume $L_n \neq 0$, and therefore that $M_n \neq 0$, and that $n > 0$. The assumptions imply that the Killing form identifies L_{-i} with the dual space of L_i. Denoting the Killing form of each algebra by B, we have for $i > 0$, $X_{-i} \in L_{-i}$, that the linear function $Y_i \mapsto B(X_{-i}, Y_i^{\Theta_i^{-1}})$ on M_i has the form $Y_i \to B(Y_{-i}, Y_i)$ for a unique $Y_{-i} \in M_{-i}$. We define $\Theta_{-i} : L_{-i} \to M_{-i}$ by $X_{-i}^{\Theta_{-i}} = Y_{-i}$ as above; thus

$$B\left(X_{-i}^{\Theta_{-i}}, X_i^{\Theta_i}\right) = B(X_{-i}, X_i), \quad 1 \leq i \leq n.$$

Thus for $X_{-i} \in L_{-i}$, $X_0 \in L_0$, $X_i \in L_i$, we have

$$B\left(X_{-i}^{\Theta_{-i}}, [X_i X_0]^{\Theta_i}\right) = B(X_{-i}, [X_i X_0]) = -B([X_{-i} X_0], X_i)$$
$$= -B\left([X_{-i} X_0]^{\Theta_{-i}}, X_i^{\Theta_i}\right).$$

On the other hand, the assumption that Θ_i is an L_0-isomorphism means that $[X_i X_0]^{\Theta_i} = [X_i^{\Theta_i} X_0^{\Theta_0}]$, from which the above is equal to

$$B\left(X_{-i}^{\Theta_{-i}}, \left[X_i^{\Theta_i} X_0^{\Theta_0}\right]\right) = -B\left(\left[X_{-i}^{\Theta_{-i}} X_0^{\Theta_0}\right], X_i^{\Theta_i}\right).$$

Since the last members of these relations are equal for all values of X_0, $X_{\pm i}$, we see that Θ_{-i} is an isomorphism of L_0-modules $L_{-i} \to M_{-i}$.

Furthermore, the restriction of the Killing form to $L_0(M_0)$ is nondegenerate, and we have for X_i, X_0 as above,

$$B\left(\left[X_i^{\theta_i} X_{-i}^{\theta_{-i}}\right], X_0^{\theta_0}\right) = B\left(X_i^{\theta_i}, \left[X_{-i}^{\theta_{-i}} X_0^{\theta_0}\right]\right)$$

$$= B\left(X_i^{\theta_i}, [X_{-i}X_0]^{\theta_{-i}}\right) = B(X_i, [X_{-i}X_0]) = B([X_i X_{-i}], X_0).$$

Now the map $L \to M$ which is given by θ_j on L_j for each j is an isomorphism of L_0-modules, as we have seen. Hence the trace forms of the two representations of L_0 are equal. For the representation of L_0 on L, this form is the restriction to L_0 of the Killing form of L; for the representation on M, it is

$B\left(X_0^{\theta_0}, X_0'^{\theta_0}\right)$, B the Killing form of M. Thus in particular,

$B\left([X_i X_{-i}]^{\theta_0}, X_0^{\theta_0}\right) = B([X_i X_{-i}], X_0)$, and it follows from the remarks above that

(14) $$[X_i X_{-i}]^{\theta_0} = [X_i^{\theta_i}, X_{-i}^{\theta_{-i}}]$$

for all $X_i \in L_i$, $X_{-i} \in L_{-i}$.

In the direct sum $L \oplus M$ (whose elements we write as ordered pairs), let A be the Lie subalgebra generated by all elements in the union of the graphs of θ_0, $\theta_{\pm 1}$, i.e., all elements $(X_0, X_0^{\theta_0})$, $(X_1, X_1^{\theta_1})$, $(X_{-1}, X_{-1}^{\theta_{-1}})$ and let N be the smallest A-submodule of $L \oplus M$ containing all $z_V = (V, V^{\theta_{-n}})$, $V \in L_{-n}$. Evidently N contains all linear combinations of elements

(15) $$z_V \, \text{ad}(W_1, Y_1) \text{ad}(W_2, Y_2) \ldots \text{ad}(W_s, Y_s),$$

where $W_j \in L_0 \cup L_1$ and $Y_j = W_j^{\theta_i}$, i having the appropriate value, 0 or 1. But the displayed elements (15) span N. To see this, it suffices to show that applying to (15) the adjoint of one of our generators for A of the form $(X_{-1}, X_{-1}^{\theta_{-1}})$ carries the element (15) into a linear combination of elements of the same type. We proceed by induction on s: for $s = 0$ the assertion is clear since $[L_{-n} L_{-1}] = 0 = [M_{-n} M_{-1}]$. Assuming the assertion for smaller values of s, first suppose $W_s \in L_0$; then

$$z_V \text{ad}(W_1, Y_1) \ldots \text{ad}(W_s, W_s^{\theta_0}) \text{ad}(X_{-1}, X_{-1}^{\theta_{-1}})$$
$$= z_V \text{ad}(W_1, Y_1) \ldots \text{ad}(X_{-1}, X_{-1}^{\theta_{-1}}) \text{ad}(W_s, W_s^{\theta_0})$$
$$+ z_V \text{ad}(W_1, Y_1) \ldots \text{ad}(W_{s-1}, W_{s-1}^{\theta_0}) \text{ad}([W_s X_{-1}], [W_s^{\theta_0} X_{-1}^{\theta_{-1}}])$$

The second term has as last factor $\mathrm{ad}([W_s X_{-1}], [W_s X_{-1}]^{\theta_{-1}})$, and hence the induction hypothesis applies to show that both terms are linear combinations of terms like (15).

When $W_s \in L_1$, we obtain two analogous terms, the last factor in the second being

$$\mathrm{ad}([W_s X_{-1}], [W_s^{\theta_1} X_{-1}^{\theta_{-1}}]) = \mathrm{ad}([W_s X_{-1}], [W_s X_{-1}]^{\theta_0}),$$

by (14). Again the inductive hypothesis yields our conclusion.

Evidently the element (15) is in $L_i \oplus M_i$, where $W_j \in L_1$ for exactly $n + i$ values of j. In particular, the element (15) is in $L_{-n} \oplus M_{-n}$ if and only if all $W_j \in L_0$, in which case it has the form

$$[[\ldots[(V, V^{\theta_{-n}}), (W_1, W_1^{\theta_0})]\ldots](W_s, W_s^{\theta_0})]$$
$$+ ([[\ldots[V, W_1]\ldots]W_s], [[\ldots[V^{\theta_{-n}} W_1^{\theta_0}]\ldots]W_s^{\theta_0}])$$
$$= ([[\ldots[V, W_1]\ldots]W_s], [[\ldots[V, W_1]\ldots]W_s]^{\theta_{-n}}).$$

The remarks above and this computation enable us to deduce that N is homogeneous with respect to the decomposition of $L \oplus M$ as the direct sum of the $L_i \oplus M_i$, and that the component of N in $L_{-n} \oplus M_{-n}$ is exactly the graph of θ_{-n}. Since this is a proper subspace of $L_{-n} \oplus M_{-n}$, N is a proper A-submodule of $L \oplus M$.

It follows that $A \neq L \oplus M$. For otherwise N would be an ideal in $L \oplus M$. Since L_{-n} is by assumption a non-trivial L_0-module, we have $[VH] \neq 0$ for some $V \in L_{-n}$, $H \in L_0$. But then $[(V, V^{\theta_{-n}}), (H, 0)] \in N$, and this is $([VH], 0)$, contrary to our conclusion that $N \cap (L_{-n} \oplus M_{-n})$ is the graph of θ_{-n}. Thus $A \neq L \oplus M$; however, it follows, by the assumption that L_0 and $L_{\pm 1}$ (M_0 and $M_{\pm 1}$) generate $L(M)$, that the projection of A on the first (resp. second) summand is all of L (resp. M). We now show that each of these projections is an *isomorphism*.

If this conclusion fails, we may assume without loss of generality that A contains a non-zero element of the form $(0, Z)$, and with it all

$$(0, Z)\mathrm{ad}(W_1, Y_1) \ldots \mathrm{ad}(W_s, Y_s), \quad W_i \in L_0 \cup L_1 \cup L_{-1},$$

$Y_i = W_i^{\theta_j}$ for the appropriate j among $0, \pm 1$. But the generation of M by $M_0, M_{\pm 1}$ and the simplicity of M show that the second components in such expressions span an ideal in M, which must be M. Thus $(0,M) \subseteq A$, and $A = L \oplus M$, since the first projection maps A onto L. Each projection is therefore an isomorphism, so that $L \cong A \cong M$ via an isomorphism $\Theta : L \to M$ derived from the projections, hence agreeing with Θ_i on L_i, $i = 0, \pm 1$. If Θ' is any isomorphism agreeing with Θ_i on L_i, $i = 0, 1$, and mapping L_{-1} to M_{-1}, the fact that Θ' preserves the Killing form yields that Θ' agrees with Θ_{-1} on L_{-1}, so that $\Theta' = \Theta$. This completes the proof of Lemma 2.

(This argument, an adaptation of the diagonal trick used in proving isomorphism of irreducible modules of the same highest weight, was suggested by the earlier modification of that trick used by Winter ([W], p. 89) in proving the isomorphism theorem for split semisimple Lie algebras. A similar trick is employed in the next lemma.)

For our second lemma on extensions, we let the notations $L, T, L_0, P, <P>, <P>^+, <P>^-, G_0, G_\pm, G$ be as in the first paragraph of §2. Then G_+ and G_- are nilpotent subalgebras of G, $[G_0, G_\pm] \subseteq G_\pm$.

<u>Lemma 4.3.</u> <u>Let</u> V <u>and</u> W <u>be</u> <u>G-modules in which</u> T <u>acts by F-diagonalizable transformations. Assume that</u> V <u>and</u> W <u>are isomorphic as</u> T<u>-modules. Let</u> V^+ <u>(resp.</u> W^+<u>) be the subspace annihilated by</u> G_+<u>, and assume further that</u>:

    a) <u>the</u> G-<u>module</u> V <u>(resp.</u> W) <u>is generated by</u> V^+ (resp. W^+);

    b) V^+ <u>is a weight-space</u> V_λ <u>for</u> T <u>in</u> V <u>and</u> $W^+ = W_\lambda$.

<u>Let</u> $\eta : V^+ \to W^+$ <u>be an isomorphism of</u> G_0-<u>modules. Then</u> η <u>has a unique extension to an isomorphism</u> $V \to W$ <u>of</u> G-<u>modules.</u>

Proof. Let U be the G-submodule of the G-module $V \oplus W$ generated by all (v, v^η), $v \in V^+$. Then U evidently contains all
$$(vX_1 X_2 \cdots X_s, v^\eta X_1 \cdots X_s), X_i \in (\cup_{\gamma \in <P>^-} L_\gamma) \subseteq G_-.$$
Let U' be the subspace of $V \oplus W$ generated by these; we show $U = U'$, for which it suffices to show that U' is a G-submodule. To do this, it is enough to show $(vX_1 \cdots X_s Y, v^\eta X_1 \cdots X_s Y) \in U'$ whenever $v, X_1, \ldots, X_s$ are as above, and when: i) $Y \in L_0$; ii) $Y = Z_\gamma \in L_\gamma$, $\gamma \in <P>^+$. In case i), if $X_j \in L_\beta$, $[X_j Y] \in L_\beta$, and we have

(16) $\quad (vX_1 \ldots X_s Y, v^n X_1 \ldots X_s Y)$

$= \sum_{j=1}^{s} (vX_1 \ldots [X_j Y] \ldots X_s, v^n X_1 \ldots [X_j Y] \ldots X_s)$

$+ (vYX_1 \ldots X_s, v^n YX_1 \ldots X_s)$.

Since $v^n Y = (vY)^n$ for $Y \in G_0 = L_0$, the last term is in U', as are all terms in the summation, and i) is done.

In case ii), we also have the relation (16), in which the last term is now zero. We argue by induction on s, the conclusion being trivial if $s = 0$. For $s > 0$, each bracket $[X_j Y]$ is either in L_0 or in some L_γ, $\gamma \in <P>$. If $[X_j Y] \in L_0$, the corresponding term in (16) is in U' by the case i); if $[X_j Y] \in L_\gamma$, $\gamma \in <P>^-$, the term is in U' by the definition of U; if $[X_j Y] \in L_\gamma$, $\gamma \in <P>^+$, the term is in U' by inductive hypothesis. Hence $U = U'$.

The same argument shows that V (resp. W) is generated as vector space by all $vX_1 \ldots X_s$, $v \in V^+$ (resp. by all $x^n X_1 \ldots X_s$, $v \in V^+$), with $X_1, \ldots, X_s$ as above, and therefore that projection on the first (resp. second) factor maps U <u>onto</u> V (resp. W). As in the preceding lemma, we show that each of these projections is an isomorphism, and the lemma follows at once (uniqueness being trivial from the mode of generation of V resp. W).

Let U^+ be the subspace of U annihilated by G_+. We show U^+ is the graph of η. For if $X_i \in L_{\gamma_i} \subseteq G_-$, $v \in V^+ = V_\lambda$ as before, we have

$$vX_1 \ldots X_s \in V_{\lambda + \gamma_1 + \ldots + \gamma_s} \text{, so that}$$

$$\Sigma \mu_{X_1, \ldots, X_s} vX_1 \ldots X_s \ (\mu_{X_1 \ldots X_s} \in F) \text{ is in } V^+ = V_\lambda$$

if and only if all components in other weight spaces cancel, <u>i.e.</u>, if and only if it is equal to $\mu_\phi v$. Now the general element of U has the form

$$\Sigma \mu_{v, X_1, \ldots, X_s} (vX_1 \ldots X_s, v^n X_1 \ldots X_s)$$

$$= \Sigma_v \Sigma_{X_1, \ldots, X_s} (\mu_{v, X_1, \ldots, X_s} vX_1 \ldots X_s, \mu_{v, X_1, \ldots, X_s} v^n X_1 \ldots X_s),$$

and is in $U^+ \subseteq V^+ \oplus W^+ = V_\lambda \oplus W_\lambda$ if and only if it is equal to $\Sigma_v (\mu_{v,\phi} v, \mu_{v,\phi} v^n) = (v_0, v_0^n)$, $v_0 = \Sigma_v \mu_{v,\phi} v$. This proves our assertion.

Finally, suppose, say, that $(0, w) \neq (0, 0)$ in U is in the kernel of the first projection. Ordering the weights of T in V and

160

W by the fact that the difference of any two is in our space X_Σ, ordered so that Π is the set of simple roots, we see that λ is the highest weight and that the actions of $Y_\gamma \in L_\gamma \subseteq G_+$ raise weights in V, W, U. Thus we can find $Y_1, \ldots, Y_s$ (possibly $s = 0$) in root-spaces $L_\gamma \subseteq G_+$ such that $wY_1 \ldots Y_s \neq 0$, $wY_1 \ldots Y_s \in W^+$. But then $(0,w)Y_1 \ldots Y_s = (0, wY_1 \ldots Y_s) \in U^+$, so that $wY_1 \ldots Y_s = 0^\eta = 0$, a contradiction. This completes the proof of Lemma 3.

§5. <u>Proof of the isomorphism theorem: Types A_1, BC_1, D, E.</u>

Lemma 2 suffices to prove the isomorphism theorem when L is of type A_1 or BC_1:

<u>Proposition 4.1.</u> <u>The isomorphism theorem is valid when T (and hence U) is of dimension one.</u>

<u>Proof.</u> The root system Σ is either $\{\pm\alpha\}$ (type A_1) or $\{\pm\alpha, \pm 2\alpha\}$ (type BC_1). Replacing α by $-\alpha$ if necessary, we may assume $\Pi = \{\alpha\}$ and that Θ_1 maps L_α to $M_{\alpha'}$. If Θ_2 maps $L_{-2\alpha}$ to $M_{-2\alpha'}$, we may use the duality of $L_{2\alpha}$ and $L_{-2\alpha}$, as in the proof of Lemma 2, to obtain an L_0-isomorphism of $L_{2\alpha}$ to $M_{2\alpha'}$. Thus we may assume Θ_2 maps $L_{2\alpha}$ to $M_{2\alpha'}$. Now let $L_i = L_{i\alpha}$, $M_i = M_{i\alpha'}$, $-2 \leq i \leq 2$. The hypotheses of Lemma 2 then follow from those of the isomorphism theorem, and from the structure of L (resp. M)- here no appeal to the clumsy "compatibility" is needed- and that Lemma yields an isomorphism $\Theta : L \to M$ extending Θ_0 and Θ_1. Clearly such an isomorphism maps $L_{-1} = L_{-\alpha}$ to $M_{-\alpha'} = M_{-1}$, and is therefore unique by Lemma 2.

For types D and E, we may prove the isomorphism theorem directly from our constructions in §III.2:

<u>Proposition 4.2.</u> <u>The isomorphism theorem is valid when L (and hence M) is of type $D_r(r \geq 4)$, E_6, E_7 or E_8.</u>

<u>Proof.</u> In Theorem 3.1 we have seen that there is a field extension A of F such that $L \cong S \otimes_F A$, where S is a split algebra of the given type over F. The torus T is identified with $T \otimes 1$, 1 the unit element of A, T a maximal (split) torus in S. Then $L_0 = T \otimes A$, and is commutative. With Θ_0, Θ_1 as in the hypotheses of the theorem, we may choose a fundamental system Π for Σ such that δ_1 is the highest root (there is only one root-length, so all roots are

conjugate). Then $U = T^{\Theta_0}$ is a maximal split torus in M, and as in §3.1 we may build a split subalgebra of M of the same type, with the image Π' of Π as fundamental system and U as maximal split torus. Calling this subalgebra S', the isomorphism theorem for split algebras yields an isomorphism $\Psi: S \to S'$ agreeing with Θ_0 on T, and mapping the root-space $S_\alpha (\alpha \in \Sigma)$ to $S'_{\alpha'}$. Using S' as we did S, we identify M with $S' \underset{F}{\otimes} B$, B another field extension of F, U with $U \otimes 1 = T^\Psi \otimes 1$, M_0 with $T^\Psi \otimes B$. The root-space L_α identifies with $S_\alpha \otimes A = x_\alpha \otimes A$ where x_α is an F-basis for S_α, and $M_{\alpha'}$ with $S'_{\alpha'} \otimes B = x_\alpha^\Psi \otimes B$. Thus if $\beta \in \Sigma$, $T \in \mathcal{T}$, we have

$$\beta'(T^\Psi) = \beta'(T^\Psi \otimes 1) = \beta'((T \otimes 1)^{\Theta_0}) = \beta(T \otimes 1) = \beta(T),$$

and therefore $H_\beta^\Psi = H_{\beta'}$ for $H_\beta, H_{\beta'}$ as usual in $\mathcal{T}, U$.

Let $T_1, \ldots, T_r$ be a basis for $\mathcal{T}$ dual to the basis $\Pi = \{\alpha_1, \ldots, \alpha_r\}$ for $\mathcal{T}^*$. Then $T_j \otimes A$ is the intersection of all Z^{α_i}, $i \neq j$, i.e., of the annihilators in L_0 of all $L_{\pm \alpha_i}$, $i \neq j$, also denoted previously by Z^{P_j}, $P_j = \Pi - \{\alpha_j\}$. By b) of Lemma 4.1, Θ_0 maps Z^{P_j} to $Z^{P_j'}$, $P_j' = \Pi - \{\alpha_j'\}$. Since $T_1^{\Theta_0} = T_1^\Psi, \ldots, T_r^\Psi$ are dual in U to Π', it follows that for $1 \leq j \leq r$ we have a map $\eta_j: A \to B$ with $(T_j \otimes a)^{\Theta_0} = T_j^\Psi \otimes a^{\eta_j}$. The maps η_j are evidently invertible and F-linear.

Furthermore, the map $\Theta_1: L_{\delta_1} \to M_{\delta_1'}$ yields a unique invertible F-linear map $\zeta: A \to B$ satisfying $(X \otimes a)^{\Theta_1} = X^\Psi \otimes a^\zeta$, where X is a basis for the subspace S_{δ_1} of S. For each j, the equivalence of the representations of L_0 on L_{δ_1} and on $M_{\delta_1'}$ via Θ_1 thus yields

$$[X \otimes a, T_j \otimes a']^{\Theta_1} = [(X \otimes a)^{\Theta_1}, (T_j \otimes a')^{\Theta_0}], \text{ or}$$

$$\delta_1(T_j) X^\Psi \otimes (aa')^\zeta = \delta'_1(T_j^\Psi) X^\Psi \otimes a^\zeta a'^{\eta_j}.$$

Since $\delta_1'(T_j^\Psi) = \delta_1(T_j)$, we have

(17) $$(aa')^\zeta = a^\zeta a'^{\eta_j}$$

for all $a, a' \in A$, whenever $\delta_1(T_j) \neq 0$. But Π was chosen so that δ_1 is the highest root, <u>viz</u>:

162

$\delta_1 = \alpha_1 + 2\alpha_2 + \ldots + 2\alpha_{r-2} + \alpha_{r-1} + \alpha_r$ for D_r; $\delta_1 = \alpha_1 + 2\alpha_2 + 3\alpha_3 + 2\alpha_4 + \alpha_5 + 2\alpha_6$ for E_6; $\delta_1 = \alpha_1 + 2\alpha_2 + 3\alpha_3 + 4\alpha_4 + 3\alpha_5 + 2\alpha_6 + 2\alpha_7$ for E_7; $\delta_1 = 2\alpha_1 + 3\alpha_6 + 4\alpha_3 + 5\alpha_4 + 6\alpha_5 + 4\alpha_6 + 2\alpha_7 + 3\alpha_8$ for E_8, in suitable labelings of Π. Thus $\delta_1(T_j) \neq 0$ for all j, and (17) holds for all j. Writing a for a', 1 for a in (17) gives $a^\zeta = c\,a^{\eta_j}$, $c = 1^\zeta$, and shows that all η_j are equal; we call the common map η, so $a^\eta = c^{-1}a^\zeta$ for all a, and $(aa')^\eta = c^{-1}(aa')^\zeta = c^{-1}a^\zeta a'^\eta = a^\eta a'^\eta$. Thus η is a homomorphism of A into B; being a linear map of A onto B, η <u>is an isomorphism of fields</u>. Thus Θ_0 sends

$$\Sigma T_j \otimes a_j \text{ to } \Sigma T_j^\psi \otimes a_j^\eta,$$ and Θ_1 sends $X \otimes a$ to $X^\psi \otimes ca^\eta$, $0 \neq c \in B$.

Now we may take δ_1 as first root in a new fundamental system $\delta_1, \ldots, \delta_r$ for S relative to T. Then $S \otimes A$ and $S' \otimes A$ are both split algebras of the same type over A, and there is an A-isomorphism ν of $S \otimes A$ onto $S' \otimes A$ sending $T_j \otimes 1$ to $T_j^\psi \otimes 1$ (this is necessarily the same as the map sending H_{δ_i} to $H_{\delta_i'}$ for all i) for all j, and $X \otimes 1$ to $X^\psi \otimes d$, where $d^\eta = c$, by the existence theorem for isomorphism in the split case. Combining ν with $\text{id} \otimes \eta : S' \otimes A \to S' \otimes B$, we have an F-isomorphism $\Theta : L \to M$ mapping $T_j \otimes a$ to $T_j^\psi \otimes a^\eta$, $X \otimes a$ to $X^\psi \otimes (ad)^\eta = X^\psi \otimes ca^\eta$. Thus Θ is the desired extension of Θ_0 and Θ_1, and the isomorphism theorem is proved in these cases.

Here the extension Θ is by no means unique (note the freedom in defining ν on $S_{\delta_2}, \ldots, S_{\delta_r}$). The field A is readily identified with the centroid of L, so it is not at all surprising that data leading to an isomorphism $L \to M$ yield an isomorphism $A \to B$.

§6. <u>Proof of the isomorphism theorem: the remaining reduced cases</u>.

Now assume L is reduced, of rank greater than one, and not of type $D_r (r \geq 4)$, E_6, E_7, E_8. Then the roots in any fundamental system Π may be so labeled that $(\alpha_i, \alpha_j) = 0$ unless $|i - j| \leq 1$, where $\Pi = \{\alpha_1, \ldots, \alpha_r\}$. We may assume α_1 is of minimal length among elements of Σ_1 and that α_r is of maximal length. With our given root δ_1 as in the hypotheses of the isomorphism theorem, we may choose our system

Π so that $\delta_1 = \alpha_1$. We prove the theorem in these cases by induction on r.

Take $P = \{\alpha_1, \ldots, \alpha_{r-1}\} \subseteq \Pi$, and note that in this case L^P is in fact <u>simple</u>: Since any non-zero ideal A in L^P is normalized by $L_0 = Z(L_0) \dotplus L_0^P$, A is the sum of its intersections with the L_γ and with L_0. The irreducibility of the action of $L_0 (= Z^P \dotplus L_0^P)$ on $L_\gamma (\gamma \in \langle P \rangle)$ implies that some $L_\gamma \subseteq A$ unless $A \subseteq L_0^P$. Once we have some $L_\gamma \subseteq A$, we may use automorphisms of L (stabilizing L^P and A) of the form $\exp(\text{ad} X_\beta)$, $\beta \in \langle P \rangle$ to assume $L_{\alpha_i} \subseteq A$ for some $\alpha_i \in P$; then the connectedness of the diagram and further such automorphisms yield $L_{\pm \alpha_j} \subseteq A$ for all $\alpha_j \in P$ of the same length as γ. But then these $H_{\alpha_j} \in A$, and the connectedness of the diagram of P yields some $L_{\alpha_i} \subseteq A$ for α_i of the <u>other</u> root-length in $\langle P \rangle$ (if such is present). In this case we conclude that $A = L^P$. On the other hand, if $A \subseteq L_0^P$, A must centralize all L_γ, $\gamma \in \langle P \rangle$, so $A \subseteq Z^P$. But we have seen that $L_0^P \cap Z^P = 0$, so $A = 0$ in this case. (Here and in what follows we use freely the notation and information developed in §2.)

<u>Proposition 4.3.</u> <u>The isomorphism theorem holds in the remaining reduced cases; with L^P as above, Θ_0 and Θ_1 extend to an isomorphism of L^P to M^P, and this in turn to an isomorphism of L to M.</u>

<u>Proof.</u> Θ_0 induces a map of P to the subset $P' = \{\alpha_1', \ldots, \alpha_{r-1}'\}$ of the fundamental system Π' for Σ'. By b) of Lemma 1, Θ_0 maps Z^P to $Z^{P'}$, and by c) of the same lemma, L_0^α goes to $M_0^{\alpha'}$ for each $\alpha \in P$, so that Θ_0 carries the sum of these, L_0^P, to $M_0^{P'}$. Thus $\Theta_{0_P}|L_0^P$ is an isomorphism onto $M_0^{P'}$ sending $T \cap L_0^P$ to $U \cap M_0^{P'}$ and inducing the map $\gamma \to \gamma'$ from $\langle P \rangle$ to $\langle P' \rangle$, the systems of roots. We check that the hypotheses of the isomorphism theorem hold for L_0^P, $M_0^{P'}$, $T \cap L_0^P$, $U \cap M_0^{P'}$, $\Theta_0 | L_0^P$, Θ_1, and (if P contains two root-lengths) for an equivalence of L_0-modules $\Theta_2: L_\beta \to M_{\beta'}$, as given by a) of Lemma 1, where $\beta \in \langle P \rangle$ has different length than has α_1. Only the compatibility presents any problems, and here we may use the fact that a minimal representation in $W = W(\Pi)$ for any element in the subgroup $(= W(P))$ generated by all w_α, $\alpha \in P$, involves only roots from P, thus is a minimal representation in $W(P)$ ([Bo], Proposition 7, p. 19). Thus, with respect to the

given representations for $w \in W(P)$, and with the maps $"\Psi"$ replaced by their restrictions to L_0^P (these being just the composites of the factors of $"\Psi"$ as restricted to L_0 or to $M_0^{P'}$), the condition ii) for compatibility is trivial, and the condition i) (now in the form $"Z(L_0^P)$ is fixed") follows since $Z(L_0^P) = L_0^P \cap Z(L_0)$ by §2. Also, if $\gamma \in <P>$, we have $Z \cap L_0^P \subseteq Z \cap L_0$, thus necessarily fixed by the $"\Psi"$ associated with a canonical representation for $\omega(w_\gamma)$ and one of $\omega(w_{\gamma'})$. The inductive hypothesis thus yields an isomorphism $\Theta_P : L^P \to M^{P'}$ extending $\Theta_0 \mid L_0$ and Θ_1.

We now use Lemmas 2 and 3 to obtain an extension of Θ_P to an isomorphism $\Theta: L \to M$, as follows: From the determination of root-systems, when each root of Σ is expressed in terms of $\Pi = \{\alpha_1, \ldots, \alpha_r\}$, the coefficient of α_r is $0, \pm 1, \pm 2$, and the coefficients ± 2 occur only in types G_2 and F_4 (recall that α_r is a long root). We obtain a grading of L: $L_{(0)} = L_0 + L^P$; for $i = \pm 1, \pm 2, L_{(i)}$ is to be the sum of those root-spaces L_γ such that the coefficient of α_r in γ is i. Then $L_{(0)}$ is a subalgebra, $[L_{(i)}, L_{(j)}] \subseteq L_{(i+j)}$, with $L_{(i+j)} = 0$ if $i + j$ is not in the indicated range, and $B(L_{(i)}, L_{(j)}) = 0$ if $i + j \neq 0$. Since $T \subseteq L_{(0)}$ acts nontrivially on each L_γ, $\gamma \neq 0$, each $L_{(i)}$ is a non-trivial $L_{(0)}$-module. All $L_{\pm\alpha_j}$ are contained in $L_{(0)} + L_{(1)} + L_{(-1)}$, so this subspace generates L. We have

$$L = \Sigma_{i=-2}^{2} \oplus L_{(i)},$$

and a corresponding decomposition for M. Now $L_0 = Z^P \oplus L_0^P$, with Z^P centralizing L^P, so we can extend Θ_P to $\Theta_{(0)} : L_{(0)} \to M_{(0)}$ by defining the extension to agree with Θ_0 on Z^P. Then $\Theta_{(0)}$ is an isomorphism of $L_{(0)}$ to $M_{(0)}$ extending Θ_0 and Θ_1. Lemma 2 will then yield the desired isomorphism $L \to M$, once we have maps $\Theta_{(i)} : L_{(i)} \to M_{(i)}$, $i = 1, 2$, satisfying the hypotheses of that Lemma.

To produce the $\Theta_{(i)}$, we appeal to Lemma 3, noting that $L_{(0)} = L_0 + \Sigma_{\gamma \in <P>} L_\gamma$ is a subalgebra G as in the hypotheses, with $G_0 = L_0$, G_+, G_- as in that Lemma (see the first paragraph of §2). Since L_γ and $M_{\gamma'}$ are isomorphic T-modules, it is clear that $L_{(i)}$ and $M_{(i)}$ are, for each i, and evidently T acts by diagonalizable transformations. We verify the hypotheses a) and b) of Lemma 3) for $L_{(i)}$ and $M_{(i)}$, $i = 1, 2$. Now $L_{(i)}^+$ is the sum of those $L_\gamma \subseteq L_{(i)}$ such that no $\gamma + \beta \in \Sigma$ for $\beta \in <P>^+$. Except when Σ is of

165

type G_2 or F_4, only $i = 1$ is admissible, so that $\gamma + \alpha_r \notin \Sigma$ for $L_\gamma \subseteq L_{(1)}$, and thus $L_\gamma \subseteq L^+_{(1)}$ if and only if no $\gamma + \alpha_i (\alpha_i \in \Pi)$ is a root, i.e., if and only if γ is the highest root α_0. Similar considerations show that $L^+_{(2)} = L_{\alpha_0}$ in types G_2 and F_4. In these latter types, one verifies by examining root-systems that $L^+_{(1)} = L_{3\alpha_1 + \alpha_2}$ for G_2, $L^+_{(1)} = L_{2\alpha_1 + 4\alpha_2 + 3\alpha_3 + \alpha_4}$ for F_4. Thus the hypothesis b) is satisfied. The hypothesis a) is verified by checking that every root γ with $L_\gamma \subseteq L_{(i)}$ is obtained by successive subtraction of roots $\alpha_j \in P$ from the root β with $L_\beta = L^+_{(i)}$. Similar observations hold in M, with $M^+_{(i)} = M_{\beta'}$. But Lemma 1, a), gives an equivalence of $L_0 (= G_0)$-modules $L_\beta \to M_{\beta'}$, for each β, in particular of $L^+_{(i)}$ and $M^+_{(i)}$, so that Lemma 3 yields $L_{(0)}$-isomorphisms $L_{(i)}$ to $M_{(i)}$. This completes the proof of the isomorphism theorem in these cases.

§7. Conclusion of the proof: the remaining non-reduced cases.

Finally, consider the case where Σ is not reduced and $\dim T \geq 2$. Here there are three root-lengths, hence three roots δ_1, δ_2, δ_3, and maps Θ_1, Θ_2, Θ_3. We must produce an isomorphism $\Theta : L \to M$ extending Θ_0, Θ_1, Θ_2. By a suitable choice of a fundamental system Π for Σ, $\Pi = \{\alpha_1, \ldots, \alpha_r\}$, we may assume $\delta_1 = \alpha_1 + \ldots + \alpha_r$, where α_r is a shortest root (so that $2\alpha_r$ is a root). Then either δ_2 is in $<\alpha_1, \ldots, \alpha_{r-1}>$, or $\delta_2 = \pm (\alpha_i + \ldots + \alpha_{j-1} + 2\alpha_j + \ldots + 2\alpha_r)$, $i < j \leq r$, by our knowledge of Σ. Since all $w_{\alpha_j} (j > 1)$ fix δ_1, we may replace Π by its image under an element of the group generated by these w_{α_j} to assume δ_1 and Π are as above, with $\delta_2 \in <P>$, $P = <\alpha_1, \ldots, \alpha_{r-1}>$.

Proposition 4.4. <u>The isomorphism theorem holds in the remaining non-reduced cases; with</u> L^P <u>defined using</u> P <u>as above,</u> Θ_0 <u>and</u> Θ_2 <u>extend to an isomorphism</u> Θ_P <u>of</u> L^P <u>to</u> $M^{P'}$, <u>with</u> $L_{(0)}$, $L_{(i)} (i = \pm 1, \pm 2)$ <u>defined as in</u> §6, Θ_1 <u>extends to an</u> L^P <u>isomorphism of</u> $L_{(1)}$ <u>to</u> $M_{(1)}$, <u>and</u> Θ_0, Θ_P, Θ_1 <u>extend to an isomorphism</u> $\Theta : L \to M$.

Proof. As in the proof of Proposition 3, we have an isomorphism $\Theta_P : L^P \to M^{P'}$, extending $\Theta_0 | L_0^P$ and Θ_2, and extend Θ_P to $L_{(0)} = L_0 + \Sigma_{\gamma \in <P>} L_\gamma = G$ by $\Theta_P | Z^P = \Theta_0 | Z^P$. As in the proof of Proposition 3, we have $L_{(i)}$, $M_{(i)}$, $i = \pm 1, \pm 2$, and one checks roots to see that $L_{(1)}^+ = L_{\delta_1}$, $L_{(2)}^+ = L_{2\delta_1}$. By Lemma 3, Θ_1 extends to an $L_{(0)}$-isomorphism $\Theta_{(1)}: L_{(1)} \to M_{(1)}$, and an L_0-equivalence $L_{2\delta_1} \to M_{2\delta_1}$, which exists by a) of Lemma 1, extends to an $L_{(0)}$-isomorphism $\Theta_{(2)}: L_{(2)} \to M_{(2)}$. Then Lemma 2 applies to extend $\Theta_{(0)}$ and $\Theta_{(1)}$ to an isomorphism $\Theta: L \to M$, thus completing the proofs of the proposition and of the isomorphism theorem.

CHAPTER V

REALIZATIONS

The present chapter connects the constructions of simple Lie algebras in Chapter III with a number of realizations of simple Lie algebras, particularly with those as derived algebras of the skew elements of simple involutorial associative algebras. In the remaining cases, those of relatively low ranks, more precise conclusions than those of Chapter III are drawn concerning the structures (and, in particular, the dimensions) of the coordinatizing spaces A, B, C, D. Relations between these and various older constructions of exceptional simple Lie algebras are explored, and our constructions are shown in fact to yield simple Lie algebras with the claimed properties.

One test as to the effectiveness of our approach is whether, combined with some classical facts, it enables us to determine the real simple Lie algebras. That this test is successfully met will be seen in §8.

§1. <u>Centroids</u>.

Let $L = D \oplus (S \otimes A) \oplus (M' \otimes B) \oplus (M'' \otimes C)$ be as in Chapter III. We assume L to be **simple**, so that the centroid K of L, i.e., the set of those F-linear transformations of L centralizing all ad X, $X \in L$, is a finite extension field of F ([J], Chapter X). From the fact that K centralizes the action of S, it is clear that each $T \in K$ stabilizes each of our four summands. If we fix $a \in A$ and a basis $\{a_i\}$ for A, then for all $x, y \in S$,

$$[x \otimes a, y \otimes 1] = [xy] \otimes f_1(a,1) = [xy] \otimes a$$

yields, with $(x \otimes a)T = \Sigma m_i(x) \otimes a_i$, that

$$\Sigma m_i([xy]) \otimes a_i = [x \otimes a, y \otimes 1]T = [(x \otimes a)T, y \otimes 1]$$
$$= \Sigma [m_i(x), y] \otimes a_i .$$

As in §III.1 it follows that $(x \otimes a)T = x \otimes aX$, where $X: A \to A$ is F-linear. In all cases but those of types A_r, C_r, $BC_r (r \geq 2)$, we have

$$[x \otimes a, y \otimes a'] = (x,y) <a,a'> + [xy] \otimes f_1(a,a')$$

for suitable products and thus from

$$[x \otimes a, y \otimes a']T = [(x \otimes a)T, y \otimes a'] \text{ we find}$$

$$<aX, a'> = <a,a'>T, \quad f_1(aX, a') = f_1(a,a')X.$$

For A_r, $r \geq 2$, the linear independence of $[xy]$ and $x \circ y$ as mappings $S \to S$ yields the above and $f_2(aX, a') = f_2(a,a')X$. Thus we see, except in cases C_r, $BC_r (r \geq 2)$, that the action $a \mapsto aX$ centralizes multiplication on the left or right in the (internal) composition determined on A. This is still the case for C_r, $BC_r (r \geq 2)$, where we have

$$[x \otimes a, y \otimes a'] = (x,y) <a,a'> + [xy] \otimes f_1(a,a') + (x \circ y) \otimes f_2(a,a'),$$

the last term in $M' \otimes B$. Here we find

$$<a,a'>T = <aX,a'> \, ; \, f_1(aX,a') = f_1(a,a')X,$$

$$((x \circ y) \otimes f_2(a,a'))T = (x \circ y) \otimes f_2(aX,a').$$

Thus the mapping $T \mapsto X$ is a homomorphism of F-algebras (with unit) from K into the centroid of (A, f_1) (of $(A, f_1 + f_2)$ in type $A_r (r \geq 2)$), and (A, f_1) may be regarded as an algebra over K.

Similar considerations applied to the remaining summands yield mappings $T \mapsto Y$ resp. $T \mapsto Z$ from K to F-linear transformations of B resp. C:

$$(s \otimes b)T = s \otimes bY, \quad (u \otimes c)T = u \otimes cZ.$$

From relations like

$$[s \otimes b, (x \otimes a)T] = [s \otimes b, x \otimes a]T = [(s \otimes b)T, x \otimes a], \text{ we find:}$$

$$g_1(b, aX) = g_1(b,a)X = g_1(bY, a),$$

$$g_2(b, aX) = g_2(b,a)Y = g_2(bY, a),$$

$$<b,b'>T = <bY, b'> \, , \, <c,c'>T = <cZ, c'> \, ,$$

$$h_1(bY, b') = h_1(b,b')X, \quad h_2(bY, b') = h_2(b,b')Y,$$

$$p_1(cZ, a) = p_1(c,a)Z = p_1(c, aX),$$

$$p_2(cZ,b) = p_2(c,b)Z = p_2(c,bY),$$
$$q_1(cZ,c') = q_1(c,c')X, \quad q_2(cZ,c') = q_2(c,c')Y,$$
$$\text{and, with} \quad f_2(a,a') \in B, \quad f_2(aX,a') = f_2(a,a')Y,$$

for all cases where the compositions result as in Chapter III (the composition h_2 being omitted for C_2, BC_2). If $\tau = 1X \in A$, it follows from the above and from $g_2(b,1) = b$, $p_1(c,1) = c$, that $bY = g_2(b,\tau)$, $cZ = p_1(c,\tau)$, as well as that $aX = f_1(a,\tau)$, and

$$<A,\tau> \ = 0; \quad g_1(B,\tau) = 0; \quad f_2(A,\tau) = 0.$$

Moreover, we must have, for all a,a', b,b', c,c':

$$<a,a'> T = <f_1(a,\tau),a'>;$$
$$f_1(f_1(a,a'),\tau) = f_1(f_1(a,\tau),a');$$
$$f_1(f_2(a,a'),\tau) = f_2(f_1(a,\tau),a') \quad \text{or}$$
$$g_2(f_2(a,a'),\tau) = f_2(f_1(a,\tau),a'), \quad \text{according}$$

as $f_2: A \times A \to A$ or $A \times A \to B$;

$$f_1(g_1(b,a),\tau) = g_1(g_2(b,\tau),a) = g_1(b,f_1(a,\tau));$$
$$f_1(h_1(b,b'),\tau) = h_1(g_2(b,\tau),b');$$
$$f_1(q_1(c,c'),\tau) = q_1(p_1(c,\tau),c');$$
$$<b,b'> T = <g_2(b,\tau),b'>; \quad <c,c'> T = <p_1(c,\tau),c'>;$$
$$g_2(g_2(b,a),\tau) = g_2(g_2(b,\tau),a) = g_2(b,f_1(a,\tau));$$
$$g_2(h_2(b,b'),\tau) = h_2(g_2(b,\tau),b');$$
$$p_1(p_1(c,\tau),a) = p_1(p_1(c,a),\tau) = p_1(c,f_1(a,\tau));$$
$$p_2(p_1(c,\tau),b) = p_1(p_2(c,b),\tau) = p_2(c,g_2(b,\tau));$$
$$g_2(q_2(c,c'),\tau) = q_2(p_1(c,\tau),c').$$

From $<A,\tau> = 0$ and the formulas for $<a,a'>$, we see that <u>any f_1-product of three factors, one of which is</u> τ, <u>is associative, and that</u> $\tau \in A$ <u>has its</u> g_2- resp. p_1-<u>action on</u> B resp. C <u>commuting with those of all other elements of</u> A. Thus B and C <u>are also K-vector spaces, on which</u> A <u>acts</u> (via g_2 or p_1) <u>as K-linear transformations</u>. Those relations listed above and not involving pointed brackets say that <u>all our compositions denoted by</u> f_i, g_i,

170

h_i, p_i, q_i are K-bilinear. Except for type BC_1, the actions of $<A,A>$, $<B,B>$, $<C,C>$ on A,B,C are given in terms of these compositions, and hence are also K-linear; the trilinearity of forms like $c<b,b'>$ over K and the above formula for quantities like $<b,b'>T$ says that the action of $T \in K$ on an element of D (regarded as endormorphism of $A \oplus B \oplus C$) is that of scalar multiplication by τ. This also holds for BC_1, as one sees from a more general class of observations: if $c \in C$, $d \in D$, and if $T \in K$ is as above, then $[u \otimes c, d]T = (u \otimes cd)T = u \otimes p_1(cd,\tau)$; but also $[u \otimes c, d]T = [u \otimes c, dT] = u \otimes c(dT)$, and $[u \otimes c, d]T = [(u \otimes c)T, d] = u \otimes p_1(c,\tau)d$. Comparing the first and third of these shows that d acts K-linearly, and the second, that the action of T on d is by scalar multiplication by τ. Then the listed relations involving pointed brackets show that these are also K-bilinear compositions. Thus all operations and parameter-spaces (A,B,C,D) admit K as ground field, and K can be identified with a subfield of (A,f_1).

The subfield in question is identified with certain $\tau \in A$ such that:

   i) $f_2(A,\tau) = 0$, $g_1(B,\tau) = 0$;
   ii) τ is in the center (i.e., associates universally) of (A,f_1);
   iii) the p_1-action of τ on C centralizes that of A.

These conditions are generally redundant; for instance, if $C \neq 0$, (A,f_1) identifies with a (special) Jordan algebra of F-endomorphisms of C via the p_1-action, so that if τ satisfies iii), it also satisfies ii). All the conditions hold for all $\tau \in A$ whenever (A,f_1) is a field, a case that we have frequently encountered (indeed the only exceptions are of types A_r, C_r, BC_r). When (A,f_1) is not a field, and $r > 1$, f_2 is just a commutator in an associative (or possible alternative) algebra, as is g_1; then i) says, for type A_r, that τ commutes with everything in the associative or alternative algebra, and ii) follows from this, while iii) is vacuous. Otherwise, when $r > 1$ and (A,f_1) is not a field, i) means that $\tau \in A$ is central in the associative division algebra $A \oplus B$. For $r = 1$, i) is vacuous, as is iii) for A_1, while ii) and iii) are consequences of iii) for BC_1.

We claim that i) - iii) characterize the centroid of L, more specifically, that:

Theorem V.1. The centroid of L identifies with:

α) A, if A is a field (with product f_1);

β) the center of $(A, f_1 + f_2)$ if L is of type A_r (if $r = 1$, $f_2 = 0$);

γ) the elements of A central in the associative division algebra $A \oplus B$ if the type is C_r or BC_r, $r > 1$, and α) does not hold;

δ) the elements of A centralizing the action of A on C, if the type is BC_1.

To complete the proof of the theorem, it suffices to show that if $\tau \in A$ satisfies the appropriate condition from α) – δ), then there is a unique element T of the centroid of L extending the map $x \otimes a \to x \otimes f_1(a, \tau)$ of $S \otimes A$ to $S \otimes A$. When (A, f_1) is a field, we have seen in our separate considerations that B, C are vector spaces over A, D acts as A-linear transformations of $B \oplus C$, and all compositions are A-multilinear; the assertion is thus trivial in these cases.

In each case, our extension T must send $s \otimes b$ to $s \otimes g_2(b, \tau)$, $u \otimes c$ to $u \otimes p_1(c, \tau)$, $<a, a'>$ to $<f_1(a, \tau), a'>$, $<b, b'>$ to $<g_2(b, \tau), b'>$ and $<c, c'>$ to $<p_1(c, \tau), c'>$, so is evidently unique, if it exists. There is no problem with extending our mapping to $(S \otimes A) \oplus (M' \otimes B) \oplus (M'' \otimes C)$ by these formulas. In case β), we have

$$[(x \otimes a)T, x' \otimes a'] = [x \otimes f_1(a, \tau), x' \otimes a']$$

$$= (x, x') <f_1(a, \tau), a'>$$

$$+ [xx'] \otimes f_1(f_1(a, \tau), a') + (x \circ x') \otimes f_2(f_1(a, \tau), a').$$

When $r = 1$, f_2 is missing and $f_1(f_1(a, \tau), a') = f_1(f_1(a, a'), \tau)$ by the fact that τ is central. It also follows that D stabilizes, hence annihilates, the center $Z(A)$ (all derivations of A do so), so that all $d \in D$ satisfy $f_1(ad, \tau) = f_1(a, \tau)d$ whenever $a \in A$, $\tau \in Z(A)$. Thus D acts $Z(A)$-linearly on A, and, moreover, $D = \text{Der}(A)$ shows that D is a $Z(A)$-vector space under scalar multiplication. This also follows from $D = <A, A>$ and

$$f_1((a'' <a,a'>),\tau) = f_1(a'',\tau) \ <a,a'> =$$

$$\lambda(f_1(f_1(f_1(a'',\tau),a),a') - f_1(f_1(f_1(a'',\tau),a'),a))$$

$$= \lambda(f_1(f_1(a'',f_1(\tau,a)),a') - f_1(f_1(f_1(a'',a'),\tau),a))$$

$$= \lambda(f_1(f_1(a'',f_1(a,\tau)),a') - f_1(f_1(a'',a'),f_1(a,\tau)))$$

$$= a'' \ < f_1(a,\tau),a' > \ , \quad \text{for suitable } \lambda \neq 0 \text{ in } F.$$

The last computation also shows that scalar multiplication of $<a,a'>$ by τ amounts to forming $<f_1(a,\tau),a'>$. Thus if we extend T to $\mathcal{D}$ by defining dT to be the composite of d and scalar multiplication by τ, we have $<a,a'>T = <f_1(a,\tau),a'>$. Thus

$$[x \otimes a, x' \otimes a']T = (x,x') \ <a,a'> \ T$$

$$+ ([xx'] \otimes f_1(a,a'))T$$

$$= (x,x') < \ f_1(a,\tau),a' > \ + [xx'] \otimes f_1(f_1(a,a'),\tau)$$

$$= [(x \otimes a)T, x' \otimes a'].$$

That $[d,d']T = [dT,d']$ follows by the fact that all $d \in \mathcal{D}$ act $Z(A)$-linearly on A. Finally, we have $[x \otimes a,d]T = (x \otimes ad)T = x \otimes f_1(ad,\tau) = x \otimes f_1(a,\tau)d = [(x \otimes a)T,d]$, and $[x \otimes a, dT] = x \otimes a(dT) = x \otimes f_1(ad,\tau) = [x \otimes a,d]T$. Thus T as defined is actually in the centroid.

When $r > 1$, $f_1(a,\tau) = a\tau$, $f_1(f_1(a,\tau),a') = \frac{1}{2}((a\tau)a' + a'(a\tau)) = \frac{1}{2}(aa' + a'a)\tau = f_1(a,a')\tau$, $f_2(f_1(a,\tau),a') = \frac{1}{2}((a\tau)a' - a'(a\tau)) = \frac{1}{2}(aa' - a'a)\tau = f_2(a,a')\tau$, juxtaposition denoting our alternative product $f_1 + f_2$. Again the derivations of A annihilate $Z(A)$, so if $g(a'',a,a')$ is $a'' <a,a'>$, g being a word in the free alternative algebra, we have $g(a'',a\tau,a') = g(a'',a,a')\tau$, so that $<a\tau,a'> = <f_1(a,\tau),a'> = <a,a'> T$, where $T: \mathcal{D} \to \mathcal{D} \subseteq \text{Der}(A)$ consists of scalar multiplication by τ. We thus have

$$[x \otimes a, x' \otimes a']T = (x,x') \ <a,a'> \ T$$

$$+ [xx'] \otimes \left(\frac{aa'+a'a}{2}\right)\tau + (x \circ x') \otimes \left(\frac{aa'-a'a}{2}\right)\tau$$

$$= [(x \otimes a)T, x' \otimes a'], \text{ and}$$

$[dd']T = [dT,d']$, $[(x \otimes a)T,d] = [x \otimes a,d]T = [x \otimes a,dT]$ as for A_1. This completes the case β).

In case γ), we have $f_1(a,a') = \frac{1}{2}(aa' + a'a)$, $f_2(a,a') = \frac{1}{2}(aa' - a'a)$, $g_1(b,a) = \frac{1}{2}(ba - ab)$, $g_2(b,a) = \frac{1}{2}(ba + ab)$, $h_1(b,b') = \frac{1}{2}(bb' + b'b)$, $h_2(b,b') = \frac{1}{2}(bb' - b'b)$ ($r > 2$).

The elements $\tau \in A$ with which we are concerned have $f_1(a,\tau) = a\tau = \tau a$, $g_2(b,\tau) = b\tau = \tau b$, so that our T sends $x \otimes a$ to $x \otimes a\tau$, $s \otimes b$ to $s \otimes b\tau$; τ evidently is annihilated by all derivations of $A \oplus B = E$, from which it follows in case C_r that an action T of scalar multiplication by τ is defined on D, identified as those derivations of E commuting with the involution. Here we have

$$((a + b) <b',b''>)\tau = ((a + b)\tau) <b',b''> =$$

$$\lambda[a\tau[b'b'']] + \mu[b\tau[b',b'']] = \lambda[a[b'\tau,b'']]$$

$$+ \mu[b[b'\tau,b'']] = (a + b) <b'\tau,b''> \text{ for}$$

fixed λ,μ, determined in §6. Thus $<b',b''> T = <b'\tau,b''>$, and likewise $<a',a''> T = <a'\tau,a''>$. It follows as before that T is in the centroid of L. (For C_r with $r = 2$, the map h_2 had to be defined otherwise, but there is no distinction to be drawn in our argument.)

In case BC_r ($r > 1$), with A not a field, we see as before that $(c\tau)d = (cd)\tau$ for all $d \in D$, $c \in C$, and τ as above. Now $(c\tau) <a,a'> = \lambda(((c\tau)a)a' - ((c\tau)a')a) = \lambda(c(\tau[a,a'])) = \lambda c[a\tau,a']$ $= c <a\tau,a'>$, and likewise for $(c\tau) <b,b'> = c <b\tau,b'>$. Thus scalar multiplication by τ maps $<A,A>$ and $<B,B>$ to themselves in an appropriate way. We also have $((a + b)\tau) <c,c'> =$ $\lambda q(c',c(a + b)\tau) + \mu q(c,c'(a + b)\tau) = \lambda q(c',c\tau(a + b)) + \mu q(c,c'(a + b))\tau$. Now $q(c,c'(a + b))\tau = \tau q(c,c'(a + b)) = \tau^* q(c,c'(a + b))$ $= q(c\tau,c'(a + b))$, so the above is $(a + b) <c\tau,c'>$. Finally, we have $(c''\tau) <c,c'> =$ $- c'q(c,c''\tau) - cq(c',c''\tau) - \frac{1}{2r}(c''\tau)(q(c,c') + q(c',c))$.

By the fact that τ is symmetric and central and q sesquilinear, it follows at once that $(c''\tau) <c,c'> = c'' <c\tau,c'>$. Thus T preserves $<C,C>$, sending $<c,c'>$ to $<c\tau,c'>$, if T is scalar multiplication (on $A \oplus B \oplus C$) by τ, and all the identities needed for showing our extension is in the centroid have been established.

Finally, in the case δ), our T must send a to $f_1(a,\tau)$, c to $p_1(c,\tau)$. Since these elements τ are contained in the field which is the centroid of A, they are annihilated by the action of D on A. As before, we define an action T of τ on D as scalar multiplication by τ (in the faithful module $A \oplus C$), and show what remains to be done by showing $<a,a'>T = <a\tau,a'>$, $<c,c'>T = <c\tau,c'>$, where $a\tau = f_1(a,\tau)$, $c\tau = p_1(c,\tau)$. That $a''(<a,a'>T) = (a''\tau)<a,a'> = a''<a\tau,a'>$ follows as in type A_1. Now we have $c(<a,a'>T) = (c\tau)<a,a'> = \lambda(((c\tau)a)a' - ((c\tau)a')a) = \lambda(((ca)\tau)a' - ((ca')a)\tau)$ by the property iii) of τ (and by (91) of §III.9). By (90) at the same stage, $2c(a\tau) = (ca)\tau + (c\tau)a = 2(ca)\tau$. Thus the above is $c<a\tau,a'>$, as required. From the same section, we find, using (94),

$$a(<c,c'>T) = (a\tau)<c,c'> = q(c,c'(a\tau)) - q(c(a\tau),c')$$
$$= q(c,(c'a)\tau) - q((c\tau)a,c') \quad \text{by the above.}$$

But $\tau<c,c'a> = 0$, which yields by (94) that $q(c,(c'a)\tau) = q(c\tau,c'a)$, and this last becomes $q(c\tau,c'a) - q((c\tau)a,c') = a<c\tau,c'>$. Finally, we must consider $(c''\tau)<c,c'>$, and show this is equal to $c''<c\tau,c'>$. By (92) of §III.9 and the fact that $<\tau,a> = <1,a>T = 0$, we find $<c\tau,c'> = <c,c'\tau>$. Now computing $c''<c\tau,c'> - c\tau<c',c''>$ and $c''<c'\tau,c> - c'\tau<c,c''>$ by (95) of §III.9, we find, since $<c'\tau,c> = <c,c'\tau> = <c\tau,c'>$, upon adding, that

$$2c''<c\tau,c'> = c\tau<c',c''> + c'\tau<c,c''>$$
$$+ 3c'\tau q(c'',c) + 3c\tau q(c'',c').$$

Likewise, we find, from $c''<c,c'> - c<c',c''>$ and $c''<c',c> - c'<c,c''>$, upon addition and using the fact that $<c,c'> = <c',c>$,

$$2c''<c,c'> = c<c',c''> + c'<c,c''>$$
$$+ 3c'q(c'',c) + 3cq(c'',c').$$

Applying τ to this latter relation and subtracting from the former yields our result.

Corollary V.1 _The identification of the theorem identifies the centroid of_ L _with the center of the Jordan algebra_ (A, f_1) _in all cases._

The corollary is evident if A is a field, and has been proved for type A_1. In all the other cases, (A, f_1) is a Jordan algebra embedded in an associative algebra (even when $(A, f_1 + f_2)$ or $A \oplus B$ is only alternative – see [JJ], p. 15), so we may write the product $f_1(a, a') = \frac{1}{2}(aa' + a'a)$. Then to say that c is central in A is to say that

$$f_1(f_1(c,a), a') = f_1(c, f_1(a, a')) \quad \text{for all } a, a',$$

or that for all a, a', $ca - ac$ commutes with a'. In particular $ca - ac$ commutes with c and a, from which it follows ([J], p. 44) that $ca - ac$ is nilpotent. But (A, f_1) is a division algebra, i.e., for every $a \neq 0$ in A there is an element $b \in A$ (its inverse in $(A, f_1 + f_2)$, or in $A \oplus B$ in those cases) such that $\frac{1}{2}(ab + ba) = 1$, $\frac{1}{2}(a^2 b + ba^2) = a$. Then $\frac{1}{2}(ab + ba)a = a = \frac{1}{2}a(ab + ba)$, from which $ba^2 = a^2 b = a$, and $ba^2 b = ab = ba$, so $ab = 1 = ba$, and b is the ordinary associative inverse of a. Since $ca - ac$ has no associative inverse, it must be zero, and c commutes (associatively) with all elements of A. In case β) this argument works with $(A, f_1 + f_2)$ even in the alternative cases, since $ac - ca$ and its inverse generate a field, and it clearly establishes the corollary in case δ). The corollary holds in case γ) if A generates the associative algebra, a case which holds except when A is itself a field [H]; this case falls under α), rather than γ), so the corollary is proved.

§2. **Algebras of Type A.**

a) _Derived algebras of simple associative algebras._

We consider here the types $A_r (r \geq 2)$, the division algebra $(A, f_1 + f_2)$ of §III. 3 being assumed **associative** (a condition that always prevails if $r \geq 3$); also the type A_1, where the Jordan division algebra (A, f_1) is assumed to arise as an associative division algebra A with $2f_1(a,b) = ab + ba$. (We write the associative product in A by juxtaposition.)

Let V be an F-vector space of dimension $r + 1$ where our split algebra S of type A_r acts as the linear mappings of trace zero, and consider the left A-vector space $A \otimes_F V$, of A-dimension $r + 1$. We show:

<u>Proposition V.1.</u>   <u>In all the cases of this subsection, the Lie algebra L as reconstructed in §§III.2,3 identifies with the derived algebra of $\text{End}_A(A \otimes_F V)$. In particular, if $r \geq 3$, each simple L of type A_r is realized as such a derived algebra for $[V:F] = r + 1$.</u>

To give the identification explicitly, note first that for each $x \in S$, $b \in A$, there is an F-endomorphism of $A \otimes V$ sending $a \otimes v$ to $ab \otimes vx$ for all $a \in A$, $v \in V$, and this is clearly an A-endomorphism. We thus obtain an F-linear mapping of $S \otimes A$ into $\text{End}_A(A \otimes V)$. Choosing a basis $v_1, \ldots, v_{r+1}$ for V over F, hence for $A \otimes V$ over A, and letting E_{ij} be the corresponding matrix units in $\text{End}_F(V)$, every element of $S \otimes A$ is uniquely expressible in the form

(*) $\sum_{i \neq j} E_{ij} \otimes b_{ij} + \sum_{i=1}^{r} (E_{i+1,i+1} - E_{ii}) \otimes b_{ii}$,

$b_{ij} \in A$. If the image of this element in $\text{End}_A(A \otimes V)$ annihilates all v_i, we see at once that all $b_{ij} = 0$, so the mapping $S \otimes A \to \text{End}_A(A \otimes V)$ is injective.

Next let D be a derivation of $(A, f_1 + f_2)$ (or of (A, f_1) in the case of type A_1. Then D is <u>inner</u> (Appendix B), which is to say in our cases that there is a unique element $b \in [AA]$, the derived algebra of the associative algebra A, such that $aD = ab - ba$. We map D to that A-endomorphism of $A \otimes V$ sending $a \otimes v$ to $ab \otimes v$ for all a, v. If this endomorphism agrees with one in the image of $S \otimes A$, say as displayed in (*), we find for each i, $2 \leq i \leq r$, that $b = b_{i-1,i-1} - b_{ii}$, and also that $b = -b_{11}$, $b = b_{rr}$ (apply to the various v_j). Starting with $b = -b_{11}$, we obtain successively $b_{ii} = -ib$, $1 \leq i \leq r$, which combines with $b = b_{rr}$ to give $(r + 1)b = 0$, and $b = 0$. It follows that we have displayed an injection of $\mathcal{D} \oplus (S \otimes A)$ into $\text{End}_A(A \otimes V)$. Each element T of the image has the property that if $v_i T = \sum_j a_{ij} \otimes v_j$, all i, then $\sum_i a_{ii} \in [A, A]$. This latter property is easily seen to characterize the derived algebra of $\text{End}_A(A \otimes V)$, so our image of $\mathcal{D} \oplus (S \otimes A)$ lies in

177

this derived algebra.

In fact, one verifies directly that every element of the derived algebra of $\text{End}_A(A \otimes V)$ differs from an element of the form (*) by an element which is in the image of $\mathcal{D}$ (just take b to be an appropriate rational multiple of Σa_{ii}, in the matrix of the last paragraph). Thus <u>our map provides an</u> F-<u>linear identification of</u> L and $\text{End}_A(A \otimes V)$. We now show this is an isomorphism of Lie algebras.

Identifying $b \in [AA]$ with that element of $\mathcal{D}$, identified in its turn with the derivations of A, sending a to [ab], we see at once that [bc] is identified with the derivation sending a to [a[bc]] = [[ab]c] − [[ac]b], the latter corresponding to the effect on $a \in A$ of the commutator of the corresponding elements of $\mathcal{D}$. The homomorphism property into $\text{End}_A(A \otimes V)$ for two such elements amounts to the relation, for all a and v,

$$((ab)c - (ac)b) \otimes v = a[bc] \otimes v,$$

which is evident from associativity. Likewise, the homomorphism property for a pair $x \otimes a \in S \otimes A$, $b \in [AA] \cong \mathcal{D}$, amounts to, for all a' and v,

$$a'ab \otimes vx - a'ba \otimes vx = a'[ab] \otimes vx,$$

which is again trivial. Finally, let $a, a' \in A$; $x, y \in S$; we still must treat the bracket $[x \otimes a, y \otimes a']$, for which it suffices to compare effects on $1 \otimes v$, i.e., to compare, for $r \geq 2$,

$$aa' \otimes vxy - a'a \otimes vyx$$

with

$$B(x,y)b \otimes v + \frac{1}{2}(aa' + a'a) \otimes v[xy]$$
$$+ \frac{1}{2}(aa' - a'a) \otimes v(x \circ y),$$

where $b \in [AA]$ corresponds to $<a,a'> \in \mathcal{D}$. Now we have $B(x,y) = 2(r+1)\text{Tr}(xy)$, $a'' <a,a'> = \frac{1}{2(r+1)^2}[a''[aa']]$ (see §III.3). Thus $2(r+1)^2 b = [aa']$, $B(x,y)b = \frac{1}{r+1}\text{Tr}(xy)[aa']$. Also, $x \circ y = xy + yx - \frac{2}{r+1}\text{Tr}(xy)I$. Thus the second member of our comparison is

$$B(x,y)b \otimes v + aa' \otimes vxy - a'a \otimes vyx - \frac{1}{(r+1)}\text{Tr}(xy)[aa'] \otimes v,$$

which agrees with the first by our computation of $B(x,y)b$.

When $r = 1$, we have $B(x,y) = 4\mathrm{Tr}(xy)$, and associative computation shows that (5) of §III.2, performed for the Jordan multiplication of our case, amounts to the associative formula $a'' <a,a'> = \frac{1}{8}[a''[a,a']]$. Here we must compare

$$aa' \otimes vxy - a'a \otimes vyx$$

with

$$B(x,y)b \otimes v + \frac{1}{2}(aa' + a'a) \otimes v[xy],$$

where $B(x,y)b = \frac{1}{2}\mathrm{Tr}(xy)[aa']$. Since in this case $xy + yx = \mathrm{Tr}(xy)I$, we may write

$$B(x,y)b \otimes v = \frac{1}{2}(aa' - a'a) \otimes v(xy + yx),$$

and our equality is an immediate consequence. This completes the proof of the Proposition.

The centralizer L_0 of $T = T \otimes 1 \subseteq S \otimes A$ is clearly $D + T \otimes A$ in this case; in our identification, L_0 consists of those elements of the derived algebra of $\mathrm{End}_A(A \otimes V)$ sending each v_i to a multiple $b_i \otimes v_i$. Thus $[L_0 L_0]$ consists of such endomorphisms with all $b_i \in [AA]$, and is isomorphic to a direct sum of $r + 1$ copies of $[AA]$; the center of L_0 identifies with all $(r+1)$-tuples $(a_1, \ldots, a_{r+1})$, a_i central in A, and with $\Sigma a_i = 0$, thus with $T \otimes Z(A)$, $Z(A)$ being the center of A, and, by §1, the centroid of L.

A customary ordering of the roots of S relative to T identifies the maximal nil subalgebra N^+ of L with all endomorphisms of $A \otimes V$ having strictly lower triangular matrices relative to the ordered basis $v_1, \ldots, v_{r+1}$, and N^- with those having strictly upper triangular matrices. By Theorem I.1, every nilpotent $X \in L$ is of the form $Y_0 \exp(\mathrm{ad}Y_1) \cdots \exp(\mathrm{ad}Y_k)$, where $Y_0 \in N^+$ and the remaining Y_i are in $N^+ \cup N^-$. Thus $\exp(\mathrm{ad}X) = \eta^{-1}\exp(\mathrm{ad}Y_0)\eta$, $\eta = \exp(\mathrm{ad}Y_1) \cdots \exp(\mathrm{ad}Y_k)$, so the group generated by all such $\exp(\mathrm{ad}X)$ coincides with that generated by all $\exp(\mathrm{ad}Y)$, $Y \in N^+ \cup N^-$, or by conjugations of the derived algebra of $\mathrm{End}_A(A \otimes V)$ by $\exp(Y)$ for such Y. Thus G' evidently contains the group generated by conjugations by all elements of $\mathrm{Aut}_A(A \otimes V)$ having matrices of the form $I + E_{ij} \otimes b$, $i \neq j$, $b \in A$, relative to the basis $v_1, \ldots, v_{r+1}$, hence by all members of the group $SL_A(A \otimes V)$ generated by these latter. But each $\exp(Y)$ as above is also in $SL_A(A \otimes V)$, so that G' consists of conjugations by $SL_A(A \otimes V)$, a group whose center is known to consist of scalars from $Z(A)$. Thus G' is isomorphic to the quotient $PSL_A(A \otimes_F V)$, and the second corollary to Theorem II.1 provides a general principle yielding the simplicity of this well-known group (for characteristic zero, $[A:Z(A)]$ finite).

If one carries out the construction of $\mathcal{D} + S \otimes A$ as above, assuming only that A is a simple associative algebra over F, a simple Lie algebra always results: The Jacobi identity uses only the associative law, the definition of $<a,a>$, and for $r = 1$ the relation $xy + yx = Tr(xy)I$ for S. For the simplicity, one notes that any ideal has the form $\mathcal{D}_0 + (S \otimes A_0)$, where A_0 is an ideal in A, hence zero or A. From $A\mathcal{D}_0 \subseteq A_0$, this ideal is zero if $A_0 = 0$, while if $A_0 = A$, we have $\mathcal{D}_0 \supseteq <A,A_0> = <A,A> = \mathcal{D}$.

b) <u>The exceptional case of rank two; Tits' second construction</u>.

When L is of type A_2 and the alternative division algebra $(A, f_1 + f_2)$ is not associative, A must be a Cayley-Dickson algebra of octonions over its center $Z = Z(A)$, which was identified in §1 with the centroid of L (see [SCH], §III.5.) The construction for L resulting in §III.3 is equivalent to a special case of a construction of Tits [TA] (see also [JE], §10, [SCH], §IV.4), other cases of which occur in connection with types G_2, C_3, F_4. One starts with a simple Jordan algebra J, all of whose elements satisfy a generic cubic polynomial - in this case, J is the 3 by 3 F-matrices, relative to $x \cdot y = \frac{1}{2}(xy + yx)$, and the cubic polynomial is the characteristic polynomial - and with a composition algebra A, <u>i.e.</u>, a simple algebra with unit (necessarily alternative), equipped with a linear form t over its center Z such that $t(1) = 1$, an involution $a \mapsto a'$ such that $a' = -a$ if and only if $t(a) = 0$, and such that every a satisfies the quadratic relation $a^2 - 2t(a)a + t(aa') = 0$ over the center Z. It will be noted that $t(a) = \frac{1}{2}(a + a')$ and that $aa' = a'a$; moreover, the symmetric Z-bilinear form β obtained by polarizing the quadratic form $t(aa')$ is non-degenerate. All of these are classical properties of octonion division algebras, thus of our A. The underlying vector space in Tits' construction is

(*) $\quad\quad\quad Der(A) \oplus (A_0 \otimes J_0) \oplus (Der(J) \otimes Z)$, where A_0 is the set of $a_0 \in A$ with $t(a_0) = 0$, and J_0 is the set of elements of J having "generic trace" - the negative of the coefficient of X^2 in the generic cubic polynomial - equal to zero. In our case this trace is the ordinary matric trace, and J_0 is the same vector space as S. The derivations of this J are all inner (Appendix B), and again identify with S, considered as acting, by bracketing, on the associative algebra of 3 by 3 F-matrices. Thus $\text{Der}(J)$ is identified with S, $\text{Der}(J) \otimes Z$ with $S \otimes Z$, therefore $(A_0 \otimes J_0) \oplus (\text{Der}(J) \otimes Z)$ with $S \otimes (A_0 \oplus Z) = S \otimes A$.

(Actually, Tits' original construction is carried out over a common field, over which A is assumed to be a composition algebra; here it is convenient to allow for a center Z of A larger than F; This necessitates extending the base field of J to Z before performing the construction, an extension which is, in effect, achieved by replacing $\text{Der}(J)$ by our $\text{Der}(J) \otimes Z$.)

The Lie product in Tits' construction is obtained by requiring that $\text{Der}(A)$ and $\text{Der}(J) \otimes Z$ commute, and that they multiply internally by brackets of derivations (and the multiplication in Z). If $D \in \text{Der}(A)$, $a \in A_0$, $x \in J_0$, one defines $[a \otimes x, D] = aD \otimes x$, and if $D \in \text{Der}(J)$, $z \in Z$, $[a \otimes x, D \otimes z] = az \otimes xD$. (One always has $AD \subseteq A_0$, $JD \subseteq J_0$, respectively.) Finally, if $a,b \in A_0$ while $x,y \in J_0$, the construction defines, in our case

(†)
$$[a \otimes x, b \otimes y] = \frac{1}{12} \text{Tr}(xy) D_{a,b}$$
$$+ (ab - \beta(a,b')) \otimes (\frac{xy + yx}{2} - \frac{1}{3} \text{Tr}(xy) I)$$
$$+ \frac{1}{4} D_{[xy]} \otimes \beta(a,b').$$

In our notation of §III.3, $D_{a,b} = 18 \langle a,b \rangle$, so the first term of (†) is $\frac{3}{2} \text{Tr}(xy) \langle a,b \rangle$. Also, $\beta(a,b') = \frac{1}{2}(ab + ba)$, so the second term of (†) is $\frac{[ab]}{2} \otimes \frac{(x \circ y)}{2}$. In our identification of S with $\text{Der}(J)$, the last term of (†) is

$$\frac{1}{4}[xy] \otimes \beta(a,b') = \frac{1}{4}[xy] \otimes \frac{ab + ba}{2}.$$

We evidently obtain a linear isomorphism (over Z) of Tits' algebra onto ours, using the map which is the identity on $\text{Der}(A)$, which is $\frac{1}{2}$

181

on $A_0 \otimes S$, and which sends $D_x \otimes Z (x \in S, z \in Z)$ to $x \otimes z \in S \otimes Z \subseteq S \otimes A$. That this map is an isomorphism of Lie algebras is immediate from the formulas in Tits' definition, except where the homomorphism property must be checked for brackets of the form (†). Here $a,b \in A_0$, $x,y \in S$, and we must verify that, as obtained in §III.3, $\frac{1}{4}[x \otimes a, y \otimes b]$ agrees with the image of the right-hand side of (†), viz. with

$$\frac{3}{2} \text{Tr}(xy) <a,b> + \frac{1}{2} \frac{(x \circ y)}{2} \otimes \frac{[ab]}{2}$$

$$+ [xy] \otimes \frac{1}{8}(ab + ba).$$

Our definition gave

$$[x \otimes a, y \otimes b] = B(x,y) <a,b> + [xy] \otimes \frac{ab + ba}{2}$$

$$+ (x \circ y) \otimes \frac{[ab]}{2}.$$

In this case $B(x,y) = 6 \text{Tr}(xy)$, from which the verification is clear.

We have not previously carried through the details of an argument to show that the construction of §III.3 actually yields a Lie algebra, let alone a simple one; this is now immediate, since all the cited references show that Tits' construction yields such a Lie algebra.

The centralizer L_0 of $T = T \otimes 1$ is evidently equal to $\mathcal{D} \oplus (T \otimes A) = \mathcal{D} \oplus (T \otimes A_0) \oplus (T \otimes Z)$, centralized by $T \otimes Z$. Since $<A_0, A_0> = \mathcal{D}$ and $A_0 \mathcal{D} = A_0$, the derived algebra of L_0 is $\mathcal{D} \oplus (T \otimes A_0)$, of dimension 28 over Z, where it identifies with the skew transformations of A with respect to the (necessarily anisotropic) symmetric bilinear form β. Namely, as one sees from [JE], §2, or from [SCH], Chapter III, elements of $\mathcal{D}$, as well as left and right multiplications by elements of A_0 (written L_a, R_a respectively) are β-skew transformations of A over Z. The set of all L_a (resp. all R_a) forms a Z-vector space of dimension <u>seven</u>, and if $D + L_b + R_a = 0$, application to 1 gives $b = -a$, $D = L_a - R_a$, $a \in A_0$, so that for $b, c \in A$,

$$a(bc) - (bc)a = (ab)c - (ba)c + b(ac) - b(ca),$$

or

$$a(bc) - (ab)c = b(ac) - (ba)c - b(ca) + (bc)a,$$

from which, by the alternative law,

$$3(a(bc) - (ab)c) = 0,$$ and hence all products of three factors with a as a factor are associative. Since $a \in A_0$, this implies $a = 0$ ([SCH], §III.8), so that $R_a = L_a = D = 0$. By comparison of dimensions ($[\mathcal{D}:Z] = 14$), we see that $\mathcal{D} \oplus L_{A_0} \oplus R_{A_0}$ exhausts the skew transformations.

An explicit isomorphism of the above with $\mathcal{D} \oplus (T \otimes A_0)$ is given by taking the basis H_i, $i = 1,2$, for T, $H_i = E_{ii} - E_{i+1,i+1} \in S$ as before. Then the desired isomorphism sends

$$D + H_1 \otimes a + H_2 \otimes b \quad \text{to} \quad D + R_a - L_b .$$

The simple group G' of Chapter II is the group of F-endomorphisms of $\mathcal{D} \oplus (S \otimes A)$ generated by all $\exp(\mathrm{ad}(E_{ij} \otimes a))$, $a \in A$, $i \neq j$. Since the map sending $a \in A$ to this element is a homomorphism into G' from the additive group of A, we may specialize to the two cases: $a = \alpha \in Z$ and $a \in A_0$. More generally, if $y \in S$, $y^2 = 0$, $\exp(\mathrm{ad}(y \otimes \alpha))$ fixes all $D \in \mathcal{D}$, and sends, for $x \in S$, $b \in A$, $x \otimes b$ to

$$x \otimes b + [xy] \otimes \alpha b - \frac{2}{3} \mathrm{Tr}(xy) y \otimes \alpha^2 b,$$

while if $a \in A_0$, $\exp(\mathrm{ad}(y \otimes a))$ sends $D \in \mathcal{D}$ to

$$D - y \otimes aD,$$

and $x \otimes b$ to

$$\frac{1}{3} \mathrm{Tr}(xy) D_{b,a} + \{ x \otimes b + [xy] \otimes \frac{ba+ab}{2} + (x \circ y) \otimes \frac{[ba]}{2}$$
$$+ \mathrm{Tr}(xy) y \otimes (-\frac{4}{3} n(a) b + \frac{10}{3} aba) \} ,$$

where $n(a) = t(aa') = - t(a^2) = - a^2 \in Z$. These formulas are obtained by straightforward computation. Our generators afford the closest available counterparts of "transvections", corresponding to the "elations", when our group G' is viewed as the "little projective group" of a certain "Cayley plane" (see [JJ], Chapter IX, where this group is treated in terms of exceptional Jordan algebras). The realization presented here is the adjoint representation, rather than the projective representation. The module L has dimension 78 over its centroid Z, and L becomes isomorphic to the split Lie algebra of type E_6 upon suitable field extension of Z

c) <u>The case of rank one</u>; <u>the Tits-Koecher construction</u>.

The construction here is one previously given by Tits [TC] and by Koecher [K]. (See also [JE], §9; [JJ], §VIII.5.) The construction produces a simple Lie algebra whenever the Jordan algebra A is simple, but, as we saw in §III.2, the F-rank of the resulting Lie algebra will be greater than one if A is not a division algebra. Conversely, whenever A is a (Jordan) division algebra, one has no nilpotent transformations of A in $R_{A_0} + \text{Der}(A)$, where A_0 is the set of elements of generic trace zero; for all these mappings are skew with respect to the (polarized) generic norm form ([JJ], §VI.9), which cannot represent zero for a division algebra ([JJ], §VI.5). But any symmetric multilinear form admitting nontrivial nilpotent skew transformations must represent zero (see, <u>e.g.</u>, [SLA II]). But we have seen that $R_{A_0} + \text{Der}(A)$ identifies with the derived algebra of $L_0 = \text{Der}(A) + T \otimes A$. Thus the latter has no nilpotents, and T is indeed a maximal split torus in L when A is a Jordan division algebra.

We shall return to the various possibilities when A is a special Jordan division algebra other than that resulting from $\frac{1}{2}(ab + ba)$ in an associative division algebra (this was covered in a) above), under our discussions of algebras of types B and C. Here we dispose, in particular, of the case where A is an <u>exceptional</u> Jordan division algebra. Then $[A:Z(A)] = 27$, $[\text{Der}(A):Z(A)] = 52$ ([JE], §3; [JJ], Chapter IX; [SCH], Chapter IV). Explicit constructions for all such A, due to Tits, are to be found in [JJ], <u>loc. cit.</u> . The field $Z(A)$ is the centroid of L, $[L:Z(A)] = 52 + 81 = 133$, and extension of the base field from $Z(A)$ results in a split Lie algebra of type E_7 ([JE], §9).

Our simple group G' of Chapter II is generated by all $\exp(\text{ad}(x_\beta \otimes a))$, $a \in A$, $\beta = \pm \alpha$, or by all $\exp(\text{ad}(y \otimes a))$, where y runs over all 2 by 2 F-matrices with $y^2 = 0$, $a \in A$. We compute at once that the latter acts on L by sending $D \in \mathcal{D}$ to

$$D - y \otimes aD,$$

and $x \otimes b$ ($x \in S$, $b \in A$) to

$$4 \, \text{Tr}(xy) \ <b,a> \ + x \otimes b + [xy] \otimes b \cdot a$$
$$- \text{Tr}(xy) \ y \otimes bU_a,$$

where $bU_a = 2(b \cdot a) \cdot a - b \cdot (a \cdot a)$, the product $a \cdot b$ being the Jordan product in A. Thus we obtain a simple group acting in a space of dimension 133 over $Z(A)$ whenever A is an exceptional Jordan division algebra.

3. Algebras of Types B and D: Skew Transformations Relative to Quadratic Forms.

a) Algebras of type $B_r (r \geq 3)$.

In this case, the structure of the Lie algebra L is given by Theorem III.4. By §1, its centroid is the field extension A. Now consider the A-vector space $(A \otimes_F M') \oplus B$, endowed with a symmetric bilinear form β relative to which $A \otimes M'$ and B are orthogonal, such that β agrees with $-f$ (f as in §III.4) on B and with the A-bilinear extension to $A \otimes_F M'$ of our original F-form of index r on the $(2r+1)$-dimensional F-space M'. We give a realization of L:

Proposition V.2. *Let L be of type $B_r (r \geq 3)$, with notations as above. Then L identifies with the Lie algebra of all A-endomorphisms of $(A \otimes_F M') \oplus B$ which are skew with respect to the form β.*

To establish the identification, we denote the Lie algebra of β-skew A-endomorphisms of each non-singular subspace U of $(A \otimes M') \oplus B$ by $K(U)$. In §III.4, we identified the subalgebra D of L with $K(B)$; we use this identification to identify D with those elements of $K(A \otimes M' + B)$ which stabilize B and annihilate $A \otimes M'$. Also, $K(A \otimes_F M') = A \otimes_F K(M') = A \otimes S = S \otimes A$, the element of $K(A \otimes M')$ corresponding to $x \otimes a$ ($x \in S$, $a \in A$) sending $a' \otimes v$ ($a \in A$, $v \in M'$) to $aa' \otimes vx$. We extend this action of $S \otimes A$ to $A \otimes M' \oplus B$ by insisting that $S \otimes A$ annihilate B. Finally, if $b \in B$, $u \in M'$, there is an A-endomorphism of $(A \otimes M') \oplus B$ sending $b' \in B$ to $f(b,b') \otimes u \in A \otimes M'$ and $a \otimes v$ to $(u,v)ab \in B$. This A-endomorphism depends F-bilinearly on b and u, and is readily seen to be a β-skew transformation $T(b,u)$. Now our isomorphism of L onto

$K(A \otimes M' \oplus B)$ is given in terms of the decomposition
$L = \mathcal{D} \underset{F}{\oplus} (S \otimes A) \oplus (M' \underset{F}{\otimes} B)$ by the above injections of $\mathcal{D}$ and of $S \otimes A$
into $K(A \otimes M' \oplus B)$, and by sending $u \otimes b$ ($u \in M'$, $b \in B$) to $T(b,u)$.

There is evidently an F-linear map ϕ of L into $K(A \otimes M' \oplus B)$ with these definitions on the summands. If $T \in K(A \otimes M' \otimes B)$, we may subtract from T an element of $\mathcal{D}^\phi + (S \otimes A)^\phi$ to assume $BT \subseteq A \otimes M'$, $(A \otimes M') T \subseteq B$. By letting b run over an orthogonal A-basis for B, and for each such b taking $u = f(b,b)^{-1} bT$, we have that $\Sigma_b T(b,u)$ matches T on a basis for B, so that we may assume $BT = 0$, $(A \otimes M')T \subseteq B$. But then the skewness of T implies that $T = 0$, and we have shown that ϕ is <u>onto</u>. That ϕ is injective follows at once by comparing the A-dimension of L : $\binom{[B:A]}{2} + \binom{[M':F]}{2} + [M':F] [B:A]$ with that of $K(A \otimes M' \oplus B)$: $\binom{[B:A] + [M':F]}{2}$. It remains only to check that ϕ preserves brackets, and for this we may assume one factor in a bracket to be of the form $u \otimes b \in M' \otimes B$. If $D \in \mathcal{D}$, we have in L, $[u \otimes b, D] = u \otimes bD$, whose image under ϕ, $T(bD,u)$, is determined by its effect on $b' \in B$, which goes to $f(bD,b') \otimes u$. On the other hand $[(u \otimes b)^\phi, D^\phi]$ sends b' to $(f(b,b') \otimes u)D^\phi - f(b,b'D) \otimes u = - f(b,b'D) \otimes u = f(bD,b') \otimes u$. If $x \in S$, $a \in A$, $[u \otimes b, x \otimes a] = ux \otimes ab$. Again the image under ϕ, $T(ab,ux)$ is determined by its effect on b', namely $f(ab,b') \otimes ux$, while $[(u \otimes b)^\phi, (x \otimes a)^\phi]$ sends b' to $(f(b,b') \otimes u)(x \otimes a)^\phi = af(b,b') \otimes ux = f(ab,b') \otimes ux$. Finally, we have $[u \otimes b, v \otimes b'] = (u,v) <b,b'> + [uv] \otimes f(b,b')$ in Theorem III.4. The image of this under ϕ sends $b'' \in B$ to $(u,v)b'' <b,b'> = (u,v)f(b'',b)b' - (u,v) f(b'',b')b$, and $a \otimes w(w \in M'$, $a \in A)$ to $af(b,b') \otimes w[uv] = af(b,b') \otimes ((w,u)v - (w,v)u)$. Meanwhile $[(u \otimes b)^\phi, (v \otimes b')^\phi]$ sends b'' to $(f(b'',b) \otimes u)(v \otimes b')^\phi - (f(b'',b') \otimes v)(u \otimes b)^\phi = f(b'',b)(u,v)b' - f(b'',b')(u,v)b$, and $a \otimes w$ to $(a(w,u)b)(v \otimes b')^\phi - (a(w,v)b')(u \otimes b)^\phi = a(w,u)f(b,b') \otimes v - a(w,v)f(b',b) \otimes u$, and the identification is complete. Since $K(A \otimes M' \oplus B)$ is known to be a simple Lie algebra (see, <u>e.g.</u>, [J], Chapter X - the A-dimension of $(A \otimes M') \oplus B$ is at least <u>seven</u>), we see that L is, also.

The centralizer L_0 of T has a component in $M' \otimes B$ equal to $u_{r+1} \otimes B$, where $(u_{r+1}, u_{r+1}) = 1$, $u_{r+1} \in M'$ being a basis for the orthogonal complement of a sum of dual maximal totally isotropic subspaces. Thus $L_0 = \mathcal{D} + T \otimes A + u_{r+1} \otimes B$, $[L_0 L_0] = \mathcal{D} + u_{r+1} \otimes B$, and this identifies as above with $K(A \otimes u_{r+1} \oplus B)$. The condition that f not represent 1, in §III.4, is equivalent to requiring that β not represent 0 on $(A \otimes u_{r+1}) \oplus B$, and amounts, as seen in §III.4, to the condition that T actually be a maximal split torus in L. This assures that the construction of Theorem 3.4 actually yields a simple Lie algebra of type B_r. The Witt index of the A-form β on $(A \otimes M') \oplus B$ is exactly r. The simple group G' of Chapter II is here generated by elements $\exp(\mathrm{ad}X)$, where $X \in L$ is a root-vector relative to T, so is one of the following: i) $x_\alpha \otimes a$, $a \in A$, $x_\alpha \in S_\alpha$, α a "long" root; ii) $x_\alpha \otimes a + u_\alpha \otimes b$, where α is <u>short</u>, $b \in B$, and $u_\alpha \in M'$ is a weight-vector belonging to α. Since $[u_\alpha \otimes b, x_\alpha \otimes a] = u_\alpha x_\alpha \otimes ab$, with $u_\alpha x_\alpha$ belonging to 2α, which is not a weight of M', we have $[u_\alpha \otimes b, x_\alpha \otimes a] = 0$, so that the case ii) reduces to the commuting automorphisms $\exp(\mathrm{ad}(x_\alpha \otimes a))$, $\exp(\mathrm{ad}(u_\alpha \otimes b))$. Here u_α is necessarily isotropic. In the canonical realization of S as acting on M' we have $x_\alpha^2 = 0$ if α is long, $x_\alpha^3 = 0$ if α is short, and in each case x_α has rank 2 as endomorphism of M' ([J], §IV.6).

For α long, the endomorphism $\exp(x_\alpha \otimes a)$ of $(A \otimes M') \oplus B$ fixes all elements of B and sends $a' \otimes u$ to $a' \otimes u + aa' \otimes ux$. When α is short, $\exp(x_\alpha \otimes a)$ also fixes B, but sends $a' \otimes u$ to $a' \otimes u + aa' \otimes ux + \frac{1}{2} a^2 a' \otimes ux^2$. In the latter case, the realization of S on M' actually yields $x_\alpha = \lambda_\alpha [u_\alpha, u_{r+1}]$, $\lambda_\alpha \in F$, where u_α, u_{r+1} are as before, with $(u_{r+1}, u_{r+1}) = 1$. From this form it follows that $\exp(x_\alpha \otimes a)$ sends $a' \otimes u$ to

$$a' \otimes u + \beta(a' \otimes u, \lambda_\alpha a \otimes u_\alpha) \otimes u_{r+1}$$
$$- \beta(a' \otimes u, 1 \otimes u_{r+1}) \lambda_\alpha a \otimes u$$
$$- \frac{1}{2} \beta(a' \otimes u, \lambda_\alpha a \otimes u_\alpha) \lambda_\alpha a \otimes u_\alpha.$$

If we set $U = \lambda_\alpha a \otimes u_\alpha$, $V = 1 \otimes u_{r+1}$, we see that for all $W \in (A \otimes M') \oplus B$, we have

$$W \exp(x_\alpha \otimes a) = W + \beta(W,U)V - \beta(W,V)U$$
$$- \frac{1}{2} \beta(V,V) \beta(W,U)U,$$

which we define to be $Ws(U,V)$. In our case, U is isotropic, V is orthogonal to U, and $\beta(V,V) = 1$.

For $(u,u) = 0$ and $b \in B$, the action of $\exp(u \otimes b)$ on $(A \otimes M') \oplus B$ is readily verified to be that of $s(U,V)$, where the isotropic vector U is now $1 \otimes u$ and the orthogonal anisotropic vector V is b. Whenever U and V are orthogonal, with U isotropic and V anisotropic, the mapping $\log(s(U,V))$: $W \to \beta(W,U)V - \beta(W,V)U$ is in $K(A \otimes M' \oplus B)$, and has $\exp(\log(s(U,V))) = s(U,V)$. Thus conjugation of $K(A \otimes M' \oplus B)$ by $s(U,V)$ is in G'. Moreover, G' is induced by β-orthogonal transformations, among which the $s(U,V)$ are a self-conjugate set. By the simplicity of G' established in §II.4, the subgroup of G' generated by conjugations by the $s(U,V)$ is equal to G' and is a simple group $\Omega(\beta)$. It is an easy matter to check that the center of $\Omega(\beta)$ is contained in $\{\pm I\}$, so that the simplicity of $\Omega(\beta)$, modulo its center, is a consequence of the second corollary to Theorem II.1 and of our identification of L. Of course, $\Omega(\beta)$ is the commutator subgroup of the orthogonal group. (See, e.g., [D], §II.9).

b) <u>Algebras of type</u> $D_r (r \geq 4)$.

In this case, we know by Theorem III.1 that $L \cong S \underset{F}{\otimes} A$, where A is a field extension of F, and where S may be assumed to be the skew transformations of a $2r$-dimensional F-space M' relative to a nondegenerate symmetric bilinear form (u,v) of Witt index r. Then $L \cong K(A \otimes M')$ with the evident extension of (u,v) to a form β which is A-bilinear on $A \otimes M'$, and G' identifies as above with conjugations of L by elements in the group $\Omega(\beta)$ (although here the $s(U,V)$ do not arise directly in the form $\exp(x_\alpha \otimes a)$ — in the action of S on M', we have $x_\alpha^2 = 0$ for all α.) The simplicity, modulo its center, of $\Omega(\beta)$ then follows from Theorem II.1 as in a).

c) <u>Algebras of type</u> B_2.

According to our convention of §III.6, an algebra of type B_2 is one of type C_2 for which the Jordan algebra (A, f_1) of that section is associative. (We have not really shown that this type and the other are exclusive, but this should not trouble us here.) In this case, Theorem III.8 gives

$$L \cong D \oplus (S \otimes A) \oplus (M' \otimes B),$$

A a field extension of F, B an A-vector space with a nondegenerate symmetric bilinear form h, not representing 1. S is the skew transformations of a 4-dimensional F-symplectic space M''', and M' is the symmetric transformations of M''' of trace zero. Also $(u,v) = \text{Tr}(uv)$ is a nondegenerate symmetric bilinear form of Witt index 2 on M', and the action of S on the 5-dimensional F-space M' is by skew transformations with respect to this form. Since $\dim. S = 10$, S identifies with the space of all such skew transformations, as Lie algebras over F. An anisotropic vector in M' orthogonal to a pair of dual 2-dimensional totally isotropic subspaces is given by our U_1 of §III.6, and $(U_1, U_1) = \text{Tr}(U_1^2) = 1$. Moreover, D is the Lie algebra of h-skew endomorphisms of B over A.

Now all the conclusions of a) apply in this case as well, where the form β is defined to be $-h$ on B instead of $-f$. That is, Prop. 1 applies without change to identify L with the β-skew A-endomorphisms of $(A \oplus M' \oplus B)$, where β has Witt index 2, and our conclusions concerning L_0 and G' are exactly as before, with u_{n+1} replaced by U_1.

d) <u>Some algebras of type</u> A_1 ("<u>type</u> B_1").

We return to the algebras of Theorem III.2 and of §2 c) of this chapter, considering here only the case where the Jordan division algebra (denoted here by "J", rather than "A", for conformity with notation of this section) has the form $J = A \oplus B$, where A is a field extension, B is an A-vector space of dimension at least 2 carrying a non-degenerate symmetric A-bilinear form f, not representing 1, and

$$(a + b) \cdot (a' + b') = (aa' + f(b,b')) + (ab' + a'b)$$

defines the Jordan product in $A \oplus B$. From [JJ], §§I.11, J is a Jordan division algebra with center A, and one readily verifies that the derivations of J are the A-linear f-skew transformations of B. From §1 it follows that A is the centroid of L, and, as vector spaces over F (or over A), one has

$$L \cong K(B,f) \oplus (S \otimes A) \oplus (S \otimes B).$$

Now we treat the adjoint S-module S as a 3-dimensional space M', with the non-degenerate symmetric bilinear form $(u,v) = \frac{1}{8}B(u,v)$, B the Killing form, of Witt index <u>one</u>, and with $(H_{\alpha_1}, H_{\alpha_1}) = 1$, H_{α_1} orthogonal to the two dually paired totally isotropic subspaces Fx_α, $Fx_{-\alpha}$. Here S identifies with the skew transformations of M'. We define the form β on the A-space $(A \otimes M') \oplus B$ exactly as before, and again obtain an isomorphism of Lie algebras $L \cong K(A \otimes M' \oplus B)$. L_0 is identified as in a), as is G'. (With regard to the simplicity of G', it will be noted that since $[B:A] \geq 2$, $(A \otimes M') \oplus B$ has A-dimension at least 5; when $[B:A] \leq 1$, J is a field, $D = 0$, and we revert to a case treated in §2.)

Thus <u>all the algebras</u> L <u>of types</u> B <u>and</u> D, <u>and certain ones of type</u> A_1, <u>are skew transformations with respect to symmetric bilinear forms over field extensions</u> A <u>of</u> F, <u>the index of the form being the F-rank of</u> L .

§4. <u>Algebras of Type</u> F_4.

A construction yielding all algebras of type F_4 is given in Theorem III.5. This is again a special case of Tits' second construction, discussed in §2b): Here we have the split exceptional Jordan algebra J, of dimension 27 over F, and a composition division algebra C over A, a field extension of F. Over A, $J \otimes_F A$ is a (split) simple Jordan algebra satisfying a generic cubic polynomial, and we apply Tits' construction with the parameter-algebras $J \otimes A$ and C, obtaining a Lie structure on

$$\text{Der}_A(J \otimes_F A) \oplus (C_0 \otimes_A (J \otimes_F A)_0) \oplus \text{Der}_A(C)$$
$$= (\text{Der}_F(J) \otimes A) \oplus (C_0 \otimes_A (J_0 \otimes_F A)) \oplus \text{Der}_A(C)$$
$$= (S \otimes A) \oplus (B \otimes_A (A \otimes_F M')) \oplus D$$
$$= (S \otimes A) \oplus (B \otimes_F M') \oplus D ,$$

in our notations. Comparison of the formulas of Theorem III.5 with those of Tits' construction yields an isomorphism of Lie algebras consistent with this decomposition as vector spaces; in fact, the numerous references to [SCH] in the proof of Theorem III.5 essentially establish this isomorphism and therefore that our construction yields a simple Lie algebra. Since $\text{Der}_A(C) = \mathcal{D}$ contains nonzero nilpotents if and only if $[C:A] > 2$ and C is not a division algebra, and since $\mathcal{D} = 0$ if $[C:A] \leq 2$, we are able to deduce that the condition that C be a division algebra is equivalent to requiring that T be a maximal F-split torus in L. By §III.5 and the second paragraph below, it suffices to show that T is maximal when $[C:A] \leq 2$, and C is A or a quadratic field extension. As we saw in §III.5, the centralizer L_0 of T is then

$$(T \otimes A) \oplus (M'_0 \otimes B).$$

If $C = A$, then $L_0 = T \otimes_F A$, and one easily sees that T is a maximal F-split torus. If C is a quadratic extension, L_0 is again commutative, and the characteristic roots of all $\text{ad}(u \otimes b)$ ($u \in M'_0, b \in B$) lie in B. From this, the linear independence of a basis for T and one for M'_0, and the diagonalizability of $\text{ad} L_0$ in C, it again follows that T is a maximal F-split torus. Thus: L is a simple Lie algebra of type F_4 over F if and only if it is given by Theorem 3.5. The centroid of L identifies with A, and $[L:A] = 52, 78, 133$ or 248, according to the possibility for C.

In general, the centralizer L_0 of T is identified with $\mathcal{D} + T \otimes A + M'_0 \otimes B$, and we have either $[L_0 L_0] = 0$ (the split resp. quasi-split cases, according as $[C:A] = 1$ or 2), or $[L_0 L_0] = \text{Der}_A(C) \oplus M'_0 \otimes B$. In the latter case, C is either a quaternion or an octonionic division algebra (cf., e.g., [SCH], §III.7) over A; $M'_0 \otimes B$ has A-dimension 6 resp. 14, while $\text{Der}_A(C)$ has A-dimension 3 resp. 14. In the octonionic case, arguments similar to those in §2b) enable us to identify $M'_0 \otimes B$ with $L_B \oplus R_B$, acting in C, and $[L_0 L_0]$ with $K(C)$, the 28-dimensional A-Lie algebra of skew transformations of C with respect to the polarized norm form. In the quaternionic case, $\text{Der}_A(C)$ identifies with $C_0 = B$ under the correspondence $b \longleftrightarrow D_b = \text{ad } b$, acting in C; M'_0 is the F-subspace of J with basis (in the canonical matrix realization) $u_1 = \text{diag}\{1,0,-1\}$ and $u_2 = \text{diag}\{1,-2,1\}$, and, as A-spaces,

$$[L_0 L_0] = D_B \oplus u_1 \otimes B \oplus u_2 \otimes B.$$

Now $[L_0 L_0]$ is identified with the direct sum of three Lie algebras isomorphic to $B = C_0$ by the following three isomorphisms of B onto

191

supplementary ideals of $[L_0 L_0]$:

i) $b \to \frac{1}{3} D_b + \frac{1}{2} u_1 \otimes b + \frac{1}{6} u_2 \otimes b$;

ii) $b \to \frac{1}{3} D_b - \frac{1}{2} u_1 \otimes b + \frac{1}{6} u_2 \otimes b$;

iii) $b \to \frac{1}{3} D_b - \frac{1}{3} u_2 \otimes b$.

(It will be noted that if D_b is formally identified with diag $\{b,b,b\}$, these are the respective maps $b \to$ diag $\{b,0,0\}$, $b \to$ diag $\{0,0,b\}$, $b \to$ diag $\{0,b,0\}$.) These calculations thus complete the proof that whenever a composition division algebra C over A is incorporated in our construction, $T = T \otimes 1$ is indeed a maximal F-split torus in L.

Finally, we give the corresponding identification of the simple group G' of §II.4. Identifying the roots of L and those of S relative to T, there are 24 long roots α, with root-spaces of the form $e_\alpha \otimes A$, $0 \neq e_\alpha \in S_\alpha$, and 24 short roots α, with root-spaces $e_\alpha \otimes A + u_\alpha \otimes B$, where u_α is a basis for the weight-space M'_α. For α long we have $e_\alpha^2 = 0$ as derivation of J, thus in its action on M' (e.g., see [SLA III]); for α short, we have $[u_\alpha \otimes b, e_\alpha \otimes a] = 0$, since 2α is not a weight, $e_\alpha^3 = 0$ on M', $u_\alpha \cdot u_\alpha = 0$, $(v \cdot u_\alpha) \cdot u_\alpha = (v, u_\alpha) u_\alpha$, $(v \cdot u_\alpha, u_\alpha) = 0 = (v \circ u_\alpha, u_\alpha) = (v \circ u_\alpha) \circ u_\alpha = \frac{1}{3}(v, u_\alpha) u_\alpha$, for all $v \in M'$ (with e_α, u_α as in [SLA III]). Thus G' is generated by the $\exp(\mathrm{ad}(e_\alpha \otimes a))$, $\alpha \in \Sigma$, $a \in A$, and by the $\exp(\mathrm{ad}(u_\alpha \otimes b))$, α short, $b \in B$, the effects of these generators on the summands of L being as follows:

$\exp(\mathrm{ad}(e_\alpha \otimes a))$: Each $D \in \mathcal{D}$ is fixed;

$$x \otimes a' \to x \otimes a' + [x e_\alpha] \otimes aa' + \frac{1}{2}[[x e_\alpha] e_\alpha] \otimes a^2 a';$$

$$u \otimes b \to u \otimes b + u e_\alpha \otimes ab + \frac{1}{2} u e_\alpha^2 \otimes a^2 b,$$

the last term in the last being zero if α is long.

$\exp(\mathrm{ad}(u_\alpha \otimes b))$: $D \to D - u_\alpha \otimes bD$;

$$x \otimes a \to x \otimes a + \frac{1}{2} D_{u_\alpha, u_\alpha x} \otimes af(b,b)$$

$$- u_\alpha x \otimes ab;$$

$$v \otimes b' \to \frac{1}{6}(v, u_\alpha) D_{b',b} + D_{v, u_\alpha} \otimes f(b', b)$$
$$+ v \otimes b' + \frac{1}{2}(v \circ u_\alpha) \otimes [b', b]$$
$$+ \frac{1}{2}(v, u_\alpha) u_\alpha \otimes (f(b,b) b' - 2 f(b, b') b).$$

That is, these generators in our spaces generate simple groups acting absolutely irreducibly in A-spaces of 52, 78, 133 and 248 dimensions. The first is the adjoint Chevalley group F_4; the second is the "twisted E_6" of Steinberg; the last two have no finite counterparts. They have been studied, in a more general context on at least two occasions by Tits: first, as central quotients of groups generated by rational unipotent elements in F-simple algebraic groups of relative type F_4 [TS]; and second, as groups of automorphisms of "buildings of type F_4" in his work on "buildings of spherical type." [TB].

§5. Algebras of Type G_2.

As in §4, all algebras of type G_2 are obtained by the construction of Theorem III.6, another special case of Tits' second construction: Here the composition algebra C is a split octonion algebra over F, $S = \text{Der}_F(C)$, $M' = C_0$, and $J = A \oplus B$ is a Jordan division algebra over its center A, satisfying a generic cubic polynomial over A, with $J_0 = B$. Our decomposition

$$\mathcal{D} \oplus (S \otimes_F A) \oplus (M' \otimes B)$$

is the same as

$$\text{Der}_A(J) \oplus \text{Der}_A(C \otimes_F A) \oplus ((C \otimes_F A)_0 \otimes_A J_0)$$

in Tits' construction; references to [SCH] in connection with Theorem III.6 establish the identification of the brackets in the two spaces, and hence that our construction yields a simple Lie algebra.

We now appeal to the classification of Jordan division algebras to sharpen our view of the possibilities for J: In a splitting extension of A, J can have at most three orthogonal idempotents, so has degree at most three ([JJ], Chapters V,VI). By [JJ], Chapter V, J must be one of the following: i) A; ii) a cubic extension field of A; iii) E^+, where E is a central associative division algebra of degree 3 over A, the operation in J being $c \cdot c' = \frac{1}{2}(cc' + c'c)$; iv) $H(E)$, where E is a central associative division algebra of degree 3 over a quadratic extension of A, having an involution of second kind fixing A, and with $H(E)$ as hermitian elements, operation as in iii); v) an exceptional Jordan

193

division algebra over A, which may be constructed as in [JJ], §IX.12;
vi) the Jordan algebra $A \oplus V$ of a nondegenerate quadratic A-form,
not representing 1, on the A-vector space V, of dimension at least
2. We eliminate the case vi) in Appendix B, by showing its incompatibility with the other conditions present in our situation. For the
admissible cases we have:

i) $\text{Der}_A(J) = 0$, $J_0 = 0$, $L = S \otimes A$, a split Lie algebra
of type G_2 over A.

ii) $\text{Der}_A(J) = 0$, $[B:A] = 2$, $L = (S \otimes A) \oplus (M' \otimes B)$ of
A-dimension 28.

iii) $\text{Der}_A(J) \cong [EE]$, of A-dimension 8, $[B:A] = 8$, and
L has A-dimension $8 + 14 + 56 = 78$.

iv) $\text{Der}_A(J) \cong [K(E), K(E)]$, where $K(E)$ is the skew elements
of E, of A-dimension 8, $[B:A] = 8$, $[L:A] = 78$;

v) $[\text{Der}_A(J):A] = 52$, $[B:A] = 26$, $[L:A] = 52 + 14 + 7 \cdot 26 = 248$.

If we let $u_0 \in M'$ be as in §III.5, the centralizer L_0 of T
in L is $D + T \otimes A + u_0 \otimes B$, and $[L_0 L_0] = 0$ in cases i), ii),
$[L_0 L_0] = D + u_0 \otimes B$ in cases iii) -v). In case ii), B is the set
of elements of trace zero in the field extension J/A and $\text{ad}(u_0 \otimes b)$
acts semisimply with characteristic roots which are rational multiples
of b; in all cases, those of $\text{ad}(t \otimes a)$ ($t \in T, a \in A$) are rational
multiples of a, provided t is one of our basis vectors with rational
eigenvalues. It follows in cases i) and ii) that $T = T \otimes 1$ is
always a maximal F-split torus in this construction, with a centralizer
of A-dimension 2 resp. 4. In the remaining cases, we have, as in §3c),
an identification of $[L_0 L_0]$ with $\text{Der}_A(J) + R_{J_0}$; since all elements
of the latter are skew with respect to the (cubic) generic norm form on
J, and since the latter does not represent zero, J therefore can
admit no nilpotent skew transformations. Since $L_0 = [L_0 L_0] \oplus (T \otimes A)$,
it follows that T is a maximal F-split torus in these cases, where
$[L_0 L_0]$ has dimensions 16, 16, resp. 78. In case v) this algebra is
known to be central simple over A ([JE], §7) and to become a split
algebra of type E_6 upon extension of A to a splitting field for J.
In case iii), if $(u_0, u_0) = -1$ (as we may assume, by [SLA II]),
one verifies directly that the maps $b \to \frac{1}{2}(D_b + u_0 \otimes b)$,
$b \to \frac{1}{2}(D_b - u_0 \otimes b)$ of $B = [EE]$ into $[L_0 L_0]$ are homomorphisms of

Lie algebras (D_b denoting the derivation $c \to [cb]$ of E), and that their images are supplementary ideals, each isomorphic to $[EE]$. Thus $[L_0 L_0] \cong [EE] \oplus [EE]$ in this case, and $[EE]$ is simple with centroid A.

In case iv), let Z be the center of E, $0 \neq \zeta \in Z \cap K(E)$. Then $H(E) = K(E)\zeta = J$, and $J_0 = B = [K(E), K(E)]\bar{\zeta}$. Furthermore, $[EE] = [K + J, K + J] = [K + K\zeta, K + K\zeta] = [KK] + [KK]\zeta = [KK] + J_0$. For $c \in [KK]$, $b \in J_0$, the map $(c + b) \to D_c + u_0 \otimes b$ is an isomorphism of the simple Lie algebra $[EE]$ onto $[L_0 L_0]$. Thus we have determined the structure of $[L_0 L_0]$, and see that there can be <u>no isomorphism between an algebra arising from</u> iii) <u>and one from</u> iv).

To identify the group G', which is an adjoint Chevalley group of type G_2 over A in case i), a "cubic-twisted D_4" in the sense of Steinberg in case ii), and without finite counterpart in the remaining cases, we give, as before, the effects of generators $\exp(\mathrm{ad}(e_\alpha \otimes a))$, $\exp(\mathrm{ad}(u_\alpha \otimes b))$, where $a \in A$, $b \in B$, e_α is a root-vector, u_α a weight-vector belonging to a short root. We assume these to be as displayed in [SLA II]. The effects are as follows:

$\exp(\mathrm{ad}(e_\alpha \otimes a))$: Fixes all $D \in \mathcal{D}$;

$x \otimes a'$ goes to $x \otimes a' + [xe_\alpha] \otimes aa' + \frac{1}{2}[[xe_\alpha]e_\alpha] \otimes a^2 a'$
$\qquad + \frac{1}{6}[[[xe_\alpha]e_\alpha]e_\alpha] \otimes a^3 a'$;

$u \otimes b$ goes to $u \otimes b + ue_\alpha \otimes ab + \frac{1}{2} ue_\alpha^2 \otimes a^2 b$

the last term in each of the last two expressions being zero if α is a long root.

$\exp(\mathrm{ad}(u_\alpha \otimes b))$: $D \in \mathcal{D}$ goes to $D - u_\alpha \otimes bD$;

$x \otimes a$ goes to $x \otimes a - \frac{1}{2} D_{u_\alpha x, u_\alpha} \otimes af(b,b)$
$\qquad - \frac{1}{6} D_{[u_\alpha x, u_\alpha], u_\alpha} \otimes af(h(b,b),b)$
$\qquad - ux \otimes ab - \frac{1}{2}[u_\alpha x, u_\alpha] \otimes ah(b,b)$;

$v \otimes b'$ goes to $(v, u_\alpha) < b', b > +$
$\qquad + D_{v, u_\alpha} \otimes f(b',b) + \frac{1}{2} D_{[v, u_\alpha], u_\alpha} \otimes f(b, h(b', b))$
$\qquad + v \otimes b' + [v u_\alpha] \otimes h(b', b) - \frac{1}{2}(v, u_\alpha) u_\alpha \otimes b < b', b >$.

We thus have explicit generators for our simple group G' in an absolutely irreducible A-representation of degree 14, 28, 78, 78 or 248.

§6. <u>Algebras of Type</u> E.

As shown in §III.2, these are all of the type $S \otimes A$, where A is an extension field and S is a split Lie algebra over F of the same type; <u>that is, they are split algebras of the same type over extension fields of</u> F. They are given a presentation by Serre's theorem. They may also be realized from Tits' constructions by using a split algebra instead of a division algebra as coordinate algebra (our $A \otimes B$, in the second construction; A, in the first): For example, use of a split exceptional Jordan algebra in the construction of Theorem III.2 yields a split algebra (over A) of type E_7; use of a split octonion algebra in the exceptional case for algebras of type A_2 yields a split algebra of type E_6. If the composition division algebra of Theorem III.5 (type F_4) is replaced by one of the three respective split composition algebras of dimension greater than one, the result is a split Lie algebra over A of type E_6, E_7, E_8, respectively. In the case of type G_2 (Theorem III.6), replacement of the cubic Jordan division algebra by a split simple Jordan algebra of degree 3:i) the 3 by 3 A-matrices; ii) the symplectic-symmetric 6 by 6 A-matrices; iii) the split exceptional Jordan algebra over A, yields split A-algebras of respective types E_6, E_7, E_8 ([JE], §10; [SCH], §IV.4).

The centralizer L_0 of T is $T \otimes A$, and is abelian, and the simple group G' is that of Chevalley associated with a system of roots of the specified type, and with the field A.

§7. <u>Algebras of Types</u> C, BC.

a) <u>General case; skew elements of simple associative algebras with involution</u>.

Under this heading, we consider all algebras of types $C_r (r \geq 4)$, all those of types $BC_r (r \geq 3)$, and all those of type C_3 for which the first case of Theorem III.7 holds ($E = A \otimes B$ is associative). Further, we consider the second case (the true "type C_2") of Theorem III.8, and the case a) of Theorem III.10 for type BC_2. Finally, in the case of types A_1 and BC_1, we include the case where the Jordan division algebra A identifies with the special Jordan algebra of symmetric elements in an associative division algebra E with involution, the Jordan algebra A not being associative, nor with $[A:Z(A)] = 6$, $[E:Z(A)] = 16$ in case BC_1.

In each of these cases, we have an involutorial associative division algebra E, with A as symmetric elements; we denote by B the set of skew elements in the cases of rank one, this being already done in the other cases. In all cases of rank greater than one, C (possibly equal to zero) carries the structure of right vector space over E. The S-module M'' is a $2r$-dimensional symplectic space over F, so that $M'' \otimes_F E$ becomes a $2r$-dimensional antihermitian space over E. In the case of types A_1 and BC_1, M'' is a 2-dimensional symplectic space over F, with S as skew transformations, and $M'' \otimes E$ is a 2-dimensional antihermitian space over E as before. The assumptions on A in these cases imply that the $Z(A)$ - linear mappings $A \mapsto \text{End}_{Z(A)}(C)$ which originally gave our A the structure of special Jordan algebra with center $Z(A)$ extend uniquely to $Z(A)$ - homomorphisms of associative algebras (with unit): $E \mapsto \text{End}_{Z(A)}(C)$ (i.e., E with its involution is a perfect involutorial associative algebra over $Z(A)$ - [JJ], §§III.4, II.4; to show that these results apply in our case, we note that since $Z(A) \neq A$, A generates E as $Z(A)$ - algebra [H], so that $Z(A)$ is the center of E as involutorial algebra, and thus that extension of the base field from $Z(A)$ results ultimately in a split involutorial algebra, simple as involutorial algebra. Theorem 6 of §III.4 of [JJ] then applies whenever this algebra is a full matrix algebra of degree at least three with involution of orthogonal type, or of degree at least six with involution of symplectic type, or whenever it is the sum of two antisomorphic full matrix algebras of degree at least three, with exchange involution. The case "$n = 2$" of that theorem applies if the split algebra is a full matrix algebra of degree two with orthogonal involution, and our restrictions rule out the case of a full matrix algebra of degree less than 6 with symplectic involution. The case $n = 2$ for the second kind of involutorial algebra was done by Jacobson and Rickart [JR]. Finally, one invokes Theorem 8 of §II.4 of [JJ] to show that perfection upon field extension implies perfection.) Thus we are assuming that C is a (right) vector space over E, the action of A (and of B, if present) on C being that of E. When $C = 0$, this condition is vacuously satisfied. Our realization of L is now as follows:

Proposition V.3. Let L be a simple Lie algebra over F of type C_r, BC_r ($r \geq 2$), or A_1, subject to the assumptions above. Then L is isomorphic to the derived Lie algebra $[K(M'' \otimes E \oplus C), K(M'' \otimes E \oplus C)]$ of the E-endomorphisms of $(M'' \otimes E) \oplus C$ skew with respect to a non-degenerate antihermitian form extending the symplectic one on M'', defined for $r \geq 2$ by $q_1 + q_2 = q$ on C, and for which C is anisotropic, $M'' \otimes E$ and C orthogonal.

Proof. For $r \geq 2$, and for A_1, we have the antihermitian E-space $(M'' \otimes E) \oplus C$ by virtue of Theorems III. 7,8,9,10 and our assumptions (with $C = 0$ for types C_r and A_1). Thus it suffices to give the desired isomorphism. Rather than writing brackets for the derived algebra, we shall denote it by a prime: thus $K'(C) = [K(C), K(C)]$.

We first assume the involution $*$ in E to be of second kind, i.e., that B contains nonzero elements in the center of E. Then these elements form a one-dimensional space B_0 over $Z = Z(A) = $ centroid of L (§1), and $Z + B_0$ is the center of E. Then if $\beta_0 \in B_0$, $\beta_0 I$ is in $K(U)$ for $U = C$, $M'' \otimes E$, or their sum, and we have the direct decompositions

$$K(U) = K'(U) \oplus B_0 I, \quad B = [BB] \oplus B_0,$$

for each $U \neq 0$ as above. Let $e_1, \ldots, e_{2r}$ be our symplectic basis for M'', identified with the basis $\{e_i \otimes 1\}$ for $M'' \otimes E$. First let $T \in K(M'' \otimes E)$; then the sum of the diagonal entries in the matrix of T relative to the above basis is

$$(\ddagger) \qquad \sum_{i=1}^{r} ((e_i, e_{2r+1-i} T) - (e_{2r+1-i}, e_i T)).$$

Since each of the r indicated terms is in B, so is the sum, which we can write as $2rb$, for $b \in B$. Now $K(M'' \otimes E)$ contains the mapping sending $u \otimes \gamma$ to $u \otimes b\gamma$ for all $u \in M''$, $\gamma \in E$, and subtracting this mapping from T leaves $T' \in K(M'' \otimes E)$ for which the diagonal-sum $(\ddagger)$ is zero. Now $S \otimes A$ (and $M' \otimes B$, if present) acts on $M'' \otimes E$, with $x \otimes a$ sending $u \otimes \gamma$ to $ux \otimes a\gamma$ ($s \otimes b$ sending $u \otimes \gamma$ to $us \otimes b\gamma$). One readily verifies that this gives an F-linear isomorphism of $(S \otimes A) \oplus (M' \otimes B)$ into $K(M'' \otimes E)$, the image being exactly those elements T' of $K(M'' \otimes E)$ for which the indicated diagonal-sum is zero (for type A_1, $S \otimes A$ already exhausts such T'). We extend

198

these elements of $S \otimes A \oplus M' \otimes B$ to elements of $K(M'' \otimes E \oplus C)$ by requiring that they annihilate C. In fact, one verifies directly that any $T' \in K(M'' \otimes E)$ satisfying (‡) is in $K'(M'' \otimes E)$, as is evidently the map $u \otimes \gamma \to u \otimes b\gamma$ whenever $b \in [BB]$. As in §2, elements of $K'(M'' \otimes E)$ have diagonal-sum in $[E,E]$, and this sum is skew, from (‡). In our case, we have $B = A\beta_0$, $A = B\beta_0$, and $[a + b, a' + b'] = [b''\beta_0 + b, b'''\beta_0 + b'] = [b'',b''']\beta_0^2 + [bb'] + ([b'',b'] + [b,b'''])\beta_0$.
Now $[b'',b''']\beta_0^2 = [b'',b'''\beta_0^2] \in [B,B]$, from which we see from considering skew and symmetric parts in sums of such elements that $[E,E] \cap B = [BB]$, and therefore that $[BB]$ identifies with a subspace of $K'(M'' \otimes E)$ in such a way that

$$K'(M'' \otimes E) = [BB] \oplus (S \otimes A) \oplus (M' \otimes B);$$

from $B = [BB] \oplus B_0$, $K(M'' \otimes E)$ identifies with the direct sum of B_0 and $K'(M'' \otimes E)$.

Next consider the subspace of $K(M'' \otimes E \oplus C)$ mapping $M'' \otimes E$ into C, C into $M'' \otimes E$. One readily verifies that any such T is determined by its effect on the $e_i (= e_i \otimes 1)$, a basis for $M'' \otimes E$, and that any set of elements $c_i \in C$ may be assigned as images for these. Thus this space is F-isomorphic to $M'' \underset{F}{\otimes} C$; to be more explicit, to $v \otimes c (v \in M'', c \in C)$ is assigned the mapping sending $u \otimes \gamma (u \in M'', \gamma \in E)$ to $c(v,u)\gamma \in C$ and $c' \in C$ to $v \otimes (c,c') \in M'' \otimes E$. By forming the commutator of such a map with one that is $\beta_0 I$ on $M'' \otimes E$ and $-\beta_0 I$ on C, we see that our map is in $K'(M'' \otimes E \oplus C)$. The result is an injection of $M'' \otimes E$ into $K'(M'' \otimes E \oplus C)$, with image the subspace defined above. We clearly have, with evident identifications, $K(M'' \otimes E \oplus C) = K(M'' \otimes E) \oplus (M'' \otimes C) \oplus K(C)$, $K'(M'' \otimes E \oplus C) \supset K'(M'' \otimes E) \oplus (M'' \otimes C) \oplus K'(C)$, the Z-codimension in the latter inclusion being <u>one if</u> $C \neq 0$, with equality if $C = 0$.

For $\beta \in B_0$, denote by $\tau(\beta)$ the mapping which is $\frac{1}{2r}\beta I$ on $M'' \otimes E$ and $-\frac{1}{m}\beta I$ on C. Then $\tau(\beta) \in K(M'' \otimes E \oplus C)$, but is not in $K'(M'' \otimes E) \oplus (M'' \otimes C) \oplus K'(C)$ (consider the diagonal subsum corresponding to the $\{e_i\}$) unless $\beta = 0$. If $C \neq 0$, we let $c_1,\ldots,c_m$ be an orthogonal basis for C over E, $(c_i, c_i) = \beta_i \in B$, and let $\mu_i = \frac{1}{2rm}\beta_i^{-1}\beta$, where $\beta \in B_0$ is fixed. Then

199

$$\Sigma_{j=1}^{r} \Sigma_{i=1}^{m} [e_j \otimes c_i, e_{2r+1-j} \otimes c_i\mu_i] \in [M'' \otimes C, M'' \otimes C] \subseteq K'(M'' \otimes E \oplus C)$$

is readily seen to be equal to $\tau(\beta)$. Thus $\tau(B_0) \subseteq K'(M'' \otimes E \oplus C)$, and

$$K'(M'' \otimes E \oplus C) = K'(M'' \otimes E) \oplus (M'' \otimes C) \oplus K'(C) \oplus \tau(B_0).$$

To establish our isomorphism between L and $K'(M'' \otimes E \oplus C)$, we define $\rho: L \to K'(M'' \otimes E \oplus C)$ by:

$\rho: S \otimes A \to K'(M'' \otimes E)$: $x \otimes a$ goes to the mapping sending $v \otimes \gamma$ to $vx \otimes a\gamma$;

$\rho: M' \otimes B \to K'(M'' \otimes E)$: $s \otimes b$ goes to the mapping sending $v \otimes \gamma$ to $vs \otimes b\gamma$;

$\rho: M'' \otimes C \to K'(M'' \otimes E \oplus C)$ as above;

$\rho: K'(C)(\subseteq D) \to K'(C)$ by the identity map;

$\rho: [BB](\subseteq D)$: $b \in [BB]$ goes to the mapping in $K'(M'' \otimes E)$ sending $v \otimes \gamma$ to $-v \otimes b\gamma$;

$\rho: B_0(\subseteq D)$: β goes to $-\dfrac{2rm}{2r+m} \tau(\beta)$.

Then ρ is clearly an isomorphism of F-vector spaces (omitting both A_1 and BC_1 for the moment) — indeed, ρ is an isomorphism of Z-vector spaces. As in §3, "piece-by-piece" verification that it is a Lie homomorphism is straightforward. Except for the following, the most complicated of the computations, we leave this to the reader.

We check the case of two factors from $M'' \otimes C$. Here we have, for $u, v, w \in M''$; $c, c' \in C$,

$$(w \otimes 1)[(u \otimes c)^\rho, (v \otimes c')^\rho] = (c(u,w))(v \otimes c')^\rho - c'(v,w)(u \otimes c)^\rho$$
$$= v \otimes (c',c)(u,w) - u \otimes (c,c')(v,w).$$

In particular, if $w = e_i$, $1 \leq i \leq r$, $u = \Sigma \lambda_j e_j$, $v = \Sigma \mu_j e_j$, the bracket above sends $e_i \otimes 1$ to

$$-v \otimes (c',c)\lambda_{2r+1-i} + u \otimes (c,c')\mu_{2r+1-i};$$

likewise, $e_{2r+1-i} \otimes 1$ goes to

$$v \otimes (c',c)\lambda_i - u \otimes (c,c')\mu_i.$$

The coefficient of e_i in the former is

$$-(c',c)\mu_i\lambda_{2r+1-i} + (c,c')\lambda_i\mu_{2r+1-i},$$

200

while that of e_{2r+1-i} in the latter is
$$(c',c)\lambda_i \mu_{2r+1-i} - (c,c')\mu_i \lambda_{2r+1-i}.$$
Thus the sum of these diagonal coefficients is
$$((c,c') + (c',c))(v,u).$$
Writing $(c,c') + (c',c) \in \mathcal{B}$ as $b' + b_0$ with $b' \in [\mathcal{B},\mathcal{B}]$, $b_0 \in \mathcal{B}_0$, we see that the effect on $w \otimes 1$ of $[(u \otimes c)^\rho, (v \otimes c')^\rho]$ is the same as that of a certain sum (implicit below) of elements in $(S \otimes A)^\rho$, $(M' \otimes B)^\rho$,
$$w \otimes 1 \to \frac{1}{2r} w \otimes b', \text{ and } w \otimes 1 \to \frac{1}{2r} w \otimes b_0.$$
In L, we have $[u \otimes c, v \otimes c'] =$
$$(u,v) <c,c'> + (u \circ v) \otimes \frac{(c,c')-(c',c)}{2} + [uv] \otimes \frac{(c,c')+(c',c)}{2}.$$
The effect on $w \otimes 1$ of the image of this under ρ is
$$(u,v)(w \otimes 1) <c,c'>^\rho - (w,u) v \otimes (c',c) + (w,v) u \otimes (c,c')$$
$$- \frac{1}{2r}(u,v) w \otimes ((c,c') + (c',c)).$$
Subtracting this from $(w \otimes 1)[(u \otimes c)^\rho, (v \otimes c')^\rho]$ leaves
$$- (u,v)(w \otimes 1) <c,c'>^\rho + \frac{1}{2r}(u,v) w \otimes ((c,c') + (c',c)).$$
Now, writing $<c,c'> = T' + b'' + b_0'$, $T' \in K'(\mathcal{C})$, $b'' \in [\mathcal{BB}]$, $b_0' \in \mathcal{B}_0$, as in §III.7, shows that the above difference is

($\ddagger$)
$$(u,v) w \otimes b'' + \frac{2rm}{2r+m} (u,v)(w \otimes 1)\tau(b_0')$$
$$+ \frac{1}{2r}(u,v) w \otimes (b' + b_0).$$

Meanwhile, applying $[(u \otimes c)^\rho, (v \otimes c')^\rho]$ to $c'' \in \mathcal{C}$ gives
$$c'(v,u)(c,c'') - c(u,v)(c',c''),$$
while applying $(u,v) <c,c>^\rho$ gives
$$(u,v) c''T' - \frac{2rm}{2r+m}(u,v) c'' \tau(b_0').$$
Here T' is obtained by taking the $K'(\mathcal{C})$-component of T in the decomposition $K(\mathcal{C}) = K'(\mathcal{C}) \oplus \mathcal{B}_0 I$: $c''T = -c(c',c'') - c'(c,c'')$. Thus $<c,c'>$ breaks down into T', $b_0''I$ (the $\mathcal{B}_0 I$ - component of T), and $-\frac{1}{2r}$ times the sum of the $[\mathcal{BB}]$ and $\mathcal{B}_0$- components of $(c,c') + (c',c)$, or $-\frac{1}{2r} b' - \frac{1}{2r} b_0$, as above. It follows that

201

$b_0' = b_0'' - \frac{1}{2r} b_0$, $b'' = -\frac{1}{2r} b'$; substituting in (‡) gives a value of

$$(u,v)w \otimes (-b_0' + b_0'') + \frac{2rm}{2r+m} (u,v)(w \otimes 1) \tau(b_0')$$

$$= (u,v)w \otimes (b_0'' - \frac{2rb_0'}{2r+m})$$

for that difference. The difference in the effects on c'' is

$$-\frac{2rm}{2r+m} (u,v)c'' \tau(b_0') - (u,v)c''b_0''$$

$$= (u,v)c'' (\frac{2rb_0'}{2r+m} - b_0'') .$$

To obtain our conclusion, we must show $(2r+m)b_0'' = 2rb_0'$, or since $b_0' = b_0'' - (2r)b_0$, that $b_0 = -mb_0''$. This amounts to showing that $T + (m^{-1})b_0 I \in K'(C)$.

Now we have $R = \text{End}_E(C)$ decomposing into symmetric and skew parts, R^+ resp. R^- ($R^- = K(C)$), so that $[RR] \cap K(C) = [R^+R^+] + [R^-,R^-]$ $= [R^-B_0, R^-B_0^{-1}] + [R^-,R^-] = [R^-,R^-] = K'(C)$, with $B_0 \in B_0$ as before. Now for $U \in R$ and any basis for C, we saw in §2 that $U \in [RR]$ if and only if the sum of the diagonal coefficients in the $e_i U$ is in $[EE]$. We apply this to $T + m^{-1}b_0 I$, the e_i being an orthogonal basis $e_1,\ldots,e_m$. By the biadditivity of T in c and c', it suffices to consider these cases: a) $c = e_i \lambda$, $c' = e_j \mu$, $i \neq j$; and b) $c = e_i \lambda$, $c' = e_i \mu$. In case a), $b_0 = 0$, $e_i T$ is a multiple of e_j, $e_j T$ a multiple of e_i, and the remaining $e_k T$ are 0; thus the diagonal-sum is zero: In case b), $b_0 \equiv \lambda^*(e_i,e_i) \mu + \mu^*(e_i,e_i)\lambda$ (mod $[EE]$), $e_j T = 0$ for $j \neq i$, and $e_i T = -e_i(\lambda\mu^*(e_i,e_i) + \mu\lambda^*(e_i,e_i))$, so that the diagonal-sum for $T + m^{-1}b_0 I$ is congruent (mod $[EE]$) to

$$-\lambda\mu^*(e_i,e_i) - \mu\lambda^*(e_i,e_i) + \lambda^*(e_i,e_i)\mu + \mu^*(e_i,e_i)\lambda$$

$$= [\mu^*(e_i,e_i),\lambda] + [\lambda^*(e_i,e_i),\mu] \in [E,E].$$

When $C = 0$ (still assuming our involution to be of second kind), the above and Theorems III.7 and III.8 ii) identify L with $K'(M'' \otimes E)$.

When the involution in E is of first kind, so that $B_0 = 0$, we have $K'(M'' \otimes E) = K(M'' \otimes E)$, $K'(M'' \otimes E \oplus C) = K(M'' \otimes E \oplus C)$, and $[BB] = B$ except in the case where E is a quaternionic algebra over its center Z, the centroid of L (§1), with $[B:Z] = 1$; in this

202

exceptional case, $B = [A,A]$. Here the ambiguous intersection of B and $K(C)$ is not present in $D = B + K(C)$.

The decomposition $K(M'' \otimes E \oplus C) = K(M'' \otimes E) + M'' \otimes C + K(C)$ still holds, and $K(M'' \otimes E)$ identifies as before with $B \oplus (S \otimes A) \oplus (M'' \otimes B) \subseteq L$, the summand "$B$" being canonically contained in D as in Theorem III.9 (See the proof of that theorem for remarks concerning the quaternionic case.) The identification of L with $K(M'' \otimes E \oplus C)$ is now straightforward.

We still must deal with the postponed cases of rank one. For type A_1, we have assumed $A = H(E,*)$, the symmetric elements of the associative division algebra E with involution $*$. (Our assumption that the Jordan product in A is not associative is not a true restriction, since in the excluded case A is a field extension of F and we have a situation already treated in §2.) The space M'' is 2-dimensional symplectic, with S as skew transformations. We let $B = K(E,*)$, the skew elements of E, and consider $M'' \otimes_F E$ as antihermitian E-space as before. With e_1, e_2 a symplectic basis for M'' over F, regarded also as E-basis for $M'' \otimes E$, we have $T \in K(M'' \otimes E)$ provided $T \in \text{End}_E(M'' \otimes E)$ sends e_1 to $-e_1 \gamma^* + e_2 \alpha_{12}$, e_2 to $e_1 \alpha_{21} + e_2 \gamma$, where $\alpha_{12}, \alpha_{21} \in A$. Now we have $S \otimes A$ embedded in $K(M'' \otimes E)$ as before, and we see at once that, working modulo $S \otimes A$, we may assume $\alpha_{12} = \alpha_{21} = 0$. Moreover, writing

$$\gamma = (\frac{\gamma+\gamma^*}{2}) + (\frac{\gamma-\gamma^*}{2}), \quad \gamma^* = (\frac{\gamma+\gamma^*}{2}) - (\frac{\gamma-\gamma^*}{2}),$$

we may assume $\gamma = \beta \in B$, so that T sends e_i to $e_i \beta$, e_2 to $e_2 \beta$. As before, the condition that $T \in K'(M'' \otimes E)$ amounts to requiring $\beta \in [EE]$, hence that $\beta \in [BB] + [A,A]$. Our assumption that A is not a field guarantees that $[BB] \subseteq [A,A]$ ([Ba]); thus if we extend the map $S \otimes A \to K'(M'' \otimes E)$ to a map $L \mapsto K'(M'' \otimes E)$ by sending $d \in D$ to the map $e_i \otimes \lambda_i \mapsto -e_i \otimes b\lambda_i (i=1,2)$, where $b \in [AA] \subseteq B$ is the unique element of $[AA]$ such that, for all $a \in A$, $ad = [a,b]$, we obtain an isomorphism of F-vector spaces of L onto $K'(M'' \otimes E)$. (The existence of such b follows from the fact that all derivations of the Jordan division algebra A are inner, and from the fact that the form taken by the inner derivation determined by a' and a'' is, to within scalar multiples, the map $a \mapsto [a,[a',a'']]$; the uniqueness of b

follows from the fact that any element of [A,A] centralizing A must be nilpotent ([J], p. 44), hence zero, in our division algebra.)

That the map is a homomorphism of Lie algebras follows at once on pairs of elements from D. For the rest of showing that $M'' \otimes E$ is an L-module under this action, we have

$$((u \otimes \lambda)(x \otimes a))D_b - ((u \otimes \lambda)D_b)(x \otimes a)$$
$$= (ux \otimes a\lambda)D_b + (u \otimes b\lambda)(x \otimes a)$$
$$= -ux \otimes ba\lambda + u \otimes ab\lambda = (u \otimes \lambda)(x \otimes [ab]),$$

and $((u \otimes \lambda)(x \otimes a))(x' \otimes a') - ((u \otimes \lambda)(x' \otimes a'))(x \otimes a)$

$$= uxx' \otimes a'a\lambda - ux'x \otimes aa'\lambda$$
$$= u[xx'] \otimes \frac{a'a+aa'}{2}\lambda - u(xx' + x'x) \otimes \frac{[aa']}{2}\lambda$$

Now $xx' + x'x = Tr(xx')I$, and, from Theorem III.2, $2a'' <a,a'> = (a \cdot a'') \cdot a' - (a'' \cdot a') \cdot a$ (Jordan multiplication), so that $8a'' <a,a'> = [a''[a,a']]$. Since $B(x,x') = 4Tr(xx')$, the second term above is $-\frac{1}{8} B(x,x') u \otimes [a,a']\lambda = B(x,x')(u \otimes \lambda) <a,a'>$. Comparing with Theorem III.2 completes the argument, and therewith the proof of Proposition V.3.

For type BC_1, we have $A = H(E,*)$ as above, and define B and the antihermitian structure on $M'' \otimes E$ as in type A_1. By virtue of the perfection of $(E,*)$, the A-module C is an E-vector space (on the right). We now have the following:

Proposition V.4. <u>The</u> <u>right</u> E-<u>module</u> C <u>carries a non-degenerate</u> anti-hermitian <u>form</u> (c,c'), <u>with</u> $2q(c,c') = (c,c') - (c',c)$, <u>and</u> L <u>identifies as before with</u> $K'((M'' \otimes E) \oplus C)$.

Before proving the proposition, we remark that Allison [AJ] has established its conclusions, assuming that A is not one-dimensional over its center, and that C is not an irreducible A-module. Moreover, he reconstructs the involutorial division algebra E without appeal to theorems concerning perfection. In more recent work [AZ] Allison has also obtained results equivalent to all those of this chapter relating to algebras of type BC_1. His methods are more purely Lie-theoretical than these, and exempt from consideration only ground fields of very small characteristics. Thus they will probably be preferable to these, in the long run.

To prove Proposition V.4, let $v_1,\ldots,v_n$ be a basis for the right E-module C. For each pair v_i, v_j of basis-vectors, let $\phi(\beta,\gamma) = q(v_i\beta, v_j\gamma)$, a bilinear mapping (over the centroid of L) from $E \times E$ to A. For $a \in A$, we have $\phi(\beta a, \gamma) + \phi(\beta, \gamma a) = q(v_i\beta a, v_j\gamma) + q(v_i\beta, v_j\gamma a) = a\phi(\beta,\gamma) + \phi(\beta,\gamma)a$ by (93) (§III.9). In Appendix B, we show that such a map $\phi: E \times E \to A$ must have the form $\phi(\beta,\gamma) = \beta^*\lambda\gamma + \gamma^*\lambda^*\beta$ for unique $\lambda \in E$. Thus we have $q(v_i\beta, v_j\gamma) = \beta^*\lambda_{ij}\gamma + \gamma^*\lambda_{ij}^*\beta$ for each i,j, where $\lambda_{ij} \in E$ is uniquely determined. The skewness of q gives

$$\gamma^*\lambda_{ji}\beta + \beta^*\lambda_{ji}^*\gamma = -\beta^*\lambda_{ij}\gamma - \gamma^*\lambda_{ij}^*\beta,$$

or $\gamma^*(\lambda_{ji} + \lambda_{ij}^*)\beta + \beta^*(\lambda_{ji} + \lambda_{ij}^*)^*\gamma = 0$

for all γ, β. It follows from our uniqueness that $\lambda_{ji} = -\lambda_{ij}^*$ for all i,j; in particular, $\lambda_{ii}^* = -\lambda_{ii}$.

Now with $c = \Sigma v_i\gamma_i$, $c' = \Sigma v_i\gamma_i'$, we define $(c,c') = \Sigma_{i,j}\gamma_i^*\lambda_{ij}\gamma_j' \in E$. Then (c,c') is clearly an antihermitian form on C over $(E,*)$, and

$$(c,c') - (c',c) = \Sigma_{i,j}(\gamma_i^*\lambda_{ij}\gamma_j' - \gamma_j'^*\lambda_{ji}\gamma_i)$$

$$= \Sigma_{i,j}(\gamma_i^*\lambda_{ij}\gamma_j' + \gamma_j'^*\lambda_{ij}^*\gamma_i)$$

$$= \Sigma_{i,j} q(v_i\gamma_i, v_j\gamma_j') = q(c,c'). \text{ From (94)}$$

we have, for $a \in A$; $c, c' \in C$ as above, that

$$a <c,c'> = q(c,c'a) - q(ca,c') = (c,c'a) - (c'a,c)$$

$$- (ca,c') + (c',ca) = (c,c')a - a(c',c)$$

$$- a(c,c') + (c',c)a = [(c,c') + (c',c), a], \text{ where}$$

$(c,c') + (c',c) \in B$, the *-skew elements of E.

For $c,c', d,d' \in C$, the derivation property of $<c,c'>$ gives

$$q(d<c,c'>,d') + q(d,d'<c,c'>) = q(d,d')<c,c'>$$

$$= [(c,c') + (c',c), q(d,d')] = [(c,c') + (c',c), (d,d')-(d',d)]$$

$$= -(d(c',c),d') - (d(c,c'),d') + (d'(c',c),d) + (d'(c,c'),d)$$

$$- (d,d'(c,c')) - (d,d'(c',c)) + (d',d(c,c'))$$

$$+ (d',d(c',c)) = q(d',d((c,c') + (c',c)))$$

$$+ q(d'((c',c) + (c,c')),d) = - q(d((c,c') + (c',c)),d')$$

$$- q(d,d'((c,c') + (c',c))), \text{ so that } <c,c'> + ((c,c') + (c',c)),$$

the latter term being simply right multiplication by $(c,c') + (c',c)$, is a q-skew transformation of C. Evidently $(c,c') = 0$ for all $c' \in C$ implies $q(c,c') = 0$ for all c', hence that $c = 0$ by §III.9. Thus our antihermitian form (c,c') is nondegenerate on C; it follows, since E is noncommutative, that C has an E-basis of elements $c_1,\ldots,c_n$ with $(c_i,c_i) \neq 0$, $(c_i,c_j) = 0$ if $i \neq j$, and that any $c \in C$ with $(c,c) \neq 0$ may be taken as the first element of such a basis.

We wish to show that for all c,c',c'',

$$c'' <c,c'> = - c''(c,c') - c''(c',c) - 2c(c',c'') - 2c'(c,c''),$$

or that $u(c,c',c'') = c'' <c,c'> + c''(c,c') + c''(c',c)$
$$+ 2c(c',c'') + 2c'(c,c'') = 0$$

for all c,c',c''. We note that $u(c,c',c'')$ is symmetric in c and c', and prove first the

<u>Lemma a</u>. <u>u is symmetric in all three variables</u>.

For this, it suffices to show $u(c,c',c'') = u(c,c'',c')$. From (95), we have $u(c,c',c'') = c' <c,c''>$
$-2cq(c',c'') + c'q(c'',c) + c''q(c,c') + c''(c,c')$
$+ c''(c',c) + 2c(c',c'') + 2c'(c,c'')$
$= c' <c,c''> - 2c(c',c'') + 2c(c'',c') + c'(c'',c) - c'(c,c'')$
$+ c''(c,c') - c''(c',c) + c''(c,c') + c''(c',c) + 2c(c',c'')$
$+ 2c'(c,c'')$
$= c' <c,c''> + c'((c'',c) + (c,c'')) + 2c(c'',c')$
$+2c''(c,c') = u(c,c'',c')$, as required.

Knowing that u is symmetric, we now can prove that u is identically zero by proving that $u(c,c,c) = 0$ for all c, and indeed (since (c,c) is not identically zero) by showing this for all c with $(c,c) \neq 0$. First we show that $c <c,c>$ is not identically zero, so that we may also assume $c <c,c> \neq 0$.

<u>Lemma b</u>. <u>It is not the case that</u> $c <c,c> = 0$ <u>for all</u> $c \in C$.

For if this were the case, partial polarization yields, for all c and c',

$$2c <c,c'> + c' <c,c> = 0,$$

Meanwhile, (95) gives

$$-c<c,c'> + c'<c,c> = -3cq(c,c').$$

Elimination of $c<c,c'>$ yields

$$c'<c,c> = -2cq(c,c') = 2c((c',c) - (c,c')).$$

Replacing c' by $c'a'$, $a' \in A$, we have

$$(c'a')<c,c> = 2c(a'(c',c) - (c,c')a').$$

Setting $c' = c$, and using that $(ca')<c,c> = (c<c,c>)a' + c(a'<c,c>) = c(a'<c,c>) = 2c((c,c)a' - a'(c,c))$, we find that $[(c,c),a'] = 0$ for all $c \in C$, $a' \in A$, hence that $A<C,C> = 0$. From (92) of §III.9 and the fact that the values of q exhaust A, we see that $<A,A> \subseteq <C,C>$, so our conclusion would imply that the Jordan algebra (A,f) is associative, which it is not.

(For later reference, it should be noted that only the non-degeneracy of (c,c') and the nonassociativity of A have been used here, and likewise that Lemma a is a consequence only of the identities of III.9, once one has the antihermitian form (c,c') with $q(c,c') = (c,c') - (c',c)$, $a<c,c'> = [(c,c') + (c',c),a].)$

Lemma c. For all $c \in C$ with $(c,c) \neq 0$, $c<c,c> = cr$, for some $r \in E$.

It suffices to show that if $(c,c') = 0$ then $(c',c<c,c>) = 0$. For such c', and for all $a \in A$, we have $a<c,c'> = [(c,c') + (c',c),a] = 0$. Thus $a<c,c><c,c'> = 0$ for all $a \in A$, so that

$$0 = a[<c,c'>,<c,c>] = a<c<c,c>,c'> + a<c,c'<c,c>>.$$

From (95) and the fact that $q(c,c') = 0 = q(c,c)$, we have $c'<c,c> = c<c,c'>$, so that $a<c<c,c>,c'> = -a<c,c'<c,c>> = -a<c,c<c,c'>>$ for all a. But we also have $A[<c,c>,<c,c'>] = 0$, and $[<c,c>,<c,c'>] = <c,c<c,c'>> + <c<c,c'>,c> = 2<c,c<c,c'>>$. Thus $a<c,c<c,c'>> = 0 = a<c',c<c,c>>$ for all $a \in A$, all $c' \in C$ with $(c,c') = 0$. Thus $(c',c<c,c>) + (c<c,c>,c') = g(c,c')$ is central (and skew) in E for all such c'. For fixed c', we have that $g(c,c'e)$ is central for every $e \in E$. But if

$(c<c,c>,c') = \lambda \neq 0$, then $g(c,c'e) = \lambda e - (\lambda e)^*$, which runs over skew elements of E as e runs over E. Since not all skew elements are central (this only happens for involutions of second kind, where A is a field, or where $E = A$ is a field), our lemma is proved.

<u>Lemma d</u>. <u>For all</u> $c \in C$ <u>with</u> $(c,c) \neq 0$ <u>and</u> $c<c,c> \neq 0$, $c<c,c> = -6c(c,c)$. <u>Hence this conclusion holds for all</u> $c \in C$.

<u>Proof</u>. We have $c<c,c> = cr \neq 0$ when c is as above. From
$0 = [<c,c>,<c,c>] = <c,c>,c> + <c,c<c,c>>$
$= 2<c,c<c,c>>$, we have $<c,cr> = 0$. Applying this to $a \in A$, we have $0 = a<c,cr> = [(c,cr) + (cr,c),a]$, from which $z = (c,c)r + r^*(c,c)$ is (skew and) central in E.

For $c' \in C$, we have $c'<c,cr> = 0$. From (95) this means that $cr<c,c'> = -2cq(c',cr) + c'q(cr,c) + crq(c,c') = -2c(c',c)r + 2cr^*(c,c') - c'(c,c)r + c'r^*(c,c) + cr(c,c') - cr(c',c)$. On the other hand, $cr<c,c'> = c<c,c'>r + c[(c,c') + (c',c),r]$, for this last holds with $r = a \in A$, and is then easily derived for $r = a_1 \ldots a_m (a_i \in A)$ by induction on m; the assertion follows since A generates E. Substitution for $cr<c,c'>$ in the previous relation gives

$$c<c,c'>r = -c(c,c')r - 3c(c',c)r$$
$$+ 2cr(c,c') + 2cr^*(c,c') - c'(c,c)r$$
$$+ c'r^*(c,c),$$

Substituting $r^*(c,c) = z - (c,c)r$ yields

$$c<c,c'> = -c(c,c') - 3c(c',c) + 2cr(c,c')r^{-1}$$
$$+ 2cr^*(c,c')r^{-1} - 2c'(c,c) + c'zr^{-1}.$$

Substitution of this in (95) gives, for all $c' \in C$,

$$c'<c,c> = c<c,c'> + 3cq(c',c)$$
$$= -4c(c,c') + 2cr(c,c')r^{-1} + 2cr^*(c,c')r^{-1}$$
$$- 2c'(c,c) + c'zr^{-1}.$$

Hence for all $c' \in C$, $a \in A$,

$$(c'a)<c,c> = -4c(c,c')a + 2cr(c,c')ar^{-1} + 2cr^*(c,c')ar^{-1}$$
$$- 2c'a(c,c) + c'azr^{-1}.$$

But $(c'a)<c,c> = (c'<c,c>)a + 2c'[(c,c),a]$. Using

the formula for $c'<c,c>$ above and subtracting our two results for $(c'a)<c,c>$ gives

$$2cr(c,c')[r^{-1},a] + 2cr^*(c,c')[r^{-1},a] + c'z[r^{-1},a] = 0.$$

It follows that either r is central in E, or that, for all $c' \in C$, $2cr(c,c') + 2cr^*(c,c') + c'z = 0$. In the former case, r^* is also central, and our formula for $c'<c,c>$ becomes

$$c'<c,c> = -2c(c,c') + 2c(c,c')r^*r^{-1} - 2c'(c,c) + c'zr^{-1}.$$

From $z = (c,c)(r+r^*)$, we conclude that (c,c) is central unless $0 = r + r^* = z$. In the latter case, we have $c'<c,c> = -4c(c,c') - 2c'(c,c)$, which is our conclusion when $c = c'$. On the other hand, for (c,c) to be central is an algebraic condition not satisfied by every $c \in C$; if $0 \neq (c,c)$ is central, as is $(ce,ce) = e^*(c,c)e = (c,c)e^*e$ for all $e \in E$, then e^*e is central for all $e \in E$. In particular, a^2 is central for all $a \in A$, so $aa' + a'a$ is central for all $a, a' \in A$, and with $a' = 1$, A is central, a contradiction. Therefore <u>it suffices to prove the conclusion of the lemma under the stronger hypothesis that</u> (c,c) <u>is not central in</u> E. By the above, this allows us to assume that r is not central in E and that

$$2cr(c,c') + 2cr^*(c,c') + c'z = 0$$

With $c' = c$, we have $z = -2r(c,c) - 2r^*(c,c)$; since $z = r^*(c,c) + (c,c)r$, we find $3z = 2[(c,c),r]$. In particular, $[(c,c),r]$ commutes with r, hence is nilpotent by a result from [J] already frequently cited. But this means $[(c,c),r] = 0$, therefore $z = 0$, $r^* = -r$, and our formula for $c'<c,c>$ now yields, as before,

$$c'<c,c> = -4c(c,c') - 2c'(c,c),$$

which is the desired conclusion when we set $c' = c$. This completes the proof of the lemma.

The proof of Proposition V.4 is now completely analogous to that of Proposition V.3. The spaces $M'' \otimes E$ and C shall be h-orthogonal, with $h(c,c') = 2(c,c')$ for $c,c' \in C$ and with $h(u \otimes e, u \otimes f) = e^*(u,v)f$ for $u,v \in M''$; $e,f \in E$. As in §III.7, $\mathcal{D} = <C,C>$ is faithfully represented on C, and consists of all the $Z(A)$-endomorphisms of C of the form $c \mapsto cb + cT$, $T \in K(C,h), b \in B$. As in that section, when the involution in E is of first kind we have $B + K(C,h) = B \oplus K'(C,h)$. When the involution is of second kind, we have $B_0 = Z(E) \cap B$, of dimension one over $Z(A)$, $B = [BB] \oplus B_0$, $K(C,h) = K'(C,h) \oplus B_0$, $K(C,h) \cap B = B_0$, so $D = B + K(C,h) = [BB] \oplus B_0 \oplus K'(C,h)$. Moreover, for involutions of first kind, $K(M'' \otimes E, h) = K'(M'' \otimes E, h)$ identifies with $(S \otimes A) \oplus B$ by identifying $x \otimes a$ ($x \in S$, $a \in A$) with the map sending $u \otimes e$ ($u \in M''$, $e \in E$) to $ux \otimes ae$, and $b \in B$ with the map sending $u \otimes e$ to $-u \otimes be$. We extend these maps to elements of $K((M'' \otimes E) \oplus C)$ by requiring that they annihilate C. For involutions of second kind, the same identification is used for $K'(M'' \otimes E) = (S \otimes A) \oplus [BB]$. For $\beta_0 \in B_0$, we define $\tau(\beta_0)$ as before, to be $\frac{1}{2}\beta_0 I$ on $M'' \otimes E$ and $-\frac{1}{m}\beta_0 I$ on C, where $[C:E] = m$, and we now define $\rho: L \to K'(M'' \otimes E \oplus C)$ as before, using $r = 1$ and h as our form on C, rather than (c,c') as in the notation for higher ranks.

That ρ acts as a Lie homomorphism on pairs $x \otimes a$, $y \otimes a'$ ($x,y \in S$; $a,a' \in A$) follows since $xy + yx = \mathrm{Tr}(xy)I$ in this case. Our identification of B (for involutions of first kind) and of $[BB]$ (for second kind) with a subspace of D results in $[x \otimes a, b] = x \otimes [ab]$ for b in this subspace, $x \in S$, $a \in A$. Likewise, $[x \otimes a, \beta_0] = 0$ for $\beta_0 \in B_0$ in the second case, and for $T \in K(C)$, the actions of T and $x \otimes a$ on $M'' \otimes C$ centralize each other, so that $[x \otimes a, T] = 0$. It follows that our map ρ has the homomorphism property for every bracket of which one factor is in $S \otimes A$. Likewise this is clear for every bracket of which one factor is in either $K'(C)$ or in $[BB]$ (in B, if the involution is of first kind). Now B_0 centralizes $[BB]$ and $K'(C)$ in L, as well as itself, and likewise $\tau(\beta_0)$ centralizes the image of all these under ρ. For $v, u \in M''$, $c \in C$, $\beta_0 \in B_0$, we have $[u \otimes c, \beta_0] = u \otimes c\beta_0$, which ρ sends to the mapping sending $v \otimes 1$ to $(u,v)c\beta_0$, $c' \in C$ to $u \otimes h(c\beta_0, c') = -u \otimes \beta_0 h(c, c')$. Meanwhile $[(u \otimes c)^\rho, \beta_0^\rho]$ sends $v \otimes 1$ to

$$\frac{-2m}{2+m}(u,v)c\tau(\beta_0) + \frac{m}{2+m}(v \otimes \beta_0)(u \otimes c)^\rho$$

$$= (u,v)c\frac{2\beta_0}{2+m} + \frac{m}{2+m}(u,v)c\beta_0 = (u,v)c\beta_0,$$

and c' to

$$(u \otimes h(c,c')) \frac{-2m}{2+m} \tau(\beta_0) - \frac{2}{2+m} u \otimes h(c,c'\beta_0)$$

$$= \frac{-m}{2+m} u \otimes \beta_0 h(c,c') \frac{-2}{2+m} u \otimes h(c,c')\beta_0$$

$$= - u \otimes \beta_0 h(c,c'),$$ verifying this homomorphism property.
There remains only the property for brackets $[u \otimes c, v \otimes c']$ from
$M'' \otimes C$. This verification is exactly as carried out for higher ranks in
connection with the proof of Proposition V.3 earlier in this section.
This completes the proof of Proposition V.4. Since $K(C)$ contains no
nilpotents, we must have h of Witt index zero on C. Thus the index
of our form h in $(M'' \otimes E) \oplus C$ is one. Conversely if we start with
an antihermitian form over an involutorial division algebra E, the
form of index $r > 0$ in at least $2r + 1$ dimensions, we obtain as derived algebra of the h-skew E-endomorphisms of our space a simple Lie
algebra of type BC_r over the fixed elements of the center of E. The
role of the space $M'' \otimes E$ is played by the sum of two dually paired
maximal totally isotropic subspaces, that of C by a complementary
anisotropic space.

In this setting, the group G', acting in $(M'' \otimes E) \oplus C$, is
contained in the unitary group and is its own commutator group. As one
sees from [D], p. 49, this means that G' is exactly the commutator
group of the unitary group, and our simple group of Chapter II is the
quotient by its center of this group.

b) The exceptional case of type C_3

This case arises when the involutorial alternative division algebra of Theorem III.7 (for $r = 3$) is not associative. In the discussion
preceding the statement of that theorem, we saw that $A \oplus B = E$ is an
octonion algebra over the center A, with $[B:A] = 7$, $E_0 = B$. Our
theorem gives rise to another special case of Tits' second construction,
namely that where the Jordan algebra J (over F) identifies with the
fixed 6 by 6 matrices with respect to the symplectic involution.
That J satisfies a cubic polynomial (which may be expressed in terms
of the Pfaffian) may be found in [JJ], §VI.4. We have $[J:F] = 15$, and
$Der_F(J)$ identifies with (bracketing with) the skew-symplectic transformations, thus with our algebra S. The elements J_0 of trace zero of
J have dimension 14 over F, and are our space M'; the decomposition

$$L = \mathcal{D} \oplus (S \underset{F}{\otimes} A) \oplus (M' \underset{F}{\otimes} B)$$

thus identifies with

$$\mathrm{Der}_A(E) \oplus \mathrm{Der}_A(J \underset{F}{\otimes} A) \oplus ((J_0 \underset{F}{\otimes} A) \otimes_A E_0),$$

displaying the three components in Tits' second construction (over A). The remarks of §4 again apply to the verification that an obvious linear isomorphism is an isomorphism of linear algebras, hence that <u>our construction yields a simple Lie algebra over</u> F. The centralizer in L of T is $L_0 = \mathcal{D} + T \otimes A + U_1 \otimes B + U_2 \otimes B$, with derived algebra $\mathcal{D} + U_1 \otimes B + U_2 \otimes B$, and this identifies as in §4 with $K(E)$, the skew A-endomorphisms of E with respect to its anisotropic norm form. Since E is a division algebra, it follows that $[L_0 L_0]$ is without nilpotent elements, and we have $Z(L_0) = T \otimes A$, in which T is clearly a maximal F-split torus. Thus the construction always results in a simple Lie algebra of relative type C_3, dimension $14 + 21 + 14 \times 7 = 133$ over its centroid A, (§1) thus an algebra which becomes the split exceptional Lie algebra of type E_7 over an extension of A. (By using a <u>split</u> octonion algebra over F in place of a division algebra, this construction may be used to show the existence of split algebras of type E_7 - cf. [JE], §10.)

We obtain generators for our simple group G' of §II.4 as the $\exp(\mathrm{ad}(x_\alpha \otimes a))$, $\exp(\mathrm{ad}(u_\alpha \otimes b))$, $(x_\alpha \in S_\alpha, u_\alpha \in M'_\alpha)$, $a \in A$, $b \in B$). Viewed as transformations of the 6-dimensional symplectic space M'', the root-vectors x_α and weight-vectors u_α are seen to have square zero, are isotropic in their respective forms, and 2α is not a root if α is. Now the effect of $\exp(\mathrm{ad}(x_\alpha \otimes a))$ on $D \in \mathcal{D}$ fixes D, sends

$$y \otimes a' \in S \otimes A \text{ to}$$
$$y \otimes a' + [[yx_\alpha] \otimes aa' + \frac{1}{2}[[yx_\alpha]x_\alpha] \otimes a^2 a',$$

the third term being zero if α is <u>long</u>, and

$$v \otimes b' \in M' \otimes B \text{ to}$$
$$v \otimes b' + [vx_\alpha] \otimes ab'.$$

Meanwhile $\exp(\mathrm{ad}(u_\alpha \otimes b))$ sends $D \in \mathcal{D}$ to $D - u_\alpha \otimes bD$,

$$y \otimes a' \text{ to } y \otimes a' - [u_\alpha y] \otimes a'b - \frac{1}{2}[[u_\alpha y]u_\alpha] \otimes a'b^2,$$

$$v \otimes b' \text{ to } v \otimes b' + [vu_\alpha] \otimes \frac{b'b+bb'}{2} + (v \circ u_\alpha) \otimes \frac{b'b-bb'}{2}$$
$$+ \mathrm{Tr}(vu_\alpha) <b',b> + \frac{1}{2}[[vu_\alpha]u_\alpha] \otimes (\frac{b'b+bb'}{2})b$$

$$-\frac{1}{2} \text{Tr}(vu_\alpha)u_\alpha \otimes b <b',b> + \frac{1}{2}(v \circ u_\alpha) \circ u_\alpha \otimes \frac{[[b'b]b]}{4} .$$

Now under these conditions $u_\alpha v u_\alpha = \text{Tr}(u_\alpha v)u_\alpha$, and
$(v \circ u_\alpha) \circ u_\alpha = \frac{4}{3} \text{Tr}(u_\alpha v)u_\alpha$, $[[vu_\alpha]u_\alpha] = -2 \text{Tr}(u_\alpha v)u_\alpha$. The last three terms thus consolidate to $-\text{Tr}(u_\alpha v)u_\alpha \otimes bb'b$, giving as our effect on $v \otimes b'$:

$$\text{Tr}(vu_\alpha) <b',b> + [vu_\alpha] \otimes \frac{b'b+bb'}{2}$$

$$+ v \otimes b' + (v \circ u_\alpha) \otimes \frac{b'b-bb'}{2} - \text{Tr}(u_\alpha v)u_\alpha \otimes bb'b.$$

We have thus given generators for the simple group G' in an absolutely irreducible representation on our 133-dimensional A-space $\mathcal{D} \oplus (S \otimes A) \oplus (M' \otimes B)$.

c) The exceptional case of type BC_2.

Under this heading, we refer to the case b) of Theorem III.10. By results of §1, the centroid of L identifies with A, so all our compositions are A-bilinear, and extension of the ground field from A is compatible with extension of these compositions, preserving all our multilinear identities, as well as the simplicity of L (but not its A-rank, in general). For the present, we suppress L and concentrate on the A-space B with its quadratic form h, and on the module C for the Clifford algebra $H(B)$, C bearing a nondegenerate alternate A-form q_1 relative to which the elements of B act skewly, the remaining operations being defined as in Theorem III.10 and satisfying the relations of §III.8. Our objectives will be to determine the constraints on B and C resulting from these conditions. We first study these in the case where B is <u>split</u> over A, which is to mean that the form h is of maximal Witt index and that, in the case where $[B:A]$ is odd, the orthogonal space to a dual pair of maximal totally isotropic spaces contains a vector b_0 with $h(b_0,b_0) = 1$ (in other words, if $[B:A] = 2r + 1$, then $(-1)^r \prod_{i=1}^{2r+1} h(b_i,b_i) \in A^2$ for every orthogonal basis $\{b_i\}$ for B).

In these cases, the structure of $H(B)$, and of its minimal right ideals, is known very explicitly [CS]: If $[B:A] = 2r$ is even, $H(B)$ is the full algebra of 2^r by 2^r matrices; with $p_1,\ldots,p_r$ and $n_1,\ldots,n_r$ dually paired bases for dually totally isotropic spaces, the

213

element $e = n_1 \ldots n_r \in H(\mathcal{B})$ generates a minimal right ideal, of dimension 2^r over A, and with A-basis the elements $ep_{i_1} \ldots p_{i_k}$, $0 \leq k \leq r$, $i_1 < i_2 \ldots < i_k$. In particular, the $H(\mathcal{B})$-module C must be a direct sum of copies of this module. Fixing an isomorphism for each summand, we denote the elements corresponding to e by $e, e', e'', \ldots$.

When $[\mathcal{B}:A] = 2r + 1$ is odd, we let $p_1, \ldots, p_r$ and $n_1, \ldots, n_r$ be as above, completing a basis with b_0 such that $h(b_0, b_0) = 1$. An orthogonal basis, say $\{b_0, p_i \pm n_i\}$, then has as its product a central element of $H(\mathcal{B})$, say z, with $z^2 \in A^2$, and 1 and z form an A-basis for the center Z of $H(\mathcal{B})$, Z being the sum of two copies of A. The (two-sided) ideals H_1, H_2 generated by these copies are full algebras of 2^r by 2^r matrices over A. One obtains minimal right ideals in $H(\mathcal{B})$ as the right ideals generated, respectively, by $e = (1 + b_0)n_1 \ldots n_r$ and by $f = (1 - b_0)n_1 \ldots n_r$. Since $eb_0 = (1)^r e$ and $fb_0 = (-1)^{r+1} f$, each of these ideals has a basis as before. In each case, the displayed element is a basis for the subspace of the right ideal annihilated by all of the n_i; combining this observation with the effects of b_0, computed above, we see that these ideals are inequivalent $H(\mathcal{B})$-modules. Thus C is a direct sum of $H(\mathcal{B})$-modules isomorphic to one or the other of these, and we denote canonically chosen corresponding generators by $e, e', \ldots$ resp. $f, f', \ldots$.

We next study the alternate form q_1 on C, beginning with the case $[\mathcal{B}:A] = 2r$. Following our conventions, we observe first that $q_1(e, e' p_{i_1} \ldots p_{i_k}) = 0$ if $\{i_1, \ldots, i_k\} \neq \{1, 2, \ldots, r\}$. For if $j \neq i_1, \ldots, i_k$, then n_j anticommutes with $p_{i_1}, \ldots, p_{i_k}$, so that $e' p_{i_1} \ldots p_{i_k} n_j = 0$. Meanwhile $p_j n_j + n_j p_j = 2$. Thus
$2q_1(e, e' p_{i_1} \ldots p_{i_k}) = q_1(e, e' p_{i_1} \ldots p_{i_k}(p_j n_j + n_j p_j)) = q_1(e, e' p_{i_1} \ldots p_{i_k} p_j n_j)$
$= - q_1(en_j, e' p_{i_1} \ldots p_{i_k} p_j) = 0$. From the skewness of the action of $\mathcal{B}$, it then follows that $q_1(ep_{i_1} \ldots p_{i_k}, e' p_{j_1} \ldots p_{j_m})$ is zero unless $\{i_1, \ldots, i_k\}$ and $\{j_1, \ldots, j_m\}$ constitute a partition of $\{1, \ldots, r\}$.

Now consider $q_2 : C \times C \to \mathcal{B}$. We have
$h(n_1, q_2(ep_{i_1} \ldots p_{i_k}, e' p_{j_1} \ldots p_{j_m})) = q_1(ep_{i_1} \ldots p_{i_k}, e' p_{j_1} \ldots p_{j_m} n_i)$
and q_2 is symmetric; thus $q_2(ep_{i_1} \ldots p_{i_k}, e' p_{j_1} \ldots p_{j_m})$ is orthogonal to all but at most one n_i, the exception occurring only when

$\{i_1,\ldots,i_k\} \cup \{j_1,\ldots,j_m\} = \{1,\ldots,r\}$ and $\{i_1,\ldots,i_k\} \cap \{j_1,\ldots,j_m\} = \{i\}$. In this case, $\pm h(n_i, q_2(ep_{i_1}\ldots p_{i_k}, e'p_{j_1}\ldots p_{j_m})) = 2q_1(e,e'p_1\ldots p_r)$, while $h(p_j, q_2(ep_{i_1}\ldots p_{i_k}, e'p_{j_1}\ldots p_{j_m})) = 0$ for all j, because of repetition of p_j with $p_j^2 = 0$. It follows that, with the indices as above,

$$q_2(ep_{i_1}\ldots p_{i_k}, e'p_{j_1}\ldots p_{j_m}) = \pm 2q_1(e,e'p_1\ldots p_r)p_i.$$

Returning to general sets of indices, we have

$$h(p_i, q_2(ep_{i_1}\ldots p_{i_k}, e'p_{j_1}\ldots p_{j_m})) = q_1(ep_{i_1}\ldots p_{i_k}, e'p_{j_1}\ldots p_{j_m} p_i),$$

which is equal to zero unless $\{i_1,\ldots,i_k\} \cap \{j_1,\ldots,j_m\} = \emptyset$, $\{i_1,\ldots,i_k\} \cup \{j_1,\ldots,j_m\} = \{1,\ldots,\hat{i},\ldots,r\}$, and in the latter case,

$$q_2(ep_{i_1}\ldots p_{i_k}, e'p_{j_1}\ldots p_{j_m}) = \pm q_1(e,e'p_1p_2\ldots p_r)n_i.$$

Now our formula of §III.8 for $b < c,c' >$ shows that $< ep_{i_1}\ldots p_{i_k}, e'p_{j_1}\ldots p_{j_m} >$ annihilates B if either $k + m < r - 2$, or $|\{i_1,\ldots,i_r\} \cap \{j_1,\ldots,j_m\}| > 2$. Since B generates $H(B)$, we see from the derivation-property that each such element must centralize the action of $H(B)$ on C. In particular, if any A-combination of such elements stabilizes $eH(B)$, an absolutely irreducible module, its action there must be as a scalar from A. Now these conditions are contradictory for $[B:A] > 4(r > 2)$, as the following shows: If $q_1(e,e'p_1\ldots p_r) = \gamma \neq 0$, $r > 4$, then $< ep_1\ldots p_{r-2}, e'p_1\ldots p_r >$ centralizes the action of $H(B)$ on C, and the formula of §III.8 for $c'' < c,c' >$, together with the above, shows that $e < ep_1\ldots p_{r-2}, e'p_1\ldots p_r > = -ep_1\ldots p_{r-2}q_1(e'p_1\ldots p_r, e) = \gamma ep_1\ldots p_{r-2}$, which is A-linearly independent of e, but in $eH(B)$, a submodule therefore stable under $< ep_1\ldots p_{r-2}, e'p_1\ldots p_r >$. By the form of q_1 and its nondegeneracy, such a pair of vectors e,e' must exist. We conclude that $[B:A] \leq 8$, since $r \leq 4$.

When $r = 4$, $< ep_1p_2p_3, e'p_1p_2p_3 >$ annihilates B and sends e to $-ep_1p_2p_3q_2(e,e'p_1p_2p_3) - e'p_1p_2p_3q_2(e,ep_1p_2p_3) = 0$, since in this case the values of q_2 are scalar multiples of n_4. On the other hand, ep_4 is sent to $-ep_1p_2p_3q_1(e'p_1p_2p_3, ep_4) - e'p_1p_2p_3q_1(ep_1p_2p_3, ep_4)$. Taking $e' = e$ if $q_1(e, ep_1p_2p_3p_4) = \gamma \neq 0$, our value is $-2\gamma ep_1p_2p_3$

or $-\gamma ep_1p_2p_3$, according as $e' = e$ or $e' \neq e$; in neither case is it zero, as it must be since e is sent to zero.

When $r = 3$, we again consider $< ep_1p_2p_3, e'p_1p_2p_3 >$, which, subject to choice of e' as above, sends e to either $2\gamma ep_1p_2p_3$ or $\gamma ep_1p_2p_3$, so is not a scalar on its stable space $eH(B)$. <u>Thus we conclude that</u> $r \leq 2$, <u>or</u> $[B:A] \leq 4$.

In these cases, we can determine the structure of C quite precisely: For $r = 2$, not every submodule $eH(B)$ can be totally isotropic; for if this is the case, we have summands $eH(B)$, $e'H(B)$, disjoint and totally isotropic, with $0 \neq \gamma = q_1(e, e'p_1p_2)$. But then $(e + e')H(B)$ is a submodule isomorphic to $eH(B)$, and $q_1(e + e', (e + e')p_1p_2) = q_1(e, e'p_1p_2) + q_1(e', ep_1p_2) = 2\gamma \neq 0$. Thus we may assume $q_1(e, ep_1p_2) = \gamma \neq 0$. By the same reasoning, the space q_1-orthogonal to $eH(B)$ is B-stable, hence $H(B)$-stable, and is either zero or has a submodule $e'H(B)$ with $q_1(e', e'p_1p_2) = \delta \neq 0$. Now suppose the latter is the case. Then $< e'p_1, e'p_1p_2 >$ is readily seen from the above to annihilate, while $e < e'p_1, e'p_1p_2 > = -\frac{1}{2} e q_2(e'p_1, e'p_1p_2) = \pm \delta ep_1$. Again we have a contradiction, and therefore conclude that $C = eH(B)$. Thus $[C:A] = 4$, a condition that therefore must hold before passing to our splitting field. That is, <u>the central simple A-algebra $H(B)$ of dimension 16, has a 4-dimensional module C, and therefore is already split</u>. It is easily seen that $B + [BB]$ is a subspace of $H(B)$ of A-dimension 10, all elements of which are q_1-skew in their action on C. Since C has dimension 4, these must be the totality of q_1-skew elements, and must contain nonzero elements of square zero.

Now $[BB]$ is readily identified with the skew transformations of B with respect to h, so contains no nonzero element of square zero. Thus our element N of square zero may be written in the form $N = b + bb' + d$, where $0 \neq b \in B, h(b,b') = 0$, and where $d = \Sigma b_i b'_i, h(b_i, b) = 0 = h(b, b'_i) = h(b_i, b'_i)$ for all i. Then the component, in the odd elements of the Clifford algebra, of N^2, is $b^2 b' + bb'b + bd + db = bd + db = 2bd$. Since b is a unit, $d = 0$, and $N = b + bb'$, $N^2 = b^2 - b^2 b'^2 = b^2(1 - b'^2)$, which cannot be zero since the form does not represent 1. <u>This eliminates the case $[B:A] = 4$</u>.

The case $[B:A] = 2$ is eliminated similarly, but more simply: It follows as above that in a splitting we have non-totally isotropic

216

modules $eH(B)$, and that if one of these fails to exhaust C we have at least two q_1-orthogonal such, say $eH(B)$, $e'H(B)$, with $\gamma = q_1(e, ep_1) \neq 0 \neq q_1(e', e'p_1)$. Now it is clear from our definitions and the q_1-orthogonality of $eH(B)$ and $e'H(B)$ that $<ep_1, e'p_1>$ annihilates B. The formula for $c''<c, c'>$ gives $e<ep_1, e'p_1> = \gamma e'p_1$, so that $<ep_1, e'p_1>$ induces an $H(B)$-isomorphism of $eH(B)$ onto $e'H(B)$. But $en_1 = 0$ while $e'p_1 n_1 = 2e' \neq 0$, a contradiction. Thus $C = eH(B)$, $[C:A] = 2$, $H(B) = \operatorname{End}_A(C)$, and this state of affairs holds over the original centroid A. Since the form h does not represent 1, $H(B)$ cannot contain idempotents $\neq 0, 1$, and we have a contradiction. Thus $[B:A]$ <u>is even only if</u> $B = 0$; <u>this is also impossible</u>, since then C is simply a symplectic A-space, necessarily of dimension at least 2, and $D = <C,C>$ is faithfully represented on C. Taking $0 \neq c \in C$, we find that $<c,c> \neq 0$ has square zero in its action on C, so has $(\operatorname{ad} <c,c>)^3 = 0$, and $<c,c>$ centralizes T. This completes the proof that $[B:A]$ <u>is odd</u>.

In this case, again assuming B split, of dimension $2r + 1$, with basis $b_0, p_1, \ldots, p_r, n_1, \ldots, n_r$ as before, and generators $e = (1 + b_0) n_1 \ldots n_r$ and $f = (1 - b_0) n_1 \ldots n_r$ for minimal right ideals, we have as before that $q_1(e, e'p_{i_1} \ldots p_{i_k}) = 0 = q_1(e, f'p_{i_1} \ldots p_{i_k}) = q_1(f, f'p_{i_1} \ldots p_{i_k})$ unless $\{i_1, \ldots, i_k\} = \{1, \ldots, r\}$, and that q_1 is completely determined on C by the values $q_1(e, e'p_1 \ldots p_r)$, $q_1(e, f'p_1 \ldots p_r)$, $q_1(f, f'p_1 \ldots p_r)$ for all e, e', f'. Here the evaluation of $q_2(ep_{i_1} \ldots p_{i_k}, e'p_{j_1} \ldots p_{j_m})$, etc., is as before, except when $\{i_1, \ldots, i_k\} \cup \{j_1, \ldots, j_m\}$ <u>is a partition of</u> $\{1, \ldots, r\}$. In that case, the element in question is orthogonal to all p_i and n_i, but $h(b_0, q_2(ep_{i_1} \ldots p_{i_k}, e'p_{j_1} \ldots p_{j_m})) = q_1(ep_{i_1} \ldots p_{i_k}, e'p_{j_1} \ldots p_{j_m} b_0) = (-1)^{m+r} q_1(ep_{i_1} \ldots p_{i_k}, e'p_{j_1} \ldots p_{j_m})$; likewise $h(b_0, q_2(ep_{i_1} \ldots p_{i_k}, f'p_{j_1} \ldots p_{j_m})) = (-1)^{m+r+1} q_1(ep_{i_1} \ldots p_{i_k}, f'p_{j_1} \ldots p_{j_m})$. Thus $q_2(ep_{i_1} \ldots p_{i_k}, e'p_{j_1} \ldots p_{j_m}) = \pm q_1(e, e'p_1 \ldots p_r) b_0$, and likewise with e' replaced by f', and/or e by f.

It follows again, with $u = e$ or f, $v = e'$ or f', that $<up_{i_1} \ldots p_{i_k}, vp_{j_1} \ldots p_{j_m}>$ <u>annihilates</u> B <u>if the indices omit at least three from</u> $1, \ldots, r$, <u>or if the two index-sets have at least three in common</u>.

217

Now suppose $r \geq 6$, so that $< up_1p_2p_3, vp_1\ldots p_r >$ and $< up_1p_2p_3p_4, vp_1\ldots p_r >$ both annihilate B. If $uH(B)$ is not totally isotropic, we take $v = u$; otherwise we take v so that $\gamma = q_1(u, vp_1\ldots p_r) \neq 0$. Then we have

$$u < up_1p_2p_3, vp_1\ldots p_r > = - up_1p_2p_3q_1(vp_1\ldots p_r, u)$$
$$- up_1p_2p_3q_2(vp_1\ldots p_r, u)$$
$$= \gamma up_1p_2p_3 - up_1p_2p_3q_2(u, vp_1\ldots p_r)$$
$$= \gamma up_1p_2p_3 - \gamma up_1p_2p_3b_0 = 2\gamma up_1p_2p_3 \text{ if}$$

$u = e, v = e'$; if $u = f, v = f'$, we have

$$up_1p_2p_3q_2(u, vp_1\ldots p_r) = - \gamma up_1p_2p_3b_0 = - \gamma up_1p_2p_3,$$

and again our value is $2\gamma up_1p_2p_3$. On the other hand, if $u = e, v = f'$, we have.

$$u < up_1p_2p_3p_4, vp_1\ldots p_r > = \gamma up_1p_2p_3p_4$$
$$- up_1p_2p_3p_4q_2(vp_1\ldots p_r, u) = \gamma up_1p_2p_3p_4$$
$$+ \gamma up_1p_2p_3p_4v_0 = 2\gamma up_1p_2p_3p_4.$$

The combination of these two cases yields an element of $< C, C >$ annihilating B and stabilizing $uH(B)$, but not acting as a scalar on this (absolutely) irreducible $H(B)$-module. This completes the proof that $[B:A] \leq 11$. It will also be noted that we have shown that $q_2(C,C) = B$ in the split case, from which this equality follows generally.

We next consider the remaining odd dimensions, deducing the precise structure of $H(B)$ and of C, and eliminating the case $[B:A] = 9$. We begin with $[B:A] = 11$, so $r = 5$, and assume we have the split case. Here we claim that any submodule of C of the form $eH(B)$ is q_1-orthogonal to any one of the form $fH(B)$: For $q_1(e, fp_1p_2p_3p_4p_5) = q_1(e, fb_0p_1\ldots p_5) = - q_1(e, fp_1\ldots p_5b_0) = q_1(eb_0, fp_1\ldots p_5) = -q_1(e, fp_1\ldots p_5)$ by the fact that $fb_0 = f, eb_0 = - e$ when r is odd, and this yields the claim. It follows that there must be a nonsingular submodule $eH(B)$ or $fH(B)$: For if $eH(B)$ is totally isotropic, we must have an e' with $\gamma = q_1(e, e'p_1p_2p_3p_4p_5) \neq 0$, and we may assume $e'H(B)$ totally isotropic. Then $q_1((e+e'), (e+e')p_1p_2p_3p_4p_5) = 2\gamma$, and $(e + e')H(B)$ is the desired submodule; the argument with f is the same.

Now suppose $uH(B)$ and $vH(B)$ ($u = e$ or f, $v = e'$ or f') are present; they may be assumed nonsingular and orthogonal with respect to q_1. We may assume $q_1(u, up_1p_2p_3p_4p_5) = \gamma \neq 0 \neq \delta = q_1(v, vp_1 \ldots p_5)$. It follows from the formulas for $b < c, c' >$ that $< up_1p_2, up_1 \ldots p_5 >$ and $< vp_1p_2, vp_1 \ldots p_5 >$ annihilate all of our standard basis for B except n_1 and n_2, with $n_1 < up_1p_2, up_1 \ldots p_5 > = - q_2(up_2, up_1 \ldots p_5) + q_2(up_1p_2, up_2p_3p_4p_5) = 4\gamma p_2$, $n_2 < up_1p_2, up_1 \ldots p_5 > = q_2(up_1, up_1 \ldots p_5) - q_2(up_1p_2, up_1p_3p_4p_5) = - 4\gamma p_1$,

and likewise with u replaced by v, γ by δ. Hence

$$\delta < up_1p_2, up_1 \ldots p_5 > - \gamma < vp_1p_2, vp_1 \ldots p_5 >$$

centralizes the action of B on C, but sends u to

$$- \delta up_1p_2 q_1(up_1p_2p_3p_4p_5, u) - \delta up_1p_2 q_2(up_1p_2p_3p_4p_5, u)$$
$$= \delta\gamma up_1p_2 + \delta\gamma\varepsilon up_1p_2 b_0, \text{ where } ub_0 = \varepsilon u,$$

and this is $\delta\gamma(1 + \varepsilon^2)up_1p_2$; $\varepsilon^2 = 1$ gives an image of $2\delta\gamma up_1p_2$ for u, and this is a contradiction as in previous cases. It follows that $[C:A] = 32$, and therefore that $H(B)$ is already the direct sum of two 32 by 32 matrix algebras over the original ground field; in particular, the center $Z(H(B))$ splits, or $-\prod_{i=1}^{11} h(b_i, b_i) \in A^2$, for any orthogonal basis $b_1, \ldots, b_{11}$ for B.

The formula for $b'' < b, b' >$ and our work in §3 show that $< B, B >$ already induces all skew transformations of B. From (79') of §III.8, and the fact that $q_2(C,C) = B$, $< B,B > \subseteq < C,C >$. Since C is now known to be an absolutely irreducible $H(B)$-module, the annihilator of B in $< C,C >$ acts on C by q_1-skew scalar transformations, hence by zero. It follows that $D = < C,C > = < B,B >$ identifies with the skew transformations of B, or with $[BB]$, in $H(B)$, and has A-dimension $\binom{11}{2} = 55$. The resulting dimension of L over A is $55 + (10 \times 1) + (5 \times 11) + (4 \times 32) = 248$.

Conversely, if we start with a quadratic form h in a space B of dimension 11 over a field extension A of F, not representing 1, and of square discriminant $((-1)^5 \prod h(b_i, b_i)$, as before), such that the Clifford algebra $H(B)$ splits completely, and with an irreducible $H(B)$-module C, then the involution in the even Clifford algebra $H^+(B)$ induced by the involution in $H(B)$ which is -1 on B has a skew space of dimension 528, and therefore results from transposition with respect to a symplectic form q_1 on C. The element $z = \prod b_i$ of Z, the center of $H(B)$, acts as a scalar on C, while for each j, $1 \leq j \leq 11$, the elements $\prod_{i \neq j} b_i \in H^+(B)$ act q_1-skewly. Since these elements are a basis for zB, the elements of B also act q_1-skewly as transformations of C. One may then define $q_2: C \times C \to B$ by $h(b, q_2(c,c')) = q_1(c, c'b)$ for all $b \in B$ and use (55"), (69'), (84'), (85') of §III.8 to define the remaining compositions $< b, b' >$, $< c, c' >$. By passing to a splitting field for h, these take the explicit forms above, and one can there verify by direct and lengthy computation the identities that assure that L, defined in terms of these, is a Lie algebra. That L is simple and that T is a maximal split torus are then easy. Thus <u>such an algebra of relative type</u> BC_2 <u>exists if and only if</u> A <u>admits a quadratic form</u> h <u>in eleven dimensions whose Clifford algebra has the prescribed structure</u>.

Next we show that $[B:A] = 9$ <u>cannot occur</u>. For, with our customary notation, $q_1(e, e'p_1p_2p_3p_4) = q_1(ep_4p_3p_2p_1, e') = q_1(ep_1p_2p_3p_4, e') = -q_1(e', ep_1p_2p_3p_4)$, showing that <u>every submodule</u> $eH(B)$ (and likewise every $fH(B)$) of C <u>must be totally</u> q_1-<u>isotropic</u>. Now assume, without loss of generality, that an $eH(B)$ is present; then there also is present a $vH(B)$ ($v = e'$ or f') with $q_1(v, ep_1p_2p_3p_4) = \gamma \neq 0$. We further have $eb_0 = \varepsilon e, \varepsilon = \pm 1$. Since $eH(B)$ is totally isotropic, $< ep_1, ep_1p_2p_3p_4 > \in < C, C >$ annihilates B, and sends v to $- ep_1 \cdot - \gamma - ep_1 \cdot q_2(ep_1p_2p_3p_4, v)$. As computed earlier, $q_2(ep_1 \ldots p_4, v) = q_2(v, ep_1 \ldots p_4) = \varepsilon \gamma b_0$. Thus the above is $\gamma ep_1 - \varepsilon \gamma ep_1 b_0 = \gamma ep_1 + \varepsilon \gamma eb_0 p_1 = 2\gamma ep_1$. Since $f'H(B)$ is not isomorphic to $eH(B)$, we must have $vH(B)$ isomorphic to $eH(B)$, with $< ep_1, ep_1p_2p_3p_4 >$ inducing the isomorphism. As before, no such isomorphism can send v to ep_1 since $vp_1 \neq 0$ while $ep_1^2 = 0$.

We next study the case $[B:A] = 7$. As with $[B:A] = 9$, we have $q_1(e, e'p_1p_2p_3) = - q_1(e', ep_1p_2p_3)$ and likewise for f, f', showing that <u>every irreducible submodule of</u> C (in the split case) <u>must be totally</u> q_1-<u>isotropic</u>. As in the last case, we assume $eH(B)$ present, $vH(B)$ present, $q_1(v, ep_1p_2p_3) = \gamma \neq 0$, $eb_0 = \varepsilon e$ (v an image of e or f under an isomorphism of $H(B)$-modules). Then $< e, ep_1p_2p_3 >$ annihilates B and sends v to $2\gamma e$. Since $< e, ep_1p_2p_3 >$ induces a homomorphism of $H(B)$-modules, $vH(B)$ must be isomorphic to $eH(B)$, or <u>non-isomorphic submodules of</u> C <u>are</u> q_1-<u>orthogonal</u>.

Thus we have C decomposed into a q_1-orthogonal direct sum of sums of q_1-dual pairs of isomorphic irreducible $H(B)$-modules, each of the irreducible modules being totally q_1-isotropic. We next show that only one such pair can be present.

Thus assume $uH(B)$, $u'H(B)$ and $vH(B)$, $v'H(B)$ are two orthogonal sets of pairs, u,u',v,v' being images of e or f, with $ub_0 = \varepsilon u$, $vb_0 = \varepsilon'v$ (hence likewise for u',v'), $\varepsilon^2 = 1 = \varepsilon'^2$, and with $q_1(u,u'p_1p_2p_3) = \gamma \neq 0$, $q_1(v,v'p_1p_2p_3) = \delta \neq 0$. Now if $\varepsilon \neq \varepsilon'$, so $\varepsilon\varepsilon' = -1$, we have that $< up_1, v'p_1p_2p_3 >$ annihilates B and sends v to $\delta up_1 - up_1 q_2(v'p_1p_2p_3, v) = \delta up_1 - up_1(-\delta\varepsilon'b_0) = \delta(1-\varepsilon\varepsilon')up_1 = 2\delta up_1 \neq 0$. As in previous cases, this is impossible, and all irreducible submodules of C are isomorphic. Thus we may assume $\varepsilon' = \varepsilon$ in the above, where we now consider $< up_1p_2, v'p_1p_2p_3 >$, which again annihilates B, and sends v to $\delta up_1p_2 - up_1p_2(-\delta\varepsilon b_0) = 2\delta\, up_1p_2$, a contradiction as before. We have thus proved that, when B is split, C is the direct sum of two isomorphic irreducible $H(B)$-modules, each totally isotropic with respect to q_1; in particular, $[C:A] = 16$.

We now deduce consequences of the above for our non-split situation. We have $[H(B):A] = 2^7$; if the center Z of $H(B)$ is a quadratic field over A, then $H(B)$ is a central simple algebra of dimension 2^6 over this field, and irreducible modules for $H(B)$ have A-dimension at least $2 \cdot 2^3 = 16$, this minimum being attained only if $H(B)$ is a full matrix algebra over Z. But in this case, the irreducible modules split into direct sums of two inequivalent irreducible ones (already inequivalent over Z) upon splitting Z. Thus Z is split over A, or the discriminant $(-1)^3 \prod_{i=1}^{7} h(b_i,b_i) \in A^2$. We therefore have $H(B) = H_1 \oplus H_2$, where the H_i are ideals, isomorphic as A-algebras, $[H_i:A] = 2^6$. (It is clear, upon projecting the even Clifford algebra $H^+(B)$ on the H_i, that each H_i is isomorphic to $H^+(B)$.) Now from $[C:A] = 2^4$, we have two possibilities: i) the H_i are split over A, and C is the direct sum of two irreducible submodules; or ii) the H_i are algebras of 8 by 8 matrices over a quaternionic division algebra Q, and C is irreducible. In the next paragraph, we show that the maximality of our split torus T rules out i).

For our analysis of the split case shows that these two submodules C_1, C_2, must be absolutely irreducible and totally q_1-isotropic. Moreover, if upon splitting, they become $eH(B)$ resp. $e'H(B)$ as before, we have

$$e < e', e'p_1p_2p_3 > = \gamma e' - e'q_2(e, e'p_1p_2p_3)$$

$= \gamma e' + \varepsilon \gamma e' b_0 = 2\gamma e' \neq 0$. Therefore it follows that $< c, c' > \neq 0$ for some $c, c' \in C_1$ (now back over the original ground field). On the other hand, it is clear from the fact that q_1 and q_2 vanish identically on the submodule C_1 that $< c, c' >$ annihilates B and C_1, while mapping C_2 into C_1. Thus $< c, c' > \neq 0$ has square zero in its action on $A \oplus B \oplus C$, and is therefore a nonzero nilpotent centralizing T. This eliminates the case i) and leaves only the possibility ii).

Again, as with $[B:A] = 11$, we have $< B, B > \subseteq < C, C > = D$, and D is the direct sum of $< B, B >$ and the annihilator of B in $< C, C >$. The elements of this annihilator act on C as $H(B)$-endomorphisms, hence as multiplications by elements of our quaternionic division algebra Q. Thus the annihilator of B has A-dimension at most 4, indeed at most 3, since multiplications by nonzero elements of A are not q_1-skew, and so cannot be included. On the other hand, we have seen in the split case that the annihilator of B in $< C, C > = D$ is non-zero; in fact, with the notations of the last paragraph, it contains $< e', e'p_1p_2p_3 >$, a scalar multiple of which sends e to e', e' to zero, and $< e, ep_1p_2p_3 >$ yields an element sending e' to e, e to zero. The bracket of these is in D and sends e to e, e' to $-e'$; the result is that the annihilator of B in D has dimension at least 3 over A, and indeed that its derived algebra has at least dimension 3 (since the above is the 3-dimensional split simple algebra). These conditions are independent of extension of A. Therefore we see that the annihilator of B in D must be exactly the multiplications on C by elements of the derived algebra $[QQ]$, and $D \cong < B, B > \oplus [QQ] \cong K_A(B) \oplus [QQ]$ has A-dimension $21 + 3$, so that L has A-dimension

$$24 + (10 \times 1) + (5 \times 7) + (4 \times 16) = 133.$$

Conversely, if we start with A, B as above, and take C to be one of our two irreducible $H(B)$-modules, then C is an irreducible $H^+(B)$-module, $H^+(B)$ being a full matrix algebra of degree 4 over the quaternionic division algebra Q. The involution in $H(B)$ acting as -1 on B induces in $H^+(B)$ an involution whose skew elements have dimension $\binom{7}{2} + \binom{7}{6} = 28$, while $[H^+(B):A] = 64$. It follows that on an irreducible $H(B)$-module C, there is a nondegenerate form q, antihermitian with respect to the standard involution in Q, such that our involution in $H^+(B)$ is transposition with respect to q. Letting q_1 be the projection of q on A (the symmetric elements of Q), q_1 is a non-degenerate symplectic form on C, relative to which the elements of the form $\prod_{i \neq j} b_i, 1 \leq j \leq 7$, $\{b_i\}$ an orthogonal basis for B, act skewly. Moreover, we may choose $\{b_i\}$ so that $z = \Pi b_i \in H(B)$ has $z^2 = 1$. Since the action of z in C centralizes that of $H^+(B)$, z acts as multiplication by an element of the division algebra Q. It follows that z acts as a scalar from A. As j runs from 1 to 7, $z \cdot \prod_{i \neq j} h_i$ runs over a basis for B; from this we see that the elements of B act q_1-skewly on C. Now define $q_2: C \times C \to B$ by $h(b, q_2(c, c')) = q_1(c, c'b)$ for all $b \in B$, and define $< b, b' >$, $< c, c' >$ using §III.8.

By passing to the split case, we get explicit formulas for these quantities as above, and can verify by computation that our construction based on these yields a Lie algebra L. That L is simple with T as maximal split torus is then easily verified. We conclude that <u>there exists such a Lie algebra of relative type</u> BC_2 <u>if and only if</u> A <u>admits a quadratic form in seven variables whose Clifford algebra has the prescribed structure</u>.

Now consider the case where $[B:A] = 5$; as before, we assume B is split to begin with. We first show that C must have q_1-non-singular irreducible submodules unless nonisomorphic ones occur in dual pairs. For if, say, $eH(B)$ and $e'H(B)$ are isomorphic, totally isotropic, and not q_1-orthogonal, we have $q_1(e, e'p_1p_2) = \gamma \neq 0$, and $q_1(e', ep_1p_2) = -q_1(e'p_1p_2, e) = \gamma$, so that as before $(e + e')H(B)$ is nonsingular. We proceed to prove that C <u>can have no nonsingular irreducible submodules</u>.

First let $eH(B)$ be such a nonsingular submodule, and assume $e'H(B)$ is a second one, canonically isomorphic, as usual, and q_1-orthogonal to $eH(B)$. We have $q_1(e, ep_1p_2) = \gamma \neq 0$, $q_2(eH(B), e'H(B)) = 0$, and $< ep_1, e'p_1p_2 >$ annihilates B, but sends e to $-e'p_1p_2q_2(ep_1, e) = -\gamma e'p_1p_2n_2 = -2\gamma e'p_1$. As before, this is impossible. Thus $(eH(B))^\perp$ contains no nonsingular module isomorphic to $eH(B)$. Likewise, if a nonsingular $fH(B) \subseteq C$, $(fH(B))^\perp$ contains no nonsingular submodule isomorphic to $fH(B)$. Now suppose our nonsingular $eH(B)$ has the property that $(eH(B))^\perp$ contains a nonsingular irreducible submodule, which must be of the form $fH(B)$. But then, as above, $< ep_1, fp_1p_2 >$ annihilates B and sends e to $-2\gamma fp_1 \neq 0$, which is impossible. Thus C <u>contains</u>

at most one irreducible nonsingular summand. Still assuming $eH(B)$ to be such, we see that if $eH(B) \neq C$, C must contain submodules $e'H(B)$, $fH(B)$, both q_1-orthogonal to $eH(B)$. But then, as above, $<ep_1, fp_1p_2>$ annihilates B and sends e to $-2\gamma fp_1 \neq 0$, a contradiction. We conclude: Either $C = eH(B)$ or $C = fH(B)$, so that $[C:A] = 4$, or C contains no irreducible q_1-nonsingular submodules.

Now we explore the latter alternative; we have seen that there must be dually paired totally isotropic submodules $eH(B)$, $fH(B)$ present, and, by adjusting f by an A-multiple, that $q_1(e, fp_1p_2) = 1$. If $eH(B) + fH(B) \neq C$, we must have a second such pair $e'H(B)$, $f'H(B)$ q_1-orthogonal to $eH(B) + fH(B)$. Then, as above, $<e'p_1, fp_1p_2>$ annihilates B and sends f' to $-fp_1p_2q_2(e'p_1, f') = -fp_1p_2n_2 = -2fp_1$, which is absurd as before. Thus $C = eH(B) + fH(B)$ in this case, and $[C:A] = 8$.

In the original case, we therefore have $[C:A] = 4$ or 8, $[H(B):A] = 2^5 = 32$. Let Z be the center of $H(B)$ as before, $[Z:A] = 2$. If Z is a field, $H(B)$ is central simple of dimension 16 over Z, and irreducible $H(B)$-modules have Z dimension at least 4, hence A-dimension at least 8, this minimum being achieved only when $H(B)$ is a full algebra of 4 by 4 Z-matrices. Thus C must be irreducible (of A-dimension 8) and $H(B)$ split over Z, provided Z is a field.

If Z is not a field, i.e., if the discriminant $(-1)^2 \prod_{i=1}^{5} h(b_i, b_i)$ (b_i an orthogonal basis) is a square in A, then, since h does not represent 1, the even Clifford algebra $H^+(B)$ is a division algebra. (I owe this observation to my colleague T. Tamagawa, and my version of his proof appears in Appendix C.) Now $H(B)$ is the direct sum of two copies of $H^+(B)$, which has A-dimension 16, as do all irreducible $H(B)$-modules; since $[C:A] = 4$ or 8, this is impossible.

We conclude: If $[B:A] = 5$, then the discriminant of h is not a square in A, while h fails to represent 1, and $H(B)$ is a full matrix algebra over its center Z. The module C for $H(B)$ is irreducible.

Again we have $B = q_2(C,C)$, hence $<B,B> \subseteq <C,C> = D$, and D is the direct sum of $<B,B>$ and the annihilator of B in $<C,C>$. Since the $H(B)$-endomorphisms of C are isomorphic to Z, this annihilator has A-dimension at most 2; since it does not contain

1, its A-dimension is at most 1, and because its elements are q_1-skew, it is non-zero if and only if it contains the map which is zero on B and acts on C by scalar multiplication by $z = b_1 b_2 b_3 b_4 b_5$, where the b_i are an orthogonal basis for B. To see that the latter actually is in D, it is enough to show that the annihilator of B in $<C,C>$ is nonzero in the split case resulting from this situation, where $C = eH(B) + fH(B)$ as above, both summands totally q_1-isotropic, where we may assume $q_1(e, fp_1 p_2) = 1$. Here one readily verifies that $<e, fp_1 p_2> - <ep_1 p_2, f>$ annihilates B and sends e to $2e$, f to $-2f$. This yields our identification of D with $<B,B> \oplus Az$, of A-dimension $10 + 1 = 11$. The resulting A-dimension of L is

$$11 + (10 \times 1) + (5 \times 5) + (4 \times 8) = 78.$$

By appealing to earlier considerations and the results in [J], Chapter X, on algebras "of types B_6 and C_6" (when split), one may see that, over a suitable extension field of A, L becomes a split algebra of type E_6.

The centralizer L_0 of T now identifies with $T \otimes A + U_0 \otimes B + <B,B> + Az$, its derived algebra with $U_0 \otimes B + <B,B>$, which contains no nonzero nilpotents as before. The elements of $T \otimes A + Az$, the center of L_0, act in $M'' \otimes C$ in such a way as to be diagonalizable over Z, the only elements to have all eigenvalues in F being those of $T = T \otimes F = T \otimes 1$.

Conversely, whenever we start with a quadratic form h over A of nonsquare discriminant in a five-dimensional space B, not representing 1, whose Clifford algebra $H(B)$ is a full matrix algebra over its center Z, then the involution in $H(B)$ which is -1 on B is of second kind, and is realized as transposition with respect to a form q on an irreducible $H(B)$-module C. This form may be taken to be antihermitian with respect to the involution induced in Z. If we take $q_1(c,c') = Tr_{Z/A} q(c,c')$, q_1 is a nondegenerate symplectic form on C relative to which the elements of B act as skew transformations. Then we may define $q_2(c,c') \in B$ for $c,c' \in C$ by $h(b, q_2(c,c')) = q_1(c, c'b)$ for all $b \in B$. In terms of these data, the remaining compositions of $L = <B,B> + <C,C> + S \otimes A + M' \otimes + M'' \otimes C$ are defined as in §III.8, using (55"), (69'), (84'), (85') of that section. By passing to a splitting field, we get explicit forms for h, q_1, q_2 as above, and verify directly from these all the identities necessary to show that L

is a Lie algebra (those of §III.8, including the derivation-property for the action of $\mathcal{D} = \langle \mathcal{B}, \mathcal{B} \rangle + \langle \mathcal{C}, \mathcal{C} \rangle = \langle \mathcal{C}, \mathcal{C} \rangle$). This Lie algebra is readily seen to be simple, and to have T as maximal split torus. Thus <u>algebras of this type exist whenever there is a quadratic form in five dimensions over</u> A <u>whose Clifford algebra has the given structure, and only then</u>.

When $[\mathcal{B}:A] = 3$, we again consider first the split case, where irreducible modules have A-dimension 2, with basis either $\{e, ep_1\}$ or $\{f, fp_1\}$. First we note that the only way for all irreducible submodules to be totally q_1-isotropic is to have all $\{e, ep_1\}$ q_1-orthogonal to one another, and likewise for the $\{f, fp_1\}$. For if, say, totally isotropic $\{e, ep_1\}$ and $\{e', e'p_1\}$ are not orthogonal, we have $q_1(e, e'p_1) = \gamma \neq 0$, and $q_1((e+e'), (e+e')p_1) = 2\gamma$ as before. No longer assuming all irreducible submodules totally isotropic, suppose now that some $\{e, ep_1\}$ and $\{f, fp_1\}$ are (present and) q_1-orthogonal. Since neither of $\{e, ep_1\}^\perp$ nor $\{f, fp_1\}^\perp$ can contain the other, we have $u = e'$ or f', $ub_0 = \varepsilon u$ ($\varepsilon = \pm 1$), with the irreducible submodule $\{u, up_1\}$ orthogonal to $\{e, ep_1\}$ but not to $\{f, fp_1\}$. Evidently $\langle e, fp_1 \rangle \in \langle \mathcal{C}, \mathcal{C} \rangle$ annihilates $\mathcal{B}$. We may assume $q_1(u, fp_1) = \gamma \neq 0$, from which it follows that $q_2(u, fp_1) = \gamma b_0$, and $u\langle e, fp_1 \rangle = -eq_1(fp_1, u) - eq_2(fp_1, u) = \gamma e + \gamma e b_0 = 2\gamma e$. It follows that $u = e'$, by the inequivalence of the two types of modules. We may also reverse the roles of e and f. From this we conclude that <u>if one irreducible submodule is nonsingular, then all irreducible submodules are isomorphic</u>.

Thus we may assume our $\{e, ep_1\}$ and $\{f, fp_1\}$ are totally isotropic, and u as above, so $u = e'$; but then $u\langle fp_1, fp_1 \rangle = -2fp_1 q_1(fp_1, u) - 2fp_1 q_2(fp_1, u) = 2\gamma fp_1 - 2\gamma fp_1 b_0 = 4\gamma fp_1$; since $\langle fp_1, fp_1 \rangle$ annihilates $\mathcal{B}$, this is impossible. Thus there can exist no orthogonal pair $\{e, ep_1\}$ and $\{f, fp_1\}$, from which it follows that <u>if all irreducible submodules are totally isotropic, then there are only two, and</u> $\mathcal{C} = \{e, ep_1\} \oplus \{f, fp_1\}$.

Returning to the non-split context, we see that the latter underlined case must hold if the center Z of $H(\mathcal{B})$ does not split. From $[\mathcal{C}:A] = 4$, we see that $H(\mathcal{B})$ must be the algebra of 2 by 2 Z-matrices. If b_1, b_2, b_3 are an orthogonal basis for $\mathcal{B}$, Z is spanned by 1 and $z = b_1 b_2 b_3$, the latter being q_1-symmetric in its action on $\mathcal{C}$:

$$q_1(c, c'b_1 b_2 b_3) = -q_1(cb_3 b_2 b_1, c') = q_1(cb_1 b_2 b_3, c').$$

Thus the q_1-skew elements of $H(B)$ are $B + zB$, of Z-dimension 3; it follows that, in the algebra of 2 by 2 Z-matrices, our involution associated with q_1 is symplectic, so has non-zero nilpotent skew elements, of which we may assume one to be $b + zb'$ $(b,b' \in B)$. Then $(b + zb')^2 = 0 = (h(b,b) + z^2 h(b',b')) + 2zh(b,b')$. By the independence of 1 and z over A, we have $h(b,b') = 0$, while evidently both are non-zero. Thus there is b'', completing an orthogonal basis containing b, b', such that $z = bb'b''$, and we have $0 = (b + zb')^2 = h(b,b) + (bb'b'')^2 h(b',b') = h(b,b)(1 + (b'b'')^2 h(b',b')) = h(b,b)(1 - h(b',b')^2 h(b'',b''))$. Thus $h(b'',b'') = h(b',b')^{-2}$, from which h would represent 1, contrary to what we know.

It follows that the center of $H(B)$ splits, so that $-h(b_1,b_1)h(b_2,b_2)h(b_3,b_3)$ is a square in A, which we may normalize to be 1: $h(b_3,b_3)^{-1} = -h(b_1,b_1)h(b_2,b_2)$. We set $z = b_1 b_2 b_3$, so that $z^2 = 1$. Now the even Clifford algebra $H^+(B)$ has A-basis 1, $b_1 b_2$, $b_1 b_3$, $b_2 b_3$, with $(b_1 b_2)^2 = h(b_1,b_1)h(b_2,b_2) = h(b_3,b_3)^{-1}$, $(b_1 b_3)^2 = h(b_2,b_2)^{-1}$, $(b_2 b_3)^2 = h(b_1,b_1)^{-1}$, $(b_1 b_2)(b_1 b_3) = -h(b_1,b_1)b_2 b_3 = -(b_1 b_3)(b_1 b_2)$, etc.; $H^+(B)$ is a quaternionic algebra over A. If $H^+(B)$ is not a division algebra, $H^+(B)$ contains nonzero elements $a_0 + a_1 b_2 b_3 + a_2 b_1 b_3 + a_3 b_1 b_2$ of square zero. The component of trace zero in the square of this is $2a_0(a_1 b_2 b_3 + a_2 b_1 b_3 + a_3 b_1 b_2)$, which evidently implies $a_0 = 0$; thus our element has square $-a_1^2 h(b_2,b_2)h(b_3,b_3) - a_2^2 h(b_1,b_1)h(b_3,b_3) - a_3^2 h(b_1,b_1)h(b_2,b_2) = 0$. But this is $\sum_{i=1}^{3} a_i^2 h(b_i,b_i)^{-1} = \sum_{i=1}^{3} (a_i h(b_i,b_i)^{-1})^2 h(b_i,b_i)$, which would force our form on B to be isotropic. Thus $H^+(B) = A + zB$ is a <u>division algebra</u>.

Now consider the map $a + b \to a + zb$ from $A \oplus B$ to $H^+(B)$. This is clearly an A-isomorphism of vector spaces, and from $(zb)^2 = z^2 b^2 = h(b,b)$ it follows that it establishes an equivalence between $H^+(B)$, with its norm form as quaternion algebra, and $A \oplus B$ as vector space with the quadratic form that is the product on A and $-h$ on B, these being orthogonal subspaces. If we use this mapping to transport the structure of quaternion algebra we obtain a multiplication on $A \oplus B$ with $(a + b) \cdot (a' + b') = (aa' + h(b,b')) + z\frac{[bb']}{2} + (ab' + a'b)$, where for $b \in B$, the square of b is $h(b,b)$, so that our norm form is the form above.

We now show that C is a right vector space for $A \oplus B$, with $c \cdot a$ as before, $c \cdot b = c(zb)$, the action of $H(B)$, for which it suffices to compare $(c \cdot b) \cdot b'$ and $c \cdot (b \cdot b')$ for $b, b' \in B$. But $(c \cdot b) \cdot b' = c(bb')$, where the product is taken in $H(B)$, and

$$bb' = \frac{bb'+b'b}{2} + \frac{bb'-b'b}{2} = h(b,b') + \frac{[bb']}{2},$$

$$c(bb') = ch(b,b') + (cz\frac{z[bb']}{2}) = c \cdot (b \cdot b').$$

From our information in the split case, z must act as the same scalar on each irreducible submodule there, and from $z^2 = 1$, this scalar ε is ± 1. Thus we have $cz = \varepsilon c$, $c \cdot b = \varepsilon cb$. We define a pairing $(c,c'): C \times C \to A \oplus B$ by $(c,c') = q_1(c,c') + \varepsilon q_2(c,c')$. We evidently have $(c',c) = -(c,c')^*$, where $*$ denotes the standard involution in our quaternion algebra $A \oplus B$. The pairing is clearly A-bilinear. For $b \in B$, we have

$$q_1(c \cdot b, c') + q_1(c, c' \cdot b) = \varepsilon(q_1(cb,c') + q_1(c,c'b)) = 0,$$

$$q_1(c,c' \cdot b) - q_1(c \cdot b, c') = 2q_1(c,c' \cdot b) = 2\varepsilon q_1(c,c'b)$$

$$= 2\varepsilon h(b, q_2(c,c')) = 2h(b, q_2'(c,c'))$$

$$= b \cdot q_2'(c,c') + q_2'(c,c') \cdot b.$$

Also, $q_2'(c \cdot b, c') - q_2'(c, c' \cdot b) = q_2(cb,c') - q_2(c,c'b)$ satisfies, for all $b' \in B$, $h(b', q_2(cb,c') - q_2(c,c'b))$

$$= q_1(cb, c'b') - q_1(c, c'bb') = q_1(c, c'(b'b + bb'))$$

$$= 2h(b,b') q_1(c,c') = 2h(b', q_1(c,c')b), \text{ so that}$$

$$q_2'(c \cdot b, c') - q_2'(c,c' \cdot b) = q_1(c,c') \cdot b + b \cdot q_1(c,c').$$

Finally, we show $q_2'(c \cdot b, c') + q_2'(c, c' \cdot b) = q_2'(c,c') \cdot b - b \cdot q_2'(c,c')$, and then it follows from the four relations that (c,c') is linear in its second variable (over $A \oplus B$). Thus let $b' \in B$; it suffices to show that the h-scalar product of b' with both members is the same and by A-bilinearity, to treat the cases $b' = b$ and $h(b',b) = 0$. We may evidently assume b and b' to be non-zero. When $b' = b$, we have $h(b, q_2'(c \cdot b, c') + q_2'(c, c' \cdot b)) = h(b, q_2'(cb,c') + q_2(c,c'b)) = q_1(cb,c'b) + q_1(c, c'b^2) = q_1(c,c'b^2) - q_1(c,c'b^2) = 0$, while that $h(b, q_2'(c,c') \cdot b - b \cdot q_2'(c,c')) = 0$ is a well-known fact about the norm form on elements of trace zero in a quaternion algebra. We may therefore assume $h(b',b) = 0$, and now treat separately the cases where b and $q_2(c,c')$ are A-linearly dependent and where these are h-orthogonal. In the latter case, we have

$$h(b', q_2(cb,c') + q_2(c,c'b)) = q_1(cb,c'b') + q_1(c,c'bb')$$
$$= q_1(c,c'(bb' - b'b)) = q_1(c,c'z(bzb' - b'zb))$$
$$= q_1(c,c'\cdot(b\cdot b' - b'\cdot b)) = 2q_1(c,c'\cdot(b\cdot b'))$$
$$= 2\varepsilon q_1(c,c'(b\cdot b')) = 2\varepsilon h(b\cdot b', q_2(c,c'))$$
$$= 2h(b\cdot b', q_2'(c,c')) = 2h(b', q_2'(c,c')\cdot b)$$
$$= h(b', q_2'(c,c')\cdot b - b\cdot q_2'(c,c')), \text{ using the orthogonality and}$$

properties of the form. In the former case, we have $h(b',q_2(cb,c') + q_2(c,c'b)) = 2h(b', b<c,c'>) = -2h(b'<c,c'>, b) = -h(q_2(cb',c') + q_2(c,c'b'),b)$, and the last argument shows this to be equal to $-h(b, q_2'(c,c')\cdot b' - b'\cdot q_2'(c,c'))$. Since b and $q_2'(c,c')$ are linearly dependent, this is zero, as is, evidently, $b\cdot q_2'(c,c') - q_2'(c,c')\cdot b$. This completes the proof.

As in other cases of type BC, we can now identify L with the skew transformations of the space $(M'' \otimes E) \oplus C$ over the quaternion division algebra $E = A \oplus B$, with respect to the antihermitian form extending (c,c') and (u,v) $(M'' \times M'' \to F)$. The A-linear inclusions $S \otimes A \hookrightarrow K(M'' \otimes E)$, $M' \otimes B \hookrightarrow K(M'' \otimes E)$, $M'' \otimes C \hookrightarrow K(M'' \otimes E \oplus C)$ go as before, as does $B \hookrightarrow K(M'' \otimes E)$, $K(C) \hookrightarrow K(M'' \otimes E \oplus C)$, and $K(M'' \otimes E \oplus C) = K(M'' \otimes E) \oplus (M'' \otimes C) \oplus K(C)$, with $K(M'' \otimes E) = B \oplus (S \otimes A) \oplus (M' \otimes B)$. It suffices therefore to show that $D = B \oplus K(C)$ in a suitable identification, as well as the homomorphism property.

For the former, we consider the action of D on C, which is faithful since D annihilates A, $q_2(C,C) = B$, and D acts faithfully on $A \oplus B \oplus C$. Here we have from (69') that $4<b,b'>$ sends $c \in C$ to $(c\cdot b)\cdot b' - (c\cdot b')\cdot b = c\cdot(b\cdot b' - b'\cdot b)$. Since $B \subseteq E$ is its own derived algebra, it follows that D contains all multiplications on C by elements of B. In our new notation, (85') reads

$$c'<c'',c> = -c\cdot(c'',c') - c''\cdot(c,c') - \tfrac{1}{2}c'\cdot q_2'(c'',c).$$

The last term is a multiplication by an element of B, so that D contains all mappings

$$c' \mapsto c\cdot(c'',c') + c''\cdot(c,c')$$

for c'',c arbitrary in C. Now A-combinations of these are easily seen to form an ideal in $K(C)$, which is simple except when $[C:E] = 1$, in which case $K(C)$ has A-dimension 1, consisting of multiplications by A-multiples of (e,e), where e is a basis for C over E. Evidently if $c \in C$ has $(c,c) \neq 0$, we have $c \cdot (c,c) \neq 0$, from which our ideal is non-zero, and must be all of $K(C)$. Thus $D = B + K(C)$ in our identifications and the sum is obviously direct. Clearly $K(C)$ is the annihilator of B in D; we have $[s \otimes b', b + T] = s \otimes [b',b]$. We prescribe the actions of $b + T$ in D on $(M'' \otimes E) \oplus C$, our antihermitian space, by: $u \otimes f \mapsto -u \otimes b \cdot f$ ($u \in M'', f \in E$), and $c \mapsto cT$ ($c \in C$). Then $(b \cdot b' - b' \cdot b)$ sends $u \otimes f$ to $u \otimes (b' \cdot b - b \cdot b') \cdot f$, and computing the commutator of the actions of b and b' gives the same result on $u \otimes f$. Since b and T commute in D, we have a homomorphism of D into $K(M'' \otimes E \oplus C)$, the image containing $K(C)$, as well as all maps $u \otimes f \mapsto u \otimes b \cdot f$, $b \in B$.

We prescribe the action of $u \otimes c \cdot (u \in M'', c \in C)$ by $v \otimes f \mapsto c \cdot (v,u)f$, $c \mapsto -u \otimes (c,c')$, as before; we let $x \otimes a (x \in S, a \in A)$ send $v \otimes f$ to $vx \otimes af$, c' to 0; and we let $s \otimes b (s \in M', b \in B)$ send $v \otimes f$ to $-\varepsilon vs \otimes b \cdot f$, c' to 0. Thus, for instance $[s \otimes b, s' \otimes b'] = Tr(ss') <b,b'> + [ss'] \otimes h(b,b')$, and $c <b,b'> = \frac{1}{4} c \cdot (b \cdot b' - b' \cdot b)$, so that $<b,b'>$ identifies with $\frac{1}{4}(b \cdot b' - b' \cdot b) \in B$. Hence $[s \otimes b, s' \otimes b']$ sends $u \otimes f$ to $-u \otimes Tr(ss') \frac{1}{4}(b \cdot b' - b' \cdot b) \cdot f + u[ss'] \otimes h(b,b')f$, and $c \in C$ to zero, by our definitions. The homomorphism property is checked by comparing the effect on $u \otimes f$ with that of the commutator of the actions of $s \otimes b$, $s' \otimes b'$, as defined above, i.e., with

$$uss' \otimes b' \cdot b \cdot f - us's \otimes b \cdot b' \cdot f$$
$$= u[ss'] \otimes \frac{b' \cdot b + b \cdot b'}{2} \cdot f + u(ss' + s's) \otimes \frac{b' \cdot b - b \cdot b'}{2} \cdot f.$$

Since $ss' + s's = \frac{1}{2} Tr(ss')I$ and $\frac{b' \cdot b + b \cdot b'}{2} = h(b,b')$, we have equality. Likewise, for $b = b + 0 \in D$, $s \otimes b' \in M' \otimes B$, we have $[s \otimes b', b] = s \otimes (b' \cdot b - b \cdot b')$, sending $u \otimes f$ to $\varepsilon us \otimes (b \cdot b' - b' \cdot b) \cdot f$, while the commutator of the two factors sends $u \otimes f$ to $\varepsilon us \otimes b \cdot b' \cdot f - \varepsilon us \otimes b' \cdot b \cdot f$, agreeing with the above. We also have $[u \otimes c, b] = u \otimes c \cdot b$, sending $v \otimes f$ to $c \cdot b(v,u) \cdot f$, c' to $-u \otimes (c \cdot b, c')$, while the commutator of the actions sends $v \otimes f$ to $-c(v,u) \cdot b \cdot f$, c' to $u \otimes b \cdot (c,c')$ (recall that $B \subseteq D$ annihilates C in our action). Since $(c \cdot b, c') = b^* \cdot (c,c') = -b \cdot (c,c')$, these agree. Finally, we check $[u \otimes c, v \otimes c'] = (u,v) <c,c'> + (u \circ v) \otimes q_1(c,c') + [u,v] \otimes q_2(c,c')$. Now $<c,c'> = T - \frac{1}{2}\epsilon q_2(c,c')$ in our $"T + b"$ - decomposition (by (85')), where $c"T = - c\cdot(c',c") - c'\cdot(c,c")$. Thus the above sends $w \otimes f$ to $\frac{1}{2}(u,v)w \otimes \epsilon q_2(c,c')\cdot f + (w,u)v \otimes q_1(c,c')f + (w,v)u \otimes q_1(c,c')f - \epsilon(w,u)v \otimes q_2(c,c')\cdot f + \epsilon(w,v)u \otimes q_2(c,c')\cdot f - \frac{1}{2}\epsilon(u,v)w \otimes q_2(c,c')\cdot f = (w,v)u \otimes (c,c')\cdot f - (w,u)v \otimes (c',c)\cdot f$, and $c" \in C$ to $- c\cdot(u,v)(c',c") - c'\cdot(u,v)(c,c")$. The commutator in question sends $w \otimes f$ to $- v \otimes (c',c(w,u)\cdot f) + u \otimes (c,c'(w,v)\cdot f)$, and $c" \in C$ to $- c'\cdot(u,v)(c,c") + c\cdot(v,u)(c',c")$, and these agree with the appropriate values. The rest of the identification is easy.

We therefore have the following situation when $[B:A] = 3$: <u>The discriminant of the form</u> h <u>on</u> B <u>is a square and a standard central element</u> $z = b_1 b_2 b_3$ (b_i <u>orthogonal in</u> B) <u>in</u> $H(B)$ <u>may be used to give</u> $A \oplus B$ <u>the structure of a quaternionic division algebra</u> E <u>over</u> A . <u>Then</u> C <u>may be given, in terms of</u> z <u>and our forms</u> q_1, q_2, <u>the structure of antihermitian</u> E-<u>vector space, and</u> L <u>is isomorphic to the Lie algebra</u> $K(M" \otimes_F E \oplus C)$ <u>of</u> E-<u>linear skew transformations, where</u> $M" \otimes_F E$ <u>inherits its antihermitian form from the symplectic form on</u> $M"$. Evidently $K(C)$ contains nilpotent elements, obtained from $<c,c>$, if C contains isotropic vectors c. Thus <u>the form on</u> C <u>must have Witt index</u> 0. Conversely, if this is the case, the centralizer of T in L is $T \otimes A + D + U_1 \otimes B$, with derived algebra $D + U_1 \otimes B$ if $[C:E] \geq 2$, $B + U_1 \otimes B$ if $[C:E] \geq 1$, and center $T \otimes A$ if $[C:E] \geq 2$, $T \otimes A + K(C)$ if $[C:E] = 1$. The adjoint action of $B + U_1 \otimes B$ on $M" \otimes C$ assigns to $b + U_1 \otimes b'$ the map sending $u \otimes c$ to $\epsilon u \otimes c\cdot b + \epsilon u U_1 \otimes c\cdot b'$. It follows that $B + U_1 \otimes B$ is isomorphic to the direct sum of two copies of the Lie algebra B , so is without nilpotent elements, as is $K(C)$. Evidently $T \otimes A$ and $K(C)$ commute, and all ad(S), $S \in T \otimes A$, have eigenvalues in A , and only in F if $S \in T$. Moreover, if $[C:E] = 1$, an element T of $K(C)$ has eigenvalue in A only if $T = 0$. It follows that T <u>is indeed a maximal split torus in</u> L <u>when</u> C <u>is anisotropic</u>. Thus <u>the form</u> $(c,c') = q_1(c,c') + q_2(c,c'z)$ <u>on</u> C <u>is antihermitian and anisotropic, and conversely, this situation and our construction give rise to a simple Lie algebra with centroid</u> A <u>and maximal</u> F-<u>split torus</u> T .

There remains the case $[B:A] = 1$. Since the form h fails to represent 1, $H(B) = A \oplus B$ is a quadratic field extension E of A , with involution $*$ fixing A and having B as skew elements. C is a right vector space over E, and we have

$$0 = 2b < c,c' > = q_2(cb,c') + q_2(c,c'b), \text{ while}$$
$$2bq_1(c,c') = q_2(c,c'b) - q_2(cb,c') = 2q_2(c,c'b)$$
$$2bq_2(c,c') = q_1(c',cb) + q_1(c,c'b)$$
$$= q_1(c,c'b) - q_1(cb,c') = 2q_1(c,c'b).$$

Setting $(c,c') = q_1(c,c') + q_2(c,c')$, we have $(c',c)^* = -(c,c')$, $(c,c'b) = (c,c')b$, and (c,c') is known to be A-bilinear. Thus (c,c') is an antihermitian E-form on C. As before, we identify D with $K(C)$, which now contains as center the scalar multiplications by elements of B, and as such identifies with $B \oplus K'(C)$. Likewise, we can identify L with a subalgebra of $K'(M'' \otimes E \otimes C)$, containing $K'(M'' \otimes E) \oplus (M'' \otimes C) \oplus K'(C)$, which is of A-codimension one in $K'(M'' \otimes E \otimes C)$. To show that L is equal to the latter one needs only check the identification of the action of $B \subseteq D$, which is done as in §7 a), since here $B = Z(E) \cap B$. Due to the similarity with these cases, we omit details. The centralizer of T is $T \otimes A + U_1 \otimes B + K(C)$, with derived algebra $K'(C)$, which is without nilpotent elements if and only if C is anisotropic, and with center $T \otimes A \oplus U_1 \otimes B \oplus B$, in which one sees as above that only elements of T have all adjoint-eigenvalues in F. Thus <u>our situation holds if and only if</u> C <u>is anisotropic, and then</u> L <u>identifies with</u> $K'(M'' \otimes E \otimes C)$, <u>the antihermitian space</u> $M'' \otimes E \otimes C$ <u>over</u> $E = A \oplus B$ <u>having Witt index</u> 2.

We sum up some of our observations on Lie algebras L of type BC_2:

<u>Theorem V.2.</u> <u>Let</u> L <u>be a simple Lie algebra of type</u> BC_2 <u>over</u> F, <u>and suppose we are in the case</u> b) <u>of Theorem III.10. With notations as in that theorem, one of the following holds:</u>

α) <u>There is a quadratic field extension</u> Z <u>of</u> A, <u>with</u> B <u>the elements of trace zero of</u> A <u>and</u> h <u>the ordinary(polarized)square form on</u> B, <u>and a</u> Z-<u>vector space</u> $(M'' \otimes Z) \oplus C$ <u>carrying an antihermitian nondegenerate form of Witt index</u> 2, <u>such that</u> $L \cong K'((M'' \otimes Z) \oplus C)$ (<u>the case</u> $[B:A] = 1$).

β) <u>There is a quaternionic division algebra</u> $Q = A \oplus B$ <u>over</u> A, <u>with</u> B <u>the elements of trace zero and</u> h <u>the (polarized) square form on</u> B, and a Q-vector space $(M'' \otimes Q) \oplus C$ carrying a nondegenerate form, antihermitian with respect to the standard involution in Q and of Witt index 2, such that $L \cong K((M'' \otimes Q) \oplus C)$ (the case $[B:A] = 3$).

γ) There is a vector space B over A with $[B:A] = 5$, carrying a quadratic form h, not representing 1, such that the center Z of the Clifford algebra $H(B)$ does not split, with $H(B)$ a full matrix algebra over Z, and an irreducible $H(B)$-module C carrying a non-degenerate A-symplectic form q_1 relative to which the elements of $B \subseteq H(B)$ act as skew transformations. The remaining operations on L are defined as in Theorem III.10, A is the centroid of L, and $[L:A] = 78$.

δ) There is a vector space B over A with $[B:A] = 7$, carrying a quadratic form h, not representing 1, such that the center of the Clifford algebra $H(B)$ splits, with the even Clifford algebra $H^+(B)$ a full matrix algebra over a quaternionic division algebra Q over A, and an irreducible $H(B)$-module C, as in Theorem III.10. The centroid of L is A, and $[L:A] = 133$.

ε) There is a vector space B over A with $[B:A] = 11$, carrying a quadratic form h, not representing 1, and such that both the center Z of the Clifford algebra $H(B)$ and the even Clifford algebra $H^+(B)$ split over A. There is an irreducible $H(B)$-module C, enjoying the properties of Theorem III.10, such that L results from that construction. The centroid of L is A, and $[L:A] = 248$.

In cases α), β), one sees as in earlier cases that the simple group G' identifies with the quotient modulo its center of the group generated by the "unitary transvections" with respect to our antihermitian form, thereby yielding another proof of the simplicity of the latter group. In the remaining cases, our decomposition of L affords explicit realizations of generators for G', thereby associating certain simple linear groups with quadratic forms having the appropriate properties.

d) <u>Algebras of type</u> BC_1; <u>the remaining cases</u>.

In §V.7. a) we have treated the case where the special Jordan division algebra A of Theorem III.11 is the set of symmetric elements of an involutorial associative division algebra, subject to certain restrictions. We have seen in §V.1 that the centroid of L is the center $Z(A)$, which consists of the elements of A centralizing the action of A on C. Regarding $Z(A)$ as the ground field for L,A,C, we note that L and A are absolutely simple, so that A becomes, upon suitable field extension, a <u>split</u> simple Jordan algebra. The split simple Jordan algebras J have the property that $1 \in J$ is a sum of a set of absolutely primitive orthogonal idempotents, the number of such idempotents being dependent only on J. This number is the <u>degree</u> of J. The <u>degree of</u> A is then defined to be the degree of the split algebra obtained from A by extending the ground field $Z(A)$ to its algebraic closure. (It may also be defined as the degree of the "generic minimum polynomial" of A - [JJ], Chapter VI.)

The classification of central simple Jordan algebras according to their degrees is known ([JJ], Chapter V); a <u>special</u> simple Jordan algebra A of degree <u>at least three</u> over $Z(A)$ is either as in §V.7.a), or else A identifies with the elements of a central simple associative division algebra E over $Z(A)$, $[E:Z(A)] \geq 9$, with operation $\frac{1}{2}(ab+ba)$. (One writes $A = E^+$.) Those A of degree <u>two</u> have the form $A = Z(A) \oplus B$, where B is a $Z(A)$-vector space carrying a non-degenerate symmetric bilinear form h, not representing 1, and

$$(\kappa+b)(\kappa'+b') = (\kappa\kappa'+h(b,b')) + (\kappa b'+\kappa'b).$$

The only A of degree <u>one</u> is $Z(A)$.

i) $A = E^+, E$ <u>an associative division algebra of degree</u> ≥ 2.

When $A = E^+$, the A-modules are completely reducible, and the irreducible ones are either one-dimensional right E-vector spaces V or one-dimensional left E-vector spaces V, with v·a = va resp. av. (Here we are only concerned with "special unital modules" - cf. [JJ], §III.4.) We show that <u>this case cannot occur in the context of</u> §III.9, for $[E:Z(A)] \geq 4$.

As in some earlier cases, we study the situation when $Z(A)$ is extended to a splitting field K; then $E_K = M_d(K)$, the d by $d(d \geq 2)$

matrices over K, $(E^+)_K = (E_K)^+ = M_d(K)^+$, and the irreducible $(E^+)_K$-modules are either the $1 \times d$ K-matrices, with right multiplication by elements of $M_d(K)$, or the $d \times 1$ K-matrices with left multiplication by elements of $M_d(K)$. We let $e_1,\ldots,e_d$ be a canonical K-basis for the former module, $e_1^*,\ldots,e_d^*$ for the latter, and have $e_i \cdot E_{jk} = e_i E_{jk} = \delta_{ij} e_k$ in the former case, $e_i^* \cdot E_{jk} = E_{jk} e_i^* = \delta_{ki} e_j^*$ in the latter, where E_{jk} is a typical matrix unit from $M_d(K)$.

With C as in §III.9, C_K admits (the extension of) q as an anti-symmetric K-bilinear map to $M_d(K)$, satisfying

(93) $\quad aq(c,c') + q(c,c')a = q(c \cdot a, c') + q(c, c' \cdot a)$.

First consider the pairing q on two isomorphic irreducible summands of C_K (possibly the same). Let them be, say, isomorphic to the $1 \times d$ K-matrices, an isomorphism pairing the basis $e_1,\ldots,e_d$ for the first, C_1, with a basis $e_1',\ldots,e_d'$ for the second, C_2. We show $q(C_1, C_2) = 0$.

From (93), $q(e_i, e_i')$ is a d by d matrix (β) satisfying $E_{jk}(\beta) + (\beta)E_{jk} = 0$ whenever $j \neq i$ and $k \neq i$. With $k = j$, it follows that $q(e_j, e_i') = \beta_i E_{ii}$. Then with $a = E_{ij}$ we find from (93) that $q(e_j, e_i') + q(e_i, e_j') = \beta_i E_{ij}$. Interchanging i and j gives $\beta_j E_{ji}$ for this same value, so that $\beta_i = \beta_j = 0$, and $q(e_i, e_i') = 0$ for all i.

Now let $i \neq j$; as above we find that $q(e_i, e_j') = (\beta)$ has zero in all positions except those corresponding to one of the four pairs involving i and j only. From $q(e_i, e_i') = 0 = q(e_j, e_j')$ and from (93) with $a = E_{ij}$ resp. E_{ji} we find $q(e_i, e_j') = \gamma_{ij}(E_{ii} - E_{jj})$. Applying (93) with $a = E_{ii}$ then gives

$$2\gamma_{ij} E_{ii} = q(e_i E_{ii}, e_j') = \gamma_{ij}(E_{ii} - E_{jj}),$$

so that $\gamma_{ij} = 0$, and $q(C_1, C_2) = 0$. The case of $d \times 1$ matrices is analogous.

Next we consider the pairing q on non-isomorphic summands C_1 (canonical basis $e_1,\ldots,e_d$) and C_2 (canonical basis $e_1^*,\ldots,e_d^*$). (The determinations here and above may be regarded as special cases of the problem treated in Appendix B 4), but seem simpler outside that context of involutorial algebras, and must be carried more thoroughly to their conclusions for present purposes.) If $(\beta) = q(e_i, e_j^*)$, we find

235

as above that $(\beta) = \beta_i E_{ii}$. Then, for $j \neq i$, we have
$E_{ij}q(e_i,e_i^*) + q(e_i,e_i^*)E_{ij} = q(e_i,E_{ij},e_i^*) + q(e_i,E_{ij}e_i^*)$, yielding
$\beta_i E_{ij} = q(e_j,e_i^*)$, while taking E_{ji} in place of E_{ij} gives
$\beta_i E_{ji} = q(e_i,e_j^*)$. Interchanging i and j gives $\beta_j E_{ji} = q(e_i,e_j^*)$
in our former relation, from which we have a fixed $\beta \in K$ with
$q(e_i,e_j^*) = \beta E_{ji}$ for all i,j. If $\beta \neq 0$, and if we replace the basis
$e_1,\ldots,e_d$ by $\beta^{-1}e_1,\ldots,\beta^{-1}e_d$, we may assume $q(e_i,e_j^*) = E_{ji}$. In this
case, $(C_1 + C_2)^\perp$ (with respect to q) is an $M_d(K)^+ (=A_K)$-submodule of
C_K, meeting $C_1 + C_2$ in zero only. It follows that C_K decomposes
into a q-orthogonal direct sum of s copies of our module $C_1 + C_2$, the
mapping q on each copy being given as above.

Suppose now that $s > 1$, i.e., $C_K \neq C_1 + C_2$, or that C_K
contains q-orthogonal $C_1 + C_2$, $C_1' + C_2'$. We take bases $\{e_i\}$, $\{e_i^*\}$,
$\{e_i'\}, \{e_i^{*'}\}$ for the four respective modules as above. From (94) of
§III.9, it is clear that $< e_i, e_j^{*'} >$ annihilates $A_K = M_d(K)$; by the
derivation-property, this element of $\mathcal{D}_K$ maps C_1 into the sum of the
irreducible submodules of C_K isomorphic to C_1'. Likewise, $< e_i', e_k >$
annihilates $M_d(K)$ and maps C_2 into the sum of the irreducible submodules isomorphic to C_2'. By (95) of §III.9, we have

$$e_k' < e_j^{*'}, e_i > - e_j^{*'} < e_i, e_k' >$$
$$= 2e_i \cdot q(e_k', e_j^{*'}) - e_k' \cdot q(e_j^{*'}, e_i) - e_j^{*'} \cdot q(e_i, e_k')$$
$$= 2e_i E_{jk} = 2\delta_{ij} e_k' \in C_1'.$$

Thus $< e_i, e_j^{*'} >$ maps C_1' to C_1', while $< e_i, e_k' >$ annihilates
C_2', and $< e_i, e_i^{*'} >$ does not annihilate C_1'. The same considerations
show that $< e_i, e_i^{*'} >$ annihilates every submodule C_3 isomorphic to
C_1, but with $q(C_3, C_2') = 0$. Therefore the annihilator of A_K in $\mathcal{D}_K$
contains all endomorphisms of C_K which act like off-diagonal matrix
units with respect to a maximal independent set of isomorphic irreducible
submodules. From the brackets of these, we see that this annihilator
contains all such acting like matrices of trace zero, a space (of

restrictions) having K-dimension s^2-1.

Upon extension of the base field, all A-modules of "right type" become sums of copies of C_1, while those of "left type" become sums of copies of C_2. It follows that the number u of summands in C is the same for each type, and that $[C:Z(A)] = 2ud^2 = 2sd$, so that $s = ud$. From the above, we conclude that the annihilator of A in $\mathcal{D}$ contains a subspace inducing in the sum $\mathcal{U}$ of those modules of right type a set of A-endomorphisms of $Z(A)$-dimension $s^2-1 = (ud)^2 - 1$, all of which have trace zero as $Z(A)$-endomorphisms. Evidently $[\mathcal{U}:E] = u$ so that the A-endomorphisms of $\mathcal{U}$ have $Z(A)$-dimension $(ud)^2$; we conclude that $\mathcal{D}$ <u>induces in</u> $\mathcal{U}$ <u>the set of all</u> A-<u>endomorphisms of</u> $Z(A)$-<u>trace zero</u>.

If $u > 1$, it follows that there is induced by $\mathcal{D}$ a non-zero diagonalizable A-endomorphism of $\mathcal{U}$ with rational eigenvalues. That is, we have E-endomorphism T of $\mathcal{U}$, induced from $D \in \mathcal{D}$, and a basis $c_1,\ldots,c_u$ for $\mathcal{U}$ over E such that $c_i T = c_i \lambda_i$ for all i, where $\lambda_i \in \mathbb{Q}$ are not all zero. Now let $t(c,c') = \mathrm{Tr}_{A/Z(A)}(q(c,c'))$, a skew $Z(A)$-bilinear form which is defined (to within a nonzero multiple) with $\mathrm{Tr}_{A/Z(A)}(a)$ either as the generic trace of [JJ], Chapter VI, or as the trace of the $Z(A)$-endomorphism of multiplication on the right by a in E. If A_0 is the set of elements of A of trace zero, to say that $t(c,c') = 0$ for all $c \in C$ is to say that $q(c,c') \in A_0$ for all c, hence that $q(ca,c') \in A_0$ for all c, all $a \in A$. By [JJ], Theorem 1, p. 224, every derivation of A maps A into A_0, so that by (94) we have also $q(c,c'a) \in A_0$ for all $c \in C$, $a \in A$. By (93) of §III.9, it follows that $q(c,c')a + aq(c,c') \in A_0$ for all a, all c. But if $q(c,c') \neq 0$, we may take $a = q(c,c')^{-1}$ to contradict this conclusion. Hence $q(c,c') = 0$ for all c, and so $c' = 0$ by §III.9; that is, $t(c,c')$ <u>is nondegenerate</u>.

Evidently $t(\mathcal{U},\mathcal{U}) = 0 = t(\mathcal{U}',\mathcal{U}')$ and our D as above is skew with respect to t in its action on C. If $a_1,\ldots,a_r (r=d^2)$ is a basis for A over $Z(A)$, we have $\{a_i c_j\}$ a basis for $\mathcal{U}$ over $Z(A)$, and $(a_i c_j)D = a_i(c_j D) = a_i c_j \lambda_j$. If $\{v_{ij}\}$ is the t-dual $Z(A)$-basis for $\mathcal{U}'$, we have $t(a_i c_j, v_{k\ell}(D+\lambda_\ell)) = -t(a_i c_j D, v_{k\ell}) + \lambda_\ell t(a_i c_j, v_{k\ell})$ $= \delta_{ik}\delta_{j\ell}(-\lambda_j+\lambda_\ell) = 0$ for all i,j,k,ℓ. Thus $v_{ij}D = -\lambda_j v_{ij}$, and D acts diagonally in $\mathcal{U}'$ with eigenvalues in $\mathbb{Q}$. But this contradicts the maximality of our one-dimensional split torus T. Thus <u>we conclude</u>

that $u = 1$.

Now we have, over the splitting field K, $C_K = U_K \oplus U'_K$, U_K being the sum of d copies of the $1 \times d$-matrices, U'_K the sum of d copies of the $d \times 1$ matrices. That is, $C_K = \sum_{i=1}^{d} \oplus (C_i \oplus C'_i)$, the C_i being copies of the $1 \times d$ matrices, C'_i of the $d \times 1$ matrices, $U_K = \Sigma C_i$, $U'_K = \Sigma C'_i$, and $q(c_1, c_2) = 0$ unless $c_1 \in C_i$, $c_2 \in C'_i$ for the same i, or vice versa. For canonical K-bases $e_1, \ldots, e_d$ for C_i; $e_1^*, \ldots, e_d^*$ for C'_i, we have $q(e_j, e_k^*) = \beta_i E_{kj}$, and $q: C_K \otimes C_K \to A_K$ is completely determined by $\beta_1, \ldots, \beta_d \in K$.

With $a = E_{mn} \in A_K$, we have

$$E_{mn} \langle e_j, e_k^* \rangle = q(e_j, E_{mn} e_k^*) - q(e_j E_{mn}, e_k^*)$$
$$= \delta_{nk} q(e_j, e_m^*) - \delta_{jm} q(e_n, e_k^*)$$
$$= \beta_i (\delta_{nk} E_{mj} - \delta_{jm} E_{kn})$$
$$= [E_{mn}, \beta_i E_{kj}] = [E_{mn}, q(e_j, e_k^*)],$$

while $a \langle c_1, c_2 \rangle = 0$ unless $c_1 \in C_i$, $c_2 \in C'_i$, or vice versa, for the same i. By K-multilinearity it follows that $a \langle c_1, c_2 \rangle = [a, q(c_1, c_2)]$ for all $a \in A_K$, $c_1, c_2 \in C_K$.

Now we may identify U_K with $M_d(K)$ by making $u^{(1)} + \ldots + u^{(d)}$ ($u^{(i)} \in C_i$) correspond to the matrix c with i-th row $u^{(i)}$, and U'_K with $M_d(K)$ by making $u'^{(1)} + \ldots + u'^{(d)}$, ($u'^{(i)} \in C'_i$) correspond to c', with i-th column $u'^{(i)}$. With these identifications, we have $q(c, c') = c'(\beta)c$, where $(\beta) = \text{diag}\{\beta_1, \ldots, \beta_d\}$ above. (It suffices to check this for $c = e_j^{(i)}$ and for each of $c' = e_k^{*(i)}$, $c' = e_k^{*(\ell)}$, $\ell \neq i$; in the former case, this formula gives $q(c, c') = E_{ki} \beta_i E_{ii} E_{ij} = \beta_i E_{kj}$, and in the latter $q(c, c') = E_{k\ell} \beta_\ell E_{\ell\ell} E_{ij} = 0$, in each case the previously determined value.)

Since q is trivial on $U_K \otimes U_K$ and on $U'_K \otimes U'_K$, we have the following properties of $q: q(c_1, c_2) = 0$ if both $c_1, c_2 \in U_K$, or if both $c_1, c_2 \in U'_K$; there is $b \in A_K$ (more precisely, a nonsingular matrix) such that for all $c_1 \in U_K$, $c_2 \in U'_K$, $q(c_1, c_2) = c_2 b c_1$, with identifications as above. If $\{u\}$ is a right E-basis for U, $\{u'\}$ a left E-basis for U', then $c = ue \longleftrightarrow e$, $c' = e'u' \longleftrightarrow e'$ make identifications of these modules with E, or of U_K resp. U'_K with $M_d(K)$, as right resp. left $M_d(K)$-modules. Combining with the identifications above, we find invertible $b_1, b_2 \in M_d(K)$ such that

238

$q(c,c') = e'b_2bb_1e = e'b_3e$. We fix these last identifications of U resp. U' with E and change notation to conclude that, by virtue of the multilinear character of these conclusions, they descend to C and A: q is trivial on $U \otimes U$ and on $U' \otimes U'$, while there is $b \neq 0$ in E such that $q(c_1,c_2) = c_2bc_1$ for all $c_1 \in U = E$ and $c_2 \in U' = E$.

Now (93) is easily verified under these conditions, and $a<c_1,c_2> = [a,c_2bc_1]$ from the above. In particular, if we take $0 \neq c_1 \in E$, $c_2 = c_1^{-1}b^{-1}$, then $<c_1,c_2>$ annihilates A, as does $<c,c_1>$ for any $c \in U(= E$; here $c_1 \in U$, $c_2 \in U'$, as above). With such c, c_1, c_2, we have, by (95),

$$c_2 <c,c_1> - c <c_1,c_2> = 2c_1q(c_2,c) - c$$
$$= -2c_1q(c,c_2) - c = -2c_1c_2bc - c$$
$$= -3c.$$

Since $c_2 <c,c_1> \in U'$ and $c <c_1,c_2> \in U$, we have $c <c_1,c_2> = 3c$ for all $c \in U$. Likewise, $c' <c_1,c_2> = -3c'$ for all $c' \in U'$. But then $<c_1,c_2>$ is non-zero, diagonalizable over F, and centralizes T, a contradiction. This completes the proof of the assertion with which we began i).

ii) The case where A is of degree two.

If Z is the center of A, we have in this case $A = Z \oplus B$, where B is a Z-vector space of dimension at least two, carrying a non-degenerate Z-bilinear form h, not representing 1, and with product $(\alpha+b)\cdot(\alpha'+b') = (\alpha\alpha'+h(b,b')) + (\alpha b'+\alpha'b)$ for $\alpha,\alpha' \in Z$; $b,b' \in B$. Using the projections associated with the decomposition $A = Z \oplus B$, we may write $q(c,c') = q_1(c,c') + q_2(c,c')$, where $q_1(c,c') \in Z$, $q_2(c,c') \in B$, and both q_1 and q_2 are skew. For $a = \alpha \in Z$, all derivations of A annihilate α; thus, from (94),

$$0 = \alpha<c,c'> = q(c,c'\alpha) - q(c\alpha,c'), \text{ or } q(c,c'\alpha) = q(c\alpha,c'),$$

Combining this with (93), which says

$$2 q(c,c') = q(c,c'\alpha) + q(c\alpha,c'),$$

we see that q, hence also q_1 and q_2, is Z-bilinear.

Since all derivations of A map B into B, we see from (94) for $b \in B$ that

(*) $\quad b <c,c'> = q_2(c,c'b) - q_2(cb,c')$,

$\quad\quad\quad\quad 0 = q_1(c,c'b) - q_1(cb,c')$.

In particular, <u>the elements of</u> B <u>are represented by</u> q_1-<u>symmetric transformations of</u> C, <u>linear over</u> Z.

In the same setting, (93) gives

$$2b \cdot (q_1(c,c') + q_2(c,c')) = q_1(cb,c') + q_2(cb,c')$$
$$+ q_1(c,c'b) + q_2(c,c'b),$$

so that, taking components and using the above,

(**) $\quad 2q_1(c,c')b = q_2(cb,c') + q_2(c,c'b)$,

$\quad\quad\quad h(b,q_2(c,c')) = q_1(c,c'b)$.

The second of these shows <u>that</u> q_2 <u>is determined by</u> h <u>and</u> q_1, as in the corresponding case of type BC_2 (§V.7.c)). From here on, many of our arguments follow the same principles as applied in that case, and some sketchiness may be excused on this basis.

From the second relation of (**) it follows that if $q_1(c,C) = 0$ then $q_2(c,C) = 0$, so that $q(c,C) = 0$, and $c = 0$. Thus q_1 <u>is non-degenerate on</u> C.

As in §V.7.c), we let $H(B)$ be the Clifford algebra of (B,h) over Z, and we let K be a field extension of Z which splits (B,h) in the sense of that section. Rather than introducing more notation, we use the modifiers "over K" or "over Z" to indicate whether or not we assume the base field to be extended. Other notations and conventions are as in §V.7.c).

First suppose $[B:Z]$ is <u>even</u>, say $2r$; then <u>we claim that</u> $r \leq 3$, <u>so</u> $[B:Z] \leq 6$. For working over K, with dual bases $p_1,\ldots,p_r; n_1,\ldots,n_r$ for totally isotropic subspaces, and generators $u,u' \longleftrightarrow n_1 \cdots n_r$ for simple $H(B)$-submodules of C, we have $q_1(u,u'p_{i_1} \cdots p_{i_t}) = 0$ for $i_1 < \cdots < i_t$ unless $\{i_1,\ldots,i_t\} = \{1,\ldots,n\}$, and $q_1(up_{i_1} \cdots p_{i_s}, u'p_{j_1} \cdots p_{j_t}) = 0$ unless the (distinct) indices form a partition of $\{1,\ldots,r\}$. Then $q_2(up_{i_1} \cdots p_{i_s}, u'p_{j_1} \cdots p_{j_t}) = 0$ unless either the two sets of distinct indices exhaust $\{1,\ldots,n\}$ and have exactly one member in common, or else are disjoint, with $s + t = r - 1$. From (*), we then find, for

$r \geq 4$, that both $\langle up_1p_2p_3, u'p_1p_2\ldots p_r\rangle$ and $\langle up_1p_2p_3p_4, u'p_1p_2\ldots p_r\rangle$ annihilate B (hence A), as do $\langle u, up_1p_2p_3\rangle$ and $\langle u, up_1p_2p_3p_4\rangle$ if $r > 6$, the former if $r > 5$. Now we may assume u' so chosen that $q_1(u, u'p_1\ldots p_r) = \gamma \neq 0$. From the above it follows that (for $r \geq 6$) $u \langle up_1p_2p_3, u'p_1\ldots p_r\rangle$ is in the subspace of C annihilated by N, the subalgebra (without 1) of $H(B)$ generated by $n_1,\ldots,n_r$, while $u'p_1\ldots p_r \langle u, up_1p_2p_3\rangle$ is in the space annihilated by P, the subalgebra generated by $p_1,\ldots,p_r$. On the other hand, (95) yields that the difference of these two is
$2up_1p_2p_3 q(u, u'p_1\ldots p_r) + uq(up_1p_2p_3, u'p_1\ldots p_r) - u'p_1\ldots p_r q(up_1p_2p_3, u) = 2\gamma up_1p_2p_3$. From the fact that $up_1\ldots p_s$ cannot be in the sum of the spaces annihilated by N and P unless either $s = 0$ or $s = r$, we have a contradiction, yielding $r \leq 6$. For $r = 4,5$, we have $q_1(u, up_1\ldots p_r) = q_1(up_r p_{r-1}\ldots p_1, u) = (-1)^{\binom{r}{2}} q_1(up_1\ldots p_r, u) = q_1(up_1\ldots p_r, u)$, using the fact that the p_i act q_1-symmetrically on C. Since q_1 is skew, this implies that $q_1(u, up_1\ldots p_r) = 0$, so that each irreducible summand of C is totally q_1-isotropic. It follows from (**) that q_2 vanishes identically on each such summand. Now let u, u', γ be as in the preceding paragraph. We again have that $\langle up_1p_2p_3, u'p_1\ldots p_r\rangle$ annihilates A, and now it follows from the vanishing of q_2 on $uH(B)$ that $\langle u, up_1p_2p_3\rangle$ annihilates A. Applying (95) as before, we obtain the same contradiction. Thus: If $[B:Z]$ is even, then $[B:Z] \leq 6$.

In the odd-dimensional case, working over K, we let $n_1,\ldots,n_r; b_0; p_1,\ldots,p_r$ be our basis for B as in §V.7.c), and let u, v, w be generators for irreducible $H(B)$-submodules of C, each generator annihilated by N, with $ub_0 = \varepsilon u$, $vb_0 = \varepsilon' v$, $\varepsilon, \varepsilon' = \pm 1$. We assume $q_1(w, vp_1\ldots p_r) = \gamma \neq 0$. If $r > 6$, let $2 < s < r - 2$, e.g., $s = 3$ or 4. Then $q(up_1\ldots p_s, vp_1\ldots p_r) = 0 = q(w, up_1\ldots p_s)$, and both $\langle up_1\ldots p_s, vp_1\ldots p_r\rangle$ and $\langle w, up_1\ldots p_s\rangle$ annihilate B. Therefore $w \langle up_1\ldots p_s, vp_1\ldots p_r\rangle$ is annihilated by N, $vp_1\ldots p_r \langle w, up_1\ldots p_s\rangle$ by P, while (95) yields the difference of these two to be $2up_1\ldots p_s q(w, vp_1\ldots p_r) = 2up_1\ldots p_s(\gamma + (-1)^r \gamma \varepsilon' b_0) = 2\gamma(1 + (-1)^{r+s}\varepsilon\varepsilon')up_1\ldots p_s$. The coefficient here is 4γ for one of the admissible choices $s = 3, 4$, and we obtain a contradiction as before. Thus if $[B:Z]$ is odd, we have $[B:Z] \leq 13$.

Next we consider some cases of low <u>odd</u> values of $[B:Z]$, following the conventions above and those of §V.7.c). Thus $e, e'\ldots$ are generators of isomorphic irreducible $H(B)$-submodules of C (over K), annihilated by N, and with $eb_0 = e$, etc.; likewise $f, f', \ldots$ satisfy the same conditions, except that $fb_0 = -f$, <u>etc</u>.

13: First let $[B:Z] = 13$, and work over K. If a submodule $eH(B)$ and submodule $fH(B)$ fail to be q_1-orthogonal, we have $q_1(e, fp_1\ldots p_6) = \gamma \ne 0$. Since b_0 acts as a symmetric transformation with respect to q_1, we have $\gamma = q_1(e, fp_1\ldots p_6) = q_1(eb_0, fp_1\ldots p_6) = q_1(e, fp_1\ldots p_6 b_0) = (-1)^6 q_1(e, fb_0 p_1\ldots p_6) = -q_1(e, fp_1\ldots p_6) = -\gamma$. Thus $q_1(eH(B), fH(B)) = 0$. Now it follows as in §V.7.c) from $q_1(e', ep_1\ldots p_6) = q_1(e'p_6\ldots p_1, e) = (-1)^{\binom{6}{2}} q_1(e'p_1\ldots p_6, e) = q_1(e, e'p_1\ldots p_6)$, and likewise for f, f', <u>that</u> C_K <u>is a direct sum of irreducible submodules, on each of which</u> q_1 <u>is nonsingular</u>.

We next show C_K <u>is irreducible</u>. First suppose two nonisomorphic q_1-nonsingular summands $eH(B)$, $fH(B)$ are present. Then we know that $0 = q_1(eH(B), fH(B))$, $0 = q_2(eH(B), fH(B))$, and $<eH(B), fH(B)>$ annihilates B. We consider the meaning of (95) for the triple ep_1, fp_2, $ep_1\ldots p_6$: Evidently $ep_1 < fp_2, ep_1\ldots p_6 >$ and $ep_1\ldots p_6 < ep_1, fp_2 >$ must both lie in the sum of the submodules of C (over K) isomorphic to $eH(B)$. But, by (95), their difference is $2fp_2 q(ep_1\ldots p_6, ep_1) = 2fp_2 q_2(ep_1\ldots p_6, ep_1)$. From the second relation of (**) and the form of q_1, we have $h(n_1, q_2(ep_1\ldots p_6, ep_1)) = q_1(ep_1\ldots p_6, ep_1 n_1) = 2q_1(ep_1\ldots p_6, e) = -2\gamma \ne 0$, while $h(u, q_2(ep_1\ldots p_6, ep_1)) = 0$ for all the remaining u in our basis. Thus $q_2(ep_1\ldots p_6, ep_1) = -2\gamma p_1$, and $-4\gamma fp_2 p_1 = 4\gamma fp_1 p_2 \in fH(B)$ is a nonzero element of a sum of submodules isomorphic to $eH(B)$. This is absurd, so <u>all irreducible submodules of</u> C_K <u>are isomorphic</u>.

Next suppose two distinct (q_1-nonsingular, q_1-orthogonal) irreducible submodules are present, which we may assume to be $uH(B)$, $vH(B)$, u and v annihilated by N, $ub_0 = \varepsilon u$, $vb_0 = \varepsilon v$, $\varepsilon^2 = 1$. Again $< uH(B), vH(B) >$ annihilates B, so that $v < uH(B), vH(B) >$ is in the subspace of C_K annihilated by N, and $vp_1\ldots p_6 < uH(B), vH(B) >$ in the subspace annihilated by P. Then we find from (95) that $v < up_1 p_2 p_3 p_4, vp_1\ldots p_6 > - vp_1\ldots p_6 < up_1 p_2 p_3 p_4, v >$ is equal, to within sign, to $2up_1 p_2 p_3 p_4 q(v, vp_1\ldots p_6) = 2up_1 p_2 p_3 p_4 (\gamma + \gamma \varepsilon b_0)$ where $0 \ne \gamma = q_1(v, vp_1\ldots p_6)$ – then $\gamma \varepsilon b_0 = q_2(v, vp_1\ldots p_6)$ from (**) – and this is in turn equal to $2(1+(-1)^4)up_1p_2p_3p_4$. As before, we have a contradiction, and therefore have shown that C_K is irreducible, or that, over Z, C is an absolutely irreducible A-module, of dimension 2^6.

As in §V.7.c), this conclusion has the implications that the discriminant of our form h on B is a square, i.e., that the center of $H(B)$ splits over Z, as well as that the even Clifford algebra $H^+(B)$ is a full matrix algebra (of dimension 2^{12}) over Z. The annihilator of B in D is seen from its action on C to be at most one-dimensional over Z, and to act in C by Z-scalar multiples of the identity. But its action in C must also be by skew transformations with respect to q_1, from which it follows that this annihilator is zero. From (**) and (92), we have $<B,B> = <B,q_2(C,C)> = <B,q(C,C)> \subseteq <C,C>$, and $<B,B>$ induces all derivations of $A = Z + B$. Together with the above, this yields $D = <C,C> = <B,B>$. Thus $c <c',c''>$ is also determined ultimately in terms of h and q_1, together with the action of B on C, as follows: Let $d \in <B,B>$ be such that $b<c',c''> = bd$ for all $b \in B$; then $c <c',c''> = cd$.

Conversely, let Z be an extension field of F. Let B be a Z-vector space of dimension 13, and let h be a nondegenerate symmetric bilinear form on B, not representing 1, such that both the center of the Clifford algebra $H(B)$ and the even Clifford algebra $H^+(B)$ are split. Then we may construct a simple Lie algebra with centroid Z, $T = T \otimes 1 \subseteq T \otimes Z$ as maximal F-split torus, and Z-dimension 248, as follows:

Let C be a minimal right ideal in $H(B)$. Then C is a simple right $H^+(B)$-module, and $H^+(B) = \text{End}_Z(C)$. The involution in $H(B)$ which is the identity on B induces an involution in $H^+(B)$ whose skew elements have dimension $\binom{13}{2} + \binom{13}{6} + \binom{13}{10} = 2^5(2 \cdot 2^5+1)$. When $H^+(B)$ is identified with $\text{End}_Z(C)$, it follows that this involution is the operation of transposing with respect to a symplectic Z-bilinear form q_1 in the space $C ([C:Z] = 2^6)$. We may take a central element $z \in H(B)$ as the product of all 13 elements in an orthogonal basis, and satisfying $z^2 = 1$. Then z centralizes the action of $H^+(B)$ on C, so acts on C as a scalar ε from Z, $\varepsilon^2 = 1$. Viewing B as contained in $H(B)$, we have $zB \subseteq H^+(B)$; in fact zB consists of linear combinations of products of 12 distinct vectors from an orthogonal basis, all of these being symmetric with respect to our involution. If B is such an element of $H^+(B)$, we have $q_1(c,c'zB) = \varepsilon q_1(c,c'B) = \varepsilon q_1(cB,c') = q_1(czB,c')$, from which we see that $q_1(c,c'b) = q_1(cb,c')$ for all $c,c' \in C$, $b \in B$.

Next one defines a map $q_2: C \times C \to B$ by $h(b,q_2(c,c')) = q_1(c,c'b)$ for all $c,c' \in C$, $b \in B$. Then $q_2(c,c') = -q_2(c',c)$, q_2 is Z-bilinear, and the mapping $b \mapsto b<c,c'> = q_2(c,c'b) - q_2(cb,c')$ is an h-skew Z-endomorphism of B. Denoting by B_2 the Z-subspace of $H^+(B)$ generated by all products $b_i b_j (i \neq j)$, where $b_1,\ldots,b_{13}$ is an h-orthogonal basis for B, one easily checks that B_2 is independent of choice of orthogonal basis, and that, for $B \in B_2$, the maps $b \mapsto [b,B]$ are h-skew Z-endomorphisms of B. By this means one obtains an identification of B_2 with $K(B,h)$; hence there is a unique $g(c,c') \in B_2$ with $b<c,c'> = 4[b,g(c,c')]$ for all $b \in B$. The mapping $g: C \times C \to B_2$ is symmetric and Z-bilinear, and one verifies from our identities that, for all $b,b' \in B$; $c,c' \in C$, one has

$$[b,g(c,c'b') - g(cb',c')] = h(b,b')q_2(c,c') - h(b,q_2(c,c'))b'.$$

For $c,c',c'' \in C$, we define $c<c',c''> = 4c\, g(c',c'')$. Then evidently $(cb)<c',c''> = (c<c',c''>)b + c(b<c',c''>)$, $q_1(c<c',c''>,c''') + q_1(c,c'''<c',c''>) = 0$, $q_2(c<c',c''>,c''') + q_2(c,c'''<c',c''>) = 4[q_2(c,c'''), g(c',c'')] = q_2(c,c''')<c',c''>$ (as defined above), $bq_2(c',c'') - q_2(c',c'')b = 4g(c'b,c'') - 4g(c'b,c'')$. From these we deduce all the identities required for §III.9 with $A = Z + B \subseteq H(B)$, $f(a,a') = \frac{1}{2}(aa'+a'a)$, $p(c,a) = ca$, $q(c,c') = q_1(c,c') + q_2(c,c')$, with the exception of (95); that is more tedious.

Under these circumstances, our dimensions over Z are as follows: $D = <B,B>$ has the dimension of $K(B)$, i.e., $\binom{13}{2} = 78$; $A = B + Z$ has dimension 14; C has dimension $2^6 = 64$. The dimension of L is $78 + 3 \times 14 + 2 \times 64 = 248$, so that L <u>must be a Z-form of the split exceptional Lie algebra</u> E_8, <u>of Z-rank one</u>.

The centralizer of T in L is $L_0 = D + T \otimes A = D + h \otimes A$, with derived algebra $L_0' = D + h \otimes B$, which is easily seen to identify with the skew transformations of $A = Z + B$ with respect to its anisotropic (norm) form $(\alpha + b, \alpha' + b') = \alpha\alpha' - h(b,b')$. Thus L_0' is an anisotropic Z-form of a split algebra of type D_7, of dimension 91.

The simple group G' is generated by all $\exp(\text{ad}(u \otimes c))$ and all $\exp(\text{ad}(v \otimes c))$, where u and v constitute canonical basis vectors for M''_1, M''_{-1}, the subscript indicating the eigenvalue of $h \in S$.

11: Next let $[B:Z] = 11$. We show that this case cannot occur. Again we work over a splitting field K, where it follows from
$q_1(e,e'p_1\ldots p_5) = q_1(eb_0,e'p_1\ldots p_5) = q_1(e,e'p_1\ldots p_5 b_0) = -q_1(e,e'b_0 p_1\ldots p_5) = -q_1(e,e'p_1\ldots p_5)$, and likewise for f, f', that all irreducible submodules of C are totally q_1-isotropic, and all isomorphic ones are q_1-orthogonal. Thus there must be present a non-isomorphic $eH(B)$, $fH(B)$, not q_1-orthogonal; we may assume $q_1(e,fp_1\ldots p_5) = 1$. Then $q(fH(B), fH(B)) = 0$, so $<fH(B), fH(B)>$ annihilates B. Moreover, if $s < 3$ we see that $q(e,fp_1\ldots p_s) = 0$ and that $<e,fp_1\ldots p_s>$ annihilates B. It follows from (95) that $fp_1\ldots p_5 <e,fp_1\ldots p_s> - e <fp_1\ldots p_s, fp_1\ldots p_5>$, which is a K-linear combination of elements annihilated by P in modules isomorphic to $fH(B)$ and of elements annihilated by N in modules isomorphic to $eH(B)$, is $-2fp_1\ldots p_s q(e,fp_1\ldots p_5)$. Since $q_2(e,fp_1\ldots p_5) = b_0$, this is $-2fp_1\ldots p_s(1+b_0) = -2(1+(-1)^{s+1})fp_1\ldots p_s$. Taking $s = 1$, we obtain a contradiction as in earlier uses. This eliminates $[B:Z] = 11$.

9: Next suppose $[B:Z] = 9$, first working over the splitting field K. Here we have $q_1(e,fp_1 p_2 p_3 p_4) = q_1(eb_0,fp_1 p_2 p_3 p_4) = q_1(e,fb_0 p_1\ldots p_4) = -q_1(e,fp_1\ldots p_4)$, from which it follows that non-isomorphic irreducible submodules are q_1-orthogonal. Moreover, if $u = e$ or f, we have $q_1(u,up_1\ldots p_4) = q_1(up_4\ldots p_1,u) = (-1)^{\binom{4}{2}} q_1(up_1\ldots p_4,u) = q_1(up_1\ldots p_4,u) = -q_1(u,up_1\ldots p_4)$, so that each irreducible submodule is totally q_1-isotropic.

Thus we have within C two distinct isomorphic irreducible submodules $uH(B)$, $u'H(B)$, with both u and u' annihilated by N, and with $ub_0 = \varepsilon u$, $u'b_0 = \varepsilon u'$, $\varepsilon^2 = 1$, and $q_1(u,u'p_1\ldots p_4) = 1$. If the sum of these modules is not C, then C contains another irreducible submodule $vH(B)$, q_1-orthogonal to both of these, with $vN = 0$, $vb_0 = \varepsilon'v$, $\varepsilon' = \pm 1$. Then both $q(uH(B), vH(B))$ and $q(u'H(B), vH(B))$ are zero, $<uH(B), vH(B)>$ and $<u'H(B), vH(B)>$ annihilate B, and for $1 \leq s \leq 4$, we have by (95) that $u <vp_1\ldots p_s, u'p_1\ldots p_4> - u'p_1\ldots p_4 <vp_1\ldots p_s, u>$ is equal to $2vp_1\ldots p_s q(u,u'p_1\ldots p_4) = 2vp_1\ldots p_s(1+b_0) = 2(1+(-1)^s \varepsilon\varepsilon')vp_1\ldots p_s$. For $s = 1$ or 2 this yields a contradiction, as before. Hence C is the sum of two isomorphic irreducible submodules, totally isotropic and dually paired by q_1. Its dimension (over K or Z) is $2 \times 2^4 = 32$.

It follows as in §V.7.c) that the center of $H(B)$ splits over Z, and that either the even Clifford algebra $H^+(B)$ is split over Z and C is the sum of two 16-dimensional $H(B)$-submodules (over Z), or $H^+(B)$ is the full algebra of endomorphisms of an 8-dimensional (left) vector space over a quaternionic division algebra Q over Z, the $H(B)$-module C being irreducible, and isomorphic to this vector space as $H^+(B)$-module. We show that the former alternative cannot occur.

For this assertion, as well as for further information, we estimate the dimension of L over Z. We have $[A:Z] = 10$, $[C:Z] = 32$, and from $D = \langle B,B \rangle \oplus D_0$, D_0 the annihilator of B in D, $[\langle B,B \rangle : Z] = \binom{9}{2} = 36$, we have $[L:Z] = 36 + [D_0:Z] + 30 + 64 = 130 + [D_0:Z]$. Now D is faithfully represented on $B \oplus C$, so D_0 is faithfully represented on C, and may be identified with a Z-subspace of the $H(B)$-endomorphisms of C. From the above it follows that $0 \leq [D_0:Z] \leq 4$. But L is a normal simple Lie algebra over its centroid Z, so that $[L:Z]$ must be the dimension of a split simple Lie algebra, $130 \leq [L:Z] \leq 134$. From the classification of split simple Lie algebras, one has $[L:Z] = 133$, so $[D_0:Z] = 3$. Since all elements of D_0 act q_1-skewly on C, they all have trace zero as Z-endomorphisms of C. Thus if C is the sum of two (necessarily absolutely irreducible) $H(B)$-submodules $C_1 \oplus C_2$, and if e, e' are corresponding generators under an $H(B)$-isomorphism, we have for $D \in D_0$, $eD = \lambda e + \mu e'$, $e'D = \nu e - \lambda e'$, and D is determined by the 2 by 2 matrix $\begin{pmatrix} \lambda & \mu \\ \nu & -\lambda \end{pmatrix}$ of trace zero with entries in Z. From $[D_0:Z] = 3$, all such matrices occur, in particular $\begin{pmatrix} 1 & 0 \\ 0 & -1 \end{pmatrix}$. But this yields a diagonalizable transformation (over F) of $B \oplus C$ in $D_0 \subseteq D$, and contradicts the maximality of T. It follows that C is an irreducible $H(B)$-module, the centralizer of the action of $H(B)$ on C being a quaternionic division algebra Q over Z.

From extending the base field to a splitting field for Q over Z, we see that D_0 identifies with the (right) multiplications of C by elements of Q_0, the quaternions of trace zero, this by the arguments above. Thus $D = \langle B,B \rangle + Q_0 \tilde{=} K(B) \oplus Q_0$. Again, the forms h and q_1, together with the structure of C as $H(B)$-module, determine all our compositions except $c \langle c',c'' \rangle$. Now we have $d \in \langle B,B \rangle \tilde{=} K(B,h)$ such that $b \langle c,c' \rangle = bd$ for all $b \in B$; in fact, $bd = [b,g(c,c')]$ for all b, where $g(c,c')$ is a well-determined element of B_2, as in the case of dimension 13. Then $c'' \mapsto c''<c,c'> - c''g(c,c')$ is an $H(B)$-endomorphism of C, so is of the form $c'' \mapsto c''\phi(c,c')$, where $\phi(c,c') \in Q$. Thus

$$c'' <c,c'> = c''g(c,c') + c''\phi(c,c'),$$

where $g(c,c') \in B_2$ is as above, $\phi(c,c') \in Q$; in fact, $\phi(c,c') \in Q_0$ by remarks above.

By passing to the split case, $g(c,c')$ can be computed explicitly, as can $c''<c,c'>$ when the variables are in our standard basis for C; for example, $<e, ep_1p_2p_3p_4>$ annihilates B, so that $g(e,ep_1p_2p_3p_4) = 0$, and likewise $g(e,e') = 0$. (Here we have $C = eH(B) \oplus e'H(B)$, $eb_0 = \varepsilon e$, $q_1(e,e'p_1p_2p_3p_4) = 1$ as before.). Thus

$$e' <e, ep_1p_2p_3p_4> - ep_1p_2p_3p_4 <e,e'>$$

has as first term an element $e'\phi(e,ep_1p_2p_3p_4)$ of the form $\lambda e' + \mu e$, and as second term, $e\phi(e,e')p_1p_2p_3p_4$, of the form $\sigma ep_1p_2p_3p_4 + \tau e'p_1p_2p_3p_4$. But (95) yields that the above expression is $2eq(e',ep_1p_2p_3p_4) = 2e(q_1(e',ep_1p_2p_3p_4) + q_2(e',ep_1p_2p_3p_4)) = 2e(-1 - \varepsilon b_0) = -4e$. It follows that $e <e,e'> = 0$, $e'<e,ep_1p_2p_3p_4> = -4e = e'\phi(e,ep_1p_2p_3p_4)$. One easily verifies that $e <e,ep_1p_2p_3p_4> = 0$, so that $\phi(e,ep_1p_2p_3p_4)$ corresponds to the 2 by 2 matrix $\begin{pmatrix} 0 & 0 \\ -4 & 0 \end{pmatrix}$ in our identification of D_0 above. By means like these, ϕ can be computed in the split case, where one verifies that, for all $d \in Q_0$, $c,c' \in C$,

$$\phi(cd,c') + \phi(c,c'd) = [\phi(c,c'),d],$$
$$\phi(c,c'd) - \phi(cd,c') = -4q_1(c,c')d,$$
$$\phi(c,c')d + d\phi(c,c') = 2(q_1(cd,c') - q_1(c,c'd)).$$

(Here all quantities are to be regarded as operators on C.) With $d \mapsto d^*$ the canonical involution in Q, the effect of these, and of the fact that $*$ is induced by q_1-transposing, is to say that $q_1 - \frac{1}{2}\phi: C \times C \to Q$ is an antihermitian form over Q. From earlier work, one sees that the transpose with respect to this form gives the same involution of $H^+(B)$ as that induced in $H^+(B)$ by the involution of $H(B)$ fixing B.

Conversely, let B be nine-dimensional Z-space carrying a non-degenerate quadratic form h, not representing 1, of square discriminant, and such that the even Clifford algebra $H^+(B)$ is a full

(4 by 4) matrix algebra over a quaternionic division algebra Q. Then we claim B can serve as the basis for construction of an algebra L as above.

For $H^+(B)$ is stable under the involution in $H(B)$ which is the identity on B. If C is a simple right $H^+(B)$-module, then C is a 4-dimensional (right) Q-space and $H^+(B) = \text{End}_Q(C)$. The involution above has in $H^+(B)$ a space of skew elements of Z-dimension $\binom{9}{2} + \binom{9}{6} < 2^7$, so may be described as the transpose with respect to an antihermitian form Θ on the Q-space C, relative to the standard involution in Q. The form Θ is non-degenerate, and we may write $\Theta = q_1 - \frac{1}{2}\phi$, where q_1 is the projection on $Z \subseteq Q$ and $-\frac{1}{2}\phi$ the projection on Q_0, the (skew) elements of trace zero. We may take z central in $H(B)$, $z \notin Z$, $z^2 = 1$, so that $H(B) = H^+(B) \oplus zH^+(B)$, and choose fixed $\varepsilon = \pm 1$. By letting $cz = \varepsilon c$ for all $c \in C$, we extend the module structure of C to that of $H(B)$-module, and the elements of B act Θ-symmetrically as Q-endomorphisms of C.

Now q_1 is a nondegenerate alternate bilinear form on C over Z, relative to which the elements of B act symmetrically and those of Q_0 act skewly. For c,c' in C, we define $q_2(c,c') \in B$ by $h(b,q_2(c,c')) = q_1(c,c'b)$ for all $b \in B$. Then q_2 is skew-symmetric and Z-bilinear. We can now define all our remaining compositions in

$$L = K(B) \oplus Q_0 \oplus (S \otimes A) \oplus (M'' \otimes C),$$

where $A = Z \oplus B$, almost as for dimension 13: The action of $K(B)$ on A is that on B, and its action on C results from the identification of $K(B)$ with B_2; $K(B)$ and Q_0 centralize one another, Q_0 annihilates A, and its action on C is that of Q; the brackets of $S \otimes A$ and $M'' \otimes C$ are defined in terms of the actions of S on M'' and A on C, as before, and for $a_1 = \alpha_1 + b_1$, $a_2 = \alpha_2 + b_2$ in A ($\alpha_i \in Z$, $b_i \in B$), then $\langle a_1, a_2 \rangle = \langle b_1, b_2 \rangle \in K(B)$ is defined as before. One verifies (e.g., by passing to a splitting field) that, for $c,c' \in C$, the map $b \mapsto q_2(c,c'b) - q_2(cb,c') = q(c,c'b) - q(cb,c')$, where $q = q_1 + q_2 : C \times C \mapsto A$, is in $K(B)$, so has the form $b \to [b,g(c,c')]$ for unique $g(c,c') \in B_2$. Then one defines $\langle c,c' \rangle = g(c,c') + \phi(c,c') \in K(B) \oplus Q_0$, and this with q defines a bracket for two elements of $M'' \otimes C$, by our formulas before. Verifying that our L is indeed a Lie algebra comes down to the remaining ones, (92) - (95) of §III.9, together with the derivation properties needed, and for this it suffices to treat the split case, where everything may be computed explicitly.

That L is simple is easily verified by using that $K(B) + Q_0 = D$, $S \otimes A$, and $M'' \otimes C$ are its homogeneous S-components. That $T = T \otimes F \subseteq T \otimes A$ is a maximal F-split torus follows from the fact that the derived algebra L_0' of its centralizer is $D + T \otimes B$, which identifies with $K(A) \oplus Q_0$, the form on A being anisotropic as in 13 dimensions. The simple group G' is again generated by all $\exp(\text{ad}(u \otimes c))$, $u \in M''$, $c \in C$, as in that case.

7: Next consider the case $[B:Z] = 7$. First working over our splitting field K, we have $q_1(e, e'p_1p_2p_3) = q_1(eb_0, e'p_1p_2p_3) = q_1(e, e'p_1p_2p_3b_0) = -q_1(e, e'b_0p_1p_2p_3) = -q_1(e, e'p_1p_2p_3)$, and likewise for f and f'. It follows that the sum of all submodules of C_K isomorphic to $eH(B)$ is totally q_1-isotropic, as is the sum of all submodules isomorphic to $fH(B)$. Thus <u>both</u> <u>types</u> <u>must</u> <u>be</u> <u>present</u>.

We may therefore assume $eH(B)$ and $fH(B)$ to be a dual pair of summands of C_K, with $q_1(e, fp_1p_2p_3) = 1$. If their sum is <u>not</u> C_K, there is a second such pair $e'H(B)$, $f'H(B)$, q_1-orthogonal to $eH(B)$ and $fH(B)$. Then evidently $<eH(B), e'>$ and $<eH(B), f'p_1p_2p_3>$ annihilate B, and $q(eH(B), e') = 0 = q(eH(B), f'H(B))$. On the other hand, $h(b_0, q_2(e', f'p_1p_2p_3)) = q_1(e', f'p_1p_2p_3b_0) = q_1(e', f'p_1p_2p_3) = 1$; this and our other information yield $q(e', f'p_1p_2p_3) = 1 + b_0$. Thus $<ep_1p_2, e'>, <ep_1p_2, f'p_1p_2p_3>$ both annihilate B, so that $f'p_1p_2p_3 <ep_1p_2, e'> - e'<ep_1p_2, f'p_1p_2p_3>$ is a combination of elements annihilated by P in submodules isomorphic to $fH(B)$ and of elements annihilated by N in submodules isomorphic to $eH(B)$. But by (95), the above is $-2ep_1p_2(1+b_0) = -4ep_1p_2$, which is impossible as before. We thus have $C_K = eH(B) \oplus fH(B)$, of K-dimension 16, so $[C:Z] = 16$. Over Z, we have that <u>either</u>: 1) C fails to be irreducible, in which case irreducible $H(B)$-modules have Z-dimension 8; then both the center of $H(B)$ and the even Clifford algebra $H^+(B)$ split; <u>or</u> 2) C is irreducible, in which case the center of $H(B)$ <u>cannot</u> split, and $H(B)$ is a full matrix algebra over its center. We eliminate the alternative 1):

In the case 1), the irreducible $H(B)$-modules are absolutely irreducible, and C is the direct sum $U \oplus U'$ of two non-isomorphic such modules. Thus the $H(B)$-endomorphisms of C are made up of scalar multiplications separately on each of U, U', and have Z-dimension 2.

Moreover, as we have seen in a splitting, U and U' are q_1-totally isotropic and q_1-dual. It follows that the q_1-skew $H(B)$-endomorphisms of C send $u \mapsto \lambda u, u' \mapsto -\lambda u'$, for λ running over Z. Now we have as before that $<B,B> \subseteq <C,C> = D$, and $D = <B,B> \oplus D_0$, D_0 the annihilator of B, $<B,B> \stackrel{\sim}{=} K(B,h)$. It follows that $[D_0:Z] \leq 1$, and that

$$[L:Z] = \binom{7}{2} + [L_0:Z] + 24 + 32 = 77 + [D_0:Z].$$

Since 77 is not the dimension of a central simple Lie algebra, we must have $[D_0:Z] = 1$. But elements of D_0 act Z-diagonally on C and trivially on A, and in particular one of them has eigenvalues 0, ± 1 in $A \oplus C$. This contradicts the maximality of the F-split torus T, and thereby eliminates 1).

We therefore have that the center of $H(B)$ is a quadratic extension field Y of Z, over which extension $H(B)$ is a full 8 by 8 matrix algebra. The $H(B)$-module C is irreducible, and $[L:Z] = 77 + [D_0:Z]$ as above, D_0 consisting of scalar multiplications in C by elements of Y. Since these must be skew with respect to q_1, they must have Z-trace zero, and this trace is in turn $[C:Y]$ times the Z-trace of the corresponding element of Y. Thus D_0 consists of scalar multiplications by elements $\eta \in Y$ with $\text{Tr}_{Y/Z}(\eta) = 0$, so that $[D_0:Z] \leq 1$ again. Our considerations of dimension now yield $[D_0:Z] = 1$, and D_0 consists exactly of scalar multiplications in C by elements of Z-trace zero in the center Y of $H(B)$. Thus $D = <B,B> \oplus Y_0 = K(B) \oplus Y_0$. Once again, q_1 and h, together with the structure of C as $H(B)$-module, determine all compositions except possibly $c <c',c''>$. The latter is of the form $cg(c',c'') + c\phi(c',c'')$ as in the case of dimension 9, where $g(c',c'') \in B_2$ is such that $[b,g(c',c'')] = b <c',c''>$ for all $b \in B$, and where $\phi(c',c'') \in Y_0$. Here it is $q_1 - \frac{1}{3}\phi$ that is an antihermitian form on C as Y-vector space, with respect to the nontrivial automorphism of Y/Z.

Conversely, let B be a seven-dimensional Z-space carrying a non-degenerate quadratic form h, not representing 1, such that the Clifford algebra $H(B)$ is a full matrix algebra (8 by 8) over its center Y, a quadratic field extension of Z. Here the involution of $H(B)$ which is the identity on B is of second kind. If C is the irreducible $H(B)$-module, so that $H(B) = \text{End}_Y(C)$, there is an antihermitian nondegenerate form Θ on C over Y such that the involution in $H(B)$ is the transpose with respect to Θ. Let q_1 be the projection of Θ on Z and $-\frac{1}{3}\phi$ be the projection of Θ on Y_0 in the decomposition $Y = Z + Y_0$ as above. Then q_1 is an alternate nondegenerate form on the Z-space C, relative to which the elements of B act symmetrically. We now define $q_2(c,c') \in B$, $g(c,c') \in B_2$, and $b < c,c' >$, $c'' < c,c' >$ as above, and use these to introduce our compositions in $L = K(B) \oplus Y_0 \oplus (S \otimes A) \oplus (M'' \otimes C)$, with $A = Z + B$ as before. Computations in the split case verify the Jacobi identity, and we see as in higher dimensions that $T = T \otimes F \subseteq S \otimes A$ is a maximal F-split torus, with centralizer $L_0 = K(B) \oplus Y_0 \oplus (T \otimes A) = K(A) \oplus Y_0 \oplus (T \otimes Z) = K(A) \oplus Y$. Once again, the simple group G' is generated by all $\exp(\mathrm{ad}(u \otimes c))$, $u \in M''$, $c \in C$.

6: Now consider the case $[B:Z] = 6$. Here $H(B)$ is central simple, and the module C is a direct sum of isomorphic irreducible modules, as is C_K when we pass to the splitting field K. Working over K, with notations as usual, we see from $q_1(e,e'p_1p_2p_3) = q_1(ep_3p_2p_1,e') = -q_1(ep_1p_2p_3,e') = q_1(e',ep_1p_2p_3)$ that C_K is a direct sum of irreducible submodules, <u>each of them</u> q_1-<u>nonsingular</u>. Let $eH(B)$ be one of these; then $q_1(e,ep_1p_2p_3) = \gamma \neq 0$. Suppose a second summand, $e'H(B)$, is present, q_1-orthogonal to $eH(B)$. Then, as before, $< e,e'p_1 >$ and $< ep_1p_2p_3,e'p_1 >$ annihilate B, so that $ep_1p_2p_3 <e,e'p_1> - e < ep_1p_2p_3,e'p_1 >$ is a sum of an element annihilated by N and one annihilated by P. But, by (95), this element is $- 2e'p_1q(e,ep_1p_2p_3) = - 2e'p_1q_1(e,ep_1p_2p_3) = - 2\gamma e'p_1$, which is not in the sum of the spaces above. That is, C_K is irreducible, necessarily of K-dimension 8, and therefore C <u>is absolutely irreducible</u> and $H(B)$ <u>is split</u> over Z. Again we have $< B,B > \subseteq < C,C > = D$, $D = D_0 + < B,B >$, and here necessarily $D_0 = 0$ (as when $[B:Z] = 13$). Thus $D = < B,B > = H(B)$, and all compositions are determined from h, q_1, and the $H(B)$-module structure of C as in the case $[B:Z] = 13$. The Z-dimension of L is

$$\binom{6}{2} + 21 + 16 = 52,$$

and L is a Z-form of the split exceptional algebra F_4. (This case for B split, as well as that with $[B:Z] = 13$, have been considered by Hein [He].)

Conversely, whenever one has a space B of dimension six over Z, with a nondegenerate quadratic form h not representing 1 and such that the Clifford algebra $H(B)$ is split over Z (of dimension 2^6), we see from the fact that the skew elements of $H(B)$ with respect to the involution which is 1 on B have dimension $\binom{6}{2} + \binom{6}{3} + \binom{6}{6} = 36 > 2^5$, that this involution results from a nondegenerate symplectic form q_1 on a simple $H(B)$-module C. The reconstruction of L from these data then proceeds as in dimension 13. Our maximal F-split torus T has centralizer $L_0 = K(B) \oplus (T \otimes A) = K(A) \oplus (T \otimes Z)$ as before, and the group G' is again generated by elements of the form $\exp(\text{ad}(u \otimes c))$.

5: We next consider the case $[B:Z] = 5$, and work at first over the splitting field K, with notations as before. Here we have $q_1(e, fp_1p_2) = q_1(eb_0, fp_1p_2) = q_1(e, fp_1p_2 b_0) = q_1(e, fb_0 p_1 p_2) = -q_1(e, fp_1p_2)$, from which <u>non-isomorphic irreducible submodules of</u> C_K <u>are q_1-orthogonal</u>. Thus the two homogeneous parts of C_K are q_1-nonsingular, and from $q_1(e, e'p_1p_2) = q_1(ep_2p_1, e') = -q_1(ep_1p_2, e') = q_1(e', ep_1p_2)$ (and likewise for f, f'), it follows that C_K <u>is a</u> q_1-<u>orthogonal sum of irreducible submodules, each of which is</u> q_1-<u>nonsingular</u>.

Next we show (still over K) that <u>all irreducible submodules of</u> C_K <u>are isomorphic</u>. For if $eH(B)$ and $fH(B)$ are both present, with $q_1(f, fp_1p_2) = \gamma \neq 0$, we know that both $<ep_1, fp_1p_2>$ and $<ep_1, f>$ annihilate B, so that $f <ep_1, fp_1p_2> - fp_1p_2 <ep_1, f>$ is in the sum of those submodules of C_K isomorphic to $fH(B)$, while by (95) this expression is $2ep_1 q(f, fp_1p_2) = 2ep_1(\gamma + q_2(f, fp_1p_2)) = 2ep_1(\gamma - \gamma b_0) = 4\gamma ep_1 \neq 0$, a contradiction. Now this implies that <u>the center of</u> $H(B)$ <u>must split over</u> Z, and then the result of Tamagawa (Appendix C) applies to determine that $H^+(B)$ <u>is a division algebra</u>.

The canonical involution of $H(B)$ fixing the elements of B maps to their negatives all elements $b_i b_j$ and fixes all elements $b_i b_j b_k b_\ell$, where the b_ν occurring in each product are orthogonal with respect to h. Thus the induced involution in $H^+(B)$ has a 6-dimensional space of fixed elements, and a 10-dimensional space of skew elements. Since the center splits over Z, we may take an orthogonal basis $b_1, \ldots, b_5$ with $\prod_{i=1}^{5} h(b_i, b_i) = 1$, so with $z^2 = 1$, $z = b_1 \ldots b_5$ central in $H(B)$. We have $cz = \varepsilon c$ for all $c \in C$, where either $\varepsilon = 1$ or $\varepsilon = -1$, and the Z-linear mapping $\alpha + b \to \alpha + zb$ identifies $A = Z \oplus B$ with the symmetric elements of $H^+(B)$ under the above involution, and <u>is</u>

moreover an isomorphism of Jordan algebras. That is, we are in the situation of §V.7.a), with $[A:Z(A)] = 6$, $[E:Z(A)] = [H^+(B):Z] = 16$. Now we do have the additional information, which could not in §V.7.a) be deduced from the Jordan theory alone, that C is a vector space over $H^+(B) = E$, and from this point on the considerations of §V.7.a) apply. That is, Proposition V.4 is proved exactly as before, and results in an identification of L with

$$K'((M'' \otimes E) \oplus C) = K((M'' \otimes E) \oplus C),$$

the involution in E being of first kind. As in that case, the simple group G' is identified with the quotient by its center of the commutator subgroup of the unitary group of $(M'' \otimes E) \oplus C$, a space over E carrying a nondegenerate antihermitian form of Witt index 1.

4: Next we note that $[B:Z] = 4$ is impossible. Working over the splitting field K, we see, from $q_1(e,e'p_1p_2) = q_1(e',ep_1p_2)$ as before, that C_K is the sum of isomorphic irreducible $H(B)$-modules, each nonsingular with respect to q_1. With e,e' as usual, we let $eH(B)$ and $e'H(B)$ be two of these, orthogonal to one another with respect to q_1. Then $q(ep_1,e') = 0 = q(ep_1,e'p_1p_2)$, and both $< ep_1,e' >$ and $< ep_1,e'p_1p_2 >$ annihilate B. With $q_1(e',e'p_1p_2) = \gamma \neq 0$, we have $\gamma = q(e',e'p_1p_2)$, and $e'p_1p_2 < ep_1,e'> - e' < ep_1,e'p_1p_2 > = 2\gamma ep_1$ as before, yielding a contradiction as before. That is, C_K is irreducible for $H(B_K)$, or C is an absolutely irreducible $H(B)$-module. Thus $[C:Z] = 4$, and $H(B)$ is the algebra of 4 by 4 Z-matrices. As before $< B,B > \subseteq < C,C >$, $< B,B >$ identifies with $K(B,h)$, and $D = < C,C > = < B,B > \oplus D_0$. But since C is absolutely irreducible for $H(B)$, $[D_0:Z] \leq 1$, and in fact $D_0 = 0$, since D_0 consists of scalar multiplications in C by elements of Z such that these operations are q_1-skew. We have thus $[D:Z] = [< B,B >:Z] = 6$, $[A:Z] = 5$, $[C:Z] = 4$, and $[L:Z] = 6 + 15 + 8 = 29$, which is not the dimension of any absolutely simple Lie algebra. The case $[B:Z] = 4$ is thus eliminated.

3: Next let $[B:Z] = 3$, where we have $q_1(e,e'p_1) = q_1(eb_0,e'p_1) = q_1(e,e'p_1b_0) = -q_1(e,e'p_1)$, and likewise for the modules $fH(B)$ (all over a splitting field K). That is, over K, we have that C is a sum of irreducible modules of both types for $H(B)$, any two of the same type being q_1-orthogonal.

253

Now suppose that more than one summand of each type is present, so that C contains a submodule $eH(B) \oplus fH(B) \oplus e'H(B) \oplus f'H(B) = U$, $q_1(eH(B), f'H(B)) = 0 = q_1(e'H(B), fH(B))$, $q_1(e,fp_1) = 1 = q_1(e',f'p_1)$. Then, as we have seen, $q_2(eH(B), eH(B)) = 0 = q_2(eH(B), e'H(B)) = q_2(eH(B), f'H(B))$, (and indeed $q_2(V,W) = 0$ for any pair of q_1-orthogonal submodules V,W. It follows that $<e,f'p_1>$ annihilates B. Now if $e''H(B)$, $f''H(B)$ are q_1-orthogonal to U, we see by (95) that

$$e'' <e,f'p_1> - f'p_1 <e,e''> = 0 = f'' <e,f'p_1> - e <f'',f'p_1>.$$

Since $<e,e''>$ and $<f'',f'p_1>$ also annihilate B, we have that $e'' <e,f'p_1> = 0 = f'' <e,f'p_1>$. Similar considerations yield $e <e,f'p_1> = 0 = f' <e,f'p_1>$, so that $<e,f'p_1>$ annihilates $U^\perp$, $eH(B)$, $f'H(B)$.

Next we have by (95) that $e' <e,f'p_1> = f'p_1 <e,e'> = 2eq(e',f'p_1) = 2e(1+b_0) = 4e$, and $<e,e'>$ annihilates B. Thus $e' <e,f'p_1> = 4e$, and likewise $fp_1 <e,f'p_1> = -4fp_1$, so that $f <e,f'p_1> = \frac{1}{2}fp_1 n_1 <e,f'p_1> = \frac{1}{2}fp_1 <e,f'p_1> n_1 = -2f'p_1 n_1 = -4f'$. Thus $<e,f'p_1>$ is an $H(B)$-endomorphism of C, annihilating $U^\perp$, $eH(B)$, $f'H(B)$, and sending e' to $4e$, f to $-4f'$. Letting $e_1, e_2, \ldots, e_n; f_1, f_2, \ldots, f_n$ be an $H(B)$-basis for C with $e_i \leftrightarrow e$, $f_i \leftrightarrow f$, we see that $<C,C>$ contains $H(B)$-endomorphisms of C having the effect on these of sending e_i to e_j, f_j to $-f_i (i \neq j)$, and the remaining elements to zero. In our previous decomposition of $D = <B,B> \oplus D_0$, this assures that (over K) D_0 contains elements having the effect on $e_1, \ldots, e_n$ of each matrix unit $E_{ij} (i \neq j)$, and therefore also of $[E_{ij} E_{ji}] = E_{ii} - E_{jj}$ $(i \neq j)$. The fact that $e_1, \ldots, e_n$ and $f_1, \ldots, f_n$ generate q_1-dual spaces assures that the effect on C of $D \in D_0$ is completely determined by $e_1 D, \ldots, e_n D$, which must be K-linear combinations of $e_1, \ldots, e_n$. That is, $[D_0:K] \leq n^2$. We have shown above that $[D_0:K] \geq n^2 - 1$, and that $[D_0:K] = n^2$ if and only if D_0 contains D <u>fixing</u> each of $e_1, \ldots, e_n$. (Actually it has been assumed that $n > 1$; but if $n = 1$, these assertions are trivial.)

Now we have $[A:K] = 4$, $[C:K] = 4n$, $[D:K] = [<B,B>:K] + [D_0:K] = 3 + n^2$, or $3 + n^2 - 1 = 2 + n^2$. Thus $[L_K:K] = n^2 + 8n + 14$, or $n^2 + 8n + 15$, according as $[D_0:K] = n^2 - 1$ or n^2. The former is impossible, since $n^2 + 8n + 14 = (n+4)^2 - 2$ <u>cannot be the dimension of a split simple Lie algebra</u>. One readily rules out the dimensions of

A_{u-1}, (viz. u^2-1), F_4, E_6, E_7, E_8, since the sum of any of these and 2 cannot be a square ≥ 16. For G_2, $14 = (n+4)^2 - 2$ only if $n = 0$, which means $C = 0$; but this is not the case here. Finally, B_u, C_u, D_u have dimensions $2u^2 + u$, $2u^2 + u$, $2u^2 - u$; replacing u by $-u$ if necessary, it suffices to show that $2u^2 + u + 2 = r^2$ has no integral solutions u, r. Now $r^2 \equiv 1$ or $0 \pmod 3$, while $2u^2 + u + 2 \equiv 2 \pmod 3$ unless $u \equiv 2 \pmod 3$, in which case $2u^2 + u + 2 \equiv 0 \pmod 3$. Thus we must have $r = 3j$, $u = 3k + 2$. But then $r^2 \equiv 0 \pmod 9$, $2u^2 + u + 2 \equiv 3 \pmod 9$, and we are done.

Thus we must have $[\mathcal{D}_0 : K] = n^2$, $\mathcal{D}_0$ contains D fixing each of $e_1, \ldots, e_n$, and $[L_K : K] = (n+4)^2 - 1$. Next we return to the ground field Z, the centroid of L, and show that <u>the center of</u> $H(B)$ <u>cannot split</u>. For if so, we should have an orthogonal basis b_1, b_2, b_3 for B, with $z^2 = 1$, where $z = b_1 b_2 b_3 \in H(B)$. The $H(B)$-module C splits into spaces where z acts as ± 1, call them C^+, C^-. In our splitting over K, z identifies with a scalar multiple of $(p_1 + n_1)b_0(p_1 - n_1) = w$, say by $\lambda \in K$, and we have $1 = z^2 = \lambda^2 w^2 = 4\lambda^2$, so that $\lambda = \frac{1}{2}$. With $\{e_i\}, \{f_i\}$ as above (over K), we find $e_i z = \frac{1}{2} e_i p_1 b_0 (p_1 - n_1) = -\frac{1}{2} e_i p_1 (p_1 - n_1) = \frac{1}{2} e_i p_1 n_1 = e_i$, $e_i p_1 z = \frac{1}{2} e_i p_1$, and likewise $f_i z = - f_i$, $f_i p_1 z = - f_i p_1$. That is, $C^+_K = \Sigma e_i H(B_K)$, $C^-_K = \Sigma f_i H(B_K)$, and the restriction to C^+_K of $\mathcal{D}_{0K}$ is the totality of $H(B_K)$-endomorphisms of C^+_K. Thus the restriction to C^+ of $\mathcal{D}_0$ is the totality of $H(B)$-endomorphisms of C^+, and contains in particular the identity. But we know now that C^+ and C^- are q_1-dual, and that $D \in \mathcal{D}_0$ has $q_1(cD, c') + q_1(c, c'D) = 0$ for $c, c' \in C$, while mapping C^+ to C^+, C^- to C^-. It follows that if $D \in \mathcal{D}_0$ induces the identity on C^+, D induces $- \text{Id}$ on C^-, and therefore that ad D has rational eigenvalues (and is diagonalizable). This is impossible by the maximality of T. Thus <u>the center of</u> $H(B)$ <u>is a quadratic extension field</u> Y <u>of</u> Z, <u>and</u> $H(B)$ <u>is a simple</u> Y<u>-algebra of dimension</u> 4.

It follows that, taking our splitting field K to be an extension of Y, each irreducible summand C_1 of C (over Z) decomposes over K into a sum of modules isomorphic to $eH(B) + fH(B)$. There are still two cases for the structure of $H(B)$: i) the algebra of 2 by 2 Y-matrices; and ii) a (quaternionic) division algebra over Y. We eliminate the case i) by showing that $H(B)$ can contain no non-zero elements of square zero. For if b_1, b_2, b_3 are an orthogonal basis for

B, and if $z = b_1 b_2 b_3 \in Y$, then $z^2 \in Z$, but is not a square, and $H(B)$ is the direct sum of the Z-subspaces Z, B, zB, Zz. Thus let $\alpha, \beta \in Z$; $b, b' \in B$, $x = \alpha + b + zb' + \beta z$, $x^2 = 0$. Then x^2 is easily seen to have as component in B, $2\alpha b + 2\beta z^2 b'$, and as component in zB, $2\alpha z b' + 2\beta z b$, so both of these are zero. Since z is invertible in $H(B)$, this gives the relations $\alpha b + \beta \gamma b' = 0 = \beta b + \alpha b'$, where $\gamma = z^2 \in Z$, $\gamma \notin Z^2$. It follows that either $b = b' = 0$ or $\alpha^2 = \beta^2 \gamma$. Since $\gamma \notin Z^2$, the latter entails $\beta = 0 = \alpha$. Thus either $b = b' = 0$ or $\alpha = \beta = 0$. In the former case, $x = \alpha + \beta z \in Y$ has square zero, from which $x = 0$ since Y is a field. We therefore have $x = b + zb'$, $x^2 = h(b,b) + z^2 h(b',b') + 2zh(b,b')$. Clearly we may assume both b, b' are non-zero, and the term in Zz above is $2zh(b,b') = 0$, so that b and b' are orthogonal, $z = bb'b''$ for b, b', b'' an orthogonal basis for B, and $0 = h(b,b) + z^2 h(b',b') = h(b,b)(1 - h(b',b')^2 h(b'',b''))$. But then $h(b'',b'') = h(b',b')^{-2}$, from which h represents 1, a contradiction.

Thus $H(B)$ is a division algebra, a quaternion algebra over Y. The involution of $H(B)$ which is the identity on B is of second kind, with $A = Z + B$ as fixed elements. Thus we are in the context of §V.7.a.

Namely, in the division algebra $E = H(B)$, the symmetric elements with respect to this involution are simply $A = Z + B$. Our C is a right E-module, and these are the conclusions resulting from perfection used in the proof of Proposition V.4. The remaining parts of that argument did not exclude the present situation, and therefore the conclusion holds, identifying L with $K'((M'' \otimes E) \oplus C)$, with respect to a non-degenerate antihermitian form of Witt index one on the E-space $(M'' \otimes E) \oplus C$, constructed as in the proof of that proposition. The identification of our simple G' with the central quotient of the commutator group of the unitary group is also valid.

2: Finally, the case $[B:Z] = 2$ is easily seen to lie in the setting of Proposition V.4. Namely, $H(B)$ is necessarily a quaternionic division algebra over Z, with $A = Z + B$ as fixed elements for the involution which is the identity on B, and C is an $H(B)$-module. The rest goes as above.

Since $[B:Z] \geq 2$ if Z is to be the center of the Jordan algebra $A = Z \oplus B$, this completes the case where A is of degree two. We thus following description for Lie algebras L of this type:

Theorem V.3: Let L be a simple Lie algebra of type BC_1 over F. Let Z be the centroid of L, and let A be a Jordan algebra associated with L as in Theorem III.11. Suppose that A is of degree 2. Then one of the following holds:

α) $[A:Z] = 3, 4$ or 6, and A is the symmetric elements of a division algebra E over Z, with respect to an involution, as follows:

If $[A:Z] = 3$, E is a quaternionic division algebra over Z;

if $[A:Z] = 4$, E is a quaternionic division algebra over a quadratic field extension Y of Z;

if $[A:Z] = 6$, E is a central division algebra of dimension 16 over Z.

The space C is a right E-module carrying an anisotropic antihermitian form, and L identifies with the derived algebra $K'((M'' \otimes E) \oplus C)$ of the skew E-endomorphisms of the antihermitian E-space $(M'' \otimes E) \oplus C$, of Witt index one.

β) $[A:Z] = 7$, and A is the Jordan algebra associated with a quadratic form in 6 dimensions over Z, not representing 1, and having split Clifford algebra; the Lie algebra L is constructed from A and from an irreducible module C for this Clifford algebra as described under "6:" above. We have $[L:Z] = 52$.

γ) $[A:Z] = 8$, and A is the Jordan algebra associated with a quadratic form in 7 dimensions over Z, not representing 1, and whose Clifford algebra is a full matrix algebra over its center Y, a quadratic field extension of Z; the Lie algebra L is constructed from A and from an irreducible module C for this Clifford algebra as described under "7:" above. We have $[L:Z] = 78$.

δ) $[A:Z] = 10$, and A is the Jordan algebra associated with a quadratic form in 9 dimensions over Z, not representing 1, and whose Clifford algebra has split center, the even Clifford algebra being a full matrix algebra over a quaternionic division algebra Q over Z; the Lie algebra L is constructed from A and from an irreducible module C for the Clifford algebra as described under "9:" above. We have $[L:Z] = 133$.

ε) $[A:Z] = 14$, and A is the Jordan algebra associated with a quadratic form in 13 dimensions over Z, not representing 1, and whose Clifford algebra has split center and split even algebra;

the Lie algebra L is constructed from A and from an irreducible module C for the Clifford algebra as described under "13:" above. We have $[L:Z] = 248$.

iii) The case where $A = Z(A)$ is of degree one.

Here we have seen in §V.1 that A is the centroid of L. The mapping $q: C \times C \to A$ is a nondegenerate symplectic A-form on C, which by (90) of §III.9 is a vector space over the field A. From (91) and (5') of Chapter III we have $<A,A> = 0$; that q is a A-bilinear then follows from (93) and (94), and from the fact that A admits no nontrivial F-derivations. The nondegeneracy of q has already been shown in §III.9. From (92) we have $<c,c'a> = <ca,c'>$, and thus the ternary composition $(c,c',c'') \mapsto c<c',c''>$ on C is A-trilinear. To be more conventional, we shall often write the operation of A on C with A on the left; thus $ac = ca$.

Prior to the general case of algebras of type BC_1, this special case had been studied by Faulkner [F], who developed a set of identities relating ternary compositions corresponding to $c<c',c''>$ and $q(c,c')c''$, and equivalent to ours of §III.9 for this case. Some discrepancies in parametrization give our versions of these a slightly different form, as follows: let $<c,c',c''> \in C$ be defined by

$$<c,c',c''> = -c<c',c''> - cq(c',c'').$$

Then, from (95), we have

(T1) $\qquad <c,c',c''> = <c',c,c''> - 2q(c,c')c''$.

The symmetry of $<c',c''>$ and skewness of q yield at once

(T2) $\qquad <c,c',c''> = <c,c'',c'> - 2q(c',c'')c$.

From the definitions, $q(<c_1,c_2,c_3>,c_4) - q(<c_1,c_2,c_4>,c_3) + 2q(c_1,c_2)q(c_3,c_4) = q(c_1<c_2,c_4>,c_3) + q(c_1,c_3)q(c_2,c_4) - q(c_1<c_2,c_3>,c_4) - q(c_1,c_4)q(c_2,c_3) + 2q(c_1,c_2)q(c_3,c_4)$. The derivation property of $<c,c'>$ here gives $q(c_1<c_2,c_4>,c_3) - q(c_1<c_2,c_3>,c_4) = q(c_1,c_4<c_2,c_3> - c_3<c_2,c_4>)$, and substitution from (95) yields

(T3) $\qquad q(<c_1,c_2,c_3>,c_4) = q(<c_1,c_2,c_4>,c_3)$
$\qquad\qquad\qquad + 2q(c_1,c_2)q(c_3,c_4).$

Finally, we consider

$$\langle\langle c_1,c_2,c_3\rangle,c_4,c_5\rangle - \langle\langle c_1,c_4,c_5\rangle,c_2,c_3\rangle$$
$$- \langle c_1,\langle c_2,c_4,c_5\rangle,c_3\rangle - \langle c_1,c_2,\langle c_3,c_5,c_4\rangle\rangle$$

$$= - \langle c_1 \langle c_2,c_3\rangle,c_4,c_5\rangle - q(c_2,c_3)\langle c_1,c_4,c_5\rangle$$
$$+ \langle c_1 \langle c_4,c_5\rangle,c_2,c_3\rangle + q(c_4,c_5)\langle c_1,c_2,c_3\rangle$$
$$+ \langle c_1,c_2 \langle c_4,c_5\rangle,c_3\rangle + q(c_4,c_5)\langle c_1,c_2,c_3\rangle$$
$$+ \langle c_1,c_2,c_3\langle c_4,c_5\rangle\rangle - q(c_4,c_5)\langle c_1,c_2,c_3\rangle$$

$$= c_1\langle c_2,c_3\rangle\langle c_4,c_5\rangle + q(c_4,c_5)c_1\langle c_2,c_3\rangle$$
$$+ q(c_2,c_3)c_1\langle c_4,c_5\rangle + q(c_2,c_3)q(c_4,c_5)c_1$$
$$- c_1\langle c_4,c_5\rangle\langle c_2,c_3\rangle - q(c_2,c_3)c_1\langle c_4,c_5\rangle$$
$$- q(c_4,c_5)c_1\langle c_2,c_3\rangle - q(c_2,c_3)q(c_4,c_5)c_1$$
$$- c_1\langle c_2 \langle c_4,c_5\rangle,c_3\rangle - q(c_2\langle c_4,c_5\rangle,c_3)c_1$$
$$- c_1\langle c_2,c_3\langle c_4,c_5\rangle\rangle - q(c_2,c_3\langle c_4,c_5\rangle)c_1.$$

The second and seventh terms cancel, as do the fourth and eighth, and the third and sixth. The fact that $d = \langle c_4,c_5\rangle$ acts q-skewly on C means that the tenth and twelfth terms cancel. Finally, from the derivation-property $[\langle c_2,c_3\rangle,d] = \langle c_2d,c_3\rangle + \langle c_2,c_3d\rangle$ we see that the first, fifth, ninth and eleventh terms add to zero. Thus our expression is zero, yielding

(T4) $$\langle\langle c_1,c_2,c_3\rangle,c_4,c_5\rangle - \langle\langle c_1,c_4,c_5\rangle,c_2,c_3\rangle$$
$$= \langle c_1,\langle c_2,c_4,c_5\rangle,c_3\rangle + \langle c_1,c_2,\langle c_3,c_5,c_4\rangle\rangle.$$

Conversely, assuming a nondegenerate symplectic form q and a ternary product $\langle c_1,c_2,c_3\rangle$ satisfying (T1-4), Faulkner constructs a simple Lie algebra along lines which are equivalent to our §III.9. To obtain a precise description of the Lie algebras of this section we must refine our information about such structures on C, and interpret in particular the meaning of our assumption that the F-rank of L is <u>one</u>.

Part of this program has been carried out by Faulkner and Ferrar [FF], in a structure theory which subsumes these structures under the name of "simple balanced symplectic algebras". The effect of their work on our question is to show that its solution lies in the study of "Freudenthal triple systems". That is, if we form the symmetric ternary

product,
$$\{c_1,c_2,c_3\} = \frac{1}{6} \Sigma_{\sigma \in S_3} <c_{\sigma(1)},c_{\sigma(2)},c_{\sigma(3)}> ,$$
we have a symmetric quartic form $q(c_1,\{c_2,c_3,c_4\})$ on C, satisfying

(T5) $\quad\quad\quad \{\{c,c,c\},c,c'\} = q(c',c)\{c,c,c\} + q(c',\{c,c,c\})c$

for all c and c' from C. From (T1) and (T2) one can recover $<c_1,c_2,c_3>$ from $\{c_1,c_2,c_3\}$ by

(T6) $\quad\quad\quad <c_1,c_2,c_3> = \{c_1,c_2,c_3\} + q(c_3,c_2)c_1 + q(c_2,c_1)c_3$
$$- q(c_1,c_3)c_2 .$$

Typically it is required (cf. [Me]) that the quartic form above be nonzero. Since q is nondegenerate, this will be the case if and only if $\{c_1,c_2,c_3\}$ is not identically zero. By our last relation above and the definition of $<c_1,c_2,c_3>$, the vanishing identically of $\{c_1,c_2,c_3\}$ would entail the relation
$$c_1<c_2,c_3> = q(c_1,c_2)c_3 - q(c_1,c_3)c_2 .$$
Thus if $0 \neq c \in C$, we have $c_1<c,c> = 2q(c_1,c)c$, not identically zero, while $C(<c,c>^2) = 0$. This means that the centralizer of T contains nilpotent elements, and violates our condition of F-rank one on L. We may therefore assume that $\{c_1,c_2,c_3\}$ is not identically zero.

Meyberg's analysis [Me] of the Freudenthal triple systems proceeds along the following lines, somewhat paraphrased: From the fact that the symmetric 4-linear form $q(c_1,\{c_2,c_3,c_4\})$ is not identically zero, it follows that $q(c,\{c,c,c\})$ is not zero for some $c \in C$. (In fact, it is a consequence of our assumption of F-rank one that $q(c,\{c,c,c\}) \neq 0$ for all $c \neq 0$; we make no use of this observation yet.) For such c, consider the mapping $L(c,c):c' \mapsto \{c',c,c\}$, of C. Now the space C has the decomposition $C = Ac \oplus A\{c,c,c\} \oplus U$, where U is q-orthogonal to both c and $\{c,c,c\} = cL(c,c)$, so the restriction of q to U is nondegenerate. From (T5), $cL(c,c)^2 = q(c,\{c,c,c\})c$, so that $L(c,c)^2$ is the scalar $q(c,\{c,c,c\})$ on $Ac + A\{c,c,c\}$. Working with a polarized form of (T5) ([Me], (1.2)), yields for $u \in U$, $3uL(c,c)^2 = q(c,\{c,c,c\})u$. Further, we have $c<c,c> = - <c,c,c> = - \{c,c,c\} = - cL(c,c)$, $cL(c,c)<c,c> = \{c,c,c\}<c,c> = 3\{c,c,c <c,c>\} = - 3cL(c,c)^2 = - 3q(c,\{c,c,c\})c$, and $u<c,c> \in U$, with $u<c,c> = - <u,c,c> = - uL(c,c)$. It follows that

260

$<c,c>^2 = \frac{1}{3} q(c,\{c,c,c\})$ on all of U, $c<c,c>^2 = 3q(c,\{c,c,c\})c$.

Departing for a moment from Meyberg's case, we note that if $\frac{1}{3}q(c,\{c,c,c\})$ were a square in A, say λ^2, then $<c,c>$ would act semisimply and diagonally in C, with eigenvalues $\pm\lambda, \pm 3\lambda$, and so $<c,\lambda^{-1}c>$ would enlarge the F-split torus T by an F-diagonalizable element of its centralizer. Thus we can conclude that $\frac{1}{3}q(c,\{c,c,c\}) = -\frac{1}{3}q(c,c<c,c>)$ is <u>not a square in</u> A.

By extending the base field, we arrive at a situation where $q(c,\{c,c,c\}) = 3$ for some $c \in C$, so that $<c,c>$ acts diagonally in U with eigenvalues ± 1. Then Meyberg shows that U^+, U^-, the subspaces of U where $<c,c> = \pm\mathrm{Id}$, are q-dually paired totally isotropic spaces. In fact, it is enough to enlarge the field A by adjoining a square root of $\frac{1}{3}q(c,\{c,c,c\})$ in order to get this decomposition. If λ is such a square root, $K = A(\lambda)$, then in $K \otimes_A C$ our element $<c,c> \in D$ acts as multiplication by λ on all elements of the form $\lambda \otimes u + 1 \otimes u<c,c>$ ($u \in U$), and as multiplication by $-\lambda$ on elements of the form $-\lambda \otimes u + 1 \otimes u<c,c>$. Likewise, for $a \in A$, $<c,c>$ acts on $\pm 3\lambda \otimes ac + 1 \otimes ac<c,c>$ as multiplication by $\pm 3\lambda$. Moreover, the mappings $u \mapsto \pm\lambda \otimes u + 1 \otimes u<c,c>$ are one-one and A-linear, and the sum of the images is $K \otimes U$. Thus the images are, respectively, the eigenspaces of $<c,c>$ in $K \otimes U$ corresponding to the eigenvalues $\pm\lambda$. Following Meyberg, we denote by $(K \otimes U)^+$ the eigenspace corresponding to λ, by $(K \otimes U)^-$ that corresponding to $-\lambda$. Then the above amounts to saying that $u \mapsto \pm\lambda \otimes u + 1 \otimes u<c,c>$ are A-linear isomorphisms of U onto $(K \otimes U)^+$ resp. $(K \otimes U)^-$.

The fact that $<c,c>^2 = \lambda^2$ on U means that if we define $\lambda u = -u<c,c>$ for $u \in U$, U itself becomes a vector space over K. The effects on λu of our mappings into $(K \otimes U)^\pm$ are as follows:

$$\lambda u \mapsto \pm\lambda \otimes \lambda u + 1 \otimes (\lambda u)<c,c> = \mp\lambda \otimes u<c,c> - \lambda^2 \otimes u$$

$$= \lambda(-\lambda \otimes u \mp 1 \otimes u<c,c>) = \mp\lambda(\pm\lambda \otimes u + 1\otimes u<c,c>).$$

That is, the map $u \mapsto -\lambda \otimes u + 1 \otimes u<c,c>$ is K-linear from U to $(K \otimes U)^-$, while $u \mapsto \lambda \otimes u + 1 \otimes u<c,c>$ is <u>semi</u>-linear with respect to the automorphism of K/A sending λ to $-\lambda$. By the same methods, $B = Ac + Ac<c,c>$ is a K-vector space with $\lambda b = -\frac{1}{3}b<c,c>$ for all $b \in B$, and $b \mapsto \mp 3\lambda \otimes b + 1 \otimes b<c,c>$ define K-linear (resp. semi-linear) isomorphisms of B onto the corresponding eigenspaces for

$< c,c >$ in $K \otimes B$.

Now Meyberg draws a distinction into cases according as $q(a,\{a,a,c\}) = 0$ for all $a \in (K \otimes U)^+$ (from which it follows that this is also the case for all $a \in (K \otimes U)^-$, and vice versa), or not. Since $q(a,c) = q(a,a) = 0$ for all such a, we have $\{a,a,c\} = -c< a,a >$, $q(a,\{a,a,c\}) = -q(a,c< a,a >) = q(a< a,a >,c) = -q(\{a,a,a\},c)$, and this vanishes for all $a \in (K \otimes U)^+$ in Meyberg's first case. Substituting $a = \lambda \otimes u + 1 \otimes u< c,c >$, $u \in U$, and expanding gives $0 = \lambda^3 \otimes q(\{u,u,u\},c) + 3\lambda \otimes q(\{u< c,c >,u< c,c >,u\},c)$ (as the component of trace zero in $K = K \otimes_A A$). From $\{u,u,u\}< c,c >^2 = 3\{u< c,c >,u,u\}$, $\{u,u,u\}< c,c >^2 = 6\{u< c,c >,u< c,c >,u\} + 3\{u< c,c >^2,u,u\} = 6\{u< c,c >,u< c,c >,u\} + 3\lambda^2\{u,u,u\}$, we may substitute for $\{u< c,c >,u< c,c >,u\}$ above and use the fact that $q(\{u,u,u\}< c,c >^2,c) = q(\{u,u,u\},c< c,c >^2) = 3\lambda^2 q(\{u,u,u\},c)$ to obtain $q(\{u,u,u\},c) = 0$ for all $u \in U$. Replacing u by $u + u< c,c >$ and polarizing yields $q(\{u< c,c >,u,u\},c) = 0$, $0 = q(\{u,u,u\}< c,c >,c)$, hence $q(\{u,u,u\},c< c,c >) = 0$. In other words, Meyberg's first condition amounts to requiring that $\{u,u,u\} \in U$ for all $u \in U$ or, by polarization, that $\{u,v,w\} \in U$ for all $u,v,w \in U$. We identify the Lie algebra L in this case with the derived algebra of the skew transformations of a K-vector space with respect to an antihermitian form of Witt index one.

Namely we define a K-valued function (x,y) on $C \times C$ by setting $(x,y) = q(x,y) + \lambda^{-1}q(x,\lambda y)$, where $\lambda y = \lambda b + \lambda u = -\frac{1}{3}b< c,c > - u< c,c >$ when $y = b + u$ ($b \in B, u \in U$). Then $(y,x) = q(y,x) + \lambda^{-1}q(y,\lambda x)$, and with $y = b + u$, $x = b' + u'$ as above $q(y,\lambda x) = -\frac{1}{3}q(b,b'< c,c >) - q(u,u'< c,c >) = \frac{1}{3}q(b< c,c >,b') + q(u< c,c >,u') = q(x,\lambda y)$. That is, $(y,x) = -q(x,y) + \lambda^{-1}q(x,\lambda y) = -(x,y)^*$, where $*$ is the non-trivial automorphism of K/A. Moreover, $(x,\lambda y) = q(x,\lambda y) + \lambda^{-1}\lambda^2 q(x,y) = \lambda(x,y)$, so that the form (x,y) is antihermitian, being linear in its second argument and semilinear in its first one. From the fact that $(x,y) = 0$ implies $q(x,y) = 0$, our form (x,y) is nondegenerate on C.

Now we show that, under Meyberg's first condition, we have for all $x,y,z \in C$,

(†) $\qquad x< y,z > = -(y,x)z - (z,x)y - \frac{1}{2}((y,z)+(z,y))x$,

the structure of C as K-vector space being that of $B \oplus U$.

When $x = y = z = c$, one checks that both sides of $(\dagger)$ are equal to $-3\lambda c$; when $y = z = c$, $x = c<c,c>$, both sides are equal to $9\lambda^2 c$; when $x \in U$, $y = z = c$, both sides are $-\lambda x$. By its Λ-linearity in x, $(\dagger)$ holds for all $x \in C$ when $y = z = c$.

Now $0 = [<c,c>,<c,c>] = 2<c<c,c>,c>$, by the derivation-property (§III.9). Thus, to verify $(\dagger)$ when $y = c, z = c<c,c>$, we must show that the right-hand side is zero for all $x \in C$. As before, it suffices to treat the cases $x = c$, $x = c<c,c>$, $x \in U$. Since $(c,c<c,c>) + (c<c,c>,c) = q(c,c<c,c>) + q(c<c,c>,c) = 0$, we need to show that $(c,x)c<c,c> + (c<c,c>,x)c = 0$ for such x. If $x \in U$, one has $0 = (c,x) = (c<c,c>,x)$, and $(\dagger)$ is immediate; if $x = c$, we have $(c,x) = \lambda^{-1}q(c,\lambda c) = -\frac{1}{3}\lambda^{-1}q(c,c<c,c>) = \lambda$, $(c<c,c>,x) = q(c<c,c>,c) = 3\lambda^2$, and $(c,x)c<c,c> = -3\lambda^2 c$, yielding $(\dagger)$. For $x = c<c,c>$, $(\dagger)$ follows from $(c,c<c,c>) = -3\lambda^2$ and from $(c<c,c>,c<c,c>) = \lambda^{-1}q(c<c,c>,c<c,c>) = -3\lambda q(c<c,c>,c) = -9\lambda^3$. Thus $(\dagger)$ holds for all $x \in C$ when $y = c, z = c<c,c>$.

When $y = z = c<c,c>$, both sides of $(\dagger)$ are equal to: $27\lambda^3 c$, if $x = c$; $-81\lambda^4 c$, if $x = c<c,c>$; $9\lambda^3 x$, if $x \in U$. It follows that $(\dagger)$ holds whenever $y, z \in B$.

Next let $y = c, z \in U$. If $x = c$, both sides of $(\dagger)$ are seen from (95) and the above to be equal to $-\lambda z$; if $x = c<c,c>$, (95) yields $x<c,z> = z<c,c<c,c>> - q(c,c<c,c>)z = 3\lambda^2 z$, which is readily seen to be the value of the right-hand side of $(\dagger)$. When $x \in U$, we must (for the first time) use the assumption that $\{UUU\} \subseteq U$. Namely, $(B,U) = 0$ and $C = B \oplus U$, from which B is seen to be the orthogonal space to U with respect to (x,y). Our relation between $\{u,v,w\}$ and $u<v,w>$ developed above then shows that $u<v,w> \in U$ whenever $u,v,w \in U$. In particular, for all $u \in U$, $u<x,z> \in U$. Now verifying $(\dagger)$ in this case amounts to showing that $x<c,z> = -(z,x)c$. From (95), $x<c,z> = c<x,z> + q(x,z)c$, so that for $u \in U$, $(u,x<c,z>) = (u,c<x,z>) + q(x,z)c) = (u,c<x,z>) = q(u,c<x,z>) + \lambda^{-1}q(u,\lambda c<x,z>)$. From the decomposition $c<x,z> = b + v$, $\lambda c<x,z> = -\frac{1}{3}b<c,c> - v<c,c>$, with the first term q-orthogonal to U, we have $q(u,\lambda c<x,z>) = -q(u,v<c,c>) = q(u<c,c>,v) = -q(\lambda u,v) = -q(\lambda u,c<x,z>)$. Thus $(u,x<c,z>) = q(u,c<x,z>) - \lambda^{-1}q(\lambda u,c<x,z>) = -q(u<x,z>,c) - \lambda^{-1}q((\lambda u)<x,z>,c)$. From the fact that $u<x,z>$ and $(\lambda u)<x,z>$ are in U, we see that

$(u, x<c,z>) = 0$ for all $u \in U$, and thus that $x<c,z> \in B$. To show our assertion, it is therefore sufficient to show that $q(x<c,z>,c) = -q((z,x)c,c)$ and that $q(x<c,z>,c<c,c>) = -q((z,x)c,c<c,c>)$. Now $q(x<c,z>,c) = -q(x,c<c,z>) = -q(x,z<c,c>) = q(x,\lambda z)$, while $-q((z,x)c,c) = -q(q(z,x)c + \lambda^{-1}q(z,x\lambda)c, c) = -q(z,x\lambda)q(\lambda^{-1}c,c) = -\lambda^{-2}q(z,x\lambda)q(\lambda c,c) = \lambda^{-2}q(z,x\lambda) \cdot \frac{1}{3}q(c<c,c>,c) = q(z,x\lambda) = -q(z,x<c,c>) = -q(x,z<c,c>) = q(x,\lambda z) = q(x<c,z>,c)$, by the above. Also, we have $q(x<c,z>,c<c,c>) = -q(x,c<c,c><c,z>) = -3\lambda^2 q(x,z)$ by the above, while $-q((z,x)c,c<c,c>) = -q(z,x)q(c,c<c,c>) - \lambda^{-2}q(z,\lambda x)q(\lambda c,c<c,c>)$; the second term is zero, and the first is $3\lambda^2 q(z,x) = -3\lambda^2 q(x,z)$, completing our verification of (†) for this case.

Next let $y = c<c,c>, z \in U$. When $x = c$, the fact that $<c,c<c,c>> = 0$ combines with (95) to give that the left-hand side of (†) is $-q(c<c,c>,c)z = -3\lambda^2 z$, and one easily checks that this is the value of the right-hand side. When $x = c<c,c>$, (95) yields that the left side of (†) is $z<c<c,c>,c<c,c>> = 9\lambda^3 z$, by the above; our earlier calculations show that this is also the value on the right. When $x \in U$, we show as in the last case that $x<c<c,c>,z> \in B$, and thereby reduce (†) to showing that $q(x<c<c,c>,z>,c) = -q((z,x)c<c,c>,c)$ and that $q(x<c<c,c>,z>,c<c,c>) = -q((z,xc<c,c>,c<c,c>)$. This proceeds as in the last case, the common value being $3\lambda^2 q(x,z)$ in the first instance and $9\lambda^2 q(\lambda x,z)$ in the second.

Finally, we assume $y,z \in U$. Then $c<y,z> = y<c,z> - q(y,z)c$ by (95), and this is by the above, $-(z,y)c - q(y,x)c = -(z,y)c - (\frac{(z,y)-(y,z)}{2})c = -(\frac{(z,y)+(y,z)}{2})c$, the right-hand side of (†) in this case. The reasoning when $x = c<c,c>$ may be carried out the same way. To complete the verification of (†), we show:

(‡)   For all $x \in U$, $\{x,x,x\} = 3(x,x)x$

Polarization of (‡) then yields

$$\{x,y,z\} = \frac{1}{2}((x,y) + (y,x))z + \frac{1}{2}((x,z) + (z,x))y$$
$$+ \frac{1}{2}((y,z) + (z,y))x, \text{ from which our relation}$$

(T.6) yields $\quad -x\langle y,z\rangle = \frac{1}{2}((y,z) - (z,y))x$

$\qquad = \frac{1}{2}((x,y) + (y,x))z + \frac{1}{2}((x,z) + (z,x))y$

$\qquad + \frac{1}{2}((y,z) + (z,y))x + \frac{1}{2}((z,y) - (y,z))x$

$\qquad + \frac{1}{2}((y,x) - (x,y))z - \frac{1}{2}((x,z) - (z,x))y,\quad$ or

$x\langle y,z\rangle = -(y,x)z - (z,x)y - \frac{1}{2}((y,z) + (z,y))x$, which is $(\dagger)$.

To show $(\ddagger)$, we have $3(x,x)x = 3q(x,\lambda x)\lambda^{-1}x = 3\lambda^{-2}q(x,\lambda x)\lambda x = 3\lambda^{-2}q(x,x\langle c,c\rangle)x\langle c,c\rangle = 3\lambda^{-2}q(x,\{x,c,c\})\{x,c,c\}$, since $\lambda x = -x\langle c,c\rangle = \{x,c,c\}$. Now complete polarization of (T5) yields an identity in $\mathcal{C}$, labelled (1.4) by Meyberg:

$\quad \{\{def\}gh\} + \{\{hef\}dg\} + \{\{deh\}fg\} + \{\{dfh\}eg\}$
$= q(g,e)\{dfh\} + q(g,f)\{deh\} + q(g,h)\{def\} + q(g,d)\{efh\}$
$+ q(g,\{dfh\})e + q(g,\{deh\})f + q(g,\{efh\})d + q(g,\{def\})h.$

Substituting x for g and d, c for f and h, $\{xcc\}$ for e, and using that $q(x,c) = 0 = q(x,x) = q(x,\{\{xcc\}cc\})$, we obtain:

(**) $\quad \begin{aligned}&\{x,x,\{\{xcc\},c,c\}\} + \{\{xcc\},\{xcc\},x\} \\ &+ 2\{\{x,\{xcc\},c\}xc\} = 2q(x,\{xcc\})\{xcc\} \\ &+ 2q(x,\{\{xcc\},x,c\})c.\end{aligned}$

From the fact that $q(c,\{xcc\}) = 0$ and the fundamental relation (T.6) between $\{c_1,c_2,c_3\}$ and $\langle c_1,c_2,c_3\rangle$, we have $\{x,\{xcc\},c\} = \langle x,\{xcc\},c\rangle + q(x,\{xcc\})c = -x\langle\{xcc\},c\rangle - q(x,x\langle c,c\rangle)c = x\langle x\langle c,c\rangle,c\rangle = q(x,x\langle c,c\rangle)c$. The formula $(\dagger)$, already verified for $y = x\langle c,c\rangle$, $z = c$, yields (since $(x,c) = 0 = (x\langle c,c\rangle,c))$ $\{x,\{xcc\},c\} = -(x\langle c,c\rangle,x)c - q(x,x\langle c,c\rangle)c = -q(x\langle c,c\rangle,x)c + \lambda^{-1}q(x\langle c,c\rangle,x\langle c,c\rangle)c - q(x,x\langle c,c\rangle)c = 0$. Thus the last term on either side of (**) vanishes, leaving

$\quad \{x,x,\{\{xcc\},c,c\}\} + \{\{xcc\},\{xcc\},x\} = 2q(x,\{xcc\})\{xcc\}.$

Since $\{\{xcc\}c,c\} = x\langle c,c\rangle^2 = \lambda^2 x$, the first term is $\lambda^2\{x,x,x\}$. For the second term, the fact that $\{x,x,x\} \in \mathcal{U}$ yields $\{x,x,x\}\langle c,c\rangle^2 = \lambda^2\{x,x,x\}$; but $\{x,x,x\}\langle c,c\rangle^2 = 3\{x\langle c,c\rangle^2,x,x\} + 6\{x\langle c,c\rangle,x\langle c,c\rangle,x\}$ by the derivation property, from which $\{x\langle c,c\rangle,x\langle c,c\rangle,x\} = -\frac{1}{3}\lambda^2\{x,x,x\}$. These substitutions in the left-hand side of (**) reduce it to

$\quad \frac{2}{3}\lambda^2\{x,x,x\} = 2q(x,\{xcc\})\{x,c,c\},$

and we have seen that $3(x,x)x = 3\lambda^{-2}q(x,\{xcc\})\{xcc\}$. The formula ($\ddagger$) is an immediate consequence, and with it ($\dagger$) is completely verified.

We now identify L with $K'(K \otimes_F M'' \oplus C)$, with $(K \otimes_F M'') \oplus C$ as antihermitian K-space relative to the form which is (x,y) on C, which has $(\mu \otimes u, \nu \otimes v)(\mu,\nu \in K; u,v \in M'')$ equal to $\mu^*(u,v)\nu$, with (u,v) the symplectic F-form on M'' relative to which S acts as skew transformations, and with C and $K \otimes M''$ orthogonal spaces. As in earlier cases, $K'(K \otimes M'' \oplus C)$ has a decomposition into $S \otimes A = K'(K \otimes M'')$, $K'(C)$, $M'' \otimes_F C$, and $A\lambda_0$, all of these being A-subspaces, where $K'(K \otimes M'')$ annihilates C and $K'(C)$ annihilates $K \otimes M''$. The action of $u \otimes z$ $(u \in M'', z \in C)$ on $\mu \otimes v(\mu \in K, v \in M'')$ is defined to give $\mu(u,v)z$, while on $y \in C$ this action produces $(z,y) \otimes u \in K \otimes M''$. The map λ_0 sends $z \in C$ to $2\lambda z$, and $\mu \otimes v \in K \otimes M''$ to $-d\lambda\mu \otimes v$, where $d = [C:K]$. The details of the identification are carried out exactly as in earlier cases; we mention only that the components of $[u \otimes y, v \otimes z] = q(y,z) \otimes (u \circ v) + (u,v)<y,z>$ according to the above decomposition of $K'(K \otimes M'' \oplus C)$ are: $q(y,z) \otimes (u \circ v) \in K'(K \otimes M'')$, $x \to -xT_{y,z} + \frac{1}{d}\text{Trace}(T_{y,z})x \in K'(C)$, where $xT_{y,z} = (y,x)z + (z,x)y$, $\text{Trace}(T_{y,z}) = (y,z) + (z,y)$, and $\beta_0 \in A\lambda_0$, sending $x \in C$ to $-\frac{1}{d}((y,z) + (z,y))x$, $\mu \otimes v \in K \otimes M''$ to $\frac{1}{2}((y,z) + (z,y))\mu \otimes v$.

The centralizer L_0 of T identifies with $T \otimes A + A\lambda_0 + K'(C)$, and consists only of semisimple elements provided C has Witt index zero, and only then. Its derived algebra $[L_0 L_0] = K'(C)$, which is simple if $[C:K] > 1$, zero if $[C:K] = 1$. In either case, one easily sees as before that, for such C, T is a maximal F-split torus of dimension one in $K'(K \otimes M'' \oplus C)$. Also as before, the simple group G' identifies with the derived group of the unitary group of $(K \otimes M'') \oplus C$, modulo its center. This completes our analysis of the case where $\{U,U,U\} \subseteq U$.

In the remaining case, which is labelled "II" by Meyberg, we shall lean more heavily on his work. This case requires the existence in $(K \otimes U)^+$ of an element a^+ with $\alpha = q(a^+,\{a^+,a^+,c\}) \neq 0$. We refer to Meyberg's formulas by his numbers, to indicate what is changed by our use of $-\lambda$ where he has used 1, and by our definition of $(K \otimes U)^\pm$. From §3, the formulas are unchanged, with the exception of the following, given in the new form:

(3.1) $\{\{u^\varepsilon u^\varepsilon c\}cc\} - \lambda\varepsilon\{u^\varepsilon u^\varepsilon c\} = 0; \{u^\varepsilon u^\varepsilon c\} \in (K \otimes U)^{-\varepsilon}$.

(3.2) $\{\{u^\varepsilon u^\varepsilon c\}cy^{-\varepsilon}\} - \varepsilon\lambda\{u^\varepsilon u^\varepsilon y^{-\varepsilon}\} = -2\varepsilon\lambda q(y^{-\varepsilon}, u^\varepsilon)u^\varepsilon$.

(3.3) $\{u^\varepsilon u^\varepsilon \{ccc\}\} - 3\lambda\varepsilon\{u^\varepsilon u^\varepsilon c\} = 0$.

(3.4) $\{u^+, y^-, \{c,c,c\}\} = 3\lambda q(u^+, y^-)c$,

$3\lambda\{u^+, y^-, c\} = q(u^+, y^-)\{c,c,c\}$.

(3.5) $\{\{u^\varepsilon u^\varepsilon c\}cy^\varepsilon\} = \frac{\varepsilon}{3\lambda}q(y^\varepsilon, \{u^\varepsilon u^\varepsilon c\})\{c,c,c\}$.

(3.6) $\{u^\varepsilon v^\varepsilon y^\varepsilon\} = -\frac{\varepsilon}{\lambda}q(u^\varepsilon, \{v^\varepsilon y^\varepsilon c\})(c - \frac{\varepsilon}{3\lambda}\{c,c,c\})$.

(3.8) $3\{\{x^\varepsilon x^\varepsilon u^{-\varepsilon}\}x^\varepsilon y^{-\varepsilon}\} = 3q(y^{-\varepsilon}, x^\varepsilon)\{x^\varepsilon x^\varepsilon u^{-\varepsilon}\}$

$+ 3q(y^{-\varepsilon}, \{x^\varepsilon x^\varepsilon u^{-\varepsilon}\})x^\varepsilon + \frac{2\varepsilon}{\lambda}q(x^\varepsilon, \{x^\varepsilon x^\varepsilon c\})\{cu^{-\varepsilon}y^{-\varepsilon}\}$.

(3.14) $2\{y^{-\varepsilon}v^\varepsilon\{y^{-\varepsilon}x^\varepsilon x^\varepsilon\}\} - \{x^\varepsilon x^\varepsilon\{y^{-\varepsilon}y^{-\varepsilon}v^\varepsilon\}\}$

$= 2q(x^\varepsilon, \{y^{-\varepsilon}y^{-\varepsilon}v^\varepsilon\})x^\varepsilon - 4q(x^\varepsilon, y^{-\varepsilon})\{y^{-\varepsilon}v^\varepsilon x^\varepsilon\}$

$+ 2\varepsilon\lambda^{-1}q(x^\varepsilon, \{x^\varepsilon v^\varepsilon c\})\{y^{-\varepsilon}y^{-\varepsilon}c\}$.

(The second term of (3.2) in [Me] is incorrect, lacking a factor of ε; otherwise one may check that our formulas agree with his when λ is set equal to -1.)

Still following [Me], a commutative product $x^- \circ y^-$ is defined on $(K \otimes U)^-$ by

(4.1) $x^- \circ y^- = \{x^- y^- a^+\} + q(a^+, x^-)y^- + q(a^+, y^-)x^-$.

This product has $e^- = \frac{3}{4}\alpha^{-1}\{a^+ a^+ c\}$ as unit element, $\sigma(x^-) = 4q(a^+, x^-)$ as an associative linear form with respect to the product, and every element x^- of $(K \otimes U)^-$ satisfies the cubic relation

(4.13) $(x^- \circ x^-)\circ x^- - \sigma(x^-)(x^- \circ x^-) + \frac{1}{2}[\sigma(x^-)^2 - \sigma(x^- \circ x^-)]x^-$

$- [\frac{1}{3}\sigma(x^- \circ (x^- \circ x^-)) - \frac{1}{2}\sigma(x^-)\sigma(x^- \circ x^-) + \frac{1}{6}\sigma(x^-)^3]e^- = 0$.

From these observations, it follows as in §4.3 of [Me] that $(K \otimes U)^-$ is a K-Jordan algebra with this product.

By working in $(K \otimes U)^+$ instead, a product is defined there by (Meyberg's numbering):

(5.2) $x^+ \odot y^+ = \{x^+ y^+ e^-\} + q(e^-, x^+)y^+ + q(e^-, y^+)x^+$.

The unit element is $-a^+$, and the map

$\psi: x^- \mapsto \{x^- a^+ a^+\} - 2q(a^+, x^-)a^+$

is an isomorphism of $(K \otimes U)^-$ onto $(K \otimes U)^+$ with respect to these structures, with inverse

$$\psi^{-1} : x^+ \mapsto -\{x^+ e^- e^-\} + 2q(e^-, x^+)e^- \ .$$

([Me], Lemmas 5.1 and 5.2.)

The rational form of our requirement on the existence of a^+ as above is the existence of $a \in U$ with $\{aaa\} \notin U$; we may then take $a^+ = \lambda \otimes a + 1 \otimes a < c,c > \in (K \otimes U)^+$. From the K-isomorphism $u \mapsto -\lambda \otimes u + 1 \otimes u < c,c >$ of U with $(K \otimes U)^-$, we have a commutative K-bilinear product $u \cdot v$ in U defined by:

$$-\lambda \otimes (u \cdot v) + 1 \otimes (u \cdot v) < c,c >$$
$$= (-\lambda \otimes u + 1 \otimes u < c,c >) \circ (-\lambda \otimes v + 1 \otimes v < c,c >).$$

From the definitions of ψ and of $x^- \circ y^-$ it follows that $q(x^-, y^{-\psi}) = q(a^+, x^- \circ y^-) = \frac{1}{4}\sigma(x^- \circ y^-)$, and that the associative bilinear form $\sigma(x^- \circ y^-)$ is <u>nondegenerate</u> on $(K \otimes U)^-$. By means of our isomorphism ψ, we may transport σ to $(K \otimes U)^+$, obtaining an analogue of (4.13) there. Since $(K \otimes U)^-$ is a Jordan algebra, it is power-associative, and we see readily from (4.13) that if $x \in (K \otimes U)^-$ is nilpotent, then $x^3 = (x \circ x) \circ x = 0$ and $\sigma(x) = 0$. Polarization shows that the cubic relation (4.13) persists upon extension (from K) of the ground field of $(K \otimes U)^-$. Thus σ is <u>normal</u> in the sense of [BK], §I.7, so that by §I.8 of [BK] we have:

i) $(K \otimes U)^-$ is a direct sum of ideals which are simple Jordan algebras.

In the presence of i), the effect of (4.13) is that:

ii) Over no field extension of K can the unit element e^- of $(K \otimes U)^-$ be written as the sum of more than three nonzero orthogonal idempotents.

The implications of i) and ii) for the structure of the Jordan algebra $(K \otimes U)^-$ are, according to the classification of simple Jordan algebras, that $(K \otimes U)^-$ must be one of the following:

α) K;

β) a commutative associative separable K-algebra of dimension 2;

γ) a commutative associative separable K-algebra of dimension 3;

δ) the Jordan algebra associated with a nondegenerate quadratic form in a K-space of dimension at least two;

ε) the direct sum of K and a Jordan algebra of δ);

ζ) the symmetric elements, of K-dimension 6, with respect to an involution of first kind in a central simple associative algebra of degree 3 over K;

η) a central simple associative algebra of degree 3 over K, with $x \circ y = \frac{1}{2}(xy+yx)$;

θ) the symmetric elements with respect to an involution of second kind in a central simple associative algebra of degree 3 over a quadratic extension of K;

ι) the symmetric elements, of K-dimension 15, with respect to an involution of first kind in a central simple associative algebra of degree 6 over K;

κ) an exceptional central simple Jordan algebra, of dimension 27 over K.

It follows that all this structure is duplicated in the K-algebra U with product $u \cdot v$. We denote by τ the K-linear function on U obtained by transporting $\sigma: \tau(u) = \sigma(-\lambda \otimes u + 1 \otimes u<c,c>)$, and by e the identity element of $U: e^- = -\lambda \otimes e + 1 \otimes e <c,c>$. For $u \in U$, we write

$$(-\lambda \otimes u + 1 \otimes u<c,c>)^\psi = \lambda \otimes u^\phi + 1 \otimes u^\phi <c,c>,$$

thereby defining an A-linear mapping $\phi: U \to U$. From the K-linearity of ψ one observes that ϕ is <u>semilinear</u> with respect to the non-trivial automorphism of K/A.

One has in $(K \otimes U)^-$ Freudenthal's cross-product

$$x^- \times y^- = x^- \circ y^- - \frac{1}{2}\sigma(x^-)y^- - \frac{1}{2}\sigma(y^-)x^-$$
$$+ \frac{1}{2}(\sigma(x^-)\sigma(y^-) - \sigma(x^- \circ y^-))e^-,$$

which we carry over to U. Thus for $u,v \in U$, we have

$$u \times v = u \cdot v - \frac{1}{2}\tau(u)v - \frac{1}{2}\tau(v)u$$
$$+ \frac{1}{2}(\tau(u)\tau(v) - \tau(u \cdot v))e.$$

Conversely, Meyberg recovers the form q and the ternary composition $\{x,y,z\}$ from the Jordan data. In the cases where these are not explicit above, one has in our setting (cf. (6.1) of [Me]):

$$q(y^-, x^+) = \frac{1}{4}\sigma((x^+)^{\psi-1} \circ y^-);$$
$$\{x^+, y^-, c\} = \frac{1}{3}\lambda^{-1} q(x^+, y^-)\{c,c,c\};$$
$$\{x^+, y^-, \{c,c,c\}\} = 3\lambda q(x^+, y^-)c;$$
$$\{x^+, c\{c,c,c\}\} = 0 = \{y^-, c, \{c,c,c\}\};$$

$$\{x^+, \{c,c,c\}, \{c,c,c\}\} = 9\lambda^3 x^+;$$

$$\{y^-, \{ccc\}, \{ccc\}\} = -9\lambda^3 y^-;$$

$$\{x^-, y^-, c\} = -\frac{3}{2}\lambda\alpha^{-1}(x^- \times y^-)^\psi,$$

where $\alpha = q(a^+, \{a^+, a^+, c\})$ is fixed. From this last and the derivation-property, $\{x^-, y^-, c < c\,c >\}$ can be determined.

As in [Me], we can derive these formulas, to be compared with the remaining ones from his (6.1):

$$\{x^-, x^-, x^-\} = q(x^-, \{x^-, x^-, c\})(\lambda^{-1} \otimes c + \frac{1}{3}\lambda^{-2} \otimes \{c,c,c\});$$

$$\{x^-, u^-, y^+\} = q(y^+, x^-)u^- + q(y^+, u^-)x^-$$
$$- \frac{9}{4}\lambda(\alpha\alpha^*)^{-1} \otimes ((u \times v)^\phi \times y)^\phi$$
$$+ \frac{9}{4}(\alpha\alpha^*)^{-1} \otimes ((u \times v)^\phi \times y)^\phi < c,c >,$$

where $\bar{x} = -\lambda \otimes x + 1 \otimes x < c,c >$, $\bar{u} = -\lambda \otimes u + 1 \otimes u < c,c >$, $y^+ = \lambda \otimes y + 1 \otimes y < c,c >$, the cross-product in U is that obtained by transporting the one in $(K \otimes U)^-$, and $\lambda \otimes u^\phi + 1 \otimes u^\phi < c,c > = (-\lambda \otimes u + 1 \otimes u < c,c >)^\psi$, for all $u \in U$.

By using the form $\pm\lambda \otimes u + 1 \otimes u < c,c >$ $(u \in U)$ for a typical element of $(K \otimes U)^\pm$, we can obtain explicit expressions, on C, for q and for the triple product $\{u,v,w\}$ in terms of the product and cross-product in U, the element $\alpha \neq 0$ of K, and the mapping ϕ. The results are as follows:

(6.2) $\quad q(u,v) = -\frac{1}{16}\lambda^{-2}(\tau(u\,v^\phi)^{-1} + \tau(u\,v^\phi)^{-1\,*})$, for $u,v \in U$, where $\kappa \mapsto \kappa^*$ is the nontrivial automorphism of K/A. Also for $u,v \in U$,

(6.3) $\quad \{u,v,c\} = -\frac{3}{8}\frac{\alpha+\alpha^*}{\alpha\alpha^*}(u \times v)^\phi + \frac{3}{8}\frac{\alpha-\alpha^*}{\lambda}\cdot\frac{1}{\alpha\alpha^*}(u \times v)^\phi < c,c >$

$$+ \frac{1}{48}\lambda^{-2}(\frac{\tau(u\cdot v^\phi)^{-1} - \tau(u\cdot v^\phi)^{-1\,*}}{\lambda})\{c,c,c\};$$

$$\{u,v,c<c,c>\} = -\frac{9}{8}\frac{\alpha+\alpha^*}{\alpha\alpha^*}(u \times v)^\phi < c,c > + \frac{9}{8}\frac{\lambda(\alpha-\alpha^*)}{\alpha\alpha^*}(u \times v)^\phi$$

$$- \frac{3}{16}\frac{\tau(u\cdot v^\phi)^{-1} - \tau(u\cdot v^\phi)^{-1\,*}}{\lambda}c.$$

For $u,v,y \in U$, we have

(6.4) $\quad \{u,v,y\} = -\frac{9\lambda^{-2}}{16}((\alpha\alpha^*)^{-1}[((u \times v)^\phi \times y)^\phi + ((v \times y)^\phi \times u)^\phi$

$$+ ((y \times u)^\phi \times v)^\phi]$$

$$+ \tau\frac{1}{16}\lambda^{-2}[\tau(u\cdot y^{\phi^{-1}})v + \tau(v\cdot y^{\phi^{-1}})u + \tau(v\cdot u^{\phi^{-1}})y$$

$$+ \tau(y\cdot u^{\phi^{-1}})v + \tau(y\cdot v^{\phi^{-1}})u + \tau(u\cdot v^{\phi^{-1}})y]$$

$$+ \frac{1}{32}\lambda^{-4}\alpha^{-1}\tau(y\cdot(u\times v))\{c,c,c\} ,$$

where left multiplication of elements of U by elements of K is defined by $(\gamma + \delta\lambda)u = \gamma u - \delta u<c,c>$ $(\gamma,\delta \in A)$, and if $x \in Ac + A\{c,c,c\}$, then $(\gamma + \delta\lambda)x = \gamma x - \frac{1}{3}\delta x<c,c>$; this is the same as our treatment of Meyberg's "Case I".

There are some additional relations present, arising from the computations in $(K \otimes U)^+$ analogous to those above where $(K \otimes U)^-$ has been emphasized. In particular, we have (cf. (6.1) of [Me]):

(6.1)b) $$\{u^+u^+c\} = -\frac{4}{3}\alpha(u^{+\psi^{-1}} \times u^{+\psi^{-1}})$$

$$= -\frac{4}{3}\alpha(u^+ \times u^+)^{\psi^{-1}}, (u^+ \in (K \otimes U)^+)$$

where we use the same symbol for the cross-products in $(K \otimes U)^{\pm}$.

Writing $u^+ = \lambda \otimes u + 1 \otimes u<c,c>$ $(u \in U)$, and likewise for $v^+ \in (K \otimes U)^+$, we may compare the term $\lambda \otimes (\{u,v<c,c>,c\} + \{u<c,c>,v,c\})$ from $\{u^+v^+c\}$ as given by (6.1)b) with the result of (6.3), to conclude that:

(6.5) $$(u \times v)^{\phi} = \frac{8}{9}\alpha\alpha^*(u^{\phi^{-1}} \times v^{\phi^{-1}}) .$$

The skewness of q and the fact that $q(u,\lambda v) = -q(\lambda u,v)$ combine with (6.2) and the semilinearity of ϕ to yield

(6.6) $$\tau(v\cdot u^{\phi^{-1}}) = -\tau(u\cdot v^{\phi^{-1}})^*, \text{ or}$$

$$\tau(u^{\phi}v) = -\tau(v^{\phi}\cdot u)^* .$$

From (6.5) and (6.6) we find, for all $u,v,w \in U$,

(6.7)
$$\tau(w^{\phi^{-1}}\cdot(u^{\phi^{-1}} \times v^{\phi^{-1}})) = -\frac{9}{8}(\alpha\alpha^*)^{-1}\tau(v\cdot(u \times v))^*$$

$$\tau(w^{\phi}\cdot(u^{\phi}\times v^{\phi})) = -\frac{8}{9}(\alpha\alpha^*)\tau(w\cdot(u \times v))^* ,$$

or that ϕ is a semi-similitude of the Jordan algebra U with respect to the "norm" form $N(u) = \frac{1}{3}\tau(u \cdot (u \times u))$, with factor of similitude $-\frac{8}{9}(\alpha\alpha^*) \in A$.

It will be noted that (6.5) permits writing the first member,
$$-\frac{9\lambda^{-2}}{16}(\alpha\alpha^*)^{-1}[((u \times v)^\phi \times y)^\phi + ((v \times y)^\phi \times u)^\phi + ((y \times u)^\phi \times v)^\phi],$$
of our expression (6.4) for $\{u,v,y\}$, as

(6.8) $\qquad -\frac{\lambda^{-2}}{2}[(u \times v) \times y^{\phi^{-1}} + (v \times y) \times u^{\phi^{-1}} + (y \times u) \times v^{\phi^{-1}}]$.

From the definition of the maps ψ and ϕ, we derive
$$\lambda \otimes u^\phi + 1 \otimes u^\phi \langle c,c \rangle = \{u^-, a^+, a^+\} - 2q(a^+, u^-)a^+$$
$$= \{-\lambda \otimes u + 1 \otimes u \langle c,c \rangle, \lambda \otimes a + 1 \otimes a \langle c,c \rangle, \lambda \otimes a + 1 \otimes a \langle c,c \rangle\}$$
$$- 2q(\lambda \otimes a + 1 \otimes a\langle c,c \rangle, -\lambda \otimes u + 1 \otimes u\langle c,c \rangle)(\lambda \otimes u + 1 \otimes a \langle c,c \rangle)$$
$$= \lambda \otimes (2(a \times a) \times u^{\phi^{-1}} - \frac{1}{2}\tau(a \cdot u^{\phi^{-1}})a)$$
$$+ 1 \otimes (2(a \times a) \times u^{\phi^{-1}} - \frac{1}{2}\tau(a \cdot u^{\phi^{-1}})a)\langle c,c \rangle$$
$$+\lambda \otimes (-\frac{1}{2}\tau(a \cdot u^{\phi^{-1}})a) + 1 \otimes (-\frac{1}{2}\tau(a \cdot u^{\phi^{-1}})a)\langle c,c \rangle,$$
so that

(6.9) $\qquad u^\phi = 2(a \times a) \times u^{\phi^{-1}} - \tau(a \cdot u^{\phi^{-1}})a,$ or
$$u^{\phi^2} = 2(a \times a) \times u - \tau(a \cdot u)a .$$

Now the operator "U_a" of Jacobson ([JJ], p. 36) in the Jordan algebra U sends u to $2(u \cdot a) \cdot a - u \cdot a^2$, and one checks from the cubic identity in U that u^{ϕ^2} is the negative of this; i.e.,

(6.10) $\qquad\qquad\qquad \phi^2 = -U_a \quad$ on U.

We next isolate those cases where the Jordan algebra U contains an idempotent e with $\tau(e) = 1$, such that $U_e = Ke$, and such that $\{u \in U | ue = \frac{1}{2}u\} = 0$. In our enumeration $\alpha) - \kappa)$, these are the cases $\epsilon)$ and those cases $\beta)$, $\gamma)$ where the algebra U is split; for the case $\epsilon)$, one takes e to be the unit element of the copy of K, for $\gamma)$ any primitive idempotent, and for $\beta)$ that one of a pair of orthogonal primitive idempotents where $\tau(e) = 1$. (One sees from the cubic relation derived from (4.13) that $\tau(e) = 1$ or 2 whenever e is an idempotent other than 0 or 1.) Since "e" has been otherwise used above, we denote our chosen element here by b.

First we note that $q(b,b<b,b>) \neq 0$, indeed that $q(v,v<v,v>) \neq 0$ for all $v \neq 0$ in C. For the embedding of $v_1 \otimes v$ ($v_1 \in M''$) in a 3-dimensional simple algebra containing $h \otimes F$ as Cartan subalgebra yields an element $v' \in C$ with $q(v,v') = 1$, $<v,v'> = 0$. From (95), we find $v'<v,v> = 2q(v',v)v - q(v,v')v = -3v$; then in particular $<v,v> \neq 0$, and the derivation-property for $<v,v>$ gives

$$0 = [<v,v'>,<v,v>] = <v<v,v>,v'> - <v,v'<v,v>>,$$

the last term being non-zero by the above. Hence the first term is also non-zero; in particular, $v<v,v> \neq 0$, and

$$v<v<v,v>,v'> = v<v,v'<v,v>> = -3v<v,v> \neq 0.$$

By (95), we have

$$v<v<v,v>,v'> - v<v,v><v,v'>$$
$$= 2q(v,v<v,v>)v' - q(v<v,v>,v')v - q(v',v)v<v,v>.$$

Now $<v,v'> = 0$, $q(v<v,v>,v') = -q(v,v'<v,v>) = 3q(v,v) = 0$. The above becomes

$$-3v<v,v> = 2q(v,v<v,v>)v' + v<v,v>.$$

Since $v<v,v> \neq 0$, we see that $q(v,v<v,v>) \neq 0$.

In particular, we conclude as for c that $<b,b>$ is an invertible transformation of C, and that $<b,b>^2$ is diagonalizable with the two eigenvalues $\frac{1}{3}q(b,\{b,b,b\})$ and $3q(b,\{b,b,b\})$, the latter having multiplicity two (over A) and eigenspace $Ab + Ab<b,b>$. The assumption that $b^2 = b$, $\tau(b) = 1$ makes $b \times b = 0$, so $\tau((b \times b) \cdot b) = 0$, and the semisimilitude property for ϕ makes $\tau(b^{\phi^{-1}} \cdot (b^{\phi^{-1}} \times b^{\phi^{-1}})) = 0$. Using the formula for $\{u,u,u\}$ derived above, we find that $b<b,b> = -\{b,b,b\} = 3\lambda^{-2}q(b,b<c,c>)b<c,c> \in Kb \subseteq U$. Furthermore, our formula (6.3) for $\{u,v,c\}$, applied with $u = b = v$ gives $\{b,b,c\} \in Kc$, and we have here that $-c<b,b> = \{b,b,c\}$. Thus

$$c<b,b> = \alpha c + \beta c<c,c>,$$

where α and β are in A. Since $<c,c<c,c>> = 0$, we have $\alpha<c,c> = <c,c<b,b>> = \frac{1}{2}[<c,c>,<b,b>]$. It would follow that $<c,c>$ is nilpotent unless $\alpha = 0$; therefore $\alpha = 0$, and $c<b,b> = \beta c<c,c>$, $0 \neq \beta \in A$.

273

We next consider the operator $<c,b>$. From (95) we have $c<c,b> = b<c,c>$, $b<c,b> = c<b,b> = \beta c<c,c>$. The space $Ac + Ac<c,c> + Ab + Ab<c,c>$ has A-dimension 4. Application of (95), the derivation-property, and what we have shown above gives

$$c<c,c><b,c> = c<b,c><c,c> + 2c<c<b,c>,c>$$
$$= b<c,c>^2 + 2c<b<c,c>,c>$$
$$= b<c,c>^2 + 2b<c,c><c,c> = 3b<c,c>^2 = 3\lambda^2 b.$$

Likewise, $b<c,c><b,c> = b<b,c><c,c> + 2b<b<c,c>,c>$

$$= \beta c<c,c>^2 + 2c<b,b<c,c>>$$
$$- 2q(b<c,c>,b)c$$
$$= \beta c<c,c>^2 + c[<b,b>,<c,c>] - 2q(c<b,c>,b)c$$
$$= \beta c<c,c>^2 - 2c<c<b,b>,c> + 2q(c,b<b,c>)c$$
$$= 9\lambda^2 \beta c - 2\beta c<c<c,c>,c> + 2\beta q(c,c<c,c>)c.$$

Since $<c<c,c>,c> = 0$ and $q(c,c<c,c>) = -3\lambda^2$, the above is $3\lambda^2 \beta c$. Thus $<b,c>$ <u>stabilizes our 4-dimensional subspace of</u> C.

From the above we see that $<b,c>^2$ is the scalar $3\lambda^2 \beta$ on this 4-dimensional subspace. We now define an A-linear operation δ_1 on (the right on) C by:

$$c\delta_1 = \tfrac{1}{3}c<c,c> = -\lambda c; \quad c<c,c>\delta_1 = \tfrac{1}{3}c<c,c>^2 = 3\lambda^2 c;$$
$$b\delta_1 = -b<c,c> = \lambda b; \quad b<c,c>\delta_1 = -b<c,c>^2 = -\lambda^2 b;$$
$$v\delta_1 = v<c,c> = -\lambda v \text{ if } v \in U, q(v,b) = 0 = q(v,b<c,c>).$$

We denote the space of such $v \in U$ by V, and observe that since $q(b,b<c,c>) = q(b,c<b,c>) = -q(b<b,c>,c) = -q(c<b,b>,c) = -\beta q(c<c,c>,c) \neq 0$, both our 4-dimensional space W and V are nonsingular with respect to q. We define a second operator, δ_2, on C to be $<b,c>$. Our calculations yield:

$$c\delta_1\delta_2 = \tfrac{1}{3}c<c,c><b,c> = \lambda^2 b.$$
$$c\delta_2\delta_1 = b<c,c>\delta_1 = -b<c,c>^2 = -\lambda^2 b,$$

from which $c<c,c>\delta_1\delta_2 = 3\lambda^2 c\delta_2 = 3\lambda^2 c<b,c> = 3\lambda^2 b<c,c>$, $c<c,c>\delta_2\delta_1 = 3\lambda^2 b\delta_1 = -3\lambda^2 b<c,c>$. Also, we have $b\delta_1\delta_2 = -b<c,c><b,c> = -3\lambda^2 \beta c$, $b\delta_2\delta_1 = \beta c<c,c>\delta_1 = 3\lambda^2 \beta c$, $b<c,c>\delta_1\delta_2 = -\lambda^2 b<b,c> = -\beta\lambda^2 c<c,c> = -b<c,c>\delta_2\delta_1$. The operators δ_1,δ_2 thus anticommute in their action on W.

This relation also holds in their action on V: For if $v \in V$, one sees from partial polarization of (T5) (cf. (1.2) of [Me]) that $3\{\{vcc\}cb\} + \{\{ccc\}vb\} = 0$, and by q-orthogonality this amounts to

($\ddagger$) $\qquad 3v<c,c><b,c> + c<c,c><v,b> = 0.$

Another partial polarization of (T5) (cf. (1.3) of [Me]) yields $2\{\{vbc\}cc\} + \{\{bcc\}vc\} + \{\{vcc\}bc\} = 2q(c,\{vbc\})c.$ But $\{vbc\}=-v<b,c>$ by q-orthogonality of the factors, and $q(c,v<b,c>) = -q(c<b,c>,v)=0$ since $v \in V$. Orthogonality (relative to q) of the factors in our relation yields

$\qquad 2v<b,c><c,c> + b<c,c><v,c> + v<c,c><b,c> = 0,$

or $\qquad 2v<b,c><c,c> + 2v<c,c><b,c>$

$\qquad = v<c,c><b,c> - b<c,c><v,c>.$

But b and v are interchangeable in ($\ddagger$) above, from which we see that the right-hand side of this last relation vanishes, and conclude:

$v\delta_1\delta_2 = -v\delta_2\delta_1$.

Evidently $\delta_1^2 = \lambda^2 \in A$, and one checks at once that, on W, $\delta_2^2 = 3\lambda^2\beta \in A$. We now show that the latter relation holds on V as well.

For $v \in V$, the fact that $v<b,c> \in V$ and that v,b,c are q-orthogonal shows that $v<b,c>^2 = \{\{vbc\}bc\}$. The partially polarized form ((1.3) of [Me]) of (T5) used above yields

$\qquad 2\{\{vbc\}bc\} + \{\{cbb\}cv\} + \{\{vbb\}cc\}$

$\qquad = 2q(c,\{vbc\})b + q(c,\{cbb\})v + q(c,\{vbb\})c.$

Since $\{vbc\} = -v<b,c> \in V$, and $q(c,\{vbb\}) = -q(c,v<b,b>) = q(c<b,b>,v) = \beta q(c<c,c>,v) = 0$, the right-hand side of the above is $q(c,\{cbb\})v = \beta q(c,\{ccc\})v = 3\lambda^2\beta v$. The left-hand side is $2v<b,c>^2 + \beta\{\{ccc\}cv\} + v<b,b><c,c>$, and the second of these terms is zero by (T5). Thus we have

(*) $\qquad 2v<b,c>^2 + v<b,b><c,c> = 3\beta\lambda^2 v.$

Furthermore, from $[<c,c>,<b,b>] = 2<c<b,b>,c> = 2\beta<c<c,c>,c> = 0$, we see that $<b,b>$ and $<c,c>$ commute (on all of C).

Now $b<b,b><c,c> = -\{b,b,b\}<c,c> = \{\{bbb\}cc\} + q(c,\{bbb\})c - q(\{bbb\},c)c = \{\{bbb\}cc\} + 2q(c,\{bbb\})c$, and $q(c,\{bbb\}) = -$

$q(c,b<b,b>) = q(c<b,b>,b) = \beta q(c<c,c>,b) = 0$, since $b \in U$. Application of the polarization ((1.2) of [Me]) of (T5) gives

$$\{\{bbb\}cc\} = -3\{\{cbb\}cb\} + 3q(c,\{cbb\})b$$
$$= 3\{b<b,c>,c,b\} + 3\beta q(c,\{ccc\})b$$
$$= 3\{c<b,b>,c,b\} + 9\beta\lambda^2 b$$
$$= 3\beta \cdot \{c<c,c>,c,b\} + 9\beta\lambda^2 b .$$

The first term is zero as before, and the conclusion is

(**) $\qquad\qquad b<b,b><c,c> = 9\beta\lambda^2 b$.

The commutativity of $<b,b>$ and $<c,c>$ gives, upon another application of $<c,c>$ to (**),

$$\lambda^2 b<b,b> = 9\beta\lambda^2 b<c,c>, \quad \text{or}$$
(***) $\qquad\qquad b<b,b> = 9\beta b<c,c>.$

From (***) we obtain $[<b,c>,<b,b>] = <b<b,b>,c> + <b,c<b,b>> = 9\beta<b<c,c>,c> + \beta<b,c<c,c>>$, while $[<b,c>,<c,c>] = <b<c,c>,c> + <b,c<c,c>>$.

Now $q(v<b,b>,c) = -\beta q(v,c<c,c>) = 0$, $q(v<b,b>,c<c,c>) = -q(v,c<c,c><b,b>) = -q(v,c<b,b><c,c>) = -9\lambda^2\beta q(v,c) = 0$, $q(v<b,b>,b) = -q(v,b<b,b>) = -9\beta q(v,b<c,c>) = 0$, $q(v<b,b>,b<c,c>) = -9\beta\lambda^2 q(v,b) = 0$. Thus $v<b,b> \in V$, and we can replace v by $v<b,b>$ in (*) to obtain

$$2v<b,b><b,c>^2 - 3\beta\lambda^2 v<b,b> = -v<b,b>^2<c,c>.$$

From the fact that $q(v,b) = 0 = q(v,b<b,b>)$ we know that $v<b,b>^2 = \frac{1}{3}q(b,\{b,b,b\})v = -3\beta q(b,b<c,c>)v$ by (***). But $q(b,b<c,c>) = q(b,c<b,c>) = -q(b<c,c>,c) = -q(c<b,b>,c) = \beta q(c,c<c,c>) = -3\beta\lambda^2$, and $v<b,b>^2 = 9\beta^2\lambda^2 v$. That is we have

$$9\beta^2\lambda^2 v<c,c> = 3\beta\lambda^2 v<b,b> - 2v<b,b><b,c>^2 .$$

On the other hand, application of $3\beta<c,c>$ to (*), the commutativity of $<b,b>$ and $<c,c>$, and the anticommutativity of $<b,c>$ and $<c,c>$ on V yield

$$9\beta^2\lambda^2 v<c,c> = 3\beta\lambda^2 v<b,b> + 6\beta v<c,c><b,c>^2 .$$

Subtraction of the last two displayed relations gives

$\xi')$ $\qquad\qquad (3\beta v<c,c> + v<b,b>)<b,c>^2 = 0$.

From the fact that $<b,c> \in <C,C>$ centralizes T, we know its action on L, and therefore on C, must be semisimple. The relation ξ) therefore has the immediate consequence

ξ') $(3\beta v<c,c> + v<b,b>)<b,c> = 0$ for all $v \in V$.

Now we have seen that $<b,b>$ and $<c,c>$ commute, and that for $v \in V$, $v<b,b>^2 = 9\beta^2\lambda^2 v = 9\beta^2 v<c,c>^2$. Thus $(<b,b> - 3\beta<c,c>)(<b,b> + 3\beta<c,c>)$ annihilates V, and both $<b,b>$ and $<c,c>$ are semisimple. It follows that V is the direct sum $V_1 \oplus V_2$ of subspaces V_1, V_2, with $<b,b> - 3\beta<c,c>$ vanishing on V_1, $<b,b> + 3\beta<c,c>$ vanishing on V_2. For $v \in V_1$, we have by ξ'), $0 = (v<b,b> + 3\beta v<c,c>)<b,c> = 6\beta v<c,c><b,c> = -6\beta v<b,c><c,c>$, from which $v<b,c> = 0$.

For $v \in V_2$, (*) becomes $2v<b,c>^2 - 3\beta v<c,c>^2 = 3\beta\lambda^2 v$, or $v<b,c>^2 = 3\beta\lambda^2 v$. It follows that, on V, $<b,c>$ <u>satisfies the polynomial</u> $X(X^2 - 3\beta\lambda^2)$ <u>over</u> A.

So far, the only property of b we have used is that $b \in U$ is idempotent with $\tau(b) = 1$. <u>Thus our considerations to this point are valid whenever the Jordan algebra U contains idempotents other than</u> 0 <u>or</u> 1. We now show that, under the further restrictions imposed at the beginning of this section, (starting after (6.10)), $<b,c>$ must annihilate no non-zero elements of V, so that $v<b,c>^2 = 3\beta\lambda^2 v$ for all $v \in V$:

In the Peirce decomposition $U = U_1 \oplus U_{\frac{1}{2}} \oplus U_0$ of U relative to the idempotent b, we have by assumption $U_1 = Kb$, $U_{\frac{1}{2}} = 0$. Since $\tau(bU_0) = \tau(0) = 0$, our formula for q in terms of τ shows that $V = U_0^\phi$. Now let $v \in U_0$, $v^\phi<b,c> = 0 = -\{v^\phi,b,c\}$. Our formula (6.3) applied to $\{v^\phi,b,c\}$, together with the fact that $\alpha \in K$ is non-zero yields $v^\phi \times b = 0$. If $v^\phi = \mu b + v_0$, $\mu \in K$, $v_0 \in U_0$, we have $0 = \mu b - \frac{1}{2}v^\phi - \frac{1}{2}\tau(v^\phi)b + \frac{1}{2}(\tau(v^\phi) - \mu)1$, $v^\phi = 2\mu b + \tau(v^\phi)(1-b) - \mu 1$, $\tau(v^\phi) = 2\tau(v^\phi) - \mu = \mu$, and $v^\phi = \mu b$. If $\mu = \xi + \eta\lambda$, $\xi, \eta \in A$, we have $v^\phi = \xi b - \eta b<c,c>$, $0 = v^\phi<b,c> = \beta\xi c<c,c> - 3\lambda^2\beta\eta c$, from the above. But this implies $\xi = \eta = 0$, and $v^\phi = 0$. This establishes the assertion, and we have $\delta_2^2 = 3\beta\lambda^2$.

Let Q be the A-algebra $A[\delta_1, \delta_2]$ of A-endomorphisms of C. We have seen that $\delta_1^2 = \lambda^2$, $\delta_2^2 = 3\beta\lambda^2$, $\delta_1\delta_2 = -\delta_2\delta_1$, so that if $1, \delta_1, \delta_2, \delta_1\delta_2$ are linearly independent, Q is a quaternion algebra over A.

But if $\alpha_0 1 + \alpha_1 \delta_1 + \alpha_2 \delta_2 + \alpha_3 \delta_1 \delta_2 = 0$, $\alpha_i \in A$, forming a commutator with δ_1 yields $2\alpha_2 \delta_1 \delta_2 + 2\alpha_3 \lambda^2 \delta_2 = 0$. Multiplying on the right by δ_2 yields $\alpha_2 \delta_1 + \alpha_3 \lambda^2 1 = 0$. Since δ_1 and 1 are linearly independent, we must have $\alpha_2 = \alpha_3 = 0$, and then $\alpha_0 = \alpha_1 = 0$.

In fact, Q is a <u>division</u> algebra; for if $\alpha_0 1 + \alpha_1 \delta_1 + \alpha_2 \delta_2 + \alpha_3 \delta_1 \delta_2$ has square zero, we see at once that $\alpha_0 = 0$. On V, we have $\delta_1 \delta_2 = \langle c,c \rangle, \langle b,c \rangle = \frac{1}{2}[\langle c,c \rangle, \langle b,c \rangle] = \langle c \langle b,c \rangle, c \rangle$, so that $\alpha_1 \delta_1 + \alpha_2 \delta_2 + \alpha_3 \delta_1 \delta_2 = \alpha_1 \langle c,c \rangle + \alpha_2 \langle b,c \rangle + \alpha_3 \langle c \langle b,c \rangle, c \rangle \in \langle C,C \rangle$, in the restrictions to V. This semisimple transformation can only have square zero if it is zero, and that implies $\alpha_1 = \alpha_2 = \alpha_3 = 0$ by our earlier considerations, applied to restrictions to $V (\neq 0)$.

We thus have shown that C <u>is a right vector space over the quaternionic division algebra</u> Q.

Now, for $u,v \in C$, define $(u,v) \in Q$ by $(u,v) = q(u,v) + q(u,v\delta_1)\delta_1^{-1} + q(u,v\delta_2)\delta_2^{-1} + q(u,v\delta_1\delta_2)\delta_2^{-1}\delta_1^{-1}$. Then (u,v) is A-bilinear, and from $q(u,v\delta_i) = -q(u\delta_i,v) = q(v,u\delta_i)$, together with $\delta_2 \delta_1 = -\delta_1 \delta_2$, it follows that $(v,u) = -(u,v)^*$, where $(\alpha_0 + \alpha_1 \delta_1 + \alpha_2 \delta_2 + \alpha_3 \delta_1 \delta_2)^* = \alpha_0 - \alpha_1 \delta_1 - \alpha_2 \delta_2 - \alpha_3 \delta_1 \delta_2$ defines the standard involution in Q. Straightforward calculations show $(u,v\delta_i) = (u,v)\delta_i$, $i=1,2$, from which it follows that (u,v) is Q-linear in v, and is thus <u>an antihermitian form on</u> C <u>over</u> Q. Since its projection $q(u,v)$ on $A \subseteq Q$ is nondegenerate, (u,v) <u>is nondegenerate</u>.

We next claim: <u>For</u> $w,x,y \in C$,

(*) $\qquad w\langle x,y \rangle = -x(y,w) - y(x,w) - \frac{1}{2}w((x,y)+(y,x))$.

By the A-trilinearity of both sides of (*), it suffices to verify (*) for x,y,w chosen arbitrarily among the following: $c, c\langle c,c \rangle, b, b\langle c,c \rangle, v \in V$.

i) $x = y = c$: If $w = c$, $w\langle x,y \rangle = c\langle c,c \rangle = 3c\delta_1$. On the other hand $q(c,c) = 0 = q(c,c\delta_2) = q(c,c\delta_1\delta_2)$, as we have seen, so that $(c,c) = q(c,c\delta_1)\delta_1^{-1} = \frac{1}{3}q(c,c\langle c,c \rangle)\delta_1^{-1} = -\lambda^2\delta_1^{-1} = -\delta_1$. Thus the right-hand side of (*), in our case, is $3c\delta_1$.

If $w = c\langle c,c \rangle = 3c\delta_1$, $w\langle x,y \rangle = 9\lambda^2 c$. Meanwhile, $(c,w) = q(c,w) = -3\lambda^2$, so the right-hand side of (*) is $6\lambda^2 c + w\delta_1 = 6\lambda^2 c + 3c\delta_1^2 = 9\lambda^2 c$.

If $w = b = \lambda^{-2} c\delta_1 \delta_2 = -\lambda^{-2} c\delta_2 \delta_1$, we have $w\langle c,c \rangle = -w\delta_1 =$

$= \lambda^{-2} c\delta_2 \delta_1^2 = c\delta_2$, while $(c,w) = q(c,w\delta_2)\delta_2^{-1} = \lambda^{-2}\delta_2^2 q(c,c\delta_1)\delta_2^{-1} = \lambda^{-2}q(c,c\delta_1)\delta_2 = -\delta_2$. The right-hand side of (*) is now $2c\delta_2 + \frac{1}{2}\lambda^{-2}c\delta_2\delta_1 \cdot (-2\lambda^2 \delta_1^{-1}) = c\delta_2$, again verifying our conclusion.

If $w = b<c,c> = -b\delta_1 = c\delta_2$, we have $w<c,c> = -c\delta_2\delta_1 = c\delta_1\delta_2$, while $(c,w) = q(c,c\delta_2\delta_1\delta_2)\delta_2^{-1}\delta_1^{-1} = -q(c,c\delta_1\delta_2^2)\delta_2^{-1}\delta_1^{-1} = -q(c,c\delta_1)\delta_2\delta_1^{-1} = \lambda^2\delta_2\delta_1^{-1} = \delta_2\delta_1$. The right-hand side of (*) is thus $-2c\delta_2\delta_1 - \frac{1}{2}c\delta_2 \cdot (-2\delta_1) = -c\delta_2\delta_1 = c\delta_1\delta_2$, and (*) holds.

Finally, if $w \in V$, $w<c,c> = w\delta_1$, $(w,c) = 0$, and $w(c,c) = -w\delta_1$; from these, * is immediate.

ii) $x = c, y = c<c,c>$. Here the left-hand side of (*) is always zero, and $(c, c<c,c>) + (c<c,c>,c) = q(c, c<c,c>) + q(c<c,c>,c) = 0$ as well. Verification of (*) comes down to showing $c(c<c,c>,w) + c<c,c>(c,w) = 0$ for all our choices of w, and this is immediate if $w \in V$.

For $w = c$, $(c<c,c>,w) = q(c<c,c>,c) = 3\lambda^2$, while $(c,c) = -\delta_1$. Our expression is $3\lambda^2 c - 3c\delta_1^2 = 0$; for $w = c<c,c>$, we have $(c<c,c>, c<c,c>) = 9(c\delta_1, c\delta_1) = 9q(c\delta_1 c\delta_1^2)\bar{\delta}^{-1} = 9q(c\delta_1, c)\delta_1 = 9\lambda^2$, $(c,w) = 3(c,c\delta_1) = -3\delta_1$, and our expression is $9\lambda^2 c - 9c\delta_1^2 = 0$. For $w = b = \lambda^{-2}c\delta_1\delta_2$, we have $(c,w) = \lambda^{-2}(c,c)\delta_1\delta_2 = -\delta_2$, $(c<c,c>,w) = 3(c\delta_1, w) = -3\delta_1\lambda^{-2}(c,c)\delta_1\delta_2 = 3\delta_1\delta_2$, so our expression is $3c\delta_1\delta_2 - 3c\delta_1\delta_2 = 0$. A similar verification applies when $w = b<c,c> = c\delta_2$, and completes ii).

The arguments of i) and ii) now show how the verification of (*) may be routinely carried out except when one of w,x,y is in V. From $<c<c,c>, c<c,c>> = [<c<c,c>,c>, <c,c>] - <c<c,c>^2, c> = -<c<c,c>^2, c> = -9\lambda^2<c,c>$, the left-hand side of (*) is, for $w \in V$, $x = y = c<c,c>$, $-9\lambda^2 w<c,c> = -9\lambda^2 w\delta_1$. Since $(w, c<c,c>) = 0$ and $(c<c,c>, c<c,c>) = -9\delta_1(c,c)\delta_1 = 9\lambda^2\delta_1^2$, the right-hand side is $9w\lambda^2\delta_1$, and (*) holds in this case.

When $w \in V$, $x = b, y = c$, the left-hand side of (*) is $w\delta_2$, while the right is $-\frac{1}{2}w((b,c) + (c,b))$. Now $(c,b) = -\delta_2$, as we have seen above, and our conclusion (*) holds. When $x = b<c,c>$, $y = c$, we see from $2<b<c,c>,c> = 2<c<b,c>,c> = [<c,c>,<b,c>] = -[<b,c>,<c,c>] = -<b<c,c>,c> -<b,c<c,c>>$ that, on V,

$3<b<c,c>,c> = -<b,c<c,c>>$ and that $2<b<c,c>,c> = \delta_1\delta_2 - \delta_2\delta_1 = 2\delta_1\delta_2$. Thus with $w \in V$, the left-hand side of (*∗*) is $w\delta_1\delta_2$, and the right-hand side is $-\frac{1}{2}w((b<c,c>,c) + (c,b<c,c>)) = -w\delta_2\delta_1 = w\delta_1\delta_2$. The relation involving $<b,c<c,c>>$ just derived also enables us to check (*∗*) for $w \in V$, $x = b, y = c<c,c>$.

Now $<b<c,c>,c<c,c>> = [<b,c<c,c>>,<c,c>] - <b,c<c,c>^2> = -\frac{3}{2}[[<c,c>,<b,c>]<c,c>] - 9\lambda^2<b,c>$, by the above; on V, this is equal to $-\frac{3}{2}[[\delta_1\delta_2]\delta_1] - 9\lambda^2\delta_2 = 6\lambda^2\delta_2 - 9\lambda^2\delta_2 = -3\lambda^2\delta_2$, and, with w as above, $w<b<c,c>,c<c,c>> = -3\lambda^2 w\delta_2$. In this case, the right-hand side of (*∗*) is

$-\frac{1}{2}w((b<c,c>,c<c,c>) + (c<c,c>,b<c,c>)) = \frac{1}{2}w((c\delta_2,3c\delta_1)+(3c\delta_1,c\delta_2))$
$= \frac{1}{2}w(3\delta_2(c,c)\delta_1 + 3\delta_1(c,c)\delta_2) = -\frac{1}{2}w(3\delta_2\delta_1^2 + 3\delta_1^2\delta_2) = -3\lambda^2 w\delta_2$,

so that (*∗*) again checks when $w \in V$, $x = b<c,c>, y = c<c,c>$. For these w and x, and with $y = b$, we have $2<b<c,c>,b> = [<b,b>,<c,c>] = -[<c,c>,<b,b>] = -2<c<b,b>,c> = -2\beta<c<c,c>,c> = 0$, while $(b<c,c>,b) = -(b\delta_1,b) = -\lambda^{-4}(c\delta_2\delta_1^2,c\delta_2\delta_1) = \lambda^{-2}\delta_2(c,c)\delta_2\delta_1 = -\lambda^{-2}\delta_2\delta_1\delta_2\delta_1 = \delta_2^2 = 3\lambda^2\beta_*$. Thus $(b<c,c>,b) + (b,b<c,c>) = (b<c,c>,b) - (b<c,c>,b)^* = 0$, and both sides of (*∗*) are zero in this case.

From $<b<c,c>,b> = 0$, we have $0 = [<b<c,c>,b>,<c,c>] = \lambda^2<b,b> + <b<c,c>,b<c,c>>$. Now $<b<c,c>,b<c,c>> = <c<c,b>,c<c,b>> = [<c,c<c,b>>,<c,b>] - <c<c,b>^2,c> = [<c,b<c,c>>,<b,c>] - \delta_2^2<c,c>$. By the computation of $<c,b<c,c>>$, this is $\frac{1}{2}[[<c,c>,<b,c>]<b,c>] - \delta_2^2<c,c>$, and is equal, on V, to $\frac{1}{2}[[\delta_1\delta_2]\delta_2] - \delta_2^2\delta_1 = \delta_2^2\delta_1$: $w<b<c,c>,b<c,c>> = w\delta_2^2\delta_1$. Meanwhile the right-hand side of (*∗*) with these values of w,x,y is $-w(b<c,c>,b<c,c>) = w\delta_1^2(b,b) = w\delta_1^2\lambda^{-4}\delta_1\delta_2(c,c)\delta_2\delta_1 = -w\delta_1^2\lambda^{-4}\delta_1\delta_2\delta_1\delta_2\delta_1 = w\delta_2^2\delta_1$. We also see that $w<b,b> = -\lambda^{-2}w\delta_2^2\delta_1$, while $-w(b,b) = -\lambda^{-4}w\delta_1\delta_2(c,c)\delta_2\delta_1 = \lambda^{-4}w\delta_1\delta_2\delta_1\delta_2\delta_1 = -\lambda^{-2}w\delta_2^2\delta_1$. <u>This completes the verification of</u> (*∗*) <u>whenever</u> $x,y \in c\mathcal{Q}$.

When $w \in V$, $x,y \in c\mathcal{Q}$, we have $q(w,x) = 0 = (w,x)$, and $x<w,y> = w<x,y> - wq(y,x)$. Since $q(y,x) = \frac{1}{2}((y,x) + (y,x)^*) = \frac{1}{2}((y,x) - (x,y))$, we see that $x<w,y> = -\frac{1}{2}w((x,y) + (y,x)) - \frac{1}{2}w((y,x) - (x,y)) = -w(y,x) = -w(y,x) - y(w,x) - \frac{1}{2}x((w,y) + (y,w))$, so (*) <u>holds whenever two of</u> x,y,w <u>are in</u> cQ.

For the remaining cases, we require a little information about the cross-product and about the mapping ϕ. Every element of our Jordan algebra $J = U$ has the form $\alpha b + u$, where $u \in J_0(b) = U_0$; $1 - b \in U_0$ is the unit element of the Jordan algebra $J_0(b)$; and $(\alpha b + u)^2 = \alpha^2 b + u^2$. From the cubic identity on J, we find by inspecting the term in U_0 with coefficient α that $u^2 - \tau(u)u + \frac{1}{2}(\tau(u)^2 - \tau(u^2))(1-b) = 0$ for all $u \in U_0$, so that $u \times u = \frac{1}{2}(\tau(u)^2 - \tau(u^2))b \in Kb$ for all such u. <u>That is</u>, $u \times v \in Kb$ <u>for all</u> $u,v \in U_0$.

Next, we note that from $b \times b = 0$ we have, from (6.5), that $b^\phi \times b^\phi = 0$. Thus if $b^\phi = \alpha b + u$, we have

$0 = (\alpha b+u) \times (\alpha b+u) = u \times u + 2\alpha(b \times u) = (u \times u) - \alpha u$
$- \alpha\tau(u)b - \alpha\tau(u)1 = (u \times u) - 2\alpha\tau(u)b - \alpha u - \alpha\tau(u)(1-b)$.

If $\alpha = 0$, we have $u \times u = 0$, or $\tau(u)^2 = \tau(u^2)$ from the above. But then the formula for $u \times u$ gives $u^2 = \tau(u)u$, and either $\tau(u) = 0$, $u^2 = 0$, or $\nu = \tau(u)^{-1} \neq 0$, $(\nu u)^2 = \nu u$, $\tau(\nu u) = 1$. if $\alpha \neq 0$, we have $u = \tau(u)(1-b)$, $\tau(u) = \tau(u) \cdot (3-1) = 2\tau(u)$, so that $\tau(u) = 0$ and $u = 0$. That is, either $b^\phi = \alpha b$, or $b^\phi = \alpha e_0$, or $b^\phi = n$, where $e_0 \in U_0$, $e_0^2 = e_0$, $\tau(e_0) = 1$; $n \in U_0$, $n^2 = 0$.

We have seen that $\{c \in J | b \times c = 0\} = Kb$, of K-dimension one. By (6.5), $\{c \in J | b^\phi \times c = 0\}$ must also have K-dimension one. In case β) this is impossible unless $b^\phi = \alpha b$ (there are no nilpotents, and $1-b$ is a primitive idempotent with $\tau(1-b) = 2$). In case γ), the same reasoning shows that $b^\phi = \alpha b$ unless $J \cong K \oplus K \oplus K$, with primitive idempotents b, e_0, e_0', each of trace 1, and $b^\phi = \alpha b$, αe_0 or $\alpha e_0' (e_0, e_0' \in J_0(b))$. In case ϵ), $U_0 = J_0(b)$ is the Jordan algebra of a quadratic form, $[U_0:K] \geq 3$, and U_0 contains non-zero nilpotents or idempotents e_0 with $\tau(e_0) = 1$ only if U_0 is split, <u>i.e.</u>, the quadratic form represents 1. Thus $b^\phi = \alpha b$ except possibly in this last instance. When U_0 is split,

$X = U_0 = Ke_0 \oplus Ke_0' \oplus (X_{\frac{1}{2}}(e_0) \cap X_{\frac{1}{2}}(e_0'))$, and $(X_{\frac{1}{2}}(e_0) \cap X_{\frac{1}{2}}(e_0')) = X_{\frac{1}{2}}(e_0)$

has K-dimension $[X:K] - 2$ whenever e_0 is a primitive idempotent in X. Since $e_0 \times X_{\frac{1}{2}}(e_0) = 0 = e_0 \times e_0$, we cannot have $b^\phi = \alpha e_0$ unless $X_{\frac{1}{2}}(e_0) = 0$, in which case $[X:K] < 3$, a contradiction. On the other hand, if $b^\phi = n \in U_0$, $n^2 = 0$, then n is in the quadratic space of codimension 1 in U_0, and is an isotropic vector. Any vector w of this space, orthogonal to n in the quadratic form, will also satisfy $n \times w = 0$, so $(n)^\perp$ must have K-dimension 1. This implies $[X:K] \leq 3$, hence $[X:K] = 3$, $[C:K] = [(cQ \oplus X):K] = 5$. But C is a right Q-vector space, and $[Q:K] = 2$, so $[C:K]$ must be even. Our conclusion is that $b^\phi = \alpha b$ except possibly when $J = F \oplus F \oplus F$.

In this case, we may assume $b^\phi = \alpha e_0$. If $e_0'^\phi = \gamma e_0'$, we may start with e_0' instead of b to assume $b^\phi = \gamma b$; therefore we may assume $e_0'^\phi = \mu b$, $e_0^\phi = \nu e_0'$. Then $b^{\phi^2} = \xi e_0'$, $e_0^{\phi^2} = \eta b$, $e_0'^{\phi^2} = \zeta e_0$. Now if $a = \alpha_1 b + \alpha_2 e_0 + \alpha_3 e_0'$, we have $a \times a = \alpha_2 \alpha_3 b + \alpha_1 \alpha_3 e_0 + \alpha_1 \alpha_2 e_0'$, $(a \times a) \times b = \alpha_1 \alpha_3 e_0' + \alpha_1 \alpha_2 e_0$, $\tau(ab) = \alpha_1$, and our formula (6.9) giving u^{ϕ^2} as a combination of $(a \times a) \times u$ and $\tau(au)a$ can only be valid for $u = b$ if $\alpha_1 = 0$. Similar tests with $u = e_0, e_0'$ force $a = 0$, a contradiction. We therefore conclude that the idempotent b with $\tau(b) = 1$ can always be chosen so that $b^\phi \in Kb$, hence so that $V = U_0^\phi = U_0$.

Our formula (6.3) now combines with the fact that $u \times v \in Kb$ and the above to show that $\{u,v,c\} \in cQ$ whenever $u,v \in V$. Since the q-skew transformations $<c,c>, <b,c>, <b<c,c>,c>$ all stabilize cQ, and hence its q-orthogonal space V, and act as derivations on $\{u,v,c\}$, we see by applying them to $\{u,v,c\}$ that $\{V,V,cQ\} \subseteq cQ$. Hence we have, for $\gamma \in Q$, $u,v \in V$, that

$$(c\gamma)<u,v> = - <c\gamma,u,v> - q(u,v)c\gamma$$
$$= - \{c\gamma,u,v\} - q(v,u)c\gamma + q(u,v)c\gamma$$
$$= - \{c\gamma,u,v\} \in cQ .$$

To verify (*) for this expression, it therefore suffices to show that, for every $\delta \in Q$,

(6.11) $\qquad q((c\gamma)<u,v>,c\delta) = -\frac{1}{2}q((c\gamma)((u,v) + (v,u)),c\delta).$

Now, by (95), $(c\gamma)<u,v> = u<c\gamma,v> - q(u,v)c\gamma$, and $q(u<c\gamma,v>,c\delta) = -q(u,(c\delta)<c\gamma,v>) = q(u,v(c\gamma,c\delta))$, by (*) for the case where two arguments are in cQ. Thus $q((c\gamma)<u,v>,c\delta) = q(u,v(c\gamma,c\delta)) - q(u,v)q(c\gamma,c\delta) = \frac{1}{2}(u,v)(c\gamma,c\delta) - \frac{1}{2}(c\gamma,c\delta)^*(v,u)$ $- \frac{1}{2}q(u,v)(c\gamma,c\delta) + \frac{1}{2}(c\delta,c\gamma)q(u,v) = \frac{1}{4}(u,v)(c\gamma,c\delta) + \frac{1}{4}(v,u)(c\delta,c\gamma) +$

$+ \frac{1}{4}(v,u)(c\gamma,c\delta) + \frac{1}{4}(c\delta,c\gamma)(u,v) = -\frac{1}{2}q((c\gamma)(u,v) + (v,u)),c\delta)$, and (6.11) is established. From (95) we now see easily that (*) holds whenever one argument is from cQ.

If $x \in cQ$ and $u,v,w \in V$, we have $q(u\langle v,w \rangle,x) = -q(u,x\langle v,w \rangle) = 0$ by the above. That is, we again have $u\langle v,w \rangle \in V$, hence $\{u,v,w\} \in V$; the proof of (*) can be completed by showing that, for $u \in V$, $\{u,u,u\} = 3u(u,u)$. This is analogous with our earlier result (‡) in the case $\{U,U,U\} \subseteq U$, and analogous arguments to those used in getting that result work: From Meyberg's identity (1.4),

$$\{u,u,\{\{ucc\},c,c\}\} + \{\{ucc\},\{ucc\},u\}$$
$$+ 2\{\{u,\{ucc\},c\},u,c\}$$
$$= 2q(u,\{ucc\})\{ucc\} + 2q(u,\{\{ucc\},u,c\})c.$$

From $\{u,u,u\}\langle c,c \rangle^2 = \lambda^2\{u,u,u\}$ and $u\langle c,c \rangle^2 = \lambda^2 u$, we have $\{u\langle c,c \rangle, u\langle c,c \rangle, u\} = -\frac{1}{3}\lambda^2\{u,u,u\}$, so that the first two terms above combine to $\frac{2}{3}\lambda^2\{u,u,u\}$. From previous cases we have $\{u,\{ucc\},c\} = \langle c,u,\{ucc\} \rangle + q(u,\{ucc\})c = -c\langle u,\{ucc\} \rangle = \frac{1}{2}c((u,\{ucc\}) + (\{ucc\},u))$ $= -\frac{1}{2}c((u,u\delta_1) + (u\delta_1,u))$. It follows that the last term in our version of Meyberg's (1.4) is zero, and that the right-hand side is $2q(u,u\delta_1)u\delta_1 = 2u\delta_1 q(u,u\delta_1) = u\delta_1(u,u)\delta_1 + \lambda^2 u(u,u)$. The last term on the left-hand side is $-\{c((u,u\delta_1) + (u\delta_1,u)),u,c\} = u\langle c,c((u\delta_1,u) + (u,u\delta_1)) \rangle^* = -\frac{1}{2}u((c,c)\delta_1^*(u,u) + (c,c)(u,u)\delta_1 - (u,u)\delta_1(c,c) - \delta_1^*(u,u)(c,c)) = -\frac{1}{2}u(\delta_1^2(u,u) - \delta_1(u,u)\delta_1 + (u,u)\delta_1^2 - \delta_1(u,u)\delta_1)$ $= -\lambda^2 u(u,u) + u\delta_1(u,u)\delta_1$. Substitution in our version of (1.4) gives $\frac{2}{3}\lambda^2\{u,u,u\} - \lambda^2 u(u,u) + u\delta_1(u,u)\delta_1 = u\delta_1(u,u)\delta_1 + \lambda^2 u(u,u)$, from which we see at once that $\{u,u,u\} = 3u(u,u)$. This completes the proof of (*).

We may now make $(M'' \otimes_F Q) \oplus C$ into an antihermitian Q-space relative to the form which is (x,y) on C and extends the symplectic form on M'' as before. Our identification of L with $K(M'' \otimes_F Q \oplus C)$ proceeds as in all previous cases of the sort. The fact that $\dim T = 1$ is reflected in the condition that (x,y) be anisotropic on C, so that our form on $M'' \otimes_F Q \oplus C$ has Witt index one. The centralizer L_0 of T is $T \otimes A + K(C)$, with derived algebra $K(C)$, a simple Lie algebra. The simple group G' identifies with the central quotient of the derived group of the unitary group of the space $M'' \otimes_F Q \oplus C$.

In our enumeration $\alpha) - \kappa)$ of possibilities for the K-Jordan algebra U, the last considerations dispose of the case $\epsilon)$, as well as of β) and γ) when U is split. In both case β) and case δ), there is a linear form t on U such that $t(1) = 2$, and such that every element $u \in U$ satisfies the relation

$$u^2 - t(u)u + \frac{1}{2}(t(u)^2 - t(u^2))1 = 0.$$

(In case β), t may be taken to be the trace of the regular representation of U; in case δ), twice the projection on $K = K \cdot 1$ in the decomposition $U = K \cdot 1 \oplus V$, V the space with the quadratic form.) We explore relations between t and τ. If $t(x) = 0$, we have $x^2 = \frac{1}{2}t(x^2)1$, $x^3 = \frac{1}{2}t(x^2)x$, and x and 1 are K-linearly independent. Now $\tau(x^2) = \frac{1}{2}t(x^2)\tau(1) = \frac{3}{2}t(x^2), \tau(x^3) = \frac{1}{2}t(x^2)\tau(x)$. Making these substitutions for x^2, $x^3, \tau(x^2), \tau(x^3)$ in the fundamental relation

$$x^3 - \tau(x)x^2 + \frac{1}{2}(\tau(x)^2 - \tau(x^2))x - (\frac{1}{3}\tau(x^3) - \frac{1}{2}\tau(x^2)\tau(x) + \frac{1}{6}\tau(x)^3)1 = 0,$$

we obtain

$$(\frac{1}{2}t(x^2) + \frac{1}{2}(\tau(x)^2 - \frac{3}{2}t(x^2)))x$$

$$= \frac{\tau(x)}{12}(t(x^2) - 2\tau(x)^2)1 .$$

From the linear independence we conclude that $t(x^2) = 2\tau(x)^2, \tau(x^2) = \frac{3}{2}t(x^2) = 3\tau(x)^2, \tau(x^3) = \frac{1}{2}t(x^2)\tau(x) = \tau(x)^3$, and $x^2 = \tau(x)^2 1 = (\tau(x)1)^2$.

Thus $x^2 = 0$ whenever $\tau(x) = 0 = t(x)$. In case δ), if $[V:K] \geq 3$, this would imply that the kernel of τ is a totally isotropic subspace of V of codimension at most one, which is impossible. Thus $[V:K] = 2$ in case δ), the quadratic form on V has Witt index one, and the intersection of the kernels of t and of τ has dimension one. In case β), we conclude that $x = \pm\tau(x)1$ (since U is not split, U is a field), $\tau(x) = \pm 3\tau(x) = 0$, and this is absurd (see also Appendix B,3)) . In the former case, with $[V:K] = 2$, along with $x \in V$ with $\tau(x) = 0$ there must be a second isotropic vector $y \neq 0$, linearly independent of x; but then $y^2 = 0$, $t(y^2) = 0 = 2\tau(y)^2$ and $\tau(y) = 0$. Therefore $\tau(V) = 0$, $z^2 = 0$ for all $z \in V$, and we contradicted the non-singularity of the quadratic form on V. This eliminates all possibilities from α) - ε) except α) and γ), when U is a cubic field extension of K. In particular, the dimension $[U:K]$ in the remaining cases is 1,3,6,9,15 or 27.

In fact, we cannot have $[U:K] = 1$. For then $U \stackrel{\sim}{=} K = K1$,

$1 \times 1 = 1$, $\tau(1) = 3$, and $a = -1^\phi = \zeta 1$, for some $\zeta \neq 0$ in K. From (6.5) with $u = v = 1$ we have $-\zeta 1 = \frac{8}{9}\alpha\alpha^*(1^{\phi^{-1}} \times 1^{\phi^{-1}})$. Now $((-\zeta^*)^{-1}1)^\phi = -\zeta^{-1}1^\phi = 1$, so $1^{\phi^{-1}} = -\zeta^{*-1}1$, so that we have $-\zeta = \frac{8}{9}\alpha\alpha^*\zeta^{*-2}$, or $-\zeta^* = \frac{8}{9}\alpha\alpha^*(\zeta\zeta^*)^{-1} \in A$, hence $\zeta^* = \zeta$. But if $u, v \in K$, we have $uU_v = uv^2$, so that (6.10) yields $1^{\phi^2} = -1U_a = -\zeta^2 1$, while we now have $1^{\phi^2} = (-\zeta 1) = -\zeta 1^\phi = \zeta^2 1$, and $\zeta = 0$, a contradiction. This eliminates the case $\alpha)$.

When our quantity $\frac{1}{3}q(c, \{c,c,c\})$ is a square in A, and when the resulting Jordan algebra U^+ (or U^-) is split, the construction yields a simple Lie algebra which is split in its turn. In the respective cases $\gamma), \zeta), \eta), \theta), \iota), \kappa)$, we have the following data for this split context:

Case	dim.C	$<C,C>$	L
$\gamma)$	8	$A_1 + A_1 + A_1$	D_4
$\zeta)$	14	C_3	F_4
$\eta)$ $\theta)$	20	A_5	E_6
$\iota)$	32	D_6	E_7
$\kappa)$	56	E_7	E_8

Since extension of the base field in these cases brings us to the split context, the dimensions yielded by this table are valid in general. (For details related to the table, see [Me], [He].)

We summarize the conclusions of §V.7.d.iii) in the following Theorem:

Theorem V.4: Let the simple Lie algebra L be of type BC_1, and such that the long root space $L_{2\alpha}$ has dimension one over the centroid A of L. Let S be the split three-dimensional algebra over F, M its two-dimensional irreducible module, (u,v) the symplectic form on M which is S-invariant, and for $u, v \in M$, $u \circ v \in S$ the mapping sending $w \in M$ to $(w,u)v + (w,v)u$. Then one of the following is the case:
a) There is a quadratic field extension K of A, a K-vector space V of dimension at least three, and a nondegenerate anti-hermitian form on V (with respect to the nontrivial automorphism of K/A) of Witt index one, such that $L \cong K'(V)$, the derived algebra of the skew K-endomor-

phisms of V.

b) There is a quaternionic division algebra Q over A, a right Q-vector space V of dimension at least three, and a nondegenerate antihermitian form on V (with respect to the unique involution of Q having A as fixed elements) of Witt index one, such that $L \stackrel{\sim}{=} K(V)$, the Lie algebra of skew Q-endomorphisms of V.

c) There is a quadratic field extension K of A and a cubic extension U of K, such that the algebra L is generated from K and U as described below; in this case, $[L:A] = 28$, and L becomes a split algebra of type D_4 upon suitable base field extension from A.

d) There is a quadratic field extension K of A and a K-involution in the algebra of 3 by 3 K-matrices, fixing a 6-dimensional K-subspace U, such that the algebra L is generated from K and U as described below; in this case, $[L:A] = 52$, and L becomes a split algebra of type F_4 upon suitable base field extension from A. $^{(1)}$

e) There is a quadratic field extension K of A and a central simple associative K-algebra U of degree 3, such that the algebra L is generated from K and U as described below; in this case, $[L:A] = 78$, and L becomes a split algebra of type E_6 upon suitable base field extension from A. $^{(1)}$

f) There is a quadratic field extension K of A, a quadratic extension L of K, and a central simple associative algebra of degree 3 over L, with an involution of second kind whose set U of fixed elements contains K, such that the algebra L is generated from K and U as described below; in this case, $[L:A] = 78$, and L becomes a split algebra of type E_6 upon suitable base field extension from A. $^{(2)}$

g) There is a quadratic field extension K of A and a central simple associative K-algebra of degree 6 with K-involution having a 15-dimensional fixed space U, such that the algebra L is generated from K and U as described below; in this case, $[L:A] = 133$, and L becomes a split algebra of type E_7 upon suitable base field extension from A. $^{(3)}$

h) There is a quadratic field extension K of A and an exceptional central simple Jordan algebra U, of dimension 27, over K, such that L is generated from K and U as described below; in this case, $[L:A] = 248$, and L becomes a split algebra of type E_8 upon suitable base field extension from A. $^{(4)}$

[The construction for c) - h):

Let $K = A(\lambda)$, $\lambda^2 \in A$, $\lambda^2 \notin A^2$, and let $\beta \mapsto \beta^*$ be the non-trivial automorphism of K/A. With Jordan multiplication, U becomes a K-Jordan algebra with unit 1 and an associative K-linear form τ, such that for all $u \in U$,

$$u^3 - \tau(u)u^2 + \frac{1}{2}(\tau(u)^2 - \tau(u^2))u =$$

$$[\frac{1}{3}\tau(u^3) - \frac{1}{2}\tau(u)\tau(u^2) + \frac{1}{6}\tau(u)^3]1 ,$$

and with $\tau(1) = 3$. For $u,v \in U$, define

$$u \times v = uv - \frac{1}{2}\tau(u)v - \frac{1}{2}\tau(v)u + \frac{1}{2}(\tau(u)\tau(v) - \tau(uv))1 .$$

Further, let there be given a mapping ϕ of U onto U, $*$-semilinear, and a nonzero element $\alpha \in K$, such that for all $u,v \in U$,

(6.5) $\qquad (u \times v)^\phi = \frac{8}{9}\alpha\alpha^* (u^{\phi^{-1}} \times v^{\phi^{-1}})$,

(6.6) $\qquad \tau(u^\phi v) = -\tau(v^\phi u)^*$.

We also require that, if $a = -1^\phi \in U$, we have for all $u \in U$,

(6.10) $\qquad u^{\phi^2} = -uU_a = 2(a \times a) \times u - \tau(au)a$.

Let Kc be a one-dimensional K-vector space with basis c, and let $C = Kc \oplus U$. Define a symplectic A-bilinear form q on C by: $q(Kc,U) = 0$; $q(u,v) = -\frac{1}{16}\lambda^{-2}(\tau(v^{\phi^{-1}}u) + \tau(v^{\phi^{-1}}u)^*)$, for $u,v \in U$; $q(c,\lambda c) = \lambda^2$. Define a symmetric ternary composition $\{x,y,z\}$ on C by A-trilinearity and:

$$\{c,c,c\} = 3\lambda c,$$

$$\{c,c,\lambda c\} = \lambda^2 c, \quad \{c,\lambda c,\lambda c\} = -\lambda^3 c,$$

$$\{\lambda c,\lambda c,\lambda c\} = -3\lambda^4 c; \quad \{u,c,c\} = \lambda u,$$

$$\{u,c,\lambda c\} = 0, \quad \{u,\lambda c,\lambda c\} = -\lambda^3 u;$$

$$\{u,u,c\} = -\frac{3}{4}\alpha^{*-1}(u \times u)^\phi$$

$$+ \frac{1}{16}\lambda^{-2}(\tau(u^{\phi^{-1}}u) - \tau(u^{\phi^{-1}}u)^*)c,$$

$$\{u,u,\lambda c\} = -\frac{3}{4}\alpha^{*-1}\lambda(u \times u)^\phi +$$

$$+ \frac{1}{16} \lambda^{-1}(\tau(u^{\phi^{-1}} \cdot u) - \tau(u^{\phi^{-1}} u)^*)c;$$

$$\{u,u,u\} = - \frac{27}{16}(\alpha\alpha^*)^{-1}\lambda^{-2}((u \times u)^{\phi} \times u)^{\phi}$$

$$+ \frac{3}{8}\lambda^{-2}\tau(u^{\phi^{-1}} u)u$$

$$+ \frac{3}{32} \lambda^{-3}\alpha^{-1}\tau(u \cdot (u \times u))c \qquad (u \in U).$$

For $x,y,z \in C$, set

$$x<y,z> = q(x,z)y + q(x,y)z - \{x,y,z\}.$$

Regarding $<y,z>$ as an operator on C, let D be the A-space spanned by these operators. Then

$$L \stackrel{\sim}{=} D \oplus (S \otimes A) \oplus (M \otimes C),$$

where: D is a subalgebra ($[<y,z>,<y',z'>] = <y<y',z'>,z> + <y,z<y',z'>>$), $[D, S \otimes A] = 0$, $[v \otimes x, d] = v \otimes xd \in M \otimes C$ if $v \in M$, $x \in C$, $d \in D$; $[s \otimes a, s' \otimes a'] = [ss'] \otimes aa'(s,s' \in S; a,a' \in A)$; $[v \otimes x, s \otimes a] = vs \otimes ax; [v \otimes x, v' \otimes y] = (v,v')<x,y> + v \circ v' \otimes q(x,y)$, for $v' \in M$, $y \in C$, where $(v,v') \in F$, $<x,y> \in D$, $v \circ v' \in S$, $q(x,y) \in A$.]

Footnotes:

(1) From the approaches using Galois cohomology, it is known that the case d) cannot occur, and it seems highly unlikely that e) can (see, e.g., [TD] for diagrams of types F_4 and E_6). However, our approach has not been carried to the point of eliminating these.

(2) It is not clear to the author whether the central simple associative algebra over L must be a division algebra; nor, indeed, is it absolutely clear whether this case actually arises. From [TD], at least one of e), f) does arise.

(3) The involutorial algebra of g) must be either the 6 by 6 K-matrices or the 3 by 3 Q-matrices, where Q is a quaternionic K-division algebra. From [TD] (diagrams for E_7) at least one of these cases actually arises. We conjecture that only the latter is possible.

(4) Tits informs the author that he has effaced the question mark opposite the diagram for E_8 in [TD] corresponding to this case. It is not clear to the author whether the exceptional Jordan algebra U must be a division algebra.

8. **Simple Lie algebras over the reals**.

   a) **Non-central simple algebras**.

   Let L be a simple Lie algebra over the real field $\mathbb{R}$. The centroid of L is a finite extension field of $\mathbb{R}$, hence either $\mathbb{R}$ or $\mathbb{C}$. In the former case, L will be referred to as "central simple"; in the latter, L carries the structure of simple Lie algebra over $\mathbb{C}$, the structure as $\mathbb{R}$-algebra being that obtained by treating L only as Lie ring and $\mathbb{R}$-module. As $\mathbb{C}$-algebra, L is split. If H is any Cartan subalgebra of the $\mathbb{C}$-algebra L, a maximal $\mathbb{R}$-split torus T in L is obtained as the $\mathbb{R}$-span of a basis $h_1, \ldots, h_r$ for H over $\mathbb{C}$, such that $\alpha(h_i) \in \mathbb{Z}$ for all roots α of L relative to H, and our construction of an $\mathbb{R}$-split subalgebra S, at the beginning of Chapter III, is such that $L \stackrel{\sim}{=} S \otimes_{\mathbb{R}} \mathbb{C}$. In the listings below, this situation yields one real simple algebra for each real split simple S, i.e., for each connected Dynkin diagram, or for each (reduced) indecomposable system of roots. We include these cases in the list without further comment, and confine all subsequent considerations to the case where the centroid of L is $\mathbb{R}$.

   b) **Compact algebras**.

   To make our classification complete, as well as for use in some cases of rank one, we give a brief resumé of some results on compact Lie algebras and on maximal compact subalgebras of real semisimple Lie algebras. It will be recalled that L is **compact** if its Killing form is negative-definite, and L will be called **anisotropic** if the only $X \in L$ with ad X nilpotent is $X = 0$. We further recall that a **real form** of a complex Lie algebra L is a real subalgebra M such that the obvious map $M \otimes_{\mathbb{R}} \mathbb{C} \to L$ is an isomorphism of complex Lie algebras, **i.e.** an $\mathbb{R}$-basis for M is a $\mathbb{C}$-basis for L.

   The following result on the existence of compact real forms of every complex semi-simple Lie algebra, is due to Weyl; a proof is given in [J], pp. 147-149.

Theorem on the Existence of Compact Real Forms: Let L be a semi-simple Lie algebra over $\mathbb{C}$. Let H be a Cartan subalgebra, $\{L_\gamma\}$ the corresponding root spaces, $H_\gamma \in [L_\gamma L_{-\gamma}]$ as usual. Then there is an automorphism σ of period two of L as Lie algebra over $\mathbb{R}$, satisfying $(\lambda x)^\sigma = \bar\lambda x^\sigma$ for all $\lambda \in \mathbb{C}$, $x \in L$, and such that the fixed elements of

σ form a compact real form M of L. Moreover, σ may be so chosen that, relative to a set of root-vectors $\{e_\gamma\}$ such that $\gamma([e_\gamma e_{-\gamma}]) = 2$ for all γ, we have $e_\gamma^\sigma = e_{-\gamma}$; thus σ stabilizes each complex 3-dimensional subalgebra $L_\gamma + \mathbb{C}H_\gamma + L_{-\gamma}$.

It is immediate that if L is compact, then L is anisotropic; for if $(\text{ad } X)^k \neq 0$, $(\text{ad } X)^{2k} = 0$, we have $0 \neq Y(\text{ad } X)^k$, but $0 = B(Y(\text{ad } X)^{2k}, Y) = \pm B(Y(\text{ad } X)^k, Y(\text{ad } X)^k)$, for suitable $Y \in L$, where B is the Killing form. That is, whenever the Killing form has Witt index 0, L is anisotropic. We now establish the converse, i.e., that an anisotropic semisimple Lie algebra L over $\mathbb{R}$ is compact, as well as the <u>uniqueness</u> of compact real M with $M \otimes_\mathbb{R} \mathbb{C} \cong_\mathbb{C} L$, where now L is a given complex semisimple algebra. The idea is due to E. Cartan; we follow S. Helgason, <u>Differential Geometry and Symmetric Spaces</u>, pp. 152-159.

Proposition a. <u>Let L be a real semisimple Lie algebra, and let σ be the complex conjugation of $L \otimes_\mathbb{R} \mathbb{C}$ fixing L. Let M be a real form of $L \otimes \mathbb{C}$ with negative-definite Killing form. Then there is an automorphism ϕ of $L \otimes \mathbb{C}$ such that $M^{\phi\sigma} \subseteq M^\phi$.</u>

Proof. By the notion of real form, there is a unique complex conjugation τ of $L \otimes \mathbb{C}$ fixing M. Then $\nu = \tau\sigma$ is an automorphism over $\mathbb{C}$ of $L_\mathbb{C} \equiv L \otimes \mathbb{C}$. Let $\{X_i\}$ be a basis for M, and let B be the Killing form of $L_\mathbb{C}$, which extends those of L and of M. Then $(B(X_i, X_j))$ is a symmetric real negative-definite matrix, and setting $g(X,Y) = B(X, Y^\tau)$ for $X, Y \in L_\mathbb{C}$ gives $(g(X_i, X_j)) = (B(X_i, X_j))$, and g is a negative-definite hermitian form on $L_\mathbb{C}$. Moreover we have $g(X^\nu, Y) = B(X^\nu, Y^\tau) = B(X, Y^{\tau\nu^{-1}}) = B(X, Y^{\tau\sigma\tau}) = B(X, Y^{\nu\tau}) = g(X, Y^\nu)$ for all $X, Y \in L_\mathbb{C}$. Thus ν is a nonsingular hermitian transformation of $L_\mathbb{C}$ with respect to g.

It follows that there is an orthonormal basis $\{Y_i\}$ (relative to g) for $L_\mathbb{C}$, and scalars $\lambda_i \neq 0$ in $\mathbb{R}$ such that $Y_i^\nu = \lambda_i Y_i$ for every i. Thus $Y_i^{(\nu^2)} = \lambda_i^2 Y_i$, $\lambda_i^2 > 0$. With $\rho = \nu^2$, we may now define ρ^t for every $t \in \mathbb{R}$ to be that endomorphism of $L_\mathbb{C}$ sending Y_i to $(\lambda_i^2)^t Y_i$ for every i, where the positive real value of $(\lambda_i^2)^t$ is taken. Clearly all ρ^t commute, and commute with ν. Moreover, if $[Y_i, Y_j] = \sum_k c_{ijk} Y_k$, we have $[Y_i^\nu, Y_j^\nu] = \sum_k c_{ijk} \lambda_i \lambda_j X_k = \sum_k c_{ijk} \lambda_k Y_k$, from

290

which $c_{ijk} = 0$ whenever $\lambda_i \lambda_j \neq \lambda_k$. It follows that ρ^t is an auto-morphism of $L_{\mathbb{C}}$ for every t; in particular $\phi = \rho^{1/4}$ is an automorphism, and M^ϕ is the set of fixed elements of $\phi^{-1}\tau\phi$.

The proposition will follow if we show that $\phi^{-1}\tau\phi$ commutes with σ. Now $Y_i^\tau = \Sigma_j d_{ij} Y_j$, $Y_i^{\phi^{-1}\tau} = |\lambda_i|^{-\frac{1}{2}} \Sigma_j d_{ij} Y_j$, $Y_i^{\tau\phi} = \Sigma_j d_{ij} |\lambda_j|^{\frac{1}{2}} Y_j$. On the other hand, $Y_i^{\nu^{-1}\tau} = Y_i^\sigma = Y_i^{\tau\nu} = \Sigma_j d_{ij} \lambda_j Y_j$, while $Y_i^{\nu^{-1}\tau} = (\lambda_i^{-1} Y_i)^\tau = \Sigma_j \lambda_i^{-1} d_{ij} Y_j$. Thus $\lambda_i^{-1} = \lambda_j$ if $d_{ij} \neq 0$, and hence $|\lambda_i|^{-\frac{1}{2}} = |\lambda_j|^{\frac{1}{2}}$ if $d_{ij} \neq 0$, from which $\phi^{-1}\tau = \tau\phi$. But then $\phi^{-1}\tau\phi\sigma = \phi^{-2}\tau\sigma = \rho^{-\frac{1}{2}}\nu$, which sends Y_i to $\text{sgn}(\lambda_i) Y_i$ for each i; therefore $\phi^{-1}\tau\phi\sigma$ is its own inverse. But $(\sigma\phi^{-1}\tau\phi)(\phi^{-1}\tau\phi\sigma) = 1$, since $\sigma^2 = \tau^2 = 1$, and therefore $\sigma\phi^{-1}\tau\phi = \phi^{-1}\tau\phi\sigma$. This completes the proof.

With L and σ as above, a <u>Cartan decomposition</u> of L is a direct sum decomposition $L = K + P$ of the real vector space L, such that for some compact real form M of $L_{\mathbb{C}}$ with $M^\sigma \subseteq M$, we have

$$K = L \cap M, \quad P = L \cap iM.$$

Such a decomposition always exists. For with M and ϕ as in Proposition a, both M^ϕ and iM^ϕ are σ-stable, and $L_{\mathbb{C}} = M + iM = M^\phi + iM^\phi$. Since $M^\phi \tilde{=}_{\mathbb{R}} M$, M^ϕ is a compact real form of $L_{\mathbb{C}}$, and the fixed elements of σ(viz. L) are the direct sum of the σ-fixed elements of M^ϕ and those of iM^ϕ. Thus M^ϕ supplies the Cartan decomposition of L.

With $L = K + P$ a Cartan decomposition of L supplied by M, as in the definition, let τ be the complex conjugation of $L_{\mathbb{C}}$ fixing M. As before, the form $g(X,Y) = -B(X,Y^\tau)$ on $L_{\mathbb{C}}$ is <u>positive</u>-definite hermitian. If $Z \in P$, then for all $X, Y \in L_{\mathbb{C}}$, $g([XZ],Y) = -B([X,Z],Y^\tau) = -B(X,[ZY^\tau]) = B(X,[Z^\tau Y^\tau]) = B(X,[ZY]^\tau) = -B(X,[YZ]^\tau) = g(X,[YZ])$ (since $Z^\tau = -Z$), or ad Z is <u>hermitian</u> with respect to g. Thus ad Z is semisimple with all characteristic roots in $\mathbb{R}$.

If $P \neq 0$, we have now seen that L contains a non-trivial split toral subalgebra, and is not anisotropic. This proves the first statement of:

Proposition b. Let L be a real semisimple Lie algebra. Then L is anisotropic if and only if L is compact. If L and M are compact real forms of the same complex semisimple algebra, then L and M are isomorphic.

Proof: (of last assertion). Take notations as above, with L, M as subalgebras of $L_{\mathbb{C}}$. With ϕ as in Proposition a, we have $L = (M^\phi \cap L) \oplus (iM^\phi \cap L)$. The remarks proving the first part of the theorem show that $M^\phi = L$.

Combined with the existence theorem, Proposition b shows that we have a one-one correspondence between anisotropic real simple Lie algebras and complex simple Lie algebras, the passage from the latter to the former by Weyl's construction. A list of realizations of compact real simple algebras is given below.

c) Real simple algebras of type $A_r (r \geq 2)$.

By remarks in a), we may concentrate on central simple algebras. By Theorem III.3 and §V.1, these are coordinatized for each $r \geq 3$ by a central real division algebra, hence either $\mathbb{R}$ or the quaternions $\mathbb{H}$, while for $r = 2$, the alternative division algebra $\mathbb{O}$ of Cayley must be allowed. Thus we have:

Proposition c. Let L be a real simple Lie algebra of type A_r. If $r > 2$, L is isomorphic to one of the following:

i) $M'_{r+1}(\mathbb{R})$; ii) $M'_{r+1}(\mathbb{C})$; iii) $M'_{r+1}(\mathbb{H})$,

where $M'_{r+1}(A)$ is the derived algebra of the $(r+1) \times (r+1)$ A-matrices. For $r = 2$, we have in addition to i), ii), iii) the algebra $D \oplus (S \otimes \mathbb{O})$ of Theorem III.3 , where $S \overset{\sim}{=} M_3'(\mathbb{R})$, $\mathbb{O}$ is the division algebra of Cayley's octonions and D is the derivation algebra of $\mathbb{O}$, a compact Lie algebra of complex type G_2.

d) Real simple algebras of type $D_r (r \geq 4)$.

By Theorem III.1 these are coordinatized by a field extension of $\mathbb{R}$, hence by $\mathbb{R}$ or $\mathbb{C}$. Since the split algebra of type D_r is realized as skew transformations with respect to a form of maximal index in $2r$ variables ([J], p 141), we have:

Proposition d). Let L be a real simple Lie algebra of type $D_r (r \geq 4)$. Then L is isomorphic to one of the following:

i) $K(\mathbb{R},2r,r)$; ii) $K(\mathbb{C},2r,r)$,

where $K(A,2r,r)$ is the Lie algebra of A-endomorphisms of a 2r-dimensional A-space which are skew with respect to a symmetric bilinear form having matrix, relative to a suitable (orthogonal) basis $\text{diag}\{I_r,-I_r\}$, where I_r is the r by r identity matrix.

e) <u>Real simple Lie algebras of type</u> $E_r (r = 6,7,8)$.

Again Theorem III.1 shows that these are coordinatized by $\mathbb{R}$ or $\mathbb{C}$. One may realize them by using Serre's presentation, one of the versions of Tits' second construction, or others of the constructions mentioned earlier. The choices in the following proposition are aimed at uniformity, rather than at maximum possible convenience.

<u>Proposition e.</u> <u>Let</u> L <u>be a real simple Lie algebra of type</u> E_r ($r = 6,7$ or 8). <u>Then</u> L <u>is isomorphic to one of the following</u>:

$r = 6$: $V(\mathbb{R} \oplus \mathbb{R}, J(\mathbb{R}))$, $V(\mathbb{C} \oplus \mathbb{C}, J(\mathbb{C}))$;

$r = 7$: $V(M_2(\mathbb{R}), J(\mathbb{R}))$, $V \cdot (M_2(\mathbb{C}), J(\mathbb{C}))$;

$r = 8$: $V(O(\mathbb{R}), J(\mathbb{R}))$, $V(O(\mathbb{C}), J(\mathbb{C}))$;

here $J(K)$ is the split exceptional Jordan algebra over the field K, and $O(K)$ is the split nonassociative alternative simple algebra over K. The algebra $V(A,J)$ is that resulting from the second construction of Tits, as in the table on p. 98 of [JE].

f) <u>Real simple Lie algebras of type</u> $B_r (r \geq 3)$.

By Proposition V.2 of §3, these are coordinatized by (finite) field extensions A of $\mathbb{R}$ and finite-dimensional A-vector spaces B carrying a nondegenerate symmetric bilinear form f, not representing 1. When $A = \mathbb{C}$, we have a split algebra of type B_r over $\mathbb{C}$, and $B = 0$. When $A = \mathbb{R}$, the form f must be negative-definite on B. With $u_{r+1} \in M'$ as in §3, the form β of Proposition V.2 is positive-definite on $\mathbb{R}u_{r+1} + B \subseteq M' \oplus B$, a space on which β has Witt index r, and such that $L \cong K(M' \oplus B)$, the Lie algebra of β-skew transformations. We give these conclusions final form:

<u>Proposition f.</u> <u>Let</u> L <u>be a real simple Lie algebra of type</u> $B_r (r \geq 3)$. <u>Then</u> L <u>is isomorphic to one of the following</u>:

i) $K(\mathbb{C},2r+1,r)$;   ii) $K(\mathbb{R},n,r)(n > 2r)$;

here $K(F,n,r)$ is the Lie algebra of all F-linear endomorphisms of an n-dimensional F-vector space with respect to a symmetric bilinear form having matrix $\text{diag}\{I_{n-r}, -I_r\}$, relative to a suitable basis.

g) Real simple Lie algebras of type F_4.

Since $\mathbb{R}$, $\mathbb{C}$, $\mathbb{H}$ and $\mathbb{O}$ are the only real composition division algebras, we see by the above and the considerations of §4 that we have:

Proposition g. Let L be a real simple Lie algebra of type F_4. Then L is isomorphic to one of the following, all resulting from Tits' second construction (see e) above):

$$V(\mathbb{C}, J(\mathbb{C})) \; ; \quad V(\mathbb{R}, J(\mathbb{R})) \; ; \quad V(\mathbb{C}, J(\mathbb{R})) \; ;$$
$$V(\mathbb{H}, J(\mathbb{R})) \; ; \quad V(\mathbb{O}, J(\mathbb{R})) \; ,$$

of respective real dimensions 104, 52, 78, 133, 248.

h) Real simple Lie algebras of type G_2.

From §5, the central ones among these are coordinatized by central Jordan division algebras J over $\mathbb{R}$ satisfying a cubic polynomial. On the other hand, no element of J can have a cubic minimum polynomial, and this eliminates the possibilities ii)-v) listed in §5. We therefore have:

Proposition h. Let L be a real simple Lie algebra of type G_2. Then L is isomorphic to one of the following cases of Tits' second construction:

$$V(O(\mathbb{C}), \mathbb{C}) \; ; \quad V(O(\mathbb{R}), \mathbb{R}) \; ;$$

i.e., L is the derivation algebra of the split octonion algebra $O(\mathbb{C})$ or $O(\mathbb{R})$, in either case regarded as $\mathbb{R}$-Lie algebra.

i) Real simple Lie algebras of type $C_r (r \geq 3)$.

Except when $r = 3$, we see from §7 a) that the central ones among these are coordinatized by giving an involutorial associative division algebra $(E, *)$ over $\mathbb{R}$ with $\mathbb{R}$ as fixed elements of the center. The Lie algebra L is then the derived algebra $K'(M'' \otimes E) = K'((E,*), 2r, r)$ of the Lie algebra of skew E-endomorphisms of the 2r-dimensional right E-vector space $M'' \otimes E$ with respect to the (unique) nondegenerate anti-hermitian form of Witt index r. The possibilities for $(E,*)$ (to within isomorphism of the Lie algebras) are: $(\mathbb{R},\mathrm{id})$; $(\mathbb{C},-)$; $(\mathbb{H},-)$; $(\mathbb{H},h \mapsto i^{-1}\bar{h}i)$, where $b \mapsto \bar{b}$ is the usual conjugation in $\mathbb{C}$ or $\mathbb{H}$, and i is one of the usual quaternionic units, $i^2 = -1$. (For the involutions in $\mathbb{H}$ are either $-$ or of the form $h \mapsto b^{-1}\bar{h}b$, $\bar{b} = -b \ne 0$. But if V is a right $\mathbb{H}$-vector space, b as above, and (v,v') a nondegenerate anti-hermitian form on V with respect to the involution $h \mapsto b^{-1}\bar{h}b$, then $< v,v' > = i^{-1}b(v,v') = -ib(v,v')$ is an antihermitian form with respect to the involution $h \mapsto i^{-1}\bar{h}i$, and $K(V,< v,v' >) = K(V,(v,v'))$.) When $r = 3$, we have the additional case of §7b), where $(E,*) = (\mathbb{O},-)$, where L results from Tits' second construction applied to $\mathbb{O}$ and to the cubic Jordan algebra $H((\mathbb{R},\mathrm{id}),6,3)$ of symmetric endomorphisms of $\mathbb{R}^6$ with respect to a nondegenerate alternate form. That is, we have:

Proposition i. Let L be a real simple Lie algebra of type $C_r (r \ge 3)$. Then L is isomorphic to one of the following:

$$K((\mathbb{C},\mathrm{id}.),2r,r); \quad K((\mathbb{R},\mathrm{id}.),2r,r) \;;$$
$$K'((\mathbb{C},-),2r,r); \quad K((\mathbb{H},-),2r,r) \;;$$
$$K((\mathbb{H},h \to i^{-1}\bar{h}i),2r,r) \;;$$
$$V(\mathbb{O},H((\mathbb{R},\mathrm{id}.),6,3)),$$

the last only for $r = 3$.

(Here we have written K rather than K' when these are known to be equal.)

j) Real simple Lie algebras of type $BC_r (r \ge 3)$.

From §7 a), we see that these are coordinatized by involutorial associative division algebras $(E,*)$ as in i), except that here the centroid cannot be $\mathbb{C}$ (see a) above); and by right E-vector spaces C of positive dimension and anisotropic antihermitian forms on these. When $(E,*) = (\mathbb{R},\mathrm{id})$, we are speaking of alternate forms, none of which is anisotropic, so only $(\mathbb{C},-)$, $(\mathbb{H},-)$ and $(\mathbb{H},h \to i^{-1}\bar{h}i)$ remain as possibilities for $(E,*)$. In the case of $(\mathbb{C},-)$, any nondegenerate antihermitian form has a diagonal matrix $\mathrm{diag}\{\lambda_1 i, \lambda_2 i, \ldots, \lambda_m i\} (\lambda_j \in \mathbb{R})$ relative to a suitable basis $v_1, \ldots, v_m$. Replacing v_j by $v_j \tau (\tau \in \mathbb{C})$ replaces $\lambda_j i = (v_j, v_j)$ by $\bar{\tau} \lambda_j i \tau = \lambda_j \bar{\tau} \tau i$, and enables us to assume $\lambda_j = \pm 1$ for every j. If two of these have different signs, the form is isotropic, so we may assume all $\lambda_j = 1$ or all $\lambda_j = -1$. Replacing the form on $(M'' \otimes E) \oplus C$ of §7 a) by its negative if necessary, we may assume all $\lambda_j = 1$.

In the case of $(\mathbb{H},-)$, the only nondegenerate anisotropic anti-hermitian space has H-dimension one (for it is a relatively easy matter to check that if $b, c \in \mathbb{H}$ are non-zero with $\bar{b} = -b$, $\bar{c} = -c$, then there is $a \in \mathbb{H}$ with $\bar{a}ba = c$; then if V is a space as above, and if u,v are linearly independent and orthogonal in V, with $(u,u) = b$, $(v,v) = -c$, then $(ua + v, ua + v) = \bar{a}ba - c = 0$.) Moreover, we may assume the matrix of the form on this space is (i). When $(E,*) = (\mathbb{H}, h \mapsto i^{-1}\bar{h}i)$, our anisotropic anti-hermitian space has an orthogonal basis $v_1, \ldots, v_m$ with $(v_j, v_j) = \lambda_j i$, $\lambda_j \in \mathbb{R}$. As before we may take $\lambda_j = \pm 1$ for all j, and the sign must be constant. Replacing the form by its negative if necessary, we may assume all $\lambda_j = 1$.

With these normalizations we have the

<u>Proposition j</u>. <u>Let</u> L <u>be a real simple Lie algebra of type</u> $BC_r (r \geq 3)$. <u>Then</u> L <u>is isomorphic to one of the following</u>:

$$K((\mathbb{C},-), \text{diag}\{iI_{m+r}, -iI_r\}), m > 0 ;$$

$$K((\mathbb{H}, h \mapsto i^{-1}\bar{h}i), \text{diag}\{iI_{m+r}, -iI_r\}), m > 0 ;$$

$$K((\mathbb{H},-), \text{diag}\{iI_{r+1}, -iI_r\}) .$$

(Here the notation $K((E,*), (\mu_{jk}))$ refers to the Lie algebra of skew transformations with respect to an antihermitian form (relative to $*$) on a right E-vector space, the matrix of that form relative to a suitable basis being (μ_{ij}).)

k) <u>Real simple Lie algebras of types</u> B_2 <u>and</u> C_2.

The bases for our assertions here are to be found in §3 c) and §7 a), from which we conclude that such an algebra is either one of those of Proposition f or one of those of Proposition i, the appropriate value of r being 2. More precisely, we have

<u>Proposition k</u>. <u>Let</u> L <u>be a real simple Lie algebra of type</u> B_2 <u>or</u> C_2. <u>Then</u> L <u>is isomorphic to one of the following</u>:

    i) $K(\mathbb{C}, 5, 2)$;   ii) $K(\mathbb{R}, n, 2)$ $(n > 4)$;
   iii) $K((\mathbb{C}, \text{id.}), 4, 2)$;   iv) $K((\mathbb{R}, \text{id}), 4, 2)$;
    v) $K'((\mathbb{C},-), 4, 2)$;   vi) $K((\mathbb{H},-), 4, 2)$;
   vii) $K((\mathbb{H}, h \mapsto i^{-1}\bar{h}i), 4, 2)$. In fact, L is isomorphic to one and

only one of i), ii), vii).

Here the notations of i), ii) are as in Proposition f, while those of iii) - vii) are as in Proposition i. It will further be noted that i) and iii) are $\mathbb{C}$-isomorphic (as split algebras of type $B_2 = C_2$), and that ii) and iv) are $\mathbb{R}$-isomorphic when $n = 5$ in ii). The algebras iii) and iv) may thus be suppressed. Furthermore, in cases v) and vi), the Jordan algebra of symmetric elements of the involutorial algebra $(\mathbb{C},-)$ or $(\mathbb{H},-)$ is associative, being $\mathbb{R}$. It follows by our breakdown into cases B_2, C_2 that these, as well, have been included under ii), the former for $n = 6$ and the latter for $n = 9$ (by dimensions). Dimensions for L do not exclude an isomorphism between vii) and ii) for $n = 8$, but the multiplicity of the long roots in ii) is one, while in vii) it is three; thus these two cannot be isomorphic. This proves the last assertion.

ℓ) <u>Real simple Lie algebras of type</u> BC_2.

Here we found our conclusions on §7 a) and §7 c). For the real field, the arguments used in suppressing the redundant cases in Proposition k show that we have only the algebras $(\mathbb{H}, h \mapsto i^{-1}\bar{h}i)$ as involutorial algebras to which §7a applies. As with Proposition j, the associated algebra L has the form $K((\mathbb{H}, h \to i^{-1}\bar{h}i), \text{diag}\{iI_{m+2}, -iI_2\}), m > 0$. For the remaining cases, we refer to Theorem V.2. of §7 c). In the cases α), β) of that theorem we are in a situation for $r = 2$ analogous to that of the first resp. third case of Proposition j, and we need only specialize those to $r = 2$ to read off our conclusions.

Finally, we consider γ) - ε) of Theorem V.2. The form h must be negative definite, since it does not represent 1. From this it follows that the center of the Clifford algebra has the required structure in each case. In particular, it is $\mathbb{C}$ in case γ), and therefore $H(B)$ must be the full algebra of 4 by 4 complex matrices in this case. The involution in $H(B)$ which maps each $b \in B$ to its negative is of second kind, hence may be associated with an antihermitian form on $\mathbb{C}^4$, whose real part affords the symplectic form q_1 on this space required in γ). In cases δ), ε), the classical tensor-decomposition results of Brauer and Weyl for the Clifford algebra, applied to a negative-definite form in 7 resp. 11 dimensions show that the even Clifford algebra is $M_8(\mathbb{R})$ resp. $M_{16}(\mathbb{H})$, contrary to our conditions. We therefore have:

Proposition ℓ. Let L be a real simple Lie algebra of type BC_2. Then L is isomorphic to one of the following:

$$K'((\mathbb{C},-), \text{diag}\{iI_{m+2},-iI_2\}), m > 0 ;$$

$$K((\mathbb{H},h \to i^{-1}\bar{h}i), \text{diag}\{iI_{m+2},-iI_2\}), m > 0 ;$$

$$K((\mathbb{H},-), \text{diag}\{iI_3,-iI_2\}) ;$$

$$E(\text{diag}\{-1,-1,-1,-1,-1\}), \text{ the 78-\underline{dimensional}}$$

exceptional Lie algebra obtained from a negative-definite form in five dimensions as in γ) of Theorem V.2.

m) Real simple algebras of type A_1.

Although it might have been more consistent with our earlier order to treat these sooner, these are treated here, followed at once by those of type BC_1, in order to collect algebras of rank one for easier reference. From §III.2, these Lie algebras are coordinatized by Jordan division algebras over $\mathbb{R}$, which we may for now assume to be central. We have considered the separate cases in §2 a), §2 c), §3 d), and §7 a). For the real field, there is no exceptional Jordan division algebra, as we observed in treating the type G_2; thus the situation of §2 c) cannot occur. In §2 a), the central associative division algebra must be $\mathbb{R}$ or $\mathbb{H}$, and our Lie algebra L identifies with $M_2'(\mathbb{R})$ or with $M_2'(\mathbb{H})$ by that section. In §3 d), $A = \mathbb{R}$, and the form f on B must be negative-definite. The resulting Lie algebra L is that described, in the notation of Proposition f, as $K(\mathbb{R},n,1)$, $n \geq 4$. Finally, in the cases covered by §7a, the only relevant one is that where the Jordan division algebra A is the fixed elements of an involution $h \mapsto b^{-1}\bar{h}b$, $\bar{b} = -b$, of $\mathbb{H}$; but the square of every element of trace zero in this algebra is in $\mathbb{R}$ and is negative. This case therefore simply repeats the case $[B:\mathbb{R}] = 2$ of §3 d). Likewise, $M_2'(\mathbb{H})$ repeats $[B:\mathbb{R}] = 3$ of §3 d). We therefore have

Proposition m. Let L be a real simple Lie algebra of type A_1. Then L is isomorphic to one of the following:

$$M_2'(\mathbb{C}) ; \qquad M_2'(\mathbb{R}) ;$$

$$K(\mathbb{R},n,1) , n \geq 4.$$

n) **Real simple Lie algebras of type** BC_1.

By a), all these must be central. By §§7 a),§7 d) and remarks of the last section, our Jordan division algebra A may be identified with $\mathbb{R}$ or with the Jordan algebra of a negative-definite quadratic form in at least two dimensions. All of these cases fall under §7 d), and in fact, under §7 d) ii), iii). The results for ii) may be read off from Theorem V.3. Under α) of that theorem, only the case $[A:Z] = 3$ can occur, in which we see from the theorem and from j) above that L is isomorphic to $K((\mathbb{H}, h \mapsto i^{-1}\bar{h}i), \mathrm{diag}\{iI_{m+1}, -iI_1\})$, $m > 0$, in the notations of that section. The cases β) – ϵ) of Theorem V.3 require information on the Clifford algebra of a negative-definite form in 6,7,9 and 13 dimensions. As we have seen in ℓ) above, the center of the Clifford algebra splits in seven dimensions, so that γ) is eliminated. The corresponding considerations show that the center of the Clifford algebra does not split in nine dimensions, thereby eliminating δ); nor does it split in thirteen dimensions, so that ϵ) is eliminated. On the other hand, the Clifford algebra of a real negative-definite form in six dimensions is $M_8(\mathbb{R})$. The main involution is of symplectic type, giving rise to the form q_1 on $\mathbb{R}^8$, and L has the structure discussed under "6:" in §7 d), ii), with underlying space $K(B) \oplus (S \otimes (\mathbb{R} \oplus B)) \oplus (M \otimes \mathbb{R}^8)$, of dimension 52. This algebra is a real form of the complex exceptional algebra F_4.

Finally, let $A = \mathbb{R}$, the case treated in §7 d), iii). Here Theorem V.4 is the basis for our conclusions. Case a) of that theorem requires $K = \mathbb{C}$, and $L \stackrel{\sim}{=} K'((\mathbb{C},-), \mathrm{diag}\{iI_{m+1}, -iI_1\})$, $m > 0$. Case b) requires $Q = \mathbb{H}$, $L \stackrel{\sim}{=} K((\mathbb{H},-), \mathrm{diag}\{iI_2, -iI_1\})$. In case c), we must have $K = \mathbb{C}$, which admits no finite nontrivial extensions U; thus this case is vacuous. In cases d) – h), we look more closely at $L \otimes \mathbb{C} = L \otimes K$, in particular at $D = <C,C>$ and its action on C in this extension, summarizing some details of the table preceding Theorem V.4, and not easily found in the literature. The one case that has been thoroughly studied [Me], [F], [FF], [Fe] is that where $(U \otimes C)^+$ is the exceptional Jordan algebra over $\mathbb{C}$; in this case, $<C,C> \otimes \mathbb{C}$ is the exceptional complex Lie algebra E_7, and the representation on $C \otimes \mathbb{C}$ is the unique 56-dimensional irreducible representation of E_7, with highest weight the fundamental weight ω_1 (in our labelling of the roots as in Appendix A). Thus, in h) of Theorem V.4, the anisotropic real algebra $<C,C>$ is the compact real form of E_7, and C is a real form of this 56-dimensional module. This would mean the existence of a complex conjugation σ in $C \otimes \mathbb{C}$ with C as fixed elements, and such that, if τ is the complex conjugation in E_7 yielding the compact form $<C,C>$, we have $(ux)^\sigma = u^\sigma x^\tau$ for all $u \in C \otimes \mathbb{C}$, $x \in <C,C> \otimes \mathbb{C}$. Now the existence theorem for compact forms and the proof of the second assertion of Proposition b) show that there is a system of roots for $<C,C> \otimes \mathbb{C}$ such that for each root α, τ sends the root-space corresponding to α to that corresponding to $-\alpha$ (and stabilizes the associated Cartan subalgebra). It follows that σ must map the weight-space in $C \otimes \mathbb{C}$ belonging to the highest weight ω_1 to that belonging to the lowest weight (with respect to the same ordering of the roots $\{\alpha\}$). Fixing a highest weight vector $v \neq 0$, we have a sequence $f_{i_1}, \ldots, f_{i_t}$ of root-vectors belonging to the negatives of fundamental roots, such that $0 \neq u = vf_{i_1}^2 \ldots f_{i_t}$ is a lowest weight vector. Then $v^\sigma = \lambda u$, $\lambda \in \mathbb{C}$, and $v = v^{\sigma^2} = \bar\lambda u^\sigma = \lambda\bar\lambda vf_{i_1} \ldots f_{i_t} e_{i_1} \ldots e_{i_t}$. We show that for a sequence $f_{i_1}, \ldots, f_{i_t}$ as above, we have $vf_{i_1} \ldots f_{i_t} e_{i_1} \ldots e_{i_t} = -v$; then no λ can give $v^\sigma = v$, and we have our contradiction. (Here $e_i = f_i^\sigma$.)

First we list all the weights of the representation in question, ordered consistently with decrease in their coefficient-sums for the $\{\alpha_i\}$; we write $\frac{1}{2}(s_1, s_2, \ldots, s_r)$ for $\frac{1}{2}\Sigma s_i \alpha_i$:

$$\omega_1 = \tfrac{1}{2}(3,4,5,6,4,2,3); \quad \tfrac{1}{2}(1,4,5,6,4,2,3);$$

$\tfrac{1}{2}(1,2,5,6,4,2,3); \quad \tfrac{1}{2}(1,2,3,6,4,2,3); \quad \tfrac{1}{2}(1,2,3,4,4,2,3);$

$\tfrac{1}{2}(1,2,3,4,2,2,3); \quad \tfrac{1}{2}(1,2,3,4,4,2,1); \quad \tfrac{1}{2}(1,2,3,4,2,0,3),$

$\tfrac{1}{2}(1,2,3,4,2,2,1); \quad \tfrac{1}{2}(1,2,3,2,2,2,1), \quad \tfrac{1}{2}(1,2,3,4,2,0,1);$

$\tfrac{1}{2}(1,2,1,2,2,2,1), \quad \tfrac{1}{2}(1,2,3,2,2,0,1); \quad \tfrac{1}{2}(1,0,1,2,2,2,1),$

$\tfrac{1}{2}(1,2,1,2,2,0,1), \quad \tfrac{1}{2}(1,2,3,2,0,0,1); \quad \tfrac{1}{2}(-1,0,1,2,2,2,1),$

$\tfrac{1}{2}(1,0,1,2,2,0,1), \quad \tfrac{1}{2}(1,2,1,2,0,0,1); \quad \tfrac{1}{2}(-1,0,1,2,2,0,1),$

$\tfrac{1}{2}(1,0,1,2,0,0,1), \quad \tfrac{1}{2}(1,2,1,0,0,0,1); \quad \tfrac{1}{2}(-1,0,1,2,0,0,1),$

$\tfrac{1}{2}(1,0,1,0,0,0,1), \quad \tfrac{1}{2}(1,2,1,0,0,0,-1); \quad \tfrac{1}{2}(-1,0,1,0,0,0,1),$

$\tfrac{1}{2}(1,0,-1,0,0,0,1), \quad \tfrac{1}{2}(1,0,1,0,0,0,-1);$ and the negatives of these.

Thus each weight has multiplicity one, and there is a sequence of two-

dimensional representations for the three-dimensional subalgebras $<e_i, f_i>$, $1 \leq i \leq 7$ showing that
$0 \neq vf_1f_2f_3f_4f_5f_6f_7f_4f_5f_3f_4f_7f_2f_3f_4f_5f_6f_1f_2f_3f_4f_5f_7f_4f_3f_2f_1 = u$ belongs to the lowest weight $-\omega_1$. From further consideration of these modules, and of the weights to which the partial products belong, we have
$ue_1 = -vf_1\ldots f_2h_1 - vf_1\ldots f_6h_1f_2\ldots f_1 - vh_1f_2\ldots f_1 = -vf_1\ldots f_2$, the second and third terms being zero. Then $ue_1e_2 = vf_1\ldots f_3h_2 +$
$vf_1\ldots f_1h_2f_3\ldots f_2 + vf_1\ldots f_4f_7h_2f_3\ldots f_2 + vf_1h_2f_3\ldots f_2 = vf_1\ldots f_3$, since the last three terms are zero. Repetition yields
$ue_1e_2e_3e_4e_5e_6e_7e_4e_5e_3e_4e_7e_2e_3e_4e_5e_6e_1e_2e_3e_4e_5e_7e_4e_3e_2e_1 = -v$, and the desired contradiction. Thus the case h) cannot occur over $\mathbb{R}$.

Likewise, one verifies in case g) that $<C,C> \otimes \mathbb{C}$ is the split algebra D_6, and that $C \otimes \mathbb{C}$ is a 32-dimensional irreducible module, corresponding to one of the fundamental weights ω_5 or ω_6 in the usual labelling. By symmetry, we may assume the highest weight is $\omega_6 = \frac{1}{2}(1,2,3,4,2,3)$. The remaining weights, ordered as before, are:

$\frac{1}{2}(1,2,3,4,2,1)$; $\frac{1}{2}(1,2,3,2,2,1)$; $\frac{1}{2}(1,2,1,2,2,1)$, $\frac{1}{2}(1,2,3,2,0,1)$;

$\frac{1}{2}(1,0,1,2,2,1)$, $\frac{1}{2}(1,2,1,2,0,1)$; $\frac{1}{2}(-1,0,1,2,2,1)$, $\frac{1}{2}(1,0,1,2,0,1)$,

$\frac{1}{2}(1,2,1,0,0,1)$; $\frac{1}{2}(-1,0,1,2,0,1)$, $\frac{1}{2}(1,0,1,0,0,1)$, $\frac{1}{2}(1,2,1,0,0,-1)$;

$\frac{1}{2}(-1,0,1,0,0,1)$, $\frac{1}{2}(1,0,-1,0,0,1)$, $\frac{1}{2}(1,0,1,0,0,-1)$;

$\frac{1}{2}(-1,0,-1,0,0,1),\frac{1}{2}(1,0,-1,0,0,-1),\frac{1}{2}(-1,0,1,0,0,-1)$;

$\frac{1}{2}(-1,-2,-1,0,0,1)$, $\frac{1}{2}(-1,0,-1,0,0,-1)$; $\frac{1}{2}(1,0,-1,-2,0,-1)$;

$\frac{1}{2}(-1,-2,-1,0,0,-1),\frac{1}{2}(-1,0,-1,-2,0,-1),\frac{1}{2}(1,0,-1,-2,-2,-1)$;

$\frac{1}{2}(-1,-2,-1,-2,0,-1)$, $\frac{1}{2}(-1,0,-1,-2,-2,-1)$; $\frac{1}{2}(-1,-2,-3,-2,0,-1)$,

$\frac{1}{2}(-1,-2,-1,-2,-2,-1)$; $\frac{1}{2}(-1,-2,-3,-2,-2,-1)$; $\frac{1}{2}(-1,-2,-3,-4,-2,-1)$;

$\frac{1}{2}(-1,-2,-3,-4,-2,-3) = -\omega_6$.

With our conjugations as before, v a vector belonging to the highest weight ω_6, we have u belonging to $-\omega_6$, where
$u = vf_6f_4f_3f_2f_1f_5f_4f_3f_2f_6f_4f_3f_5f_4f_6$, and
$ue_6e_4e_3e_2e_1e_5e_4e_3e_2e_6e_4e_3e_5e_4e_6 = -v$. As in case h), we now see that g) is impossible over $\mathbb{R}$.

Since K must be the complex numbers, no extension L of K as in f) can exist; thus f) is impossible. In e), we must have $<C,C> \otimes \mathbb{C}$ a complex Lie algebra of type A_5, with $C \otimes \mathbb{C}$ an irreducible module of dimension 20. One easily sees that this module must be the one with highest weight ω_3 in the usual labelling, and weights

$\omega_3 = \frac{1}{2}(1,2,3,2,1); \frac{1}{2}(1,2,1,2,1); \frac{1}{2}(1,0,1,2,1), \frac{1}{2}(1,2,1,0,1);$
$\frac{1}{2}(-1,0,1,2,1), \frac{1}{2}(1,0,1,0,1), \frac{1}{2}(1,2,1,0,-1); \frac{1}{2}(-1,0,1,0,1), \frac{1}{2}(1,0,-1,0,1),$
$\frac{1}{2}(1,0,1,0,-1)$; and their negatives. The lowest weight is $-\omega_3$, and if v belongs to ω_3, then $vf_3f_2f_1f_4f_5f_3f_2f_4f_3 = u$ belongs to $-\omega_3$. But $ue_3e_2e_1e_4e_5e_3e_2e_4e_3 = -v$, and we see as in cases g) and h) that e) is impossible over $\mathbb{R}$.

Finally, in d) one checks that $<C,C> \otimes \mathbb{C}$ is a 21-dimensional simple Lie algebra, hence a B_3 or C_3, and that $C \otimes \mathbb{C}$ is a 14-dimensional irreducible module. But B_3 has no such module, so we must have C_3; both the irreducible module with highest weight ω_2 and that with highest weight ω_3 have dimension 14; explicit computation of the structure in the complex case shows that our module is that with highest weight $\omega_3 = (1,2,\frac{3}{2})$. Letting v belong to this weight, we find that $u = vf_3f_2f_1f_2f_3f_2f_1f_2f_3$ belongs to the lowest weight $-\omega_3$, and that $ue_3e_2e_1e_2e_3e_2e_1e_2e_3 = -4v$. It now follows as before that it is impossible for $<C,C>$ to be anisotropic in this case. Thus d) is eliminated, and only cases a) and b) of Theorem V.4 remain over the reals. We therefore have

<u>Proposition n</u>. <u>Let</u> L <u>be a real simple Lie algebra of type</u> BC_1. <u>Then</u> L <u>is isomorphic to one of the following</u>:

$K((\mathbb{H}, h \mapsto i^{-1}\overline{h}i), \text{diag}\{iI_{m+1}, -iI_1\}), m > 0;$

$K'((\mathbb{C},-), \text{diag}\{iI_{m+1}, -iI_1\}), m > 0;$

$K((\mathbb{H},-), \text{diag}\{i,i,-i\});$

<u>the</u> 52-<u>dimensional</u> $K(\mathcal{B}) \oplus (S \otimes (\mathbb{R} \oplus \mathcal{B})) \oplus (M'' \otimes \mathbb{R}^8)$ <u>of Theorem</u> V.3 β), <u>where</u> $A = \mathbb{R} \oplus \mathcal{B}$ <u>and the quadratic form on the</u> 6-<u>dimensional space</u> $\mathcal{B}$ <u>is negative-definite</u>.

o) <u>Compact real simple algebras</u> ("type ϕ")

In the spirit of the above, we complete the list of real simple Lie algebras by giving more "constructive" descriptions of the simple ones than those given in a).

<u>Proposition o</u>. <u>Let</u> L <u>be a compact real simple Lie algebra</u>. <u>Then</u> L <u>is isomorphic to exactly one of the following</u>:

$A_r(r \geq 1)$: $K'((\mathbb{C},-), iI_{r+1})$;

$B_r(r \geq 2)$: $K(\mathbb{R}, 2r+1, 0)$;

$C_r(r \geq 3)$: $K((\mathbb{H}, h \mapsto i^{-1}\bar{h}i), iI_r)$;

$D_r(r \geq 4)$: $K(\mathbb{R}, 2r, 0)$;

G_2 : $Der(\mathbb{O})$, the derivations of Cayley's octonions;

F_4 : $Der(\mathbb{J})$, $\mathbb{J}$ the formally real exceptional Jordan algebra discovered by Jordan, von Neumann and Wigner;

E_6 : $V(\mathbb{C}, \mathbb{J})$;

E_7 : $V(\mathbb{H}, \mathbb{J})$;

E_8 : $V(\mathbb{O}, \mathbb{J})$.

The labels at the left refer to the complex type. The algebras $V(\mathbb{C},\mathbb{J})$, etc., apply to Tits' second construction as in [JE], where verifications that these are the compact forms in the exceptional cases can be found. The notations for the classical cases are as previously established; thus an antihermitian form is indicated in A, C and a symmetric anisotropic bilinear real form in B, D.

APPENDIXES

Appendix A. Computation of Multiplicities.

The principal tool will be Freudenthal's formula, giving the multiplicity of a weight in an irreducible representation of a split semi-simple Lie algebra in terms of the multiplicities of the weights lying above it. We follow a program for computer given by Krusemeyer [BIT = Nordisk Tidskrift for Infromationsbehandling 11 (1971),310-316] for determining the multiplicities of the dominant weights using this formula. We also use a routine of Krusemeyer for finding the multiplicity of a general weight by showing that it is congruent under the Weyl group to a dominant weight above it, whose multiplicity is already known, or to a linear function which is a non-weight by virtue of lying above the highest weight (in at least one coefficient of a fundamental root).

Thus let L be a split simple Lie algebra over F, H a splitting Cartan subalgebra, and $\{\alpha_1,\ldots,\alpha_r\} = \Pi$ a fundamental system of roots relative to H. We label Π as follows, according to its diagram:

A_r: •—•—···—•—• ; for $\Sigma \xi_i \alpha_i \in H^*$, write $(\xi_1,\ldots,\xi_r)$;
     $\alpha_1\ \alpha_2\ \ \alpha_{r-1}\ \alpha_r$

B_r: •—•—···—•⇒• ; for $\Sigma \xi_i \alpha_i \in H^*$, write $(\xi_1,\ldots,\xi_r)$;
     $\alpha_1\ \alpha_2\ \ \alpha_{r-1}\ \alpha_r$

C_r: •—•—···—•⇐• ; for $\Sigma \xi_i \alpha_i \in H^*$, write $(\xi_1,\ldots,\xi_r)$;
     $\alpha_1\ \alpha_2\ \ \alpha_{r-1}\ \alpha_r$

$D_r (r \geq 4)$ •—•—···—•⟨ α_{r-1} / α_r ; for $\Sigma \xi_i \alpha_i$, write $(\xi_1,\ldots,\xi_{r-1},\xi_r)$
     $\alpha_1\ \alpha_2\ \ \alpha_{r-2}$

G_2: •⇛• ; for $\xi_1 \alpha_1 + \xi_2 \alpha_2$ write (ξ_1,ξ_2);
     $\alpha_1\ \alpha_2$

F_4: •—•⇒•—• ; for $\Sigma \xi_i \alpha_i$, write $(\xi_1,\xi_2,\xi_3,\xi_4)$;
     $\alpha_1\ \alpha_2\ \alpha_3\ \alpha_4$

$E_r (r = 6,7,8)$

•—•—···—•—•—• for $\Sigma \xi_i \alpha_i$,
$\alpha_1\ \alpha_2\ \ \alpha_{r-3}\ \alpha_{r-2}\ \alpha_{r-1}$
    with α_r branching up from α_{r-2}

write $(\xi_1, \ldots, \xi_r)$.

We assume a positive definite scalar product on the rational span $\mathbb{Q}\Pi$ of the roots, denoted by (λ,μ), invariant under the Weyl group, and with $(\alpha_i,\alpha_i) = 1$ if α_i is a <u>short</u> root in the above; thus $(\alpha_r,\alpha_r) = 1$ except in case C_r, where $(\alpha_1,\alpha_1) = 1$. With this normalization, the whole scalar product can be read off from the diagram.

The formula of Freudenthal is

$$2 \sum_{\substack{n=1 \\ \alpha > 0}}^{\infty} m_{\lambda+n\alpha} (\lambda+n\alpha,\alpha) = ((\hat{\lambda} + \delta, \hat{\lambda} +\delta) - (\lambda + \delta,\lambda + \delta))m_\lambda,$$

where m_μ is the multiplicity of the weight $\mu(\neq \hat{\lambda})$ in the irreducible representation of L (now assumed simple) of highest weight $\hat{\lambda}$ (relative to our system Π); the scalar product is as above; "$\alpha > 0$" means α is a non-negative integral combination of Π, and $\delta = \frac{1}{2} \sum_{\alpha > 0} \alpha$ is the unique element of $\mathbb{Q}\Pi$ satisfying $2(\delta,\alpha_i) = (\alpha_i,\alpha_i)$ for all i. We always have $m_{\hat{\lambda}} = 1$.

An element $\mu \in \mathbb{Q}\Pi$ is <u>dominant</u> if $(\mu,\alpha_i) \geq 0$ for all i; a dominant μ is a weight of the representation of highest weight $\hat{\lambda}$ only if μ is of the form $\hat{\lambda} - \Sigma m_i \alpha_i$, m_i non-negative integers. Indeed, this is true for <u>every</u> $\mu \in \mathbb{Q}\Pi$, while for <u>dominant</u> $\mu \in \mathbb{Q}\Pi$ the condition is also sufficient. (We do not really use the latter fact here, and therefore omit a proof.) Now let $\mu = \hat{\lambda} - \Sigma m_i \alpha_i$ as above; if $(\mu,\alpha_i) \geq 0$ for all i, then μ is dominant; if not, let i be the first index with $(\mu,\alpha_i) < 0$. Then the Weyl reflection $w_i = w_{\alpha_i}$ sends μ to $\mu - \frac{2(\mu,\alpha_i)}{(\alpha_i,\alpha_i)} \alpha_i$, a weight if and only if μ is, and with all coefficients the same as in μ, except for that of α_i; we repeat this process starting with μ^{w_i}, eventually either arriving at a dominant λ of our form, or at an element of $\mathbb{Q}\Pi$ with some coefficient larger than that in $\hat{\lambda}$. In the latter case, this element is not a weight, hence μ is not, and $m_\mu = 0$. In the former case, the multiplicity of λ as a weight is equal to that of μ, and the computation of m_μ reduces to that of m_λ.

§A-1. **Multiplicities of dominant weights**.

In the case of type $A_r (r \geq 1)$, we are interested only, for $r \geq 3$, in $\lambda = 2(\omega_1 + \omega_r) = (2,2,\ldots,2); \lambda = \omega_2 + 2\omega_r = (1,2,\ldots,2);$ $\lambda = 2\omega_1 + \omega_{r-1} = (2,2,\ldots,2,1); \lambda = \omega_2 + \omega_{r-1} = (1,2,\ldots,2,1);$ $\lambda = \omega_1 + \omega_r = (1,1,\ldots,1); \lambda = 0.$ For $r = 2$, we omit $\omega_2 + \omega_{r-1}$ from this list; for $r = 1$, we consider $\lambda = j\omega_1 = \frac{j}{2}\alpha_1$, $0 \leq j \leq 8$. In each case, it is easily seen (probably most easily by consulting Planche I, particularly (VI), of Bourbaki [Bo], p. 250) that for each $\hat{\lambda}$ in this list, the list contains all those dominant functions λ of the form $\hat{\lambda} - \Sigma m_i \alpha_i$ as above. In the tables below, the row with index $\hat{\lambda}$ meets the column with index λ in an entry which is the multiplicity $m_\lambda(\hat{\lambda})$ of λ as a weight in the irreducible representation of highest weight $\hat{\lambda}$. The number $d_{\hat{\lambda}}$ is the degree of the representation with highest weight $\hat{\lambda}$, computed directly from Weyl's formula as in [J], pp. 257 ff. The number $c_{\hat{\lambda}}$ is $(\hat{\lambda} + \delta, \hat{\lambda} + \delta) - (\delta, \delta)$.

$r \geq 3$:

$\hat{\lambda}$	$d_{\hat{\lambda}}$	$c_{\hat{\lambda}}$	0	$\omega_1+\omega_r$	$\omega_2+\omega_{r-1}$	$2\omega_1+\omega_{r-1}$	$\omega_2+2\omega_r$	$2(\omega_1+\omega_r)$
$2(\omega_1+\omega_r)$	$\frac{r(r+1)^2(r+4)}{4}$	$2(r+2)$	$\frac{r(r+1)}{2}$	r	1	1	1	1
$\omega_2+2\omega_r$	$\frac{(r-1)r(r+2)(r+3)}{4}$	$2(r+1)$	$\frac{(r-1)r}{2}$	$r-1$	1	0	1	0
$2\omega_1+\omega_{r-1}$			$\frac{(r-1)r}{2}$	$r-1$	1	1	0	0
$\omega_2+\omega_{r-1}$	$\frac{(r-2)(r+1)^2(r+2)}{4}$	$2r$	$\frac{(r-2)(r+1)}{2}$	$r-2$	1	0	0	0
$\omega_1+\omega_r$	$r(r+2)$	$r+1$	r	1	0	0	0	0
0	1	0	1	0	0	0	0	0

r = 2:

$\hat{\lambda}$	$d_{\hat{\lambda}}$	$c_{\hat{\lambda}}$	0	$\omega_1+\omega_2$	$3\omega_1$	$3\omega_2$	$2(\omega_1+\omega_2)$
$2(\omega_1+\omega_2)$	27	8	3	2	1	1	1
$3\omega_2$	10	6	1	1	0	1	0
$3\omega_1$	10	6	1	1	1	0	0
$\omega_1+\omega_2$	8	3	2	1	0	0	0
0	1	0	1	0	0	0	0

r = 1

$\hat{\lambda}$	$d_{\hat{\lambda}}$	$c_{\hat{\lambda}}$	0	ω_1	$2\omega_1$	$3\omega_1$	$4\omega_1$	$5\omega_1$	$6\omega_1$	$7\omega_1$	$8\omega_1$
$8\omega_1$	9	20	1	0	1	0	1	0	1	0	1
$7\omega_1$	8	$\frac{63}{4}$	0	1	0	1	0	1	0	1	0
$6\omega_1$	7	12	1	0	1	0	1	0	1	0	0
$5\omega_1$	6	$\frac{35}{4}$	0	1	0	1	0	1	0	0	0
$4\omega_1$	5	6	1	0	1	0	1	0	0	0	0
$3\omega_1$	4	$\frac{15}{4}$	0	1	0	1	0	0	0	0	0
$2\omega_1$	3	2	1	0	1	0	0	0	0	0	0
ω_1	2	$\frac{3}{4}$	0	1	0	0	0	0	0	0	0
0	1	0	1	0	0	0	0	0	0	0	0

Actually as will be explained below, we shall need only the multiplicities in the columns corresponding to $2(\omega_1+\omega_r)$, $2\omega_1+\omega_{r-1}$, $\omega_2+2\omega_r$, $\omega_2+\omega_{r-1}$ for $r \geq 3$, and to the first three of these for $r = 2$. That the multiplicities are 1 or 0 as listed for $r = 1$ is well known ([J], pp. 114-116). With $\hat{\lambda} = 2\omega_1+\omega_{r-1} = (2,2,\ldots,2,1)$, $\lambda = \omega_2+\omega_{r-1} = (1,2,\ldots,2,1)$, the only root $\alpha > 0$ such that

$\lambda + \alpha$ is a weight of the representation of highest weight $\hat{\lambda}$ is α_1, in which case $\lambda + \alpha_1 = \hat{\lambda}$, $m_{\lambda + \alpha_1} = 1$, $(\lambda + \alpha_1, \alpha_1) = (\hat{\lambda}, \alpha_1) = 2$, $c_{\hat{\lambda}} - c_\lambda = 2$, and $m_\lambda = 1$ from Freudenthal's formula. The same considerations show $m_\lambda = 1$ when $(\hat{\lambda}, \lambda) = : (\omega_2 + 2\omega_r, \omega_2 + \omega_{r-1}), (2(\omega_1 + \omega_r), \omega_2 + 2\omega_r), (2(\omega_1 + \omega_r), 2\omega_1 + \omega_{r-1})$. For the pair $(2(\omega_1 + \omega_r), \omega_2 + \omega_{r-1})$ we have $\lambda + \alpha$ a weight for $\alpha = \alpha_1$, $\alpha = \alpha_r$ only, and $m_{\lambda + \alpha_1} = 1 = m_{\lambda + \alpha_r}$ by the above, while $(\lambda + \alpha_1, \alpha_1) = 2 = (\lambda + \alpha_r, \alpha_r)$, $c_{\hat{\lambda}} - c_\lambda = 4$. Freudenthal's formula gives $4m_\lambda = 4$, or $m_\lambda = 1$. Verification of the remaining multiplicities can be a little more laborious, but all we shall use is the last four columns of our table for $r \geq 3$, the last three for $r = 2$.

Next consider type $D_r (r \geq 4)$, where we are interested only in $\hat{\lambda} = 2\omega_2 = 2(1, 2, \ldots, 2, 1, 1)$ and those dominant functions λ of the form $\hat{\lambda} - \Sigma m_i \alpha_i$, m_i non-negative integers. We consider in particular $\lambda = \hat{\lambda} - \alpha_2 = \omega_1 + \omega_3$ $(r \geq 5)$, $\lambda = \hat{\lambda} - \alpha_2 - \alpha_1 - \alpha_3 - \alpha_2 = \omega_4$ $(r \geq 6)$, $\lambda = \hat{\lambda} - 2\alpha_2 - 2\alpha_3 - \ldots - 2\alpha_{r-2} - \alpha_{r-1} - \alpha_r = 2\omega_1$, $\lambda = \omega_2$, $\lambda = 0$; for $r = 5$, we also consider $\lambda = \hat{\lambda} - \alpha_2 - \alpha_1 - \alpha_3 - \alpha_2 = \omega_4 + \omega_5$, and for $r = 4$, $\lambda = \hat{\lambda} - \alpha_2 = \omega_1 + \omega_3 + \omega_4$, $\hat{\lambda} - \alpha_2 - \alpha_1 - \alpha_3 - \alpha_2 = 2\omega_4$, $\hat{\lambda} - \alpha_2 - \alpha_1 - \alpha_4 - \alpha_2 = 2\omega_3$. As before, we have the following tables:

$r \geq 6$:

$\hat{\lambda}$	$d_{\hat{\lambda}}$	$c_{\hat{\lambda}}$	ω_4	$\omega_1 + \omega_3$	$2\omega_2$
$2\omega_2$	$\dfrac{r(r+1)(2r-3)(2r+1)}{3}$	$4r-2$	2	1	1
$\omega_1 + \omega_3$	$\dfrac{r(r+1)(2r-3)(2r-1)}{2}$	$4(r-1)$	3	1	0
ω_4	$\dfrac{(r-1)r(2r-3)(2r-1)}{6}$	$4(r-2)$	1	0	0
$2\omega_1$	$(r+1)(2r-1)$	$2r$	0	0	0
ω_2	$r(2r-1)$	$2(r-1)$	0	0	0
0	1	0	0	0	0

$r = 5$:

$\hat{\lambda}$	$d_{\hat{\lambda}}$	$c_{\hat{\lambda}}$	$\omega_4 + \omega_5$	$\omega_1 + \omega_3$	$2\omega_2$
$2\omega_2$	770	18	2	1	1
$\omega_1 + \omega_3$	945	16	3	1	0
$\omega_4 + \omega_5$	210	12	1	0	0
$2\omega_1$	54	10	0	0	0
ω_2	45	8	0	0	0
0	1	0	0	0	0

$r = 4$:

$\hat{\lambda}$	$d_{\hat{\lambda}}$	$c_{\hat{\lambda}}$	$2\omega_4$	$2\omega_3$	$\omega_1 + \omega_3 + \omega_4$	$2\omega_2$
$2\omega_2$	300	14	2	2	1	1
$\omega_1 + \omega_3 + \omega_4$	350	12	3	3	1	0
$2\omega_3$	35	8	0	1	0	0
$2\omega_4$	35	8	1	0	0	0
$2\omega_1$	35	8	0	0	0	0
ω_2	28	6	0	0	0	0
0	1	0	0	0	0	0

The multiplicities m_λ listed are just those we shall need; for their computations, we consider the example of $\omega_4 = \lambda$ when $\hat{\lambda} = \omega_1 + \omega_3$ and when $\hat{\lambda} = 2\omega_2$, with $r \geq 6$. For either choice of $\hat{\lambda}$, those positive roots α for which $\lambda + \alpha$ is a weight of the representation with highest weight $\hat{\lambda}$ are seen from Krusemeyer's test routine to be: $\alpha_1, \alpha_2, \alpha_3, \alpha_1 + \alpha_2, \alpha_2 + \alpha_3, \alpha_1 + \alpha_2 + \alpha_3$, and in each case $m_{\lambda + \alpha} = 1$, $2(\lambda+\alpha, \alpha) = 2$. For $\hat{\lambda} = \omega_1 + \omega_3$, we have $c_{\hat{\lambda}} - c_\lambda = 4$, and for $\hat{\lambda} = 2\omega_2$, $c_{\hat{\lambda}} - c_\lambda = 6$. Freudenthal's formula gives $12 = (c_{\hat{\lambda}} - c_\lambda)m_\lambda$, resulting in our entries.

In the case of E_6, we consider the following:

$2\omega_6 = (2\ 4\ 6\ 4\ 2\ 4)$, $\omega_3 = (2\ 4\ 6\ 4\ 2\ 3)$, $\omega_1 + \omega_5 = (2\ 3\ 4\ 3\ 2\ 2)$, $\omega_6, 0$.

From Planche V, p. 261 of Bourbaki [Bo], these are the only possible dominant weights of the form $2\omega_6 - \Sigma m_i \alpha_i$. (Our labelling of the system is somewhat at variance with theirs.)

We shall use the following table:

$\hat{\lambda}$	$d_{\hat{\lambda}}$	$c_{\hat{\lambda}}$	$\omega_1 + \omega_5$	ω_3	$2\omega_6$
$2\omega_6$	2430	26	3	1	1
ω_3	2925	24	4	1	0
$\omega_1 + \omega_5$	650	18	1	0	0
ω_6	78	12	0	0	0
0	1	0	0	0	0

With $\lambda = \omega_1 + \omega_5$ and either $\hat{\lambda} = \omega_3$ or $2\omega_6$, we find $2 \sum_{\alpha > 0} m_{\lambda + n} (\lambda + n\alpha, \alpha) = 2 \sum_{\alpha > 0} m_{\lambda + \alpha} (\lambda + \alpha, \alpha)$, with $m_{\lambda + \alpha} \neq 0$ only for $n \geq 1$ those α which are combinations of $\alpha_2, \alpha_3, \alpha_4, \alpha_6$ (there are 12 of these), and for each of these, $m_{\lambda + \alpha} = 1$, $2(\lambda + \alpha, \alpha) = 2$, so the above is equal to 24. Since $c_{\hat{\lambda}} - c_\lambda$ is respectively 6 or 8, we obtain the first two entries in our column under $\omega_1 + \omega_5$.

In the case of E_7, we start with $\hat{\lambda} = 2\omega_6 = (2\ 4\ 6\ 8\ 6\ 4\ 4)$, and find, using Planche VI, p. 265 of Bourbaki [Bo], that the only dominant weights of the form $\hat{\lambda} - \Sigma m_i \alpha_i$ are:

$\omega_5 = (2\ 4\ 6\ 8\ 6\ 3\ 4)$, $\omega_2 = (2\ 4\ 5\ 6\ 4\ 2\ 3)$, ω_6, 0.

We shall use the following table:

$\hat{\lambda}$	$d_{\hat{\lambda}}$	$c_{\hat{\lambda}}$	ω_2	ω_5	$2\omega_6$
$2\omega_6$	7371	38	4	1	1
ω_5	8645	36	5	1	0
ω_2	1539	28	1	0	0
ω_6	133	18	0	0	0
0	1	0	0	0	0

The first two entries in the column headed by ω_2 result from the fact that with $\hat{\lambda} = \omega_5$ or $2\omega_6$, $\lambda = \omega_2$, we have
$2 \sum_{\substack{n \geq 1 \\ \alpha > 0}} m_{\lambda+n\alpha}(\lambda+n\alpha,\alpha) = 2 \sum_{\alpha > 0} m_{\lambda+\alpha}(\lambda+\alpha,\alpha)$, and $m_{\lambda+\alpha} \neq 0$ exactly when $\alpha > 0$ is one of the 20 positive roots which are combinations of $\alpha_3, \alpha_4, \alpha_5, \alpha_6, \alpha_7$, in which cases $m_{\lambda+\alpha} = 1$, $2(\lambda+\alpha,\alpha) = 2$ and our sum is 40.

In the case of E_8, we start with $\hat{\lambda} = 2\omega_1 = (4,6,8,10,12,8,4,6)$, and find, using Planche VII, p. 269 of Bourbaki [Bo], that the only dominant weights of the form $\hat{\lambda} - \Sigma m_i \alpha_i$ are:

$\omega_2 = (3,6,8,10,12,8,4,6)$, $\omega_7 = (2,4,6,8,10,7,4,5)$, ω_1, 0.

We use the following table:

$\hat{\lambda}$	$d_{\hat{\lambda}}$	$c_{\hat{\lambda}}$	ω_7	ω_2	$2\omega_1$
$2\omega_1$	27,000	62	6	1	1
ω_2	30,380	60	7	1	0
ω_7	3875	48	1	0	0
ω_1	248	30	0	0	0
0	1	0	0	0	0

Again, the first two entries in the column headed by ω_7 result from the fact that the member $2\sum_{n \geq 1 \atop \alpha > 0} m_{\omega_7+n\alpha}(\hat{\omega_7}+n\alpha,\alpha)$ of Freudenthal's formula is equal, for $\hat\lambda = \omega_2$ or $2\omega_1$, to the sum over all 42 roots $\alpha > 0$ which are combinations of $\alpha_1,\ldots,\alpha_6,\alpha_8$ of the quantities $2(\omega_7+\alpha,\alpha) = 2(\alpha,\alpha) = 2$, or has the value 84.

When the system is of type $B_r (r \geq 3)$, we have <u>three</u> dominant weights to emphasize: With $\omega_2 = (1,2,\ldots,2)$ and $\omega_1 = (1,1,\ldots,1)$, we consider $2\omega_2$, $\omega_1+\omega_2$ and $2\omega_1$. Since the latter two are obtained from the first by subtractions of α_i, we list all possible dominant weights so obtained from the first; the completeness of the list may be checked by using (VI) of Planche II, p. 253 of Bourbaki [Bo]:

For $r > 4$: $2\omega_2$, $\omega_1+\omega_3$, ω_4, $\omega_1+\omega_2$, ω_3, $2\omega_1$, ω_2, ω_1, 0.
For $r = 4$: $2\omega_2$, $\omega_1+\omega_3$, $2\omega_4$, $\omega_1+\omega_2$, ω_3, $2\omega_1$, ω_2, ω_1, 0.
For $r = 3$: $2\omega_2$, $\omega_1+2\omega_3$, $\omega_1+\omega_2$, $2\omega_3$, $2\omega_1$, ω_2, ω_1, 0.

As before, we have the table, first for $r > 4$:

$\hat\lambda$	$d_{\hat\lambda}$	$c_{\hat\lambda}$	ω_3	$\omega_1+\omega_2$	ω_4	$\omega_1+\omega_3$	$2\omega_2$
$2\omega_2$	$\frac{(r-1)(r+1)(2r+1)(2r+3)}{3}$	$8r$	2	1	2	1	1
$\omega_1+\omega_3$	$\frac{(r-1)r(2r+1)(2r+3)}{2}$	$4(2r-1)$	3	1	3	1	0
ω_4	$\frac{(r-1)r(2r-1)(2r+1)}{6}$	$4(2r-3)$	1	0	1	0	0
$\omega_1+\omega_2$	$\frac{(2r-1)(2r+1)(2r+3)}{3}$	$6r$	2	1	0	0	0
ω_3	$\frac{r(2r-1)(2r+1)}{3}$	$6(r-1)$	1	0	0	0	0
$2\omega_1$	$r(2r+3)$	$2(2r+1)$	0	0	0	0	0
ω_2	$r(2r+1)$	$2(2r-1)$	0	0	0	0	0
ω_1	$2r+1$	$2r$	0	0	0	0	0
0	1	0	0	0	0	0	0

For r = 4:

$\hat{\lambda}$	$d_{\hat{\lambda}}$	$c_{\hat{\lambda}}$	ω_3	$\omega_1+\omega_2$	$2\omega_4$	$\omega_1+\omega_3$	$2\omega_2$
$2\omega_2$	495	32	2	1	2	1	1
$\omega_1+\omega_3$	594	28	3	1	3	1	0
$2\omega_4$	154	20	1	0	1	0	0
$\omega_1+\omega_2$	231	24	2	1	0	0	0
ω_3	84	18	1	0	0	0	0
$2\omega_1$	44	18	0	0	0	0	0
ω_2	36	14	0	0	0	0	0
ω_1	9	8	0	0	0	0	0
0	1	0	0	0	0	0	0

For r = 3:

$\hat{\lambda}$	$d_{\hat{\lambda}}$	$c_{\hat{\lambda}}$	$2\omega_3$	$\omega_1+\omega_2$	$\omega_1+2\omega_3$	$2\omega_2$
$2\omega_2$	168	24	2	1	1	1
$\omega_1+2\omega_3$	189	20	3	1	1	0
$\omega_1+\omega_2$	105	18	2	1	0	0
$2\omega_3$	35	12	1	0	0	0
$2\omega_1$	27	14	0	0	0	0
ω_2	21	10	0	0	0	0
ω_1	7	6	0	0	0	0
0	1	0	0	0	0	0

The entries in the column headed by $\omega_4 (r > 4)$ or $2\omega_4$ $(r = 4)$ result from the fact that for this λ, and for $\hat{\lambda} = \omega_1 + \omega_3$ or $2\omega_2$, we have $2 \sum_{\substack{n \geq 1 \\ \alpha > 0}} m_{\lambda+n\alpha} (\lambda+n\alpha, \alpha) = 2 \sum_{\alpha > 0} m_{\lambda+\alpha} (\lambda+\alpha, \alpha) =$
$2 \sum (\lambda+\alpha, \alpha) = 24$.
α positive combination of $\alpha_1, \alpha_2, \alpha_3$

From the coefficient of α_1, we see that $\lambda = \omega_1 + \omega_2$ cannot be a weight in the representation of highest weight ω_4 (or $2\omega_4$, if $r = 4$). The remaining entries in the column of $\lambda = \omega_1 + \omega_2$, those for $\hat{\lambda} = 2\omega_2, \omega_1 + \omega_3$ ($\omega_1 + 2\omega_3$, if $r = 3$), result from the computation of $2 \sum_{\substack{n \geq 1 \\ \alpha > 0}} m_{\lambda+n\alpha} (\lambda+n\alpha, \alpha)$. In this sum, the only $\lambda + n\alpha$ which can occur with positive multiplicities have $n = 1$, $\alpha = \alpha_i + \ldots + \alpha_r$, where $i \geq 3$ if $\hat{\lambda} = \omega_1 + \omega_3$ (or $\omega_1 + 2\omega_3$), $i \geq 2$ if $\hat{\lambda} = 2\omega_2$, and for each of these, $2(\lambda+\alpha, \alpha) = 2(\alpha, \alpha) = 2$ if $i \geq 3$, while $\hat{\lambda} = 2\omega_2$, $2(\lambda+\alpha, \alpha) = 2(2\alpha, \alpha) = 4$. Thus our sum has respective values $2(r-2)$ and $2r$, yielding the indicated multiplicities (1 in each case). For $\lambda = \omega_3$ ($2\omega_3$ if $r = 3$) and $\hat{\lambda} = \omega_1 + \omega_2$, our sum has the value 12, resulting in the multiplicity 2 for this case as entered. For $\lambda = \omega_3$ and $\hat{\lambda} = \omega_4 (r > 4)$ or $\lambda = 2\omega_4 (r = 4)$, the value of the sum is $2r - 6$, resulting in the multiplicity 1, as entered. For $\lambda = \omega_3 (2\omega_3$ if $r = 3$) and $\hat{\lambda} = \omega_1 + \omega_3$ (or $\omega_1 + 2\omega_3$), we have $r - 3$ terms in our sum of the form $2 m_{\lambda+\alpha} (\lambda+\alpha, \alpha)$, $\alpha = \alpha_i + \ldots + \alpha_r$, $i \geq 4$, with $m_{\lambda+\alpha} = 3$ (see ω_4 - column), $2(\lambda+\alpha, \alpha) = 2$, three terms of the same form with $1 \leq i \leq 3$, having $m_{\lambda+\alpha} = 1$, $2(\lambda+\alpha, \alpha) = 4$, and three terms of the same type with $\alpha = \alpha_1, \alpha_2, \alpha_1 + \alpha_2$, $m_{\lambda+\alpha} = 1$, $2(\lambda+\alpha, \alpha) = 4$, giving the sum a value of $6r + 6$, and resulting in the entry "3". For the same λ with $\hat{\lambda} = 2\omega_2$, the terms are the same, except that the $m_{\lambda+\alpha}$ which were 3 previously are now 2, giving a sum of $4r + 12$, and resulting in the entry "2".

When the system is of type F_4, we consider dominant weights obtained by subtracting sequences of positive roots from $2\omega_1 = (4,6,8,4)$. Again one may consult Bourbaki [Bo], Planche VIII, p. 273, (VI), to see that the only possibilities are $2\omega_1$, $\omega_2 = (3,6,8,4)$, $\omega_1 + \omega_4 = (3,5,7,4)$, $2\omega_4 = (2,4,6,4)$, $\omega_3 = (2,4,6,3)$, $\omega_1, \omega_4, 0$.

Our table:

$\hat{\lambda}$	$d_{\hat{\lambda}}$	$c_{\hat{\lambda}}$	ω_3	$2\omega_4$	$\omega_1 + \omega_4$	ω_2	$2\omega_1$
$2\omega_1$	1053	40	3	3	1	1	1
ω_2	1274	36	4	3	1	1	0
$\omega_1 + \omega_4$	1053	32	4	1	1	0	0
$2\omega_4$	324	26	1	1	0	0	0
ω_3	273	24	1	0	0	0	0
ω_1	52	18	0	0	0	0	0
ω_4	26	12	0	0	0	0	0
0	1	0	0	0	0	0	0

We suppress remarks as to the calculations, which go as before, and which duplicate those of Veldkamp (Jour. of Algebra 16 (1970), p. 328).

When the system is of type G_2, we consider dominant weights obtained by subtracting sequences of positive roots from $2\omega_1 = (4,6)$, obtaining (cf. Bourbaki, op. cit., p. 274) $3\omega_2 = (3,6)$, $\omega_1 + \omega_2 = (3,5)$, $2\omega_2 = (2,4)$, ω_1, ω_2, 0. Our table:

$\hat{\lambda}$	$d_{\hat\lambda}$	$c_{\hat\lambda}$	$2\omega_2$	$\omega_1+\omega_2$	$3\omega_2$	$2\omega_1$
$2\omega_1$	77	30	2	1	1	1
$3\omega_2$	77	24	2	1	1	0
$\omega_1+\omega_2$	64	21	2	1	0	0
$2\omega_2$	27	14	1	0	0	0
ω_1	14	12	0	0	0	0
ω_2	7	6	0	0	0	0
0	1	0	0	0	0	0

The multiplicities are computed as before, starting with the right-hand column. When $\lambda = 2\omega_2$, $\hat\lambda = \omega_1+\omega_2$; $3\omega_2$; $2\omega_1$, the set of weights $\lambda+n\alpha$, $\alpha > 0$, is: $\lambda+\alpha_1$, $\lambda+\alpha_2$, $\lambda+\alpha_1+\alpha_2$; $\lambda+\alpha_1$, $\lambda+\alpha_2$, $\lambda+\alpha_1+\alpha_2$, $\lambda+\alpha_1+2\alpha_2$; and the latter set together with $\lambda+2\alpha_2$, $\lambda+2(\alpha_1+\alpha_2)$, according the respective case; the respective values of $2\sum_{\substack{n\geq 1\\ \alpha>0}} m_{\lambda+n\alpha}(\lambda+n\alpha,\alpha)$ are 14, 20, 32.

When the system is of type $C_r (r \geq 2)$, we consider dominant weights obtained by sequences of positive roots from $4\omega_1 = (4,4,\ldots,4,2)$ and from $3\omega_1 = (3,3,\ldots,3,\frac{3}{2})$; one may check the completeness of our list from Planche III, (VI), p. 254 of Bourbaki, op. cit. It is: $4\omega_1$, $2\omega_1+\omega_2 = (3,4,\ldots,4,2)$, $2\omega_2 = (2,4,4,\ldots,4,2)$, $\omega_1+\omega_3 = (2,3,4,\ldots,4,2)$, $\omega_4 = (1,2,3,4,\ldots,4,2)$, $2\omega_1$, ω_2, 0; $3\omega_1$, $\omega_1+\omega_2 = (2,3,3,\ldots,3,\frac{3}{2})$, $\omega_3 = (1,2,3,\ldots,3,\frac{3}{2})$, ω_1 (this for $r \geq 5$). For $r = 4$, we have $4\omega_1$, $2\omega_1+\omega_2$, $2\omega_2$, $\omega_1+\omega_3$, $\omega_4 = (1,2,3,2)$, $2\omega_1$, ω_2, 0; $3\omega_1$, $\omega_1+\omega_2$, ω_3, ω_1; for $r = 3$: $4\omega_1$, $2\omega_1+\omega_2$, $2\omega_2$, $\omega_1+\omega_3 = (2,3,2)$, $2\omega_1$, ω_2, 0; $3\omega_1$, $\omega_1+\omega_2$, $\omega_3 = (1,2,\frac{3}{2})$, ω_1; for $r = 2$: $4\omega_1$, $2\omega_1+\omega_2 = (3,2)$, $2\omega_2 = (2,2)$, $2\omega_1 = (2,1)$, ω_2, 0; $3\omega_1$, $\omega_1+\omega_2 = (2,\frac{3}{2})$, $\omega_1 = (1,\frac{1}{2})$. Our tables in this case:

$(r \geq 4)$:

$\hat{\lambda}$	$\hat{d}_\lambda$	$\hat{c}_\lambda$	ω_3	$\omega_1+\omega_2$	$3\omega_1$	ω_4	$\omega_1+\omega_3$	$2\omega_2$	$2\omega_1+\omega_2$	$4\omega_1$
$4\omega_1$	$\dfrac{r(r+1)(2r+1)(2r+3)}{6}$	$4(r+2)$	0	0	0	1	1	1	1	1
$2\omega_1+\omega_2$	$\dfrac{(r-1)r(2r+1)(2r+3)}{2}$	$4(r+1)$	0	0	0	3	2	1	1	0
$2\omega_2$	$\dfrac{(r-1)r(2r-1)(2r+3)}{3}$	$4r+2$	0	0	0	2	1	1	0	0
$\omega_1+\omega_3$	$\dfrac{(r-2)(r+1)(2r-1)(2r+1)}{2}$	$4r$	0	0	0	3	1	0	0	0
ω_4	$\dfrac{(r-3)r(2r-1)(2r+1)}{6}$	$4(r-1)$	0	0	0	1	0	0	0	0
$3\omega_1$	$\dfrac{2}{3}r(r+1)(2r+1)$	$\dfrac{3}{2}(2r+3)$	1	1	1	0	0	0	0	0
$\omega_1+\omega_2$	$\dfrac{8}{3}(r-1)r(r+1)$	$\dfrac{3}{2}(2r+1)$	2	1	0	0	0	0	0	0
ω_3	$\dfrac{2}{3}(r-2)r(2r+1)$	$\dfrac{3}{2}(2r-1)$	1	0	0	0	0	0	0	0
$2\omega_1$	$r(2r+1)$	$2r+2$	0	0	0	0	0	0	0	0
ω_2	$(r-1)(2r+1)$	$2r$	0	0	0	0	0	0	0	0
ω_1	$2r$	$\dfrac{1}{2}(2r+1)$	0	0	0	0	0	0	0	0
0	1	0	0	0	0	0	0	0	0	0

320

$(r = 3):$ $\hat{\lambda}$	$d_{\hat{\lambda}}$	$c_{\hat{\lambda}}$	ω_3	$\omega_1+\omega_2$	$3\omega_1$	$\omega_1+\omega_3$	$2\omega_2$	$2\omega_1+\omega_2$	$4\omega_1$
$4\omega_1$	126	20	0	0	0	1	1	1	1
$2\omega_1+\omega_2$	189	16	0	0	0	2	1	1	0
$2\omega_2$	90	14	0	0	0	1	1	0	0
$\omega_1+\omega_3$	70	12	0	0	0	1	0	0	0
$3\omega_1$	56	$\frac{27}{2}$	1	1	1	0	0	0	0
$\omega_1+\omega_2$	64	$\frac{21}{2}$	2	1	0	0	0	0	0
ω_3	14	$\frac{15}{2}$	1	0	0	0	0	0	0
$2\omega_1$	21	8	0	0	0	0	0	0	0
ω_2	14	6	0	0	0	0	0	0	0
ω_1	6	$\frac{7}{2}$	0	0	0	0	0	0	0
0	1	0	0	0	0	0	0	0	0

(r = 2):

$\hat{\lambda}$	$d_{\hat{\lambda}}$	$c_{\hat{\lambda}}$	$\omega_1 + \omega_2$	$3\omega_1$	$2\omega_2$	$2\omega_1 + \omega_2$	$4\omega_1$
$4\omega_1$	35	16	0	0	1	1	1
$2\omega_1 + \omega_2$	35	12	0	0	1	1	0
$2\omega_2$	14	10	0	0	1	0	0
$3\omega_1$	20	$\frac{21}{2}$	1	1	0	0	0
$\omega_1 + \omega_2$	16	$\frac{15}{2}$	1	0	0	0	0
$2\omega_1$	10	6	0	0	0	0	0
ω_2	5	4	0	0	0	0	0
ω_1	4	$\frac{5}{2}$	0	0	0	0	0
0	1	0	0	0	0	0	0

With $\lambda = \omega_1 + \omega_3$, the only weight of the form $\lambda + n\alpha$ when $\hat{\lambda} = 2\omega_2$ is $\lambda + \alpha_2 = \lambda$; for $\hat{\lambda} = 2\omega_1 + \omega_2$ or $4\omega_1$, the only such weights are $\lambda + \alpha_1$, $\lambda + \alpha_2$, $\lambda + \alpha_1 + \alpha_2$. Thus the quantity $2\sum_{n \geq 1} m_{\lambda+n\alpha} (\lambda+n\alpha, \alpha)$ has the values $2, 8, 8$, giving our values in the $\alpha > 0$ column under $\omega_1 + \omega_3$. In the column of ω_4, the corresponding sum has the values $12, 12, 24, 12$, according as $\hat{\lambda} = \omega_1 + \omega_3$, $2\omega_2$, $2\omega_1 + \omega_2$, $4\omega_1$ — the same $\lambda + n\alpha$ are considered in each case, but have multiplicity 2 in the third case, as conjugates of $\omega_1 + \omega_3$. Freudenthal's formula yields the entries. Finally, none of ω_3, $\omega_1 + \omega_2$, $3\omega_1$ is congruent to any of the elements heading the first five rows (four if $r = 3$, three if $r = 2$) modulo the group generated by the roots, accounting for the zeros in these five rows and three columns, and in each case where $\lambda = \omega_3$ is present ($\hat{\lambda} = \omega_1 + \omega_2$, $\hat{\lambda} = 3\omega_1$), the sum $2\Sigma\, m_{\lambda+n\alpha}(\lambda+n\alpha, \alpha)$ has three terms, and value 6, enabling us to complete the table.

§A-2. **Decomposition of Certain Tensor Products.**

Evidently the multiplicity of the weight ν in the tensor product $M_{\hat{\lambda}} \otimes M_{\hat{\mu}}$ of irreducible representations is $\sum_{\lambda+\mu=\nu} m(\lambda,\hat{\lambda}) m(\mu,\hat{\mu})$, where $m(\lambda, \hat{\lambda})$ is the multiplicity of λ in $M_{\hat{\lambda}}$. Also, the only ν which can occur as highest weights of irreducible summands of $M_{\hat{\lambda}} \otimes M_{\hat{\mu}}$ are of the form $\nu = \hat{\lambda} + \hat{\mu} - \Sigma\, m_i \alpha_i$, the m_i non-negative integers. Based on these considerations and using our tables, we conclude the presence in certain $M_{\hat{\lambda}} \otimes M_{\hat{\mu}}$ of certain irreducible summands, as follows:

In any $M_{\hat{\lambda}} \otimes M_{\hat{\mu}}$, the weight $\hat{\lambda} + \hat{\mu}$ occurs with multiplicity one; hence the summand $M_{\hat{\lambda}+\hat{\mu}}$ is present <u>once</u>.
We now consider cases.

A_r: We consider $M_{\omega_1+\omega_r} \otimes M_{\omega_1+\omega_r}$ for $r > 1$, the tensor product with itself of the adjoint representation. We claim
$$M_{\omega_1+\omega_r} \otimes M_{\omega_1+\omega_r} = M_{2(\omega_1+\omega_r)} + M_{\omega_2+2\omega_r} + M_{2\omega_1+\omega_{r-1}} + M_{\omega_2+\omega_{r-1}} + 2M_{\omega_1+\omega_r} + M_0$$
for all $r \geq 4$; for $r = 3$, we claim
$$M_{\omega_1+\omega_3} \otimes M_{\omega_1+\omega_3} = M_{2(\omega_1+\omega_3)} + M_{\omega_2+2\omega_3} + M_{2\omega_1+\omega_2} + M_{2\omega_2} + 2M_{\omega_1+\omega_3} + M_0,$$
while, for $r = 2$,
$$M_{\omega_1+\omega_2} \otimes M_{\omega_1+\omega_2} = M_{2(\omega_1+\omega_2)} + M_{2\omega_1} + M_{3\omega_1} + 2M_{\omega_1+\omega_2} + M_0.$$

It is readily verified from the tables of §1 that the sum of the dimensions of the summands on the right is equal to the dimension of the tensor product. Thus it suffices to show that each summand on the right is <u>present</u> in the tensor product. The invariance of the Killing form assures the presence of (at least one) M_0 as a summand, and the linearly independent compositions [xy] and x o y of §1 of Chapter III assure the presence of (at least) <u>two</u> summands equivalent to the adjoint module $M_{\omega_1+\omega_r}$.

For the rest, we know that $M_{2\omega_1+2\omega_r}$ is present exactly once and, from our tables, that $2\omega_1+\omega_{r-1}$ and $\omega_2+2\omega_r$ (for r = 2, $3\omega_1$ and $3\omega_2$) have multiplicity <u>one</u> as weights of $M_{2\omega_1+2\omega_r}$. Moreover, since these are $2(\omega_1+\omega_r) - \alpha_i$, i = r,1, they can occur as weights of no other summands than $M_{2\omega_1+2\omega_r}$, $M_{2\omega_1+\omega_{r-1}}$ (resp. $M_{\omega_2+2\omega_r}$). But their multiplicity in the tensor product is 2 in each case, since the ordered ways $2\omega_1+\omega_{r-1}$ can be written as a sum of two roots are $(\alpha_1+...+\alpha_r) + (\alpha_1+...+\alpha_{r-1})$, and the reverse order. We conclude that $M_{2\omega_1+\omega_{r-1}}$ and $M_{\omega_2+2\omega_r}$ are both present, each with multiplicity <u>one</u>. This completes the case r = 2. When r ≥ 3, we consider the multiplicity of $\omega_2+\omega_{r-1} = \alpha_1+2\alpha_2+...+2\alpha_{r-1}+\alpha_r$. The only possible summands having this as a weight are the three so far discussed in this paragraph, together with $M_{\omega_2+\omega_{r-1}}$. Our tables show the multiplicity of $\omega_2+\omega_{r-1}$ to be <u>one</u> in each of the former summands, while in the tensor product it occurs with multiplicity <u>four</u> [(1,...,1)+(0,1,...,1,0) and the reverse, (1,1,...,1,0)+(0,1,...,1) and the reverse]. Thus $M_{\omega_2+\omega_{r-1}}$ must be present, and our decompositions are established.

For A_1, verification of the following is an easy (and classical) matter (one may use our tables and the above reasoning):

$M_{\omega_1} \otimes M_{\omega_1} = M_{2\omega_1} + M_0$; $M_{\omega_1} \otimes M_{2\omega_1} = M_{3\omega_1} + M_{\omega_1}$;

$M_{2\omega_1} \otimes M_{2\omega_1} = M_{4\omega_1} + M_{2\omega_1} + M_0$; $M_{2\omega_1} \otimes M_{4\omega_1} = M_{6\omega_1} + M_{4\omega_1} + M_{2\omega_1}$;

$M_{4\omega_1} \otimes M_{4\omega_1} = M_{8\omega_1} + M_{6\omega_1} + M_{4\omega_1} + M_{2\omega_1} + M_0$.

The argument for A_r above is used in each of the other cases, with reference to our tables. For type D_r, the adjoint representation is M_{ω_2}. Recalling that our (split) L in this case is the skew transformations of a 2r-dimensional F-space V with respect to a symmetric bilinear form of Witt index r, and noting that the symmetric transformations of V form an L-submodule of $\text{Hom}_F(V,V)$ which decomposes into $M_{2\omega_1} + M_0$, $M_{2\omega_1}$ being the symmetric transformations of trace zero, we again have that $x,y \mapsto xy + yx - \frac{1}{r}\text{Tr}(xy)I$ defines an element of $\text{Hom}_L(M_{\omega_2} \otimes M_{\omega_2}, M_{2\omega_1})$, so that we know that each of the following is present (at least once) in $M_{\omega_2} \otimes M_{\omega_2}$: $M_{2\omega_1}$, M_{ω_2}, M_0. We claim that, for $r > 5$,

$$M_{\omega_2} \otimes M_{\omega_2} = M_{2\omega_2} + M_{\omega_1+\omega_3} + M_{\omega_4} + M_{2\omega_1} + M_{\omega_2} + M_0.$$

For $r = 5$, we claim

$$M_{\omega_2} \otimes M_{\omega_2} = M_{2\omega_2} + M_{\omega_1+\omega_3} + M_{\omega_4+\omega_5} + M_{2\omega_1} + M_{\omega_2} + M_0,$$

and, for $r = 4$, that

$$M_{\omega_2} \otimes M_{\omega_2} = M_{2\omega_2} + M_{\omega_1+\omega_3+\omega_4} + M_{2\omega_3} + M_{2\omega_4} + M_{2\omega_1} + M_{\omega_2} + M_0.$$

Again, it follows by dimensions and the remarks above that we need only show the first three summands (the first four if $r = 4$) are present. Now $M_{2\omega_2}$ is clearly present exactly once, and $\omega_1 + \omega_3 = 2\omega_2 - \alpha_2$ ($\omega_1 + \omega_3 + \omega_4$ if $r = 4$) can only be a weight of a summand $M_{2\omega_2}$ or with itself as highest weight. It has multiplicity 1 in $M_{2\omega_2}$ (see tables), while it is equal to $\omega_2 + (\omega_2 - \alpha_2)$ and to $(\omega_2 - \alpha_2) + \omega_2$, giving it multiplicity 2 in the tensor product. Thus $M_{\omega_1+\omega_3}$ ($M_{\omega_1+\omega_3+\omega_4}$ if $r = 4$) is present, <u>once</u>. Passing to ω_4 (if $r > 5$; $\omega_4 + \omega_5$, if $r = 5$; $2\omega_3$, $2\omega_4$ if $r = 4$), we see that each of these can only occur in $M_{\omega_2} \otimes M_{\omega_2}$ with multiplicity equal to the sum of multiplicity of the representation with itself as highest weight, plus 5. On the other hand, in the tensor product, we realize this weight by the partitions $\omega_r = (1,2,3,4,4,\ldots,4,2,2) = (1,2,\ldots,2,1,1) + (0,1,1,2,\ldots,2,1,1) = (1,1,2,\ldots,2,1,1) + (0,1,1,2,\ldots2,1,1) = (0,1,2,\ldots,2,1,1) + (1,1,1,2,\ldots,2,1,1)$ and the reverses of these partitions ($r > 5$); $\omega_4 + \omega_5 = (1,2,3,2,2) = (1,2,2,1,1)+(0,0,1,1,1) =$

$(1,1,2,1,1) + (0,1,1,1,1) = (1,1,1,1,1) + (0,1,2,1,1)$ and their reverses, for $r = 5$; $2\omega_3 = (1,2,2,1) = (1,2,1,1) + (0,0,1,0) = (1,1,1,1) + (0,1,1,0) = (1,1,1,0) + (0,1,1,1)$ and their reverses, for $r = 4$, and likewise for $2\omega_4$, showing that this weight has multiplicity at least 6 in the tensor product, and thereby the presence of a summand having this highest weight. Our decomposition is thereby established.

For E_6, the adjoint representation is M_{ω_6}; for E_7, M_{ω_6}; for E_8, M_{ω_1}; clearly the adjoint representation and M_0 are present in the tensor product of the adjoint representation with itself, which we claim decomposes as follows:

E_6: $M_{\omega_6} \otimes M_{\omega_6} = M_{2\omega_6} + M_{\omega_3} + M_{\omega_1 + \omega_5} + M_{\omega_6} + M_0$.

E_7: $M_{\omega_6} \otimes M_{\omega_6} = M_{2\omega_6} + M_{\omega_5} + M_{\omega_2} + M_{\omega_6} + M_0$.

E_8: $M_{\omega_1} \otimes M_{\omega_1} = M_{2\omega_1} + M_{\omega_2} + M_{\omega_7} + M_{\omega_1} + M_0$.

As before, the dimension-relations are correct, and $M_{2\omega_6}$, $M_{2\omega_6}$, $M_{2\omega_1}$, respectively, are present with multiplicity 1, containing ω_3, ω_5, ω_2, respectively with multiplicity 1, $\omega_3 = 2\omega_6 - \alpha_6$, $\omega_5 = 2\omega_6 - \alpha_6$, $\omega_2 = 2\omega_1 - \alpha_1$, respectively, being present in the tensor product with multiplicity 2, hence establishing the presence of M_{ω_3}, M_{ω_5}, M_{ω_2}, respectively, with multiplicity <u>one</u>. Finally, the same analysis applied to $\omega_1 + \omega_5$, ω_2, ω_7 respectively, yields multiplicities of 8,10,14 (necessarily even, since none of the three is twice a root) in the tensor product, and of 7,9,13 in the two higher summands found thus far. The presence of the remaining summand follows, and the decompositions are established.

Proceeding to $B_r (r \geq 3)$, the adjoint representation has highest weight ω_2; the "defining representation" on V, a vector space of dimension $2r + 1$ with a symmetric bilinear form of maximal Witt index, has highest weight ω_1, and $V \otimes V \cong V \otimes V^* \cong \mathrm{Hom}_F(V,V)$ decomposes into skew transformations (M_{ω_2}), symmetric transformations of trace zero ($M_{2\omega_1}$) and scalar multiples of the identity (M_0). Thus

$$M_{\omega_1} \otimes M_{\omega_1} = M_{2\omega_1} + M_{\omega_2} + M_0 .$$

We also consider $M_{\omega_1} \otimes M_{\omega_2}$ and $M_{\omega_2} \otimes M_{\omega_2}$. In the former case, we know by §1 of Chapter III that M_{ω_1} is a summand of the tensor product. In the latter case, we see by the remarks above and the map $x,y \mapsto xy + yx - \frac{2}{2r+1} \mathrm{Tr}(xy) I$ of $M_{\omega_2} \otimes M_{\omega_2}$ into $M_{2\omega_1}$, that each of $M_{2\omega_1}$, M_{ω_2}, M_0 is present at least once. We show:

$$M_{\omega_1} \otimes M_{\omega_2} = M_{\omega_1+\omega_2} + M_{\omega_3} + M_{\omega_1} \quad (r > 3),$$

$$M_{\omega_1} \otimes M_{\omega_2} = M_{\omega_1+\omega_2} + M_{2\omega_3} + M_{\omega_1} \quad (r = 3),$$

$$M_{\omega_2} \otimes M_{\omega_2} = M_{2\omega_2} + M_{\omega_1+\omega_3} + M_{\omega_4} + M_{2\omega_1} + M_{\omega_2} + M_0 \quad (r > 4),$$

$$M_{\omega_2} \otimes M_{\omega_2} = M_{2\omega_2} + M_{\omega_1+\omega_3} + M_{2\omega_4} + M_{2\omega_1} + M_{\omega_2} + M_0 \quad (r = 4),$$

$$M_{\omega_2} \otimes M_{\omega_2} = M_{2\omega_2} + M_{\omega_1+2\omega_3} + M_{2\omega_3} + M_{2\omega_1} + M_{\omega_2} + M_0 \quad (r = 3).$$

For the first two, it suffices to show that the multiplicity of the weight ω_3 (or $2\omega_3$) in the tensor product is at least <u>three</u>, which it is, because it has the three representations $(1,\ldots,1) + (0,1,2,\ldots,2) = \omega_1 + (\omega_2 - \alpha_2 - \alpha_1)$, $(0,1,\ldots,1) + (1,1,2,\ldots,2) = (\omega_1 - \alpha_1) + (\omega_2 - \alpha_2)$, $(0,0,1,\ldots,1) + (1,2,\ldots,2) = (\omega_1 - \alpha_1 - \alpha_2) + \omega_2$. For the rest, we first note, as before, that $\omega_1 + \omega_3$ (or $\omega_1 + 2\omega_3$, if $r = 3$) is $2\omega_2 - \alpha_2$, and so has multiplicity <u>two</u> in the tensor product, but <u>one</u> in $M_{2\omega_2}$ (see tables). Thus $M_{\omega_1+\omega_3}$ (or $M_{\omega_1+2\omega_3}$) is present, with multiplicity <u>one</u>. The combined multiplicity of ω_4 (of $2\omega_4$, of $2\omega_3$) as a weight of the representations thus far accounted for ($M_{2\omega_2}$ and $M_{\omega_1+\omega_3}$ or $M_{\omega_1+2\omega_3}$, together with $M_{2\omega_1}$, M_{ω_2}, M_0) is <u>five</u>, and that of $\omega_1 + \omega_2$ is <u>two</u>. If $r \geq 4$, $\omega_4 (2\omega_4$, if $r = 4)$ is not a weight of $M_{\omega_1+\omega_2}$, nor of M_{ω_3}, so any excess of its multiplicity in $M_{\omega_2} \otimes M_{\omega_2}$ over five must be accounted for by the presence of summands M_{ω_4} (or $M_{2\omega_4}$). We have $(1,2,3,4,\ldots,4) = (0,1,2,\ldots,2) + (1,1,1,2,\ldots,2) = (0,0,1,2,\ldots,2) + (1,2,2,\ldots,2) = (0,1,1,2,\ldots,2) + (1,1,2,\ldots,2)$, and doubling these shows a multiplicity of ω_4 (or $2\omega_4$) of at least <u>six</u> in

$M_{\omega_2} \otimes M_{\omega_2}$, establishing our decomposition of this module except for r = 3. In that case, we have $\omega_1 + \omega_2 = (2,3,3)$, whose only representation as sum of two roots (weights of M_{ω_2}) is as $(1,2,2) + (1,1,1)$ and the reverse. It follows from remarks above that $M_{\omega_1+\omega_2}$ can <u>not</u> occur in $M_{\omega_2} \otimes M_{\omega_2}$, and that $M_{2\omega_3}$ will occur if we show $2\omega_3$ is a weight of the tensor product of multiplicity at least 6. Now $2\omega_3 = (1,2,3) = (1,1,1) + (0,1,2) = (1,1,2) + (0,1,1) = (1,2,2) + (0,0,1)$; these and the three reversed decompositions show that this multiplicity is indeed at least 6, and establish our decompositions.

When L is of type F_4, we claim the following:

$$M_{\omega_4} \otimes M_{\omega_4} = M_{2\omega_4} + M_{\omega_3} + M_{\omega_1} + M_{\omega_4} + M_0,$$

$$M_{\omega_4} \otimes M_{\omega_1} = M_{\omega_1+\omega_4} + M_{\omega_3} + M_{\omega_4},$$

$$M_{\omega_1} \otimes M_{\omega_1} = M_{2\omega_1} + M_{\omega_2} + M_{2\omega_4} + M_{\omega_1} + M_0,$$

From the tables, the dimension-relations are correct; M_{ω_1} is the adjoint representation, and M_{ω_4} is the representation on elements of trace zero in a split exceptional Jordan algebra, as discussed in §1 of Chapter III. From the same discussion, it follows that M_{ω_1}, M_{ω_4}, M_0 are present in the above tensor products with at least the multiplicities indicated. Clearly $M_{2\omega_4}$ is present in $M_{\omega_4} \otimes M_{\omega_4}$ with multiplicity one, and has the weight $\omega_3 = 2\omega_4 - \alpha_4$ of multiplicity one, while $\omega_3 = (2,4,6,3) = (1,2,3,2) + (1,2,3,1) = (1,2,3,1) + (1,2,3,2)$ is present in $M_{\omega_4} \otimes M_{\omega_4}$ with multiplicity at least <u>two</u>. Thus M_{ω_3} occurs in $M_{\omega_4} \otimes M_{\omega_4}$, establishing our first decomposition. Likewise ω_3 occurs with multiplicity 4 in $M_{\omega_1+\omega_4}$, $\omega_1+\omega_4 = (3,5,7,4)$, while $2\omega_1$ and ω_2 do not occur in the tensor product $M_{\omega_4} \otimes M_{\omega_1}$, and $2\omega_4 = (2,4,6,4)$ occurs only from $(1,2,3,2) + (1,2,3,2)$, thus with the same multiplicity, 1, as in $M_{\omega_1+\omega_4}$, while $\omega_3=(1,2,3,2) + (1,2,3,1) = (1,2,3,1) + (1,2,3,2) = (1,2,2,1) + (1,2,4,2) = (1,1,2,1) + (1,3,4,2) = (0,1,2,1) + (2,3,4,2)$ occurs in the tensor product with multiplicity at least 5. The second decomposition is thereby

328

established. For the third, it is clear as above that M_{ω_2} is present once in the tensor product, and together $M_{2\omega_1}$ and M_{ω_2} account for the multiplicity 2 of $\omega_1 + \omega_4$. However $(2,3,4,2) + (1,2,3,2)$ and the reverse are the only ways of writing $\omega_1 + \omega_4$ as a sum of two roots, so that $M_{\omega_1+\omega_4}$ is <u>not</u> present. The multiplicities of $2\omega_4$ in $M_{2\omega_1}$ and M_{ω_2} add up to 6, while in the tensor product $M_{\omega_1} \otimes M_{\omega_1}$, we have definitely an <u>odd</u> multiplicity, since ω_4 is a root; indeed, $2\omega_4 = (1,2,3,2) + (1,2,3,2) = (1,2,2,2) + (1,2,4,2) = (1,1,2,2) + (1,3,4,2) = (0,1,2,2) + (2,3,4,2)$ shows the latter to be at least 7. The presence of $M_{2\omega_4}$ is thus established, and we are done.

For type G_2, we have

$$M_{\omega_2} \otimes M_{\omega_2} = M_{2\omega_2} + M_{\omega_1} + M_{\omega_2} + M_0,$$

$$M_{\omega_2} \otimes M_{\omega_1} = M_{\omega_1+\omega_2} + M_{2\omega_2} + M_{\omega_2},$$

$$M_{\omega_1} \otimes M_{\omega_1} = M_{2\omega_1} + M_{3\omega_2} + M_{2\omega_2} + M_{\omega_1} + M_0.$$

In each case, the dimension-relation is correct. The presence of the last three terms in the first decomposition has been shown in §1 of Chapter III, and the first term is clearly present. Thus the first decomposition holds. Likewise the last term in the second case, and the last two in the third case, are present, as is the first in each of these cases. In $M_{\omega_2} \otimes M_{\omega_1}$, $3\omega_2 = (3,6)$ cannot occur, nor can $2\omega_1$, while $2\omega_2 = (2,4) = (1,2) + (1,2) = (1,1) + (1,3) = (0,1) + (2,3)$ occurs with multiplicity at least 3, thereby exceeding its multiplicity in $M_{\omega_1+\omega_2}$. The second decomposition follows. For the third, the presence of $M_{2\omega_1}$ and $M_{3\omega_2}$, each with multiplicity one, is seen as before. Together these account for a total multiplicity 2 for $\omega_1 + \omega_2 = (3,5)$. The multiplicity of $(3,5)$ in the tensor product is also 2, since $(2,3) + (1,2)$ and its reverse are the only ways of writing $(3,5)$ as a sum of roots. Thus $M_{\omega_1+\omega_2}$ is <u>not</u> present in the tensor product. The multiplicity of $2\omega_2 = (2,4)$ is at least 5: $2\omega_2 = (1,2) + (1,2) = (1,3) + (1,1) = (2,3) + (0,1)$, of which multiplicity $M_{2\omega_1}$ and $M_{3\omega_2}$ only account for 4 units. Thus $M_{2\omega_2}$

is present in the third decomposition, and we are done.

Finally, we consider type $C_r (r \geq 2)$ where we claim:

$$M_{\omega_1} \otimes M_{\omega_1} = M_{2\omega_1} + M_{\omega_2} + M_0;$$

$$M_{\omega_1} \otimes M_{\omega_2} = M_{\omega_1+\omega_2} + M_{\omega_3} + M_{\omega_1} \quad (r \geq 3; \ M_{\omega_3} \text{ is absent if } r = 2);$$

$$M_{\omega_1} \otimes M_{2\omega_1} = M_{3\omega_1} + M_{\omega_1+\omega_2} + M_{\omega_1};$$

$$M_{\omega_2} \otimes M_{\omega_2} = M_{2\omega_2} + M_{\omega_1+\omega_3} + M_{\omega_4} + M_{2\omega_1} + M_{\omega_2} + M_0$$

($r \geq 4$); M_{ω_4} is absent if $r = 3$, and $M_{\omega_1+\omega_3}$, M_{ω_4}, M_{ω_2} are absent if $r = 2$;

$$M_{\omega_2} \otimes M_{2\omega_1} = M_{2\omega_1+\omega_2} + M_{\omega_1+\omega_3} + M_{2\omega_1} + M_{\omega_2}$$

($r \geq 3$; $M_{\omega_1+\omega_3}$ is absent if $r = 3$);

$$M_{2\omega_1} \otimes M_{2\omega_1} = M_{4\omega_1} + M_{2\omega_1+\omega_2} + M_{2\omega_2} + M_{2\omega_1} + M_{\omega_2} + M_0.$$

In each case, the dimensions are correct. M_{ω_1} is the representation on a $2r$-dimensional symplectic space V, $M_{2\omega_1}$ is the adjoint representation on the symplectic-skew transformations of V, and M_{ω_2} that on the symplectic-symmetric transformations of trace zero. From these remarks and those of §1 of Chapter III, we see that the decomposition of $M_{\omega_1} \otimes M_{\omega_1}$ is as indicated, and that M_{ω_1}, $M_{2\omega_1}$, M_{ω_2}, M_0 have at least the indicated multiplicities in all cases. For $r \geq 3$, the only way ω_3 can occur as a weight of $M_{\omega_1} \otimes M_{\omega_2}$ is as a highest weight of a summand, or as a weight of the unique summand $M_{\omega_1+\omega_2}$, where ω_3 has multiplicity 2. But $\omega_3 = (1,2,3,\ldots,3,\frac{3}{2}) = (1,1,\ldots,1,\frac{1}{2}) + (0,1,2,\ldots,2,1) = (0,1,1,\ldots,1,\frac{1}{2}) + (1,1,2,\ldots,2,1) = (0,0,1,\ldots,1,\frac{1}{2}) + (1,2,\ldots,2,1)$ shows the multiplicity in $M_{\omega_1} \otimes M_{\omega_2}$ to be at least 3, and establishes the second decomposition. Likewise, $\omega_1 + \omega_2$ had multiplicity 1 in $M_{3\omega_1}$, but at least 2 in $M_{\omega_1} \otimes M_{2\omega_1}$ [$\omega_1 + \omega_2 = (1,1,\ldots,1,\frac{1}{2}) + (1,2,\ldots,2,1) = (0,1,1,\ldots,1,\frac{1}{2}) + (2,2,\ldots,2,1)$], yielding the third decomposition.

For the fourth, we see that neither $4\omega_1$ nor $2\omega_1 + \omega_2$ is a weight of $M_{\omega_2} \otimes M_{\omega_2}$, so that $M_{2\omega_2}$ occurs with multiplicity 1, as does $M_{\omega_1+\omega_3}$ ($r \geq 3$), by an argument analogous to other cases. This establishes the fourth decomposition for $r \leq 3$, while for $r \geq 4$, the total multiplicity of ω_4 accounted for by $M_{2\omega_2}$ and $M_{\omega_1+\omega_3}$ is 5. On the other hand, $\omega_4 = (1,2,3,4,\ldots,4,2)$ [$= (1,2,3,2)$ if $r = 4$] $= (1,2,\ldots,2,1) + (0,0,1,2,\ldots,2,1) = (1,1,2,\ldots,2,1) + (0,1,1,2,\ldots,2,1) = (1,1,1,2,\ldots,2,1) + (0,1,2,\ldots,2,1)$ and the reverses show the multiplicity of ω_4 in $M_{\omega_2} \otimes M_{\omega_2}$ to be at least 6, thus establishing the fourth decomposition. The fifth is now clear for $r = 2$, and follows for $r \geq 2$ by the fact that $4\omega_1$ is not present in $M_{\omega_2} \otimes M_{2\omega_1}$, while $2\omega_2$ is present with multiplicity 1 (only as $\omega_2 + \omega_2$), its multiplicity in $M_{2\omega_1+\omega_2}$, so that $M_{2\omega_2}$ is not a constituent; meanwhile, $\omega_1 + \omega_3 = (2,3,4,\ldots,4,2) = (1,2,\ldots,2,1) + (1,1,2,\ldots,2,1) = (1,1,2,\ldots,2,1) + (1,2,\ldots,2,1) = (0,1,2,\ldots,2,1) + (2,2,\ldots,2,1)$ has multiplicity at least 3 in the tensor product, but only 2 in $M_{2\omega_1+\omega_2}$.

For the last decomposition, we see as before that $M_{4\omega_1}$ and $M_{2\omega_1+\omega_2}$ are present, each with multiplicity 1, together accounting for a multiplicity 2 for $2\omega_2$, which has in $M_{2\omega_1} \otimes M_{2\omega_1}$ a multiplicity of at least 3 (from $2\omega_2 = \omega_2 + \omega_2 = 2\omega_1 + (2\omega_1 - 2\alpha_1) = (2\omega_1 - 2\alpha_1) + 2\omega_1$). As a consequence, we have our sixth and final decomposition.

Appendix B. Some facts about Jordan algebras.

1) The radical is derivation-stable.

In the work of Albert (see [AA]; [JJ], Chap. V) on the structure of Jordan algebras, one finds a result analogous to that of Wedderburn-Artin in the associative case, namely that every non-nilpotent ideal contains a non-zero idempotent. The notion of nilpotent ideal N here is that, for some m, every product of m elements of N, in whatever association, is zero. One also has a Peirce decomposition relative to an idempotent e: the operator R_e of multiplication by e is diagonalizable with eigenvalues $0, \frac{1}{2}, 1$, and 1 is necessarily present if $e \neq 0$. Hence $Tr(R_{e \cdot e}) \neq 0$ if the characteristic is zero. It follows that if N is an ideal in the Jordan algebra A such that $Tr(R_{z \cdot a}) = 0$ for all $z \in N$, $a \in A$, then N is nilpotent.

Conversely, if N is a nilpotent ideal, we have $z \cdot a \in N$ for $z \in N$, $a \in A$, $R_{z \cdot a}$ maps A into N and acts nilpotently on N. Hence $R_{z \cdot a}$ acts nilpotently on A, and $Tr(R_{z \cdot a}) = 0$. Now consider the symmetric bilinear form $(a,b) = Tr(R_{a \cdot b})$ on the Jordan algebra A. For $c \in A$, we have $(a \cdot c, b) = (a, c \cdot b)$; this follows since polarizing the basic Jordan identity $(a \cdot a) \cdot (b \cdot a) = a \cdot (b \cdot (a \cdot a))$ yields
$(a \cdot b) \cdot (c \cdot d) + (a \cdot d) \cdot (b \cdot c) + (a \cdot c) \cdot (b \cdot d) = (a \cdot (c \cdot d)) \cdot b + (a \cdot (b \cdot c)) \cdot d + (a \cdot (b \cdot d)) \cdot c$, or, comparing as operators on d,

$$R_{a \cdot (b \cdot c)} = R_c R_{a \cdot b} + R_a R_{b \cdot c} + R_b R_{a \cdot c} - R_c R_a R_b - R_b R_a R_c.$$

Interchanging a and b gives

$$R_{(a \cdot c) \cdot b} = R_c R_{b \cdot a} + R_b R_{a \cdot c} + R_a R_{b \cdot c} - R_c R_b R_a - R_a R_b R_c.$$

The difference of the two operators is

$[R_c, R_b R_a] + [R_a R_b, R_c]$, which evidently has trace zero, and yields the identity.

It follows that the radical of the form (a,b) is an ideal in A, which is nilpotent by the above, and which contains every nilpotent ideal. We denote this ideal by N, calling it the radical of A (with which it coincides in a more comprehensive structure theory -see [AA] or [JJ]).

This radical N has the fundamental property that if $N = 0$, A is a direct sum of simple ideals. This follows from the quite elementary remark of Cartan-Dieudonné (see [J], p. 71) on algebras with associative form and no ideals of square zero. Namely, if $N = 0$ the symmetric bilinear form (a,b) is non-degenerate; it satisfies the associative condition $(a \cdot c, b) = (a, c \cdot b)$, and any ideal M with $M \cdot M = 0$ must be contained in N, hence zero. For Jordan algebras, these are exactly the hypotheses of the theorem of Cartan-Dieudonné.

Moreover, if D is a derivation of A, we have $(aD,b) + (a,bD) = \text{Tr}(R_{aD \cdot b} + R_{a \cdot bD}) = \text{Tr}(R_{(a \cdot b)D})$, and $DR_c + R_{cD} = R_c D$ gives $R_{cD} = [R_c, D]$, which is of trace zero. Thus $(aD,b) + (a,bD) = 0$, from which it follows that $ND \subseteq N$, i.e., the radical of a Jordan algebra A (in characteristic zero) is stable under all derivations of A.

2) When all derivations are inner.

Always assuming a ground field of characteristic zero, let A be a finite-dimensional linear algebra (not necessarily associative) with unit element 1. Assume that A is completely reducible as a module for the associative algebra E generated by all the left and right multiplications L_a resp. R_a by elements a of A. Let D be a derivation of A; we claim D is inner, in the sense that $D \in E$. Here we follow [JD].

It follows from the basic derivation-relations $[R_a, D] = R_{aD}$, $[L_a, D] = L_{aD}$, that the derivation $d: X \mapsto [X, D]$ of the ring of linear transformations of A stabilizes E, as well as L, the Lie subalgebra of E generated by all L_a, R_a ($a \in A$). By assumption E (resp. L) is a completely reducible associative (resp. Lie) algebra of endomorphisms of A. By classical theory (resp. the Jacobson theory of completely reducible Lie algebras), E is a semisimple associative algebra and L has the form $L = [LL] \oplus Z(L)$, where $[LL]$ is a semisimple Lie algebra and the center $Z(L)$ of L consists of semisimple endomorphisms of A. In particular, $Z(L) \subseteq Z(E)$, the center of E, a sum of fields and stable under d. Since the characteristic is zero, the derivation d annihilates each of these fields, from which $[Z(L), D] = 0$. Moreover, d stabilizes $[LL]$, where we know from Lie theory that every derivation is inner. That is, we have $E \in [LL]$ such that for all $X \in [LL]$, $[XD] = [XE]$. From $[Z(L), D] = 0 = [Z(L), E]$, this relation holds for all $X \in L$, in particular for all $X = R_a$, $X = L_a$ ($a \in A$).

Thus $R_{aD} = [R_a, D] = [R_a, E]$; applying to $1 \in A$ gives

333

$aD = aE - (1E)a$, or $D = E - L_{(1E)} \in L \subseteq E$. That is, D is inner (in fact, $D \in L$).

When A is a Jordan algebra, our argument of 1) above for the associativity of the form (a,b) used the relation $R_{(a \cdot (c \cdot b) - (a \cdot c) \cdot b)} = [R_c, R_b R_a - R_a R_b] = [R_c, [R_b R_a]]$, from which it follows that L consists of endomorphisms of A of the form $R_a + \Sigma_i [R_{b_i} R_{c_i}]$, hence that any derivation D of A has this form. From $0 = 1D = 1[R_b R_c]$ for all b,c, it follows that $a = 0$ in the expression for D as an element of this form, so that $D = \Sigma_i [R_{b_i} R_{c_i}]$.

In case the radical N of A, as defined in 1) above, is zero, these considerations apply. In particular, if A is a Jordan algebra with unit and no derivation-stable ideal other than A and 0, we must have $N = 0$ by 1); hence A satisfies the hypotheses of this section, and every derivation is inner. Thus all ideals are derivation-stable, and such an A is a simple Jordan algebra.

3) **Jordan division algebras arising in connection with Lie algebras of type G_2.**

We refer to §III.5, where $J = A \oplus B$ was a Jordan algebra in which every element x satisfies a cubic polynomial $p_3(x;X) = X^3 - t_3(x)X^2 + s_3(x)X - n_3(x)$ over A, where if $x = a + b$ $(a \in A, b \in B)$, $t_3(x) = 3a$, $s_3(x) = 3a^2 - \frac{3}{2} f(b,b)$, $n_3(x) = a^3 - \frac{3}{2} af(b,b) + f(h(b,b),b)$. Here A is the center of J, and polarizing the identity in J given by this relation yields a trilinear identity which is preserved under field extension, starting from A. When such an extension splits $1 \in J$ to the maximum possible number of orthogonal idempotents, it follows that this maximum is at most three (since $\lambda_1 e_1 + \lambda_2 e_2 + \lambda_3 e_3 + \lambda_4 e_4$ would satisfy no polynomial not having all λ_i as roots, if the e_i are four non-trivial orthogonal idempotents). We wish to show that this number of idempotents (the degree of J) is not two. The effect of this conclusion is to constrain J to be one of the Jordan division algebras considered in §V.5; in particular, when $F = \mathbb{R}$ we conclude that $J = A$, and that L is a split (real or complex) Lie algebra of type G_2.

By [JJ], §V.7, what we must eliminate is the case where J contains an A-subspace V, of dimension at least two, such that

$J = A \oplus V$, and where every $x \in J$ satisfies a generic polynomial over A of the form $p_2(x;X) = X^2 - t_2(x)X + n_2(x)$, where $t_2(x)$ is a linear form with kernel V and $n_2(x)$ is a nondegenerate quadratic form, not representing zero, with $n_2(1) = 1$, hence with $t_2(1) = 2$. If $x \notin A$, the fact that J is a division algebra implies that $p_2(x;X)$ is irreducible, hence that $p_2(x;X)$ divides $p_3(x;X)$, as a polynomial in X. Thus $p_3(x;X) = p_2(x;X)(X - \phi(x))$, where $\phi(x) \in A$, and $t_3(x) = t_2(x) + \phi(x)$, $s_3(x) = n_2(x) + t_2(x)\phi(x)$, $n_3(x) = n_2(x)\phi(x)$.

Now the subspace B of J is seen, from our construction, to contain JD, the image of every (inner) derivation of J. On the other hand, the derivations of $J = A \oplus V$ are just the skew transformations of V with respect to the restriction to V of the polarized n_2, so that J Der $J = V \subseteq B$. Since both spaces are of A-codimension one in J, we have $V = B$, i.e., t_2 and t_3 have the same kernel, so that $t_3 = \lambda t_2$ for some $\lambda \neq 0$ in A. Evaluation at 1 shows that $\lambda = \frac{3}{2}$. It follows that $\phi(x) = \frac{1}{2}t_2(x)$ for all x, $n_3(x) = \frac{1}{2}n_2(x)t_2(x)$; that is, $n_3(x) = 0$ whenever $t_2(x) = 0$, i.e., whenever $x \in B$. In other words, $f(h(b,b),b)$ <u>must be identically zero</u>. Now it follows from our considerations in §III.5 that $f(h(b,b'),b'')$ is symmetric in its three variables, so that the above implies $f(h(b,b'),b'') = 0$ for all b,b',b'', hence $h(b,b') = 0$ for all b,b', by the nondegeneracy of f (cf. §III.5). We show that this is impossible in our case $(B \neq 0)$.

Namely, the short roots $\pm\alpha_2$, $\pm(\alpha_1+\alpha_2)$, $\pm(\alpha_1+2\alpha_2)$ satisfy $(\alpha_1+2\alpha_2)(h_{\alpha_2}) = 1 > 0$. By Lemma I.7, we have $[L_{\alpha_1+\alpha_2}, X_{\alpha_2}] = L_{\alpha_1+2\alpha_2}$ for each non-zero $X_{\alpha_2} \in L_{\alpha_2}$. Now α_2 is a weight of M', so that if $u \in M'$ is a basis for M'_{α_2}, and if $0 \neq b \in B$, then $u \otimes b = X_{\alpha_2}$ is non-zero in L_{α_2}, and $[L_{\alpha_1+\alpha_2}, u \otimes b] = L_{\alpha_1+2\alpha_2}$. Since $[B:A] \geq 2$, if v is a basis for $M'_{\alpha_1+2\alpha_2}$, $L_{\alpha_1+2\alpha_2}$ contains an element of the form $v \otimes b'$, where b' and b are linearly independent over A. Thus we have, for e a basis for $S_{\alpha_1+\alpha_2}$, w for $M'_{\alpha_1+\alpha_2}$, elements $a \in A$, $c \in B$, with

$$[u \otimes b, e \otimes a + w \otimes c] = v \otimes b'.$$

The formulas of §III.5 show that this is impossible unless $h(b,c) \neq 0$, and therefore that J <u>cannot be of degree two</u>.

4) **A Theorem on Some Mappings of Involutorial Algebras.**

In connection with Proposition V.4 (§V.7.a), we have required the following result:

<u>Theorem.</u> <u>Let</u> $(E,*)$ <u>be an involutorial associative division algebra, of finite dimension over its center. Assume that</u> E <u>is generated by its symmetric elements,</u> A, <u>and that</u> A <u>contains elements of trace zero whose squares are not central. Let</u> f <u>be a pairing of</u> $E \times E$ <u>to</u> A, <u>bilinear over the fixed elements of the center, and such that for all</u> $e,e' \in E$, $a \in A$, $f(ea,e') + f(e,e'a) = f(e,e')a + af(e,e')$. <u>Then there is a unique</u> $b \in E$ <u>with, for all</u> e,e',

$$f(e,e') = e^*be' + e'^*b^*e.$$

<u>Proof.</u> The uniqueness of b follows at once from the fact that if $e^*be' = -e'^*b^*e$ for all e,e', then we have for $e' = 1$, $e^*b = -b^*e$ for all e, i.e., e^*b is skew for all e; but e^*b runs over E as e does, unless $b = 0$. Therefore $b = 0$. Maps of the form $f(e,e') = e^*be' + e'^*b^*e$ evidently satisfy the specified relation, which may be reinterpreted to say that a certain system of linear equations with coefficients in the field F of fixed elements of the center has a solution in F. As such, the problem is solved if we can solve it when F is algebraically closed, where E is no longer a division algebra, but rather $(E,*)$ is one of the following:

i) $(M_{2r}(F),*)$, the involution $*$ being that associated with a nondegenerate symplectic form;

ii) $(M_n(F),*)$, the involution $*$ being associated with a nondegenerate symmetric form;

iii) $(M_r(F) \oplus M_r(F),*)$, the involution $*$ sending $a \oplus b$ to $b^t \oplus a^t$, a^t being the transpose of a.

In case i), let $u_1,\ldots,u_{2r}$ be a symplectic basis for F^{2r}, with $(u_i,u_{2r+1-i}) = 1$ if $i \leq r$, -1 if $i > r$, and all other scalar products zero. A complete set of primitive idempotents in the Jordan algebra A are then the $E_i = E_{ii} + E_{2r+1-i,2r+1-i}$, $1 \leq i \leq r$, the $E_{k,\ell}$ being matrix-units relative to our basis. One easily verifies that the only $a \in A$ such that $aE_i + E_i a = 2a$ are the scalar multiples of E_i. Taking $e = E_{2r+1-j,2r+1-j}$, $e' = E_{ij}$ for arbitrary i,j, we have $f(eE_j,e') + f(e,e'E_j) = 2f(e,e')$, hence $f(e,e') = \lambda_{ji}E_j$ for some

336

$\lambda_{ji} \in F$. We set $b = \Sigma_{i,j} \lambda_{ji} E_{ji}$, and show

$$f(d,d') = d^* b d' + d'^* b^* d$$

for all d,d'. First with $d = E_{2r+1-j, 2r+1-j}$, $d' = E_{ij}$, we have $d^* = E_{jj}$, $d'^* = \varepsilon E_{2r+1-j, 2r+1-i}$, the sign ε being 1 if i and j belong to the same one of the intervals $[1,r]$, $[r+1,2r]$, and -1 if they do not. Then $d^* b d' + d'^* b^* d$

$$= \lambda_{ji} E_{jj} + \varepsilon E_{2r+1-j, 2r+1-i} \varepsilon \lambda_{ji} E_{2r+1-i, 2r+1-j}$$
$$= \lambda_{ji} (E_{jj} + E_{2r+1-j, 2r+1-j}) = \lambda_{ji} E_j.$$ Now it

suffices to show that the values of a mapping f, satisfying our conditions, at the pairs d,d' as above, completely determine f.

What are assumed known are all $f(E_{jj}, E_{i, 2r+1-j})$. Next we note that these determine all $f(E_{jj}, E_{i\ell})$, $\ell \neq j$, for which we may assume $\ell \neq 2r+1-j$. Taking $a = E_{2r+1-j, \ell} + E^*_{2r+1-j, \ell} = E_{2r+1-j, \ell} + \varepsilon' E_{2r+1-\ell, j}$, we have $f(E_{jj} a, E_{i, 2r+1-j}) + f(E_{jj}, E_{i, 2r+1-j} a) = f(E_{jj}, E_{i\ell})$ if $\ell \neq j$, and this is $f(E_{jj}, E_{i, 2r+1-j}) a + a f(E_{jj}, E_{i, 2r+1-j})$, so is determined. Since A generates E, we have $r > 1$, so may take $\ell \neq j$, $2r+1-j$ in the above. Then, with $a = E_{\ell j} + E^*_{\ell j} = E_{\ell j} + \varepsilon E_{2r+1-j, 2r+1-\ell}$, we have $f(E_{jj} a, E_{i\ell}) + f(E_{jj}, E_{i\ell} a) = f(E_{jj}, E_{ij})$, from which we see that all $f(E_{jj}, E_{i\ell})$ are determined. Then with $k \neq \ell$, $2r+1-\ell$, we take $a = E_{jk} + E^*_{jk} = E_{jk} + \varepsilon E_{2r+1-k, 2r+1-j}$ to see that $f(E_{jk}, E_{i\ell})$ are all determined if $k \neq \ell$, $2r+1-\ell$, $2r+1-j$. Starting from such a k and taking $a = E_{k\ell} + E^*_{k\ell} = E_{k\ell} + \varepsilon E_{2r+1-\ell, 2r+1-k}$, we see that $f(E_{j\ell}, E_{i\ell})$ are also determined, so that $f(E_{jk}, E_{i\ell})$ are all determined if $k \neq 2r+1-j$, $2r+1-\ell$. The assumption that some squares of elements of A are not central forces $r \geq 3$, therefore that we have such $k \neq j, \ell$. Taking $a = E_{k, 2r+1-j} + E^*_{k, 2r+1-j}$, we see that $f(E_{j, 2r+1-j}, E_{i\ell})$ are determined whenever $\ell \neq j$. This leaves only $f(E_{j, 2r+1-\ell}, E_{i\ell})$ to show determined for all $i, j, \ell \neq 2r+j-1$ (since in this case, we have one of the original pairs). From earlier remarks,

$f(E_{j, 2r+1-\ell}, E_{i\ell}) = \lambda_\ell (E_{\ell\ell} + E_{2r+1-\ell, 2r+1-\ell})$ for some λ_ℓ. With $k \neq \ell$, $2r+1-\ell$, $a = E_{\ell k} + E^*_{\ell k} = E_{\ell k} + \varepsilon E_{2r+1-k, 2r+1-\ell} \neq 0$, we find

$$f(E_{j, 2r+1-k} a, E_{i\ell}) + f(E_{j, 2r+1-k}, E_{i\ell} a)$$
$$= \varepsilon f(E_{j, 2r+1-\ell}, E_{i\ell}) + f(E_{j, 2r+1-k}, E_{ik})$$
$$= \varepsilon \lambda_\ell (E_{\ell\ell} + E_{2r+1-\ell, 2r+1-\ell}) + \lambda_k (E_{kk} + E_{2r+1-k, 2r+1-k}),$$

and this is determined by previously found values of f and our relation. Since the two terms are linearly independent, λ_k and λ_ℓ are determined, and case i) is complete.

The case iii) is actually a consequence of i); namely, we may identify $(E,*)$ with the involutorial subalgebra of i) generated by the E_{ij}, where i and j are in the same block, either $[1,r]$ or $[r+1,2r]$. With this restriction on the indices, we define b exactly as before, with $\lambda_{ij} = 0$ if i,j are in different blocks. The conditions on A assure that $r > 1$, and we start with all $f(E_{jj}, E_{i,2r+1-j})$, where now i and j are in opposite blocks. Then with $\ell \neq 2r+1-j$ in the same block as i, we see with a as before that $f(E_{jj}, E_{i\ell})$ are determined. Then with $k \neq 2r+1-\ell$ in the same block as j, we see that all $f(E_{jk}, E_{i\ell})$ are determined for E_{jk}, $E_{i\ell}$ in opposite blocks, except when $\ell = 2r+1-k$, and this must be a scalar multiple of $E_{kk} + E_{2r+1-k, 2r+1-k}$, which multiple we determine as before. Finally, if E_{jk}, $E_{i\ell}$ are in the same block, we have for $k = \ell$, $f(E_{j\ell}, E_{i\ell}) = \lambda(E_{\ell\ell} + E_{2r+1-\ell, 2r+1-\ell})$, while for $a = E_{2r+1-\ell, m} + E^*_{2r+1-\ell, m}$, m and ℓ in opposite blocks, $m \neq \ell$, we have

$$0 = f(E_{j\ell}a, E_{i\ell}) + f(E_{j\ell}, E_{i\ell}a) = \lambda a. \text{ Thus}$$

$f(E_{j\ell}, E_{i\ell}) = 0$. Now if $k \neq \ell$, taking $a = E_{k\ell} + E^*_{k\ell}$ gives

$f(E_{jk}a, E_{i\ell}) + f(E_{jk}, E_{i\ell}a) = f(E_{j\ell}, E_{i\ell}) = 0$, so

$f(E_{jk}, E_{i\ell})a + af(E_{jk}, E_{i\ell}) = 0$, while

$f(E_{jk}, E_{i\ell}) = f(E_{jk}(E_{kk} + E^*_{kk}), E_{i\ell}) + f(E_{jk}, E_{i\ell}(E_{kk} + E^*_{kk}))$
$= f(E_{jk}, E_{i\ell})(E_{kk} + E^*_{kk}) + (E_{kk} + E^*_{kk})f(E_{jk}, E_{i\ell})$.

The last says that all non-zero entries of $f(E_{jk}, E_{i\ell})$ lie in the k-th row or the k-th column, or in the 2r+1-k-th row or column, but that the diagonal entries are zero. Thus the ℓ-th column of $f(E_{jk}, E_{i\ell})a + af(E_{jk}, E_{i\ell})$ is the same as the k-th column of $f(E_{jk}, E_{i\ell})$, and the latter must be zero, as must the 2r+1-k-th row. Applying similar considerations with $E_{\ell k} + E^*_{\ell k}$ and with $E_{\ell\ell} + E^*_{\ell\ell}$ shows that the only non-zero entries in the block of $f(E_{jk}, E_{i\ell})$ containing the indices i,j,k,ℓ are in the (k,ℓ) and (ℓ,k) positions, and the ℓ-th column must be zero. It follows that this block of $f(E_{jk}, E_{i\ell}) \in A$ must be zero, so the opposite block is also zero. This completes the case iii).

ii) Here we have a basis $u_1,\ldots,u_n$, with $(u_i,u_{n+1-i}) = 1$, other scalar products being zero. We can only conclude at first that $f(E_{n+1-j,n+1-j},E_{ij})$ is a combination of $E_{jj} + E_{jj}^*$, $E_{j,n+1-j}$, $E_{n+1-j,j}$. When $n = 2r+1$, $j = r+1$, these three are all multiples of E_{jj}^*. Otherwise, if this quantity is $\alpha(E_{jj} + E_{jj}^*) + \beta E_{j,n+1-j} + \gamma E_{n+1-j,j}$, taking $a = E_{n+1-j,j}$ gives $f(E_{n+1-j,j},E_{ij}) = 2\alpha a + (\gamma+\beta)(E_{jj} + E_{jj}^*)$, and repeating with the same "a" gives $0 = 2\beta a$, so $\beta = 0$; likewise, with $a = E_{j,n+1-j}$, we find $\gamma = 0$, so that $f(E_{n+1-j,n+1-j},E_{ij}) = \lambda_{ji}E_j$ as before, $E_j = E_{jj} + E_{jj}^*$, enabling us to define $b = \Sigma \lambda_{ji} E_{ji}$.

In case $n = 2r$ is even, one completes the argument as in the symplectic case. Here no restriction on r is necessary, since we have at our disposal in A the elements $E_{j,n+1-j}$ which were not available in the symplectic case. When $n = 2r+1$ is odd (≥ 3), the same argument shows that the matrix b determines all $f(E_{ij},E_{k\ell})$ when none of the indices i,j,k,ℓ is $r+1$. Taking one such, with $j \neq \ell$, we set
$a = E_{j,r+1} + E_{j,r+1}^* = E_{j,r+1} + E_{r+1,2r+1-j} \in A$ and find
$f(E_{i,r+1},E_{k\ell}) = f(E_{ij}a,E_{k\ell}) + f(E_{ij},E_k a) = f(E_{ij},E_{k\ell})a + af(E_{ij},E_{k\ell})$ is determined, as is $f(E_{ij},E_{k,r+1})$. Now $f(E_{i,r+1},E_{k,r+1})$ is a scalar multiple of $E_{r+1,r+1}$, as before, and with $a = E_{j,r+1} + E_{j,r+1}^*$, $j \neq r+1$ we have $f(E_{ij}a,E_{k,r+1}) + f(E_{ij},E_{k,r+1}a)$ to be determined. The first term is $f(E_{i,r+1},E_{k,r+1})$, while the second is a combination of E_j^*, $E_{j,2r+2-j}$, $E_{2r+2-j,j}$. By linear independence we see that $f(E_{i,r+1},E_{k,r+1})$ is determined. Finally, we have that $f(E_{r+1,r+1},E_{i,r+1})$ is determined for all i by our b as required. Next, for $j \neq r+1$, $f(E_{r+1,r+1},E_{ij})$ is seen as above to be a combination of $E_{r+1,j}$, $E_{j,r+1}$, $E_{2r+2-j,r+1}$, $E_{r+1,2r+2-j}$, hence of $E_{j,r+1} + E_{j,r+1}^*$ and $E_{r+1,j} + E_{r+1,j}^*$. With $k \neq j, r+1$ taking $a = E_{jk} + E_{jk}^*$ shows that $f(E_{r+1,r+1},E_{ik}) = \alpha(E_{r+1,k} + E_{r+1,k}^*)$, yielding this form of relation for all i,k. Then using $a = E_{k,r+1} + E_{k,r+1}^*$, we see that α is determined by $f(E_{r+1,r+1},E_{i,r+1})$, so that all $f(E_{r+1,r+1},E_{ik})$ are determined. Then for $k \neq 2r+2-j$, taking $a = E_{r+1,j} + E_{r+1,j}^*$ shows that all $f(E_{r+1,j},E_{ik})$ are determined for $k + j \neq 2r + 2$, and finally an argument using $a = E_{j,2r+2-k} + E_{j,2r+2-k}^*$, $j \neq k, 2r+2-k$ and taking components as before shows that this last is determined. The theorem now follows by checking the fundamental relation $f(e,e') = e^*be' + e'^*b^*e$ for the pairs e,e' used to define b. This is completely straightforward.

Appendix C. A Theorem of Tamagawa.

In §§V. 7c),d), we have used the following result, due to Tamagawa:

Theorem. Let A be a field (here assumed of characteristic different from 2), h nondegenerate symmetric bilinear form on a five-dimensional A-vector space B. If h does not represent 1, and if the discriminant of h is a square in A, then the even Clifford algebra $H^+(B)$ is a division algebra.

Proof. The center of the full Clifford algebra $H(B)$ contains by assumption a central element $z = \lambda b_1 b_2 b_3 b_4 b_5$, where $b_1, \ldots, b_5$ is an orthogonal basis, $\lambda^{-2} = \Pi h(b_i, b_i)$, so that $z^2 = 1$. The subspace $zB \subseteq H^+(B)$ is 5-dimensional, with $(zb)^2 = z^2 b^2 = b^2 = h(b,b)$ for all $b \in B$. Thus zB carries a quadratic form $(zb)^2$ equivalent to h under $zb \leftrightarrow b$.

The involution of $H(B)$ which is the identity on B stabilizes $H^+(B)$, where it has a 10-dimensional space of skew elements and a 6-dimensional space of fixed elements, the latter being $A1 + zB$. A basis for the space of skew elements consists of the products $b_i b_j$, $1 \leq i < j \leq 5$. Now consider the trace form t of the right regular representation of $H^+(B)$; one checks directly that zB is a non-singular subspace, where $t((zb)(zb')) = t(bb') = 0$ if $h(b,b') = 0$, while $t((zb)^2) = t(b^2) = 16h(b,b)$. Thus the restriction of t to zB is also equivalent to h under $zb \leftrightarrow 4b$. Moreover zB is characterized as the set of fixed elements where t vanishes.

Since $H^+(B)$ is a central simple algebra of degree four over A, it is either a division algebra, the algebra of 2 by 2 matrices over a quaternionic division algebra, or the full algebra of 4 by 4 matrices over A. In the last case, the fact that our involution in $H^+(B)$ has 10 and 6 as the dimensions of its skew resp. fixed elements assures that the involution in $H^+(B)$ is realized as the transpose with respect to a symplectic form in a four-dimensional space over A. But in such a case there are nonzero symplectic-symmetric elements of square 0 (e.g., $E_{12} + E_{34}$, in the notation of §III.6). These have trace zero, and afford elements $u \neq 0$ in zB with $t(u) = 0 = t(u^2)$. But this would mean that h represents zero, and therefore represents 1 as well.

We may therefore assume that $H^+(B) \cong \mathcal{Q}_2, \mathcal{Q}$ a quaternionic division algebra over A, and that the involution is that of forming the adjoint with respect to a nondegenerate hermitian form, associated with the standard involution in $\mathcal{Q}$, in the 2-dimensional $\mathcal{Q}$-space U where $H^+(B) = \text{End}_{\mathcal{Q}}(U)$. (The other case, where the form is anti-hermitian with respect to an involution $\alpha \mapsto \nu^{-1}\alpha^*\nu$, $\alpha \mapsto \alpha^*$ the standard involution, $\nu^* = -\nu \neq 0$ in $\mathcal{Q}$, is seen by a standard device to have the same skew and fixed elements as one of our type.). Thus U has an orthogonal $\mathcal{Q}$-basis e_1, e_2 with respect to our form (u,v), with $(e_1,e_1) = \lambda \neq 0$, $(e_2,e_2) = \mu \neq 0$, $\lambda^* = \lambda, \mu^* = \mu$.

But then if $T \in \text{End}_{\mathcal{Q}}(U)$ sends e_1 to e_1, e_2 to $-e_2$, we have $T \in H^+(B)$, $T^2 = \text{Id.}$, and the trace of T in the regular representation is zero, while that of T^2 is 16. Thus T and our isomorphism provide us with an element $b \in B$ with $h(b,b) = 1$, a contradiction. This completes the proof that $H^+(B)$ must be a division algebra.

BIBLIOGRAPHY

[AA] Albert, A. A., A structure theory for Jordan algebras. Annals of Math. 48 (1947), 546-567.

[A] Allison, B. N., Isomorphism of simple Lie algebras. Transactions A.M.S. 117 (1973), 173-190.

[AJ] _____, A construction of Lie algebras from J-ternary algebras. Amer. J. Math., to appear.

[AL] _____, Lie algebras of type BC_1. Transactions A.M.S., to appear.

[AZ] _____, J-ternary algebras without zero divisors. Journal of Algebra, to appear.

[Ba] Baxter, W. E., Lie simplicity of a class of associative rings. II. Transactions A.M.S. 87 (1958), 63-75.

[B],[BO],[Bo] Bourbaki, N., Groupes et algèbres de Lie, Chaps. 4-6. Herrmann, Paris, 1968.

[BK] Braun, H. and Koecher, M., Jordan-algebren. Springer-Verlag, Berlin 1966.

[CS] Chevalley, C., The Algebraic Theory of Spinors. Columbia Univ. Press, New York, 1954.

[C II] _____, Théorie des groupes de Lie. Tome II, Groupes algébriques. Herrmann, Paris, 1951.

[C III] _____, ibid. Tome III, Théorèmes généraux sur les algèbres de Lie. Herrmann, Paris, 1955.

[D] Dieudonné, J., La Géométrie des Groupes Classiques. Springer-Verlag, Berlin, 1955, 1963, 1971.

[F],[FT] Faulkner, J. R., A construction of Lie algebras from a class of ternary algebras. Transactions A.M.S. 155 (1971), 397-408.

[FF] _____ and Ferrar, J. C., On the structure of symplectic ternary algebras. Indagationes Math. (Proc. Ned. Ak. Wet. Series A) 34 (1972), 247-256.

[He] Hein, W., A construction of Lie algebras by triple systems. Transactions A.M.S. 205 (1975), 79-95.

[H] Herstein, I. N., Lie and Jordan systems in simple rings with involution. Amer. J. Math. 78 (1956), 629-649.

[HU] Humphreys, J. E., Introduction to Lie Algebras and Representation Theory. Springer-Verlag, New York, 1972.

[J] Jacobson, N., Lie Algebras. Interscience, New York, 1962.

[JC] _____, Composition algebras and their automorphisms. Rend. Circ. Mat. Palermo 7 (1958), 55-80.

[JD] _____, Derivation algebras and multiplication algebras of semi-simple Jordan algebras. Annals of Math. 50 (1949), 866-874.

[JE] _____ , *Exceptional Lie Algebras*. Marcel Dekker, New York, 1971.

[JJ] _____ , *Structure and Representations of Jordan Algebras*. Amer. Math. Society, Providence, 1968.

[JR] _____ and Rickart, C. E., Homomorphisms of Jordan rings of self-adjoint elements. Transactions A.M.S. 72 (1952), 310-322.

[K] Koecher, M., Imbedding of Jordan algebras in Lie algebras. I. Amer. J. Math. 89 (1967), 787-815.

[Ma] Malcev, A., Analytic loops (Russian). Mat. Sbornik 78 (1955), 569-576.

[Me] Meyberg, K., Eine Theorie der Freudenthalschen Tripelsysteme. Indag. Math. 30 (1968), 162-190.

[M] Mostow, G. D., Fully reducible subgroups of algebraic groups. Amer. J. Math. 78 (1956), 200-221.

[SAG] Sagle, A. A., Malcev algebras. Transactions A.M.S. 101 (1961), 426-458.

[SC] Satake, I., *Classification Theory of Semi-Simple Algebraic Groups*. Marcel Dekker, New York, 1971.

[SG] _____ , On the theory of reductive algebraic groups over a perfect field. J. Math. Soc. Japan 15 (1963), 210-235.

[SCH] Schafer, R. D., *An Introduction to Nonassociative Algebras*. Academic Press, New York, 1966.

[SLA II] Seligman, G. B., On automorphisms of Lie algebras of classical type. II. Transactions A.M.S. 94 (1960), 452-482.

[SLA III] _____ , *ibid*. III. Transactions A.M.S. 97 (1960), 286-316.

[SR] Serre, J-P., *Algebres de Lie semi-simples complexes*. W. A. Benjamin, New York, 1966.

[TA] Tits, J., Algèbres alternatives, algèbres de Jordan et algèbres de Lie exceptionnelles. I. Construction. Indag. Math. 28 (1966), 223-237.

[TB] _____ , *Buildings of Spherical Type and Finite* (B,N)-*Pairs*. Springer-Verlag, Berlin, 1974.

[TC] _____ , Une classe d'algèbres de Lie en relation avec les algèbres de Jordan. Indag. Math. 24 (1962), 530-535.

[TD] _____ , Classification of algebraic semisimple groups. Proc. Symposia Pure Math. IX (1966), 33-62. (Amer. Math. Society, Providence).

[TG] _____ , Groupes semi-simples isotropes. Colloque sur la théorie des groupes algébriques. C.B.R.M. Brussels, 1962, pp. 137-147.

[TS] _____ , Algebraic and abstract simple groups. Annals of Math. 80 (1964), 313-329.

[W] Winter, D. J., *Abstract Lie Algebras*. M.I.T. Press, Cambridge (MA), 1972.

INDEX

Algebraic completely reducible Lie algebra	1
Alternative algebra	67
Anisotropic simple Lie algebra	289
Balanced symplectic algebra	259
(B,N)-pairs (Tits systems)	43
Cartan decomposition of real simple Lie algebra	291
Center of Jordan algebra	176
Centroid of simple Lie algebra	167
Clifford algebras in 5 dimensions	224, 252, C2
Closed set of roots	40
Compact real simple Lie algebra	289
Composition algebra	281, 282
Derivation of Jordan algebra into bimodule	132
Dominant weight	A2
Dynkin diagram	15
Exceptional Jordan division algebra	16
Freudenthal's cross-product	269
Freudenthal's formula	A2
Freudenthal triple system	259
Inner derivation of linear algebra	B2
Involution in alternative algebra	97
Involution in simple associative algebra	99
Involutorial algebra, simple	97
Jordan algebra of quadratic form	189
Jordan bimodule	132, 136
Jordan division algebra	64
Malcev algebra	67
Modules for Clifford algebra	213, 214
Ordering (of relative roots)	7
Perfect involutorial associative algebra	197
Primitive roots	11, 12
Radical of Jordan algebra	B1
Rational isomorphism (of reductive Lie algebras)	31
Real form of complex simple Lie algebra	289
Reduced system of roots	112

Replica of diagonal matrix	19
Representations of split 3-dimensional Lie algebra	4,5
Root spaces (relative to maximal split torus)	3
Semi-similitude of Jordan algebra	272
Serre's theorem	54,55
Simple roots	8
Simple system of roots	11
Split quadratic form	213
Split toral subalgebra ("torus")	2
Splittable ("almost algebraic") Lie algebra	3
System of roots	10
Tits' second construction	84
Toral subalgebra ("torus")	2
Weyl group	10,11